MOLECULAR INTERACTIONS ON TWO-DIMENSIONAL MATERIALS

MOLECULAR INTERACTIONS ON TWO-DIMENSIONAL MATERIALS

Editors

Konstantin S. Novoselov
Andrew T. S. Wee
Arramel

National University of Singapore, Singapore

World Scientific

NEW JERSEY • LONDON • SINGAPORE • BEIJING • SHANGHAI • HONG KONG • TAIPEI • CHENNAI • TOKYO

Published by

World Scientific Publishing Co. Pte. Ltd.
5 Toh Tuck Link, Singapore 596224
USA office: 27 Warren Street, Suite 401-402, Hackensack, NJ 07601
UK office: 57 Shelton Street, Covent Garden, London WC2H 9HE

British Library Cataloguing-in-Publication Data
A catalogue record for this book is available from the British Library.

MOLECULAR INTERACTIONS ON TWO-DIMENSIONAL MATERIALS

ISBN 978-981-124-784-2 (hardcover)
ISBN 978-981-124-785-9 (ebook for institutions)
ISBN 978-981-124-786-6 (ebook for individuals)

For any available supplementary material, please visit
https://www.worldscientific.com/worldscibooks/10.1142/12574#t=suppl

Desk Editor: Nur Syarfeena Binte Mohd Fauzi

Typeset by Stallion Press
Email: enquiries@stallionpress.com

https://doi.org/10.1142/9789811247859_fmatter

Foreword

Molecular Interactions on Two-Dimensional Materials

Two-dimensional (2D) materials became an active field of research with the isolation of graphene in 2004 by Andre Geim and Konstantin Novoselov at the University of Manchester. This work resulted in the winning the Nobel Prize in Physics in 2010 "for groundbreaking experiments regarding the 2D material graphene." Today, research in 2D materials is conducted in many laboratories internationally, driven by their physical properties as well as potential applications in diverse fields ranging from electronics, photonics, spintronics, catalysis, membrane science, and so on.

This book addresses the fundamental and applied topics in molecular interactions on 2D materials. It begins with an introduction of 2D materials by Novoselov himself. Various experts then introduce on-surface molecular interactions, and then specifically on the molecular functionalization of 2D materials. The electronic structure at molecular-2D material heterointerfaces is presented, followed by related screening effects at these heterointerfaces. Analytical techniques used to study such heterointerfaces are then described, such as synchrotron techniques, scanning probe microscopy, and electron holography. The final chapters cover related topics such as the molecule-2D perovskite heterointerface, and graphene oxide membranes.

This book will be useful to graduate students and researchers in the field of molecular interactions on 2D materials. It lays the foundation on many aspects of physics and chemistry at this molecule-2D material heterointerface, in anticipation of useful applications that will emerge in the next decade or so.

Editors

Andrew Wee
National University of Singapore, Singapore

Kostya S. Novoselov
National University of Singapore, Singapore

Arramel
National University of Singapore, Singapore

https://doi.org/10.1142/9789811247859_fmatter

Contents

https://doi.org/10.1142/9789811247859_0001

Chapter 1

Introduction to Two-Dimensional Materials

Xuanye Leng[*,§], Siyu Chen[*,§], Kou Yang[*,§], Musen Chen[*,§], Majid Shaker[*,†], Evgenii E. Vdovin[‡], Qi Ge[†], Kostya S. Novoselov[*,†,§] and Daria V. Andreeva[*,§,¶]

[*]*Department of Materials Science and Engineering, National University of Singapore, 9 Engineering Drive 1, Singapore 117575, Singapore*

[†]*Chongqing 2D Materials Institute, Liangjiang New Area, Chongqing 400714, P. R. China*

[‡]*Institute of Microelectronics Technology and High Purity Materials, Russian Academy of Sciences, 6 Academician Ossipyan Str., Chernogolovka, 142432 Moscow, Russia*

[§]*Institute for Functional Intelligent Materials, National University of Singapore, 117544, Singapore*

[¶]*daria@nus.edu.sg*

The application areas of two-dimensional (2D) materials, their heterostructures and composites are rapidly expanding — thanks to their extraordinary mechanical, optical, electronic and thermal properties. The family of 2D materials, which are finding a use in applications, is rapidly growing, and now includes graphene, graphene oxide, hexagonal boron nitride and transition-metal dichalcogenide, as well as the recently obtained chemical derivatives, physically modified hybrids and allotropes. New applications also trigger the development of new and more efficient methods for the synthesis of 2D materials and the formulation of composites. In this paper, we focus on the relation between the structure of the 2D materials and their heterostructures on the one hand and their properties on the other,

[¶]Corresponding author.

summarize recent improvements in the synthesis approaches and discuss the perspectives in applications. We highlight existing applications in electronics, optoelectronics, electrocatalysis and energy storage and discuss new potential areas such as functional coatings and biomedical materials and devices.

Keywords: Graphene; 2D materials; van der Waals heterostructures.

1. Background

Conventional theory predicts the absence of long-range ordering in dimensionality two due to logarithmic divergencies.[1–5] However, it has been demonstrated experimentally that finite-size two-dimensional (2D) crystals are stable (often stabilized by substrate or ripples) and can be of very high quality.[6–10] Furthermore, such crystals can be produced in large quantities, making them suitable for a number of exciting applications.[11–14]

These days, the family of 2D materials includes dozens of crystals beyond graphene[15]: graphene oxide and reduced graphene oxide (GO/rGO),[16–18] hexagonal boron nitrite (*h*-BN),[19,20] graphane,[21] fluorographene,[22] transition-metal dichalcogenides (TMDCs; e.g. MoS_2),[23] 2D metal oxides (e.g. $Bi_2Sr_2CaCu_2O_x$),[7] graphitic carbon nitrite (*g*-C_3N_4),[24] black phosphorus (BP),[25–27] layered double hydroxides (LDHs),[28] silicene,[29] germanene,[30] 2D metal carbides and nitrides (MXenes),[31] 2D metal–organic frameworks (2D MOFs),[32] 2D covalent–organic frameworks (2D COFs)[33] and 2D perovskite.[7,34–37] The schematic illustration of chemical structures of some of 2D materials is shown in Fig. 1.

Despite such a variety, graphene still stays at the focus of attention even these days. Graphene has a planar honeycomb lattice structure with the carbon atoms bonded with σ bonds with the length of 1.42 Å,[34,38] and the π-bonds being responsible for the linear dispersion spectrum, making graphene a semimetal with zero band overlap.[39] GO and rGO are two most common chemically modified graphene derivates. GO possesses oxygen-contained functional groups (–OH, –COOH and –O–) attached on graphene backbone,[17,40] which makes it an insulator. Also, rGO is a reduced derivative of GO and, comparing to GO, it is conductive, has less oxygen-contained radicals and maintains better pristine frame structure of graphene.[41]

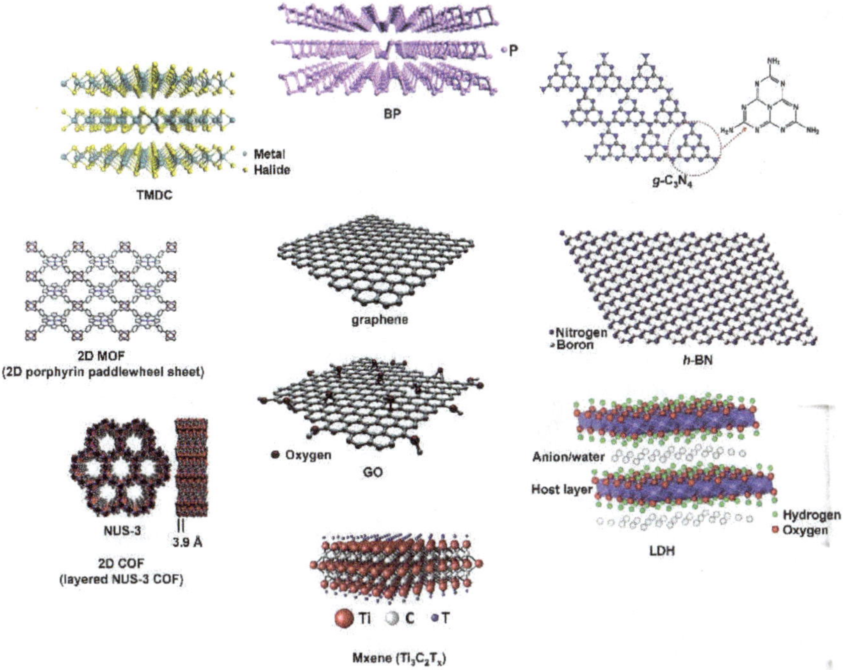

Fig. 1. Schematic structures of typical 2D materials.

Source: Graphene and GO adapted with permission from Ref. 63. Published by the Royal Society of Chemistry. The *h*-BN adapted with permission from Ref. 43. Copyright © 2015 Elsevier, Ltd. TMDC adapted with permission from Ref. 64. Copyright © 2018 Wiley-VCH Verlag GmbH & Co. KGaA, Weinheim. BP adapted with permission from Ref. 65. Copyright © 2016 The Royal Society of Chemistry. The g-C_3N_4 adapted with permission from Ref. 66. Copyright © 2014 American Chemical Society. LDH adapted with permission from Ref. 28. Copyright © 2006 The Royal Society of Chemistry. The 2D MOF adapted with permission from Ref. 32. Copyright © 2009 The Royal Society of Chemistry. The 2D COF adapted with permission from Ref. 33. Copyright © 2016 American Chemical Society. MXene adapted with permission from Ref. 67. Copyright © 2016 Wiley-VCH Verlag GmbH & Co. KGaA, Weinheim.

Beyond graphene and its derivatives, *h*-BN is probably the most important member of the family of 2D crystals. Being a very good insulator, it is often used as a high-quality substrate for graphene devices.[42] In *h*-BN, the boron atoms and nitrogen atoms appear alternatively bonded with covalent bonds, where the bond length of B–N bonds in *h*-BN is 1.44 Å.[43] TMDCs are 2D semiconductors, and have a layered X–M–X structure, where X is a chalcogen and M is transition metal element.[44,45] For instance, MoS_2, a typical TMDC, has a layer-stacking structure where the spacing of two adjacent MoS_2 layers is reported to be 6.5 Å.[46] BP is an anisotropic layered 2D semiconductor. Single-layer BP is a puckered honeycomb-structured 2D material where each phosphorus atom is linked with three phosphorus atoms by covalent bonds.[47]

The 2D materials exhibit outstanding characteristics in many aspects. One obvious example is the high specific surface area. For instance, the specific surface area of graphene is as high as 2630 m^2/g.[48] This justifies the use of graphene as additives in batteries and supercapacitors. Other properties are extraordinary as well: For instance, graphene has a Young's modulus as high as ~ 1 TPa,[49] an intrinsic mobility of up to 200,000 $cm^2 \cdot V^{-1} \cdot s^{-1}$ at room temperature,[50] a high light transmittance of $\sim 97.7\%$ (Ref. 51) and a superior thermal conductivity of about 5000 $W \cdot m^{-1} \cdot K^{-1}$ at room temperature.[16,52] Besides, the unconventional superconductivity observed in twisted (under 1.1°, so-called magic-angle) bilayer graphene has also drawn extensive attention recently.[53] Additionally, beyond the thermal fluctuations limit of Meimin–Wagner theorem,[3,54] 2D van der Waals (vdW) materials have also been revealed with intriguing magnetic properties.[55–57] Below, we will discuss the properties in detail.

With tolerable structure, exotic electronic property, tunable capability as well as superior specific area, 2D materials have exhibited promising potential in a wide range of fields such as electronic/optoelectronic devices,[58,59] catalyst,[60] purification and separation engineering[61] and battery/capacitor systems.[62] However, till now, the majority of applications are limited to laboratory scale. Thus, an increasing number of lab-to-industry practices is expected. To this end, easy and affordable fabrication methods for 2D materials with satisfactory quality are also of great significance. This review provides an overview of basic information and properties of the common 2D materials as well as the general synthetic approaches and applications.

2. Properties of 2D Materials

This section will review some basic properties of most widely studied 2D materials such as graphene, GO, *h*-BN, TMDCs and BP.

Young's modulus and breaking strength are two key parameters to describe the mechanical properties of materials. Regarding the 2D materials, a 2D elastic constant E^{2D} with unit of force/length has been used to indicate their elastic stiffness due to the dimensionality, where E^{2D} = Young's modulus× thickness.[49,68] Lee *et al.*[49] measured the elastic properties of monolayer graphene by atomic force microscopy (AFM) and obtained the Young's modulus, E^{2D} and breaking strength as 1 ± 0.1 TPa, $342\pm$ 30 N/m and 130 ± 10 GPa, respectively. Suk *et al.*[68] investigated the mechanical properties of monolayer GO and obtained the Young's modulus of $207.6\pm$ 23.4 GPa and the corresponding E^{2D} of $145.3\pm$ 16.4 N/m. Such a degradation of the elastic properties of GO in comparison with graphene is attributed to the significant number of defects. The mechanical properties of BP are strongly anisotropic along the zigzag and armchair directions.[25,27] Tao *et al.*[26] tested and obtained the Young's modulus of BP to be 27.2 ± 4.1 GPa and 58.6 ± 11.7 GPa in the armchair and zigzag directions, respectively. The experimentally measured mechanical properties of several typical 2D materials are summarized in Table 1 (we also include those for stainless steel for comparison).

We are now turning to the electronic properties of 2D materials. A schematic band structure of graphene is shown in Fig. 2(a). Carbon atoms in graphene are sp^2-hybridized, leaving a free valence electron

Table 1. Experimentally measured mechanical properties of several typical 2D materials versus stainless steel.

Material	Young's modulus (GPa)	E^{2D} (N/m)	Breaking strength (GPa)	Source
Graphene (single layer)	~ 1000	~ 342	~ 130	Ref. 49
GO (single layer)	~ 207.6	~ 145.3	—	Ref. 68
h-BN (two layers)	~ 216.5	~ 223	~ 8.5	Ref. 69
MoS_2 (single layer)	~ 270	~ 180	~ 23	Ref. 70
BP (few layers), zigzag	~ 58.6	—	~ 4.8	Ref. 26
BP (few layers), armchair	~ 27.2	—	~ 2.3	Ref. 26
Stainless steel ASTM-A514	205	—	0.9	Ref. 70

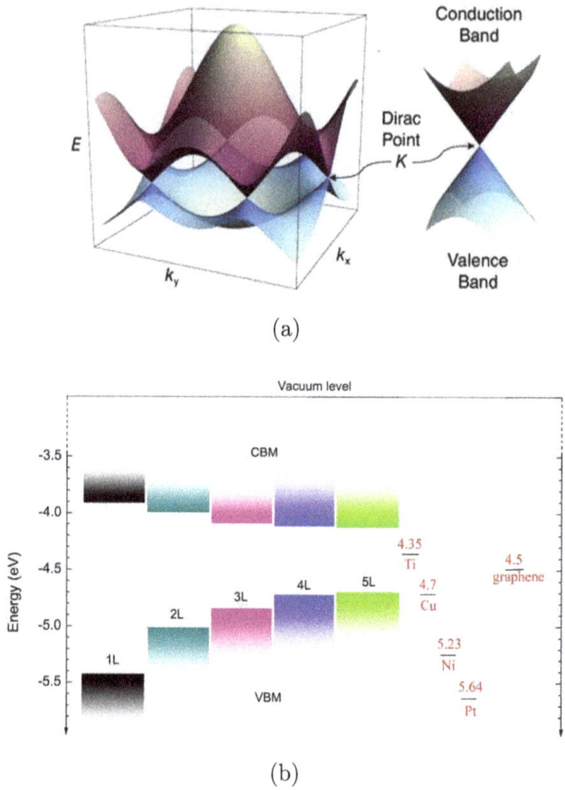

Fig. 2. (a) The schematic band structure of graphene. (b) The variations of conduction band minimun (CBM) and valence band maximum (VBM) in BP of difference thickness.

Source: Panel (a) adapted with permission from Ref. 73. Copyright © 2012 Wiley-VCH Verlag GmbH & Co. KGaA, Weinheim. Panel (b) adapted with permission from Ref. 74. Copyright © 2014 The author(s).

at each carbon atom. Valence and conduction bands touch each other at two inequivalent K- and K′-points of the Brillouin zone, forming linear dispersive conical low-energy bands.[16,71] Thus, the low-energy electronic excitations in graphene can be described by Dirac equation rendering massless chiral quasirelativistic particles.[8,39,72] The conical band structure of graphene enables observation of both hole and electron conductivities, which leads to ambipolar field effect.[8] High quality of graphene crystals, combined with the low effective mass (zero at the Dirac point), leads to observation of ballistic transport and electron mobilities as high as 200,000 $cm^2 \cdot V^{-1} \cdot s^{-1}$ even at room temperatures.[50]

Unlike graphene, BP is a semiconductor with a $\sim$ 2-eV direct bandgap at the Γ-points of first Brillouin Zone.[47,75] The bandgap in BP strongly depends on the number of layers. The mobility of charges in BP was also found to be thickness-dependent, and was estimated to be 1000 $cm^2 \cdot V^{-1} \cdot s^{-1}$ for BP with a thickness of $\sim$ 10 nm.[47] As for MoS_2, a typical TMDC, the charge carrier mobility of its monolayer form can be as high as 200 $cm^2 \cdot V^{-1} \cdot s^{-1}$.[46] Bandgaps, corresponding transistors' on/off ratios, carrier mobilities as well as thermal conductivities of some 2D crystals in mono- and few-layer forms are summarized in Table 2.

Owing to monolayer thickness, 2D materials usually have high optical transmittances. For instance, the light transmittances of graphene and GO can be, respectively, up to 97.7% and 95%,[51,92] and scale with the number of layers. Differently, the optical properties of semiconducting 2D crystals change dramatically with the number of layers. As have been illustrated, most TMDCs are direct bandgap semiconductors, whereas their 3D counterparts are typically indirect ones.[45,93] In contrast, BP exhibits direct bandgap independently of the thickness.[91] Besides, due to the quasiparticle bandgap feature and strong Coulomb interaction between charge carriers,[94] the bandgap values of monolayer 2D materials vary noticeably with the number of layers. The optical bandgap information of several common 2D materials and the bulk materials are shown in Table 2. The layer-number-dependent bandgap value of BP is presented in Fig. 2(b).

Apart from the exceptional mechanical, electronic and optical properties, graphene is also reported to have a superior thermal conductivity as high as 5300 $W \cdot m^{-1} \cdot K^{-1}$ at room temperature.[52] However, unlike graphene, the thermal conductivities of other 2D materials appear to be significantly lower comparing to their bulk counterparts. This could be explained by the interfaces-induced thermal resistance or newly formed phonon bands in these 2D materials, and on this basis, defects, mechanical strains, thickness and size can make a difference on the thermal conductivities of 2D materials.[99]

Table 2. The charge carrier mobilities, bandgaps, on/off ratios of corresponding transistors and thermal conductivities of typical 2D materials and their bulk materials.

Material	Mobility ($cm^2 \cdot V^{-1} \cdot s^{-1}$)	Bandgap (eV)	Transistor I_{on}/I_{off}	Thermal conductivity ($W \cdot m^{-1} \cdot K^{-1}$)	Source
Graphene (monolayer)	200,000	0	—	5300	Refs. 50,52 and 76
GO (multilayer)	—	0.72	—	3.77	Refs. 77 and 78
Graphite	6.7	0	2–7	2600	Refs. 43,79 and 80
h-BN (few layers)	—	5.97	—	227–280	Refs. 36,43 and 81
Bulk *h*-BN	—	5.92	—	400	Refs. 36 and 43
MoS_2 (monolayer)	200	1.9, D^c	10^8	131	Refs. 45,46 and 82
Bulk MoS_2	200–500	1.3, I^c	—	18.06 (in-plane) 4.17 (out-of-plane)	Refs. 46,83 and 84
$MoSe_2$ (monolayer)	50	1.55, D	10^6 for e^- 10^6 for h^+	54	Refs. 85 and 86
Bulk $MoSe_2$	100–160	1.1, I	—	—	Refs. 85 and 87
WSe_2 (monolayer)	200[a]	1.7, D	10^7	53	Refs. 86,88 and 89
Bulk WSe_2	500	1.2, I	—	9.7 (in-plane) 2 (out-of-plane)	Refs. 89 and 90
BP (few layers)	1000[b]	0.8–1.5, D	10^5	10 (armchair) 20 (zigzag)	Refs. 47 and 91
Bulk BP	—	0.3, D	—	20[d] (armchair) 40[d] (zigzag)	Ref. 91

Notes: [a]at 160 K; [b]the thickness of corresponding BP is about 10 nm; [c]D means direct bandgap and I means indirect bandgap; and [d]thickness is larger than 15 nm.

For instance, the thermal conductivity of GO is significantly reduced and is measured to be $3.77 \pm 0.09\,\mathrm{W \cdot m^{-1} \cdot K^{-1}}$.[78] Thermal conductivity of puckered 2D materials such as BP and TMDCs could differ in different directions (armchair or zigzag). As an example, the thermal conductivities of few-layer BP are $\sim 10\,\mathrm{W \cdot m^{-1} \cdot K^{-1}}$ and $\sim 20\,\mathrm{W \cdot m^{-1} \cdot K^{-1}}$ for armchair direction and zigzag direction, respectively.[100] Table 2 summarizes the thermal conductivities of a number of 2D materials and their bulk counterparts.

Raman spectroscopy is one of the most reliable and fast methods to characterize 2D materials. It can render information on the amount and type of defects, doping as well as the layer numbers of

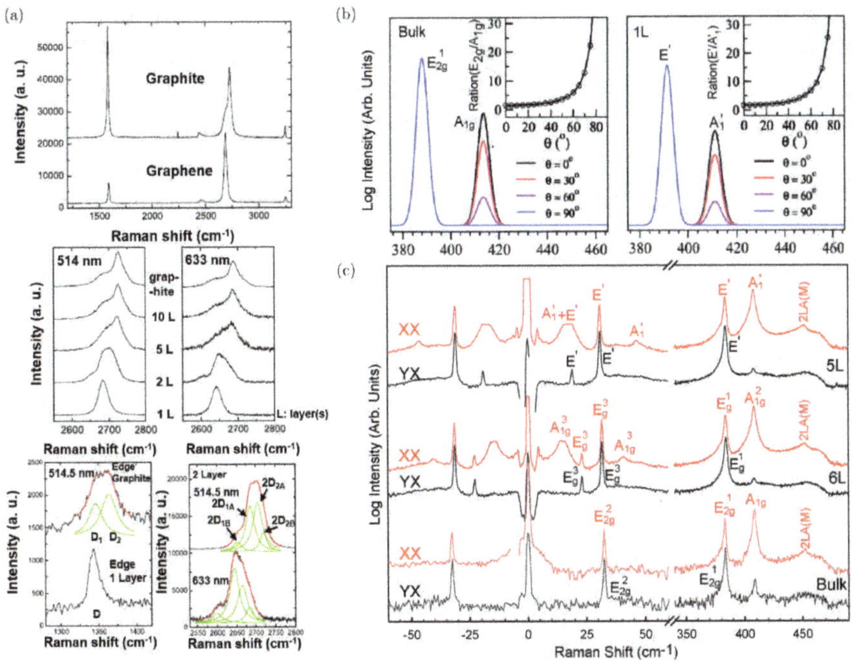

Fig. 3. (a) Raman shifts of graphite and graphene. (b) Computational and (c) experimental Raman shifts of monolayer, few layers and bulk MoS_2.[96–98]

Source: Panel (a) adapted with permission from Ref. 95. Copyright © 2006 The American Physical Society. Panels (b) and (c) adapted with permissions from Ref. 96, Copyright © 2014 The Royal Society of Chemistry; Ref. 97, Copyright © 2013 American Physical Society; and Ref. 98, Copyright © 2015 The Royal Society of Chemistry.

corresponding 2D crystals.[95,101,102] Graphene has two intense characteristic Raman peaks at $\sim 1580\,\mathrm{cm}^{-1}$ and $\sim 2700\,\mathrm{cm}^{-1}$ in its Raman spectra, which are called G- and $2D$-peaks, respectively.[95] Therein, G-peak is related to the zone center E_{2g} mode while $2D$ denotes the second order of D-peak, which is not active in the pristine samples. The D-peak at $\sim 1350\,\mathrm{cm}^{-1}$ is activated when the hexagonal symmetry is broken and can be caused by defects.[95] The Raman spectra of mono- and few-layer graphene and graphite are shown in Fig. 3(a). As the number of layers increases, the $2D$-peak grows wider while its intensity decreases and the peak position moves towards higher Raman shift. In graphite, the G-peak position is 3–$5\,\mathrm{cm}^{-1}$ lower than that in monolayer graphene, and the $2D$-peak could split into two subpeaks namely $2D_1$ and $2D_2$ with the intensities of $1/4$ and $1/2$ of G-peak, respectively. Taking advantage of this Raman shift feature, we can characterize the quality of the crystal as well as the number of layers. Similarly, other 2D materials such as TMDCs have also exhibited thickness-dependent Raman shifts. The theoretical and experimental Raman shifts of monolayer and bulk MoS_2 are exhibited in Figs. 3(b) and 3(c), respectively.[96–98] Interestingly, owing to their abundant density of states near the Fermi level, 2D materials such as graphene and NbS_2 have been utilized as efficient substrates for plasmon-free surface-enhanced Raman scattering.[103]

Recently, the magnetic properties of 2D materials have also attracted intensive attention. The 2D vdW magnet families, such as 2D halides (e.g. CrI_3, $CrBr_3$) and 2D chalcogenides (e.g. $Fe_2P_2S_6$, $Cr_2Ge_2Te_6$ and Fe_3GeTe_2),[54–56,105–108] have been experimentally and theoretically excavated.[109] In the above 2D magnets, magnetic anisotropy, by producing a nonzero excitation gap in magnon spectrum, is thought to play the key role in preserving the 2D long-range magnetic order and thus counteracts the thermal fluctuation limits.[54,105,109] Different models have been explored to explain the origin of magnetic anisotropy in 2D vdW magnets, such as the Heisenberg model (e.g. in $Cr_2Ge_2Te_6$),[110] Ising model (e.g. in CrI_3)[111] and Stoner model (e.g. in Fe_3GeTe_2).[54,107] Generally, the spin–orbit interaction as well as single-ion anisotropy of the atoms (e.g. spin–orbit interaction of I ligand and the magnetic moment carried by Cr^{3+}

in CrI_3) and strong electron correlation (e.g. in Fe_3GeTe_2) are accounted for the magnetic anisotropy in 2D magnets.[54,55,105,107] Compared to their 3D counterparts, the Curie temperatures of 2D vdW magnets usually appear lower. For instance, the Curie temperature of bulk CrI_3 is 61 K while that of monolayer CrI_3 is 45 K,[111] and the Curie temperature of 3D Fe_3GeTe_2 is 207 K while that of 2D Fe_3GeTe_2 is 130 K.[107] Moreover, room-temperature ferromagnetism in 2D Fe_3GeTe_2 has also been realized with the help of gate control.[57] Interestingly, the magnetic properties of 2D vdW magnets can be controlled by adjustable factors such as electron doping,[55,112] layer number,[111] stacking[113] and light.[108] To illustrate, the layer-number-dependent Curie temperature of 2D vdW magnet is shown in Fig. 4. With the intrinsic and controllable ferromagnetic/antiferromagnetic properties, 2D magnets are promising candidates for spintronic devices and memory devices, and exceptional platforms for studying the magnetic and electronic properties. Additionally, the applications such as photodetector,[108] oxygen evolution reaction (OER)[114] and adsorption with 2D vdW magnets have also been investigated.[115]

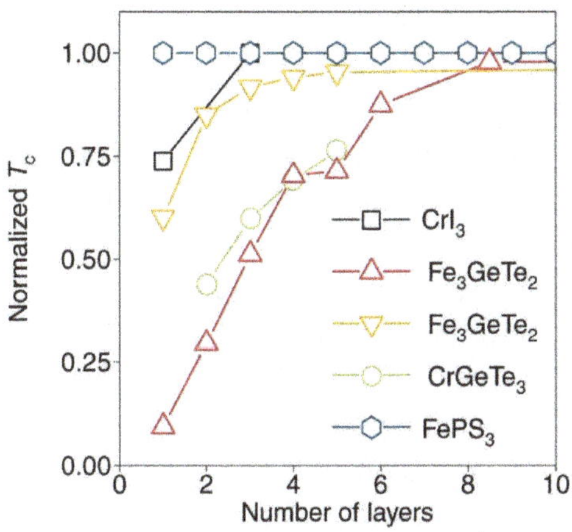

Fig. 4. Variations of Curie temperatures (T_c) of 2D magnets with the layer number. *Source*: Figure adapted with permission from Ref. 104. Copyright © 2019 Springer Nature Limited.

3. Synthetic Approaches to Produce 2D Materials

At the moment, there is a large variety of mass-production methods of 2D materials. The choice of the particular method depends on the desired quality and the form factor of the particular application. Till now, mechanical exfoliation (e.g. repeated peeling), liquid-phase exfoliation, chemical vapor deposition (CVD) and physical vapor deposition (PVD) have been reported as effective approaches for the fabrication of 2D materials.[6,7,116–119] Traditionally, mechanical exfoliation is the method used at the laboratory scale.[6,7] The other production methods can be divided into top–down (e.g. solution-phase exfoliation) and bottom–up (e.g. CVD).

Bulk layered materials are formed by van der Waals attractions among the adjacent layered units. To obtain the monolayer 2D materials, one approach is to break the van der Waals attractions in bulk layered materials with mechanical force. Mechanical exfoliation was first used to produce graphene and is still a very popular method which allows easy production of high-quality 2D crystals of few microns (up to hundreds of microns) in size.[6,7] Figure 5(c) exhibits the optical image of mechanical exfoliated MoS_2 with different layers on Si/SiO_2 substrate.[124] Even though the monolayer 2D materials obtained from mechanical exfoliation are of good quality, the mechanical exfoliation methods are inefficient for massive production. The bulk versions of micromechanical cleavage are shear force-assisted exfoliation, sonication-assisted exfoliation and ball-milling exfoliation. All these methods have been adapted to produce 2D materials such as graphene, *h*-BN and TMDCs.[35,125–127]

A more efficient top–down approach is the chemical-assisted exfoliation. It was reported that strong oxidative agents can intercalate into the adjacent layers of the bulk materials and thus expand the interlayer spacing to exfoliate 2D materials from the bulk materials.[35] In the case of graphene, this method includes oxidation of hydrophobic graphite into hydrophilic graphene oxide. Generally, the Hummers method is used, where the oxidative radicals NO_3^-, SO_4^{2-}, MnO_4^- and H_2O_2 in an acid environment are employed to oxidize graphene and foster exfoliating GO from stacked graphite, forming stable water suspension of GO.[128] In addition, reduction-based

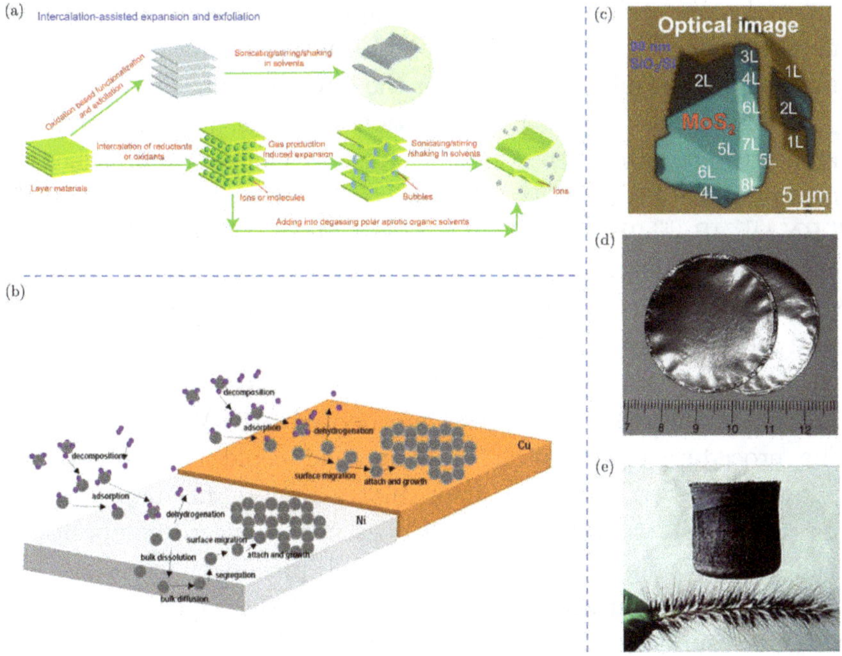

Fig. 5. (a) A schematic of chemical-assisted exfoliation. (b) The schematic CVD growth mechanism of graphene on Cu and Ni with CH_4. (c) Optical image of single-layer crystal of mechanically cleaved 2D MoS_2. (d) GO membrane (GO paper) obtained by filtration. (e) Ultrathin graphene-based aerogel.

Source: Panel (a) adapted with permission from Ref. 35. Copyright © 2018 The Royal Society of Chemistry. Panel (b) adapted with permission from Ref. 120. Copyright © 2013 Wiley-VCH Verlag GmbH & Co. KGaA, Weinheim. Panel (c) adapted with permission from Ref. 121. Copyright © 2013 American Chemical Society. Panel (d) adapted with permission from Ref. 122. Copyright © 2008 Wiley-VCH Verlag GmbH & Co. KGaA, Weinheim. Panel (e) adapted with permission from Ref. 123. Copyright © 2013 Wiley-VCH Verlag GmbH & Co. KGaA, Weinheim.

exfoliation, ion-exchange-assisted exfoliation and etching-assisted exfoliation are effective chemical-assisted exfoliation methods which have been explored to obtain TMDCs, LDHs and MXenes.[31,116,129] A schematic procedure of chemical-assisted intercalation exfoliation is shown in Fig. 5(a). In some cases, chemical-assisted exfoliation can functionalize the 2D materials with oxygen-contained radicals (e.g. GO from graphite), which provides active reaction sites for multiple applications such as catalyst. Generally, the chemically processed 2D materials are of inferior quality compared with the intrinsic 2D materials. Besides, the chemical solution waste produced during the processing steps might be a risk for the environment.

In addition to top–down exfoliation methods, bottom–up CVD methods have also been widely studied as efficient approaches to mass-produce high-quality 2D materials such as graphene,[117,130] *h*-BN,[131] BP,[132] 2D TMDCs (e.g. MoS_2, $MoSe_2$, $MoTe_2$, WS_2, WSe_2 and $HfSe_2$)[133,134] and 2D heterostructures (e.g. 2H-$MoTe_2$–1T′-$MoTe_2$ and MoS_2–$NbSe_2$).[135] Generally, the precursor materials, a catalytic substrate such as Cu, Ni and sapphire (where the 2D material is formed) and a tube furnace (accommodating and controlling the reaction process) are needed for CVD fabrications. The procedure of a typical CVD method could be described as follows[136,137]: (i) The precursor materials are decomposed (or precursor gas can be used, which does not require decomposition, such as methane); (ii) the decomposed precursor materials are delivered to the surface of the substrate; (iii) the vapor decomposes at the hot surface of the substrate (catalytically or noncatalytically); (iv) atoms reorganize onto the surface of the substrate and form 2D materials; and (v) 2D materials are separated from the substrate and thus the freestanding 2D materials are obtained.

The graphene growth by CVD can be divided into two large classes, depending on the solubility of carbon in the metal. Thus, growth on Ni or Fe, where carbon solubility is large, goes through precipitation mechanism.[117] CH_4:H_2:Ar mixture was applied as the precursor and 300-nm Ni-coated SiO_2/Si was used as the substrate. By heating the sample up to 1000°C, a large quantity of carbon was dissolved in Ni. Upon cooling of the substrate, graphene layer was grown on the surface of Ni by precipitation. After an etching process, a floating or stamped graphene was finally fabricated. The disadvantage of the method is the uncontrollable thickness of graphene growth.

Alternatively, growth on copper, where solubility is very low, develops under a different scenario.[138] Upon catalytic cracking on the hot surface of copper, carbon atoms do not penetrate inside the substrate, but instead arrange themselves into graphene crystal.

Copper surface covered by graphene largely loses its catalytic properties, making the growth process self-limiting, resulting in high-quality graphene crystals. schematic of CVD growth mechanisms of graphene on Ni and Cu substrates are depicted in Fig. 5(b).

CVD methods have been widely employed for both graphene and other 2D materials such as *h*-BN and TMDCs.[131,133]

In the CVD growth of *h*-BN, there are normally a thermally active BN precursor (e.g. ammonia borane) which provides the raw source in the upstream, and a substrate (e.g. Cu, Ni, Fe and metal alloy) which is the platform for growth of *h*-BN in the middle of the tube furnace chamber.[139–142] High temperature leads to the decomposition of the precursor, which is brought by the flow of the gas carrier to the substrate, on which the *h*-BN can be grown.[131,143] Both atmosphere pressure CVD (APCVD) and low-pressure CVD (LPCVD) have been utilized for the growth of *h*-BN; while the CVD reaction of former is recognized by mass transport control, that of latter is done by surface reaction control.[131] The size of the aimed *h*-BN film by CVD relies on the substrate.

As for fabricating 2D TMDCs with CVD methods, various approaches have been implemented such as direct metal sulfurization,[144] thermolysis of thiosalts[145] and sulfurization of metal oxide.[133,146] Zhan *et al.* utilized pre-deposited Mo on SiO_2 as the substrate for growing MoS_2 by CVD, where sulfur acted as the S source in the upstream ceramic boat.[144] In this direct sulfurization method, they found that the size and thickness of MoS_2 film were dependent on the size of SiO_2 and the thickness of Mo. Sang *et al.* reported an investigation on thermolysis growth of 2D MoS_2.[145] The 2D MoS_2 flakes were synthesized by thermally decomposing $(NH_4)_2MoS_4$ under different temperatures of 300°C, 600°C and 900°C on different substrates (carbon film or freestanding). They found that grain size, defined edge structure and layer number increased as the temperature increased. Furthermore, substrate and morphology of the precursor were proved to impact the grain coalescence in the process of forming MoS_2 flakes. Huang *et al.* applied Se to react with WO_3 under 925°C and achieved large-size monolayer WSe_2 upon the sapphire under

Ar/H_2 gas carrier atmosphere,[146] where H_2 was thought to activate the reaction between WO_3 and Se. Overall, the source-to-substrate distance, temperature, gas carrier and the precursors and substates can influence the nucleation density and grain in CVD growth for TMDCs, and thus impact the structure of obtained TMDCs.[134] Intriguingly, 2D in-plane heterostructure 2H-$MoTe_2$–1T′-$MoTe_2$ and out-of-plane heterostructure MoS_2–$NbSe_2$ have been successfully realized by molten-salt-assisted CVD.[147] Other than TMDCs, CVD for the growth of 2D BP is also reported.[132]

One of the advantages of the bottom–up methods is the ability to control the doping (e.g. doping N into graphene[148]) as well as the fabrication of alloys. The bottom–up CVD growth also makes it possible to produce designed structure and size of 2D materials by adjusting the variables such as the precursors, substrates, pressure, rate and composition of the carrier gas, temperature, precursor-to-substrate distance and heating process. Recently, massive production of graphene films via CVD has also been realized.[137] However, uniformity of the 2D materials' thickness and the relatively high energy cost (as the samples are heated from hundreds up to 1000°C) remain as the challenges for CVD methods.[131] To this end, modified CVD methods have also been reported such as metal–organic CVD (MOCVD, for high uniformity and large area production) and plasmon-enhanced CVD (for contaminant-free and low-temperature production).[119,149,150]

Besides CVD, PVD is also a well-explored bottom–up method for fabricating 2D materials such as graphene,[151] h-BN,[152] TMDCs (e.g. MoS_2),[153] 2D vdW magnets (e.g. Fe_3GeTe_2)[109] and 2D vdW heterostructure (e.g. lateral heterostructure $MoSe_2$–WSe_2).[119,154,155] Generally, PVD employs a heated atomic source to deposit related 2D crystals on the substrate.[155] Compared to CVD, ultrahigh vacuum (UHV) conditions and high quality of atomic sources (well calibrated and of high purity) are usually required for PVD. Molecular beam epitaxy (MBE) and pulsed laser deposition (PLD) are the two main PVD approaches.[119] Remarkably, MBE method has a good reputation for concise control of thickness and stoichiometric growth, while PLD provides fast deposition capability but with uncontrollable thickness.[118,119,156] In addition, atomic layer deposition (ALD) with analogous CVD process also has been reported for low-temperature, scalable, thickness-controllable and highly crystalline production.[157,158] In practice, 2D materials can be processed into macro-assemblies for particular usage. Among them, filtered GO membranes and 3D hybrid graphene sponges have been widely explored, as shown in Fig. 5(d) and Fig. 5(e) respectively.

The 2D materials are often combined with some target substrates and appear as composites to fulfill specific functions. For instance, graphene composite coatings have a bright perspective for different surface-related applications including anticorrosion and biomedical applications. To fabricate the composite coatings, the existing conventional industrial production techniques can be used for both manufacturing and up-scaling. During the coating engineering process, graphene undergoes noncovalent modifications through the interaction of π–π bonds.[159–163] This type of modification is highly beneficial to the applications requiring high surface area and the electrical conductivity of graphene.[164] Noncovalently functionalized composites of graphene nanosheets and polymers can be produced via π–π interactions.[165] Such materials exhibit excellent conductivity and durability. It was shown that the coatings performed their functions even after losing integrity. Illustration of graphene-based composites is made in Fig. 6.

In another attempt, Shen *et al.*[166] synthesized a more complex composite composed of graphene, metallic zinc particles and polyaniline (PANI). The optimum coatings showed a combination of cathodic protection and physical barrier corrosion protection mechanisms. The samples with high ratios of zinc showed a significant cathodic protection mechanism due to the large proportions of zinc particles and the presence of graphene that improved electrical conductivity between the steel substrate and zinc particles. Richards *et al.*[167] dispersed graphene nanoplatelets (GNPs) in polyvinyl butyral (PVB) and coated the as-fabricated composite on hot-dip-galvanized steel coupons. The GNPs formed a galvanic couple with the metallic substrate in which the GNP played the role of the cathode and protected the substrate. Furthermore, the PVB–GNP

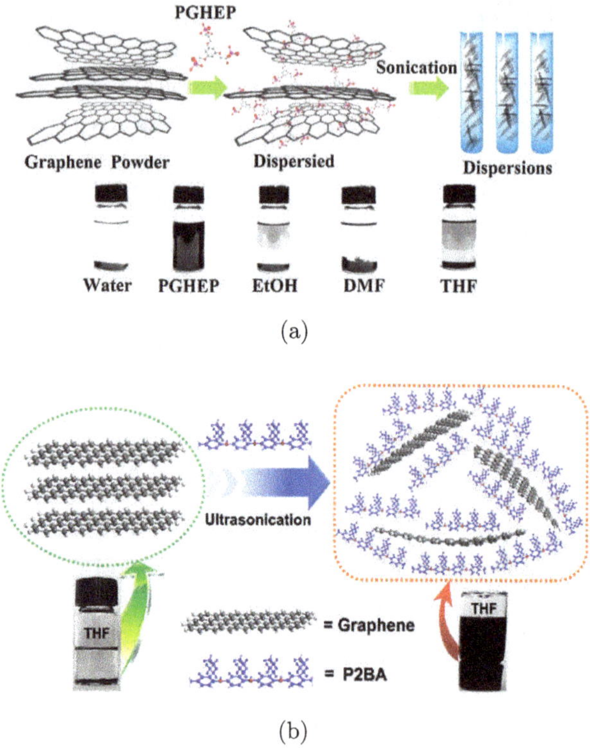

Fig. 6. Graphene composites and the fabrication process. (a) Graphene and hydroxyl epoxy phosphate monomer (PGHEP) composite. (b) Graphene and poly (2-butylaniline) (P2BA) composite.

Source: Panel (a) adapted with permission from Ref. 159. Copyright © 2017 Elsevier B.V. Panel (b) adapted with permission from Ref. 160. Copyright © 2016 Elsevier, Ltd.

composite coating successfully reduced the oxygen permeation rate over an order of magnitude.

As mentioned above, numerous methods including both top–down and bottom–up approaches have been illustrated. Still, efforts might need to be spared to pave the way for high-quality and industrially applicable production of 2D materials, as the laboratory synthesis and industrial production of 2D materials and composites are essential for their applications. Within the exceptional properties, 2D materials exhibit a wide range of applications. In the next section, we highlight the versatile applications in multiple fields including electronics,[58] magnetics,[109] optoelectronics,[59] catalyst,[60] ion/molecular purification and separation,[61] protective coatings and battery and capacitor system.[168–170]

4. Applications of 2D Materials

The promising versatility of 2D materials gives them a conspicuous place in the construction of functional electronic devices. With a majority of 2D materials such as TMDCs and BP being semiconductors with satisfactory charge carrier mobility, 2D materials-based channel transistors such as field-effect transistor (FET) have been intensely studied for multiple usage. Therein, sensors and nonvolatile memories (NVMs) appear to be the two main aspects.[171,172]

Various 2D materials such as rGO, BP and MoS_2 have been investigated as charge channels of FET-based sensors for gas (e.g. CO, H_2 and NH_3) and bio molecules in solutions (human immunoglobulin G).[171,173] Li *et al.*[174] deposited MoS_2 films on Si/SiO_2 wafer to construct FET devices for NO detection. The device configuration is depicted in Fig. 7(a). They checked the thickness-dependent characteristic of this MoS_2-based FET and found that monolayer MoS_2 FET showed a fast, dramatic but unstable response for NO while two-layer, three-layer and four-layer MoS_2 FETs exhibited stable as well as sensitive responses with a detection limit of 0.8 ppm for NO at room temperature.

In addition, NVM is another promising application for 2D materials-based channel transistor. NVMs are gradually replacing hard disk drives in the global storage market, and 2D materials exhibit good potential for nonvolatile memories to take the place of their conventional basis material — silica, which is approaching its physical limit.[172] Many 2D materials-based NVMs have been reported including resistive NVMs, ferroelectric NVMs and flash memories, where MoS_2,[176] BP,[177] WSe_2[178] and graphene are mainly involved.[172] Graphene is thought to act as ultrathin, transparent and flexible electrodes or as a protective interfacial layer for decreasing the power consumption in resistive NVMs.[179] Beyond graphene, TMDCs offer more properties such as on-demand

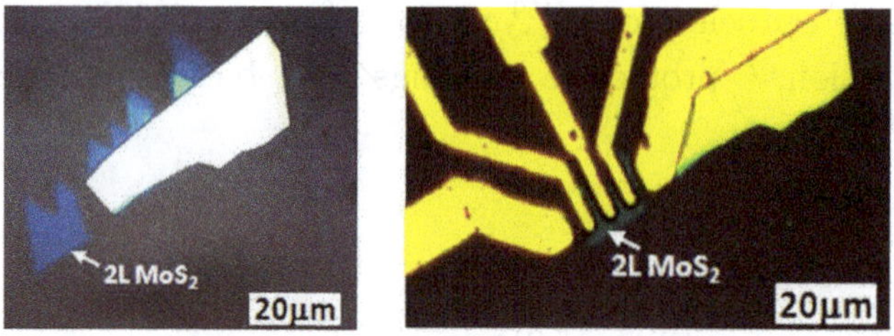

(a)

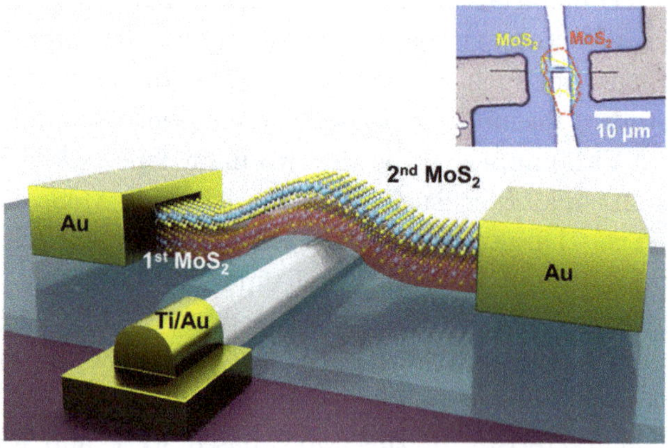

(b)

Fig. 7. (a) MoS_2-based FET device for NO sensor. (b) MoS_2-based channel device for NVM.

Source: Panel (a) adapted with permission from Ref. 174. Copyright © 2012 Wiley-VCH Verlag GmbH & Co. KGaA, Weinheim. Panel (b) adapted with permission from Ref. 175. Copyright © 2020 Wiley-VCH Verlag GmbH & Co. KGaA, Weinheim.

bandgap, tunable oxidation states and chemical reactive surface.[172,176,178] Recently, Kim *et al.*[175] fabricated a novel 2D TMDC-based channel transistor for NVM application with 1D ZnO nanowire as the gate. In this 1D–2D mixed structure, two MoS_2 nanoflakes were stacked on 1D ZnO nanowire, as illustrated in Fig. 7(b). It was revealed that the interface between first and second MoS_2 flakes could produce deep-level electron traps which contribute to long retention time of the NVM over 10^4 s. ZnO nanowire here acted as an n-type channel in this architecture. Apart from graphene and 2D semiconductors, h-BN also serves as exceptional dielectric gate due to its large bandgap.[43]

As mentioned previously that existing 2D materials play important roles in channel transistors, a much wealthy source named vdW heterostructures based on 2D materials would open up abundant applicable potential beyond the NVM.[15,156,172,181] Since the atomic surface and bandgap differ among the 2D materials, ranging from zero-bandgap semimetallic graphene, semiconductor TMDCs, to insulating h-BN, the vdW heterostructures composed of alternative stacks of different 2D materials or other-dimensional materials are destined to produce plenty of fascinating science as well as larger potential for wide applications including in electronics and optoelectronics.[15,156] The methods for obtaining 2D van der Waals heterostructures can be divided into mechanically assembled stacks and physical or chemical epitaxy (PVD or CVD),[15] as shown in Fig. 8(a). According to the dimensionality of the composition, van der

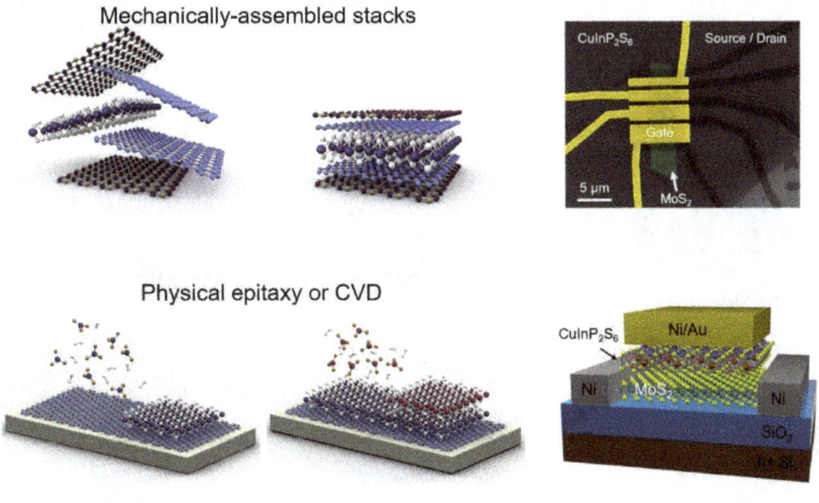

Fig. 8. (a) Two approaches for 2D/2D vdW heterostructures. (b) Top-view (top) and schematic (bottom) structures of 2D vdW NVM device.

Source: Panel (a) adapted with permission from Ref. 15. Copyright © 2016 American Association for the Advancement of Science. Panel (b) adapted with permission from Ref. 180. Copyright © 2018 American Chemical Society.

Waals heterostructures can be divided into 0D/2D, 1D/2D, 2D/2D, 3D/2D and 3D/3D structures.[156] With these distinct 2D materials-based vdW heterostructures, various structures and electronic devices can be constructed such as FET, giant magnetoresistance (GMR), photodiode, superlattice and photoconductor. An example of $CuInP_2S_6$–MoS_2-based 2D vdW heterostructure device for ferromagnetic NVM is exhibited in Fig. 8(b).[180] Moreover, by integrating the basic heterostructures into designed systems, exciting and functional devices with super-characteristic can be realized such as complementary metal–oxide–semiconductor (CMOS) circuits and more complicated systems, which could possess the potential to be next-generation computing technologies, substituting the current Si-based chips.[156,181] An integrated stimulus-responsive smart device based on 1D V_2O_5–2D graphene microsupercapacitor is exhibited in Fig. 9.

With exotic properties of 2D van der Waals heterostructures mentioned above, another practical usage of these heterostructures is for catalyst. For 2D van der Waals heterostructures, they can have a greater catalytic activity compared to single 2D crystals, which might originate from their changes in work function, conduction or valence

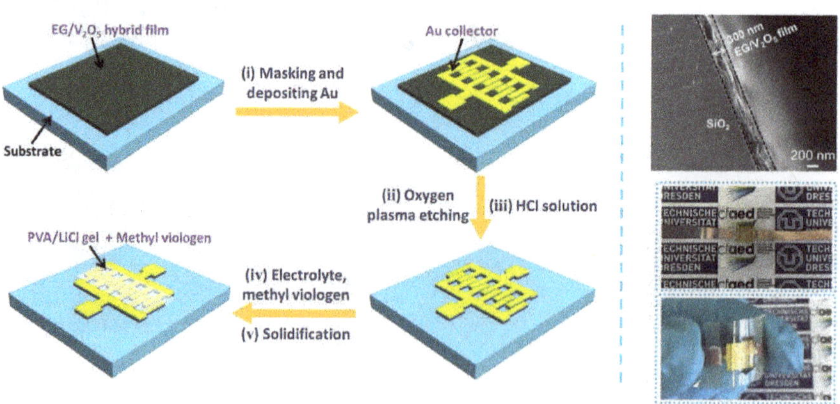

Fig. 9. Fabrication (left) and device structure (right) of 2D graphene–1D V_2O_5 microcapacitor-based smart stimulus-responsive device.

Source: Figure adapted with permission from Ref. 182. Copyright © 2016 Wiley-VCH Verlag GmbH & Co. KGaA, Weinheim.

band position and the density of states.[60,183] Zhang *et al.*[184] fabricated a 2D NiS/graphene heterostructure for efficient electrocatalytic water splitting. The as-prepared 2D NiS/graphene heterostructures performed water splitting at a low cell voltage of 1.54 V with a current density of 10 mA/cm^2, the result of which surpassed the conventional Pt/C and IrO_2 couples.

Based on the atomic magnetic properties of 2D vdW magnets, related spintronic and optoelectronic devices have emerged. Both 2D halide and 2D chalcogenide magnets-based spintronics have been explored such as graphite–CrI_3–graphite and Fe_3GeTe_2–BN–Fe_3GeTe_2 magnetic tunnel junctions.[109,185,186] With the atomic thickness of 2D magnets, large tunneling magnetoresistance is observed on the spintronics. Wang *et al.*[186] reported a tunneling resistance behavior based on Fe_3GeTe_2–BN–Fe_3GeTe_2 heterostructure device. A magnetoresistance of 160% was obtained with this magnetic tunnel junction at 4.2 K, where the spin polarization was 66%. Moreover, they found that the spin polarization was proportional to the applied magnetization with increasing temperature. A much high magnetoresistance of 10,000% has been realized with 2D CrI_3.[185] Utilizing exfoliated CrI_3, Wang *et al.*[185] established the vertical sandwiched graphite–CrI_3–graphite FET. Within this heterojunction device, the maximum magnetoresistance was observed at 10 K under an applied magnetic field of 2 T. Magnetic states transition in 2D CrI_3 was found to account for this GMR, which also indicated the strong coupling between the transport and magnetism in 2D vdW magnets. Though large and tunneling magnetoresistance has been realized with 2D vdW spintronics, it remains to be further developed since most of the demonstrated devices operate at low-temperature conditions from several K to dozens of K. With the photoresponsive magnetism of 2D magnets, Xie *et al.*[108] reported a 2D $Cr_2Ge_2Te_6$-based FET photodetector. In this detector, a negative/positive photoconductivity switching was revealed along with the change of the light intensity and a detection limit of light intensity as low as 0.04 pW accompanied with a responsivity as high as 340 A$\cdot$W^{-2} was achieved.

Theoretical simulations have also indicated the applicable potential for 2D magnets in other areas such as adsorption and

OER, due to their full potential for surface chemical functionalization.[114,115] The computational simulation on adsorption capability of 2D ferromagnetic Fe_3GeTe_2 was reported by Cui *et al.*[115] In their work, various gas molecules such as CO, CO_2, NO, H_2, H_2O and O_2 were tested, and found that H_2O, CO_2 and H_2 were physisorbed and that CO, NO and O_2 were chemisorbed on the surface. Remarkably, the gas molecule adsorption capacity of 2D Fe_3GeTe_2 was found to be superior to that of well-studied 2D MoS_2 and MXenes. Zhao *et al.*[114] revealed the capability of 2D Fe_3GeTe_2 for OER catalysis. The preserved metallic character and abundant exposed active sites of 2D Fe_3GeTe_2 were thought to be the key points for good OER performance. In general, with the intriguing properties of 2D magnets, more promising applications in compact nanoelectronic or magnonic devices and other fields can be expected.

Back to pristine 2D materials including graphene, GO/rGO, TMDCs, *h*-BN and *g*-C_3N_4, huge specific area, robust structure, tunable bandgap and density of states and good thermal and electric conductivities have made them excellent candidates for various catalysts such as oxygen reduction reaction (ORR),[148,187] OER,[188] photoinduced water-splitting,[189] CO_2 activation[190] and hydrogen evolution reaction (HER).[60,191] Defects engineering and doping are two effective methods to control the catalytic performance of 2D materials, since defects or edges can act as active sites for catalytic reaction while doping with the external atoms such as N and S can change the corresponding electronic states and the doped atoms also can serve as active sites. Particularly, due to the suitable bandgap value of $\sim 2.7\,\text{eV}$, abundant density of states at the band edge and appropriate chemical and structural properties, *g*-C_3N_4 has been widely explored for photo-, electro- and hetero-catalysts.[60,66,189]

The exceptional device performances, high flexibility, good compatibility and excellent mechanical properties of 2D materials also attract active attention to the fabrication of graphene-based flexible solid-state devices,[169,170,192–194] ranging from secondary batteries, fuel cells, supercapacitors to solar cells. Surprisingly, 2D materials can play important roles in almost all aspects from electrodes to solid-state electrolytes in these applications.

This makes the design of all-graphene-based devices achievable,[195–197] which is indeed the new trend (e.g. portable, wearable and lightweight) in the development of modern electronics.

Besides, 2D materials have also been applied in solar cells, owing to their exceptional electronic property, transparency, good mechanical strength and chemical stability.[169,170] As reported previously, 2D materials have been engaged in different types of solar cells including organic solar cells, dye-sensitized cells and perovskite solar cells.[122,124,127] Moreover, with the flexibility of 2D materials, flexible cells are also promising fields for 2D materials to realize their applicable potential.[116]

The first GO-based 2D membrane applied in sieving water and ions was reported by Nair *et al.*[198] They found that by controlling the interlayer *d*-spacing, this membrane can permeate water but impede ions. This phenomenon can be explained as follows: As graphene nanochannels were formed in the membrane, water can go through the membrane by frictionless capillary action while ions are blocked by the nanochannels.[199] Among other 2D materials, GO demonstrates expanding utilization as an indispensable component in membrane applications for various ionic and molecular separations.[200]

Apart from water/ion separation, gas separation with 2D materials has also attracted numerous interests. It was found that the nanopores in 2D materials play important roles in their gas sieving properties, where both the size and shape of the pore and the interaction between the gas and pore matter.[61] Jiang *et al.*[201] proved that pore size explicitly impacts the permeability and selectivity of gas. They also found that the interaction between pore and gas makes a difference since the selectivity values for H_2/CH_4 mixture are 10^8 and 10^{23} when the pores were passivated with nitrogen and hydrogen, respectively. Moreover, theoretical calculation also agrees with this finding.[61,202]

Recently, electromagnetic invisibility cloaking[203–205] and adaptive thermal camouflage were developed using 2D materials.[206–209] In such devices, the modification of the optical properties is achieved due to doping of few-layer graphene films by ion intercalation or electrostatic doping of monolayer graphene, which leads to strong changes in the refractive index and optical adsorption.

An atomically thin graphene-nanopatch cloaking metasurface with widely tunable scattering cancellation in the terahertz spectrum was demonstrated to have potential applications in low observable nanostructures and for efficient terahertz detection, sensing and routing.[204] Balci *et al.*[205] successfully designed large-area atomically thin graphene electrodes for electrically switchable radar-absorbing surface. The electrical control of absorption, reflection and transmission of microwaves is enabled by electrostatic tuning of the charge density on the electrode.

Thermal camouflage is also a type of electromagnetic cloaking, as it is realized by the control of thermal emission over infrared spectrum. Salihoglu *et al.*[207] developed a new class of adaptive thermal surfaces by reversible intercalation of ionic liquid into graphene multilayers to achieve electro-modulation of infrared absorptivity and emissivity by tuning the charge density and Fermi energy, as illustrated in Fig. 10. The fabricated device is light ($30\,g/m^2$), thin ($< 50\,\mu m$), ultra-flexible and more importantly, can reconfigure its thermal appearance (disguise hot objects as cold and cold ones as hot) and blend itself with the varying thermal background in a few seconds when combined with a feedback mechanism.

The 2D photocatalysis, although being widely researched, suffers from the photocorrosion of semiconductors, which decreases both the efficiency and lifetime of the photocatalysts. Among various approaches to overcome photocorrosion, hybridizing and/or co-catalysis with different 2D semiconductors and 2D carbon materials is a practical strategy. The 2D materials such as WS_2 and MoS_2 have demonstrated a favorable capability to overcome photocorrosion and provide durable photocatalysis.[210–214] For instance, Zhong and co-workers utilized WS_2 nanosheets as hydrogen-evolution active sites, electron transporters and a scaffold for the inaction of ZnS nanoparticles for photocatalytic H_2 production.[215] The composite of ZnS nanoparticles with WS_2 nanosheets could maintain its high photostability after 50 h of radiation. Ma *et al.*[212] fabricated an antiphotocorrosion 0D/2D nanocomposite of CdS and MoS_2 for stable and efficient generation of hydrogen under visible light. Indeed, MoS_2 nanosheets protected CdS from photocorrosion by reducing S atoms

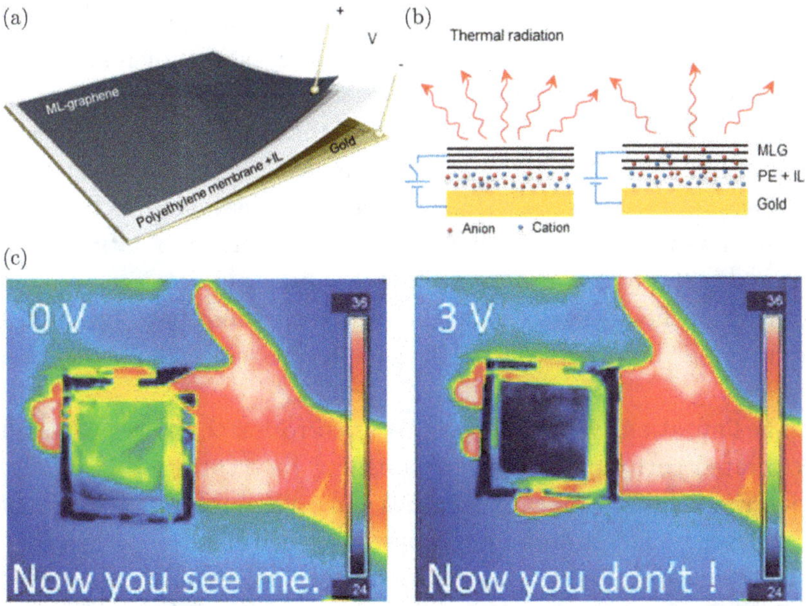

Fig. 10. Schematic of adaptive thermal surfaces: (a) Schematic drawing of the fabricated surface consisting of multilayer graphene (MLG) electrode, room-temperature ionic liquid (IL)-soaked porous polyethylene (PE) membrane and back gold electrode coated on heat-resistive nylon. (b) Representation of the working principle, suppression of the surface emissivity by intercalation of anions into MLG. (c) Thermal camera images of the device placed on the hand under respective voltage bias.

Source: Figures adapted with permission from Ref. 207. Copyright © 2018 American Chemical Society.

to S^- while interacting with lactic acid sacrificial agent and providing suitable sites for recrystallization of CdS.[216]

The enticing features of 2D materials, such as large specific surface area and good mechanical property, promote the expanding utilization of 2D materials as indispensable components in membrane applications and anticorrosion protection. Graphene composite materials can have a broader application area than that of pure graphene. Shortcomings of graphene and 2D materials such as their structural defects and unfavorable dispersibility in the coating have led scientists to search for substitutes such as GO and rGO with similar properties but better compatibility.

The 2D materials have revealed excellent performance in anticorrosion coatings for preserving a wide variety of materials in corrosive conditions, without affecting the physical properties of the substrates.[217–220] Interestingly, 2D materials are able to protect a wide range of different materials, including implants and 2D semiconductor materials, which suffer from dissimilar biocorrosion and photocorrosion, respectively.[221–223] However, the relatively high prices of 2D materials have limited their application to some momentous fields such as automotive and aerospace industries, biomaterials and electronics, in which corrosion results in great losses.

The limitations of pure graphene anticorrosion coatings such as the accelerated corrosion rate after damaging the coating, scale-up and industrialization technical difficulties and high preparation costs have urged researchers to apply graphene in combination with other suitable materials and fabricate graphene composite coatings since 2012.[224] To fabricate barrier anticorrosion coating taking advantage of both the film-constituting feature of the coating matrix and the firm adhesion properties of graphene, graphene is dispersed as filler particles inside the coating matrix. Graphene holds excellent barrier properties due to an extended diffusion path length for the corrosive compounds in graphene-reinforced coatings.

The appearance of graphene-like materials with insulating properties has opened a new way for applications of 2D materials. Nowadays, *h*-BN became the substrate of choice for many electronic devices, thanks to its insulating nature, distinguishable impermeability, thermal stability and desirable mechanical properties.[225] Several research groups have confirmed the excellent anticorrosion properties of *h*-BN,[220,226–231] even though its synthesis is costlier than graphene.[225] The *h*-BN has also demonstrated potential application for the protection of biomaterials from uncontrollable degradation.[232–234] As a proof, Tozar and Karahan employed *h*-BN because of its superb corrosion protection, tribological and mechanical properties as an inorganic reinforcing phase in combination with hydroxyapatite/chitosan (HA/CTS) and coated the fabricated biocomposite alloy.[168]

Furthermore, 2D MXenes became a research hotspot due to their chemical stability, high modulus and strength, good workability and excellent heat and electrical conductivities. However, the anticorrosion and catalytic properties of MXenes are relatively unexploited research areas.[235–237] The 2D MXenes have been used as components in anticorrosion coatings.[238,239] For instance, Yan *et al.*[240] showed that Ti_3C_2 nanosheets acted as a physical barrier in the coating, impeding the penetration of corrosive species. Besides, Ti_3C_2 nanosheets diminished the water uptake of pure epoxy to one fourth.

A well-known strategy to solve *in-vivo* interfacial problems of biomaterials is the surface modification, which can improve the bioactivity of metallic devices and interface biocompatibility simultaneously.[241] The 2D materials have shown a propitious contribution to protective coatings on biomaterials in terms of their *in-vivo* biodegradation resistance properties.[242–244] To illustrate, Zhang *et al.*[245] corroborated that graphene coating has the ability to efficaciously protect Cu surface from degradation in multiple biological aqueous mediums via inhibiting the corrosion of Cu substrates. Surface functionalization was used to prevent the permeation of corrosive species through graphene's defects, thereby improving the durability of biomaterials. Cell viability tests exhibited that graphene can significantly remove copper's toxicity by inhibiting leakage of Cu^{2+} ions. Moreover, they illustrated for the first time that the graphene coating relieved the response of the immune system to copper metal in clinical conditions by conducting the lymphocyte transformation test. Eventually, an animal experiment demonstrated the beneficial protection of graphene provided to Cu under *in-vivo* conditions.

5. Outlook

The appearance of tunable 2D crystals significantly broadens the scientific and technological opportunities, thanks to their exceptional electronic, mechanical, optical, magnetic, thermal and chemical properties. However, there are still a number of exciting challenges remaining for this class of materials.

For instance, fabrication of 2D materials largely depends on the particular application and needs to be further tuned as we find new areas of utilization for such crystals. Existing methods to fabricate 2D materials have their drawbacks. The mechanical exfoliation can easily produce high-quality 2D materials, but the scale of the obtained 2D materials is usually limited and the efficiency of this method is not satisfactory. The chemical exfoliation can massively produce 2D materials, however, there are only certain technologies which can benefit from such a production. The CVD method can produce high-quality 2D materials on a large scale, but the cost is relatively high and the growth temperatures need to be compatible with the post-processing requirements. It is indeed urgent to optimize or develop approaches for the mass production of high-quality 2D crystals which would be cost-effective, environmentally friendly and compatible with other technologies.

In addition, although 2D materials hold a plethora of exciting properties which are already utilized in applications, their full potential will be developed when such materials are combined into heterostructures or composites. The 2D materials offer an excellent platform for developing materials on demand with predetermined properties.

Acknowledgments

We thank NRF (Singapore) for financial support through the project Medium-sized centre programme R-723-000-001-281. KSN also acknowledges support from EU Flagship Programs (Graphene CNECTICT-604391 and 2D-SIPC Quantum Technology), European Research Council Synergy Grant Hetero2D, the Royal Society, EPSRC grants EP/N010345/1, EP/P026850/1, EP/S030719/1, FLAG-ERA project TRANS2DTMD. This project was supported by the Ministry of Education (Singapore) through the Research Centre of Excellence program (grant EDUN C-33-18-279-V12, I-FIM). Evgenii E. Vdovin was supported by the State Task No. 075-00355-21-00.

References

1. L. D. Landau, *Phys. Z. Sowjetunion* **11** (1937) 26.
2. R. Peierls, *Ann. Inst. Henri Poincaré* **5** (1935) 177.
3. N. D. Mermin and H. Wagner, *Phys. Rev. Lett.* **17** (1966) 1133.
4. N. D. Mermin, *Phys. Rev.* **176** (1968) 250.
5. A. O'Hare, F. V. Kusmartsev and K. I. Kugel, *Nano Lett.* **12** (2012) 1045.
6. K. S. Novoselov, A. K. Geim, S. V. Morozov, D. Jiang, Y. Zhang, S. V. Dubonos, I. V. Grigorieva and A. A. Firsov, *Science* **306** (2004) 666.
7. K. S. Novoselov, D. Jiang, F. Schedin, T. J. Booth, V. V. Khotkevich, S. V. Morozov and A. K. Geim, *Proc. Natl. Acad. Sci. USA* **102** (2005) 10451.
8. A. K. Geim and K. S. Novoselov, *Nature Mater.* **6** (2007) 183.
9. J. C. Meyer, A. K. Geim, M. I. Katsnelson, K. S. Novoselov, D. Obergfell, S. Roth, C. Girit and A. Zettl, *Solid State Commun.* **143** (2007) 101.
10. J. C. Meyer, A. K. Geim, M. I. Katsnelson, K. S. Novoselov, T. J. Booth and S. Roth, *Nature* **446** (2007) 60.
11. A. K. Geim, *Science* **324** (2009) 1530.
12. A. K. Geim and I. V. Grigorieva, *Nature* **499** (2013) 419.
13. K. S. Novoselov, V. I. Fal'ko, L. Colombo, P. R. Gellert, M. G. Schwab and K. Kim, *Nature* **490** (2012) 192.
14. A. C. Ferrari, F. Bonaccorso, V. Fal'ko, K. S. Novoselov, S. Roche, P. Boggild, S. Borini, F. H. L. Koppens, V. Palermo, N. Pugno, J. A. Garrido, R. Sordan, A. Bianco, L. Ballerini, M. Prato, E. Lidorikis, J. Kivioja, C. Marinelli, T. Ryhanen, A. Morpurgo, J. N. Coleman, V. Nicolosi, L. Colombo, A. Fert, M. Garcia-Hernandez, A. Bachtold, G. F. Schneider, F. Guinea, C. Dekker, M. Barbone, Z. P. Sun, C. Galiotis, A. N. Grigorenko, G. Konstantatos, A. Kis, M. Katsnelson, L. Vandersypen, A. Loiseau, V. Morandi, D. Neumaier, E. Treossi, V. Pellegrini, M. Polini, A. Tredicucci, G. M. Williams, B. H. Hong, J. H. Ahn, J. M. Kim, H. Zirath, B. J. van Wees, H. van der Zant, L. Occhipinti, A. Di Matteo, I. A. Kinloch, T. Seyller, E. Quesnel, X. L. Feng, K. Teo, N. Rupesinghe, P. Hakonen, S. R. T. Neil, Q. Tannock, T. Lofwandera and J. Kinaret, *Nanoscale* **7** (2015) 4598.
15. K. S. Novoselov, A. Mishchenko, A. Carvalho and A. H. Castro Neto, *Science* **353** (2016) aac9439.
16. Y. Zhu, S. Murali, W. Cai, X. Li, J. W. Suk, J. R. Potts and R. S. Ruoff, *Adv. Mater.* **22** (2010) 3906.
17. D. R. Dreyer, S. Park, C. W. Bielawski and R. S. Ruoff, *Chem. Soc. Rev.* **39** (2010) 228.
18. J. Yan, T. Wei, B. Shao, Z. Fan, W. Qian, M. Zhang and F. Wei, *Carbon* **48** (2010) 487.
19. S. J. Cartamil-Bueno, M. Cavalieri, R. Wang, S. Houri, S. Hofmann and H. S. J. van der Zant, *npj 2D Mater. Appl.* **1** (2017) 16.
20. R. V. Gorbachev, I. Riaz, R. R. Nair, R. Jalil, L. Britnell, B. D. Belle, E. W. Hill, K. S. Novoselov, K. Watanabe, T. Taniguchi, A. K. Geim and P. Blake, *Small* **7** (2011) 465.

21. D. C. Elias, R. R. Nair, T. M. G. Mohiuddin, S. V. Morozov, P. Blake, M. P. Halsall, A. C. Ferrari, D. W. Boukhvalov, M. I. Katsnelson, A. K. Geim and K. S. Novoselov, *Science* **323** (2009) 610.
22. R. R. Nair, W. Ren, R. Jalil, I. Riaz, V. G. Kravets, L. Britnell, P. Blake, F. Schedin, A. S. Mayorov, S. Yuan, M. I. Katsnelson, H.-M. Cheng, W. Strupinski, L. G. Bulusheva, A. V. Okotrub, I. V. Grigorieva, A. N. Grigorenko, K. S. Novoselov and A. K. Geim, *Small* **6** (2010) 2877.
23. X. Li and H. Zhu, *J. Materiomics* **1** (2015) 33.
24. G. Dong, Y. Zhang, Q. Pan and J. Qiu, *J. Photochem. Photobiol. C, Photochem. Rev.* **20** (2014) 33.
25. J. W. Jiang and H. S. Park, *J. Phys. D, Appl. Phys.* **47** (2014) 385304.
26. J. Tao, W. Shen, S. Wu, L. Liu, Z. Feng, C. Wang, C. Hu, P. Yao, H. Zhang, W. Pang, X. Duan, J. Liu, C. Zhou and D. Zhang, *ACS Nano* **9** (2015) 11362.
27. Q. Wei and X. Peng, *Appl. Phys. Lett.* **104** (2014) 251915.
28. R. Ma, Z. Liu, L. Li, N. Iyi and T. Sasaki, *J. Mater. Chem.* **16** (2006) 3809.
29. P. Vogt, P. De Padova, C. Quaresima, J. Avila, E. Frantzeskakis, M. C. Asensio, A. Resta, B. Ealet and G. Le Lay, *Phys. Rev. Lett.* **108** (2012) 155501.
30. Z. Ni, Q. Liu, K. Tang, J. Zheng, J. Zhou, R. Qin, Z. Gao, D. Yu and J. Lu, *Nano Lett.* **12** (2012) 113.
31. B. Anasori, M. R. Lukatskaya and Y. Gogotsi, *Nat. Rev. Mater.* **2** (2017) 16098.
32. E.-Y. Choi, C. A. Wray, C. Hu and W. Choe, *CrystEngComm* **11** (2009) 553.
33. Z. Kang, Y. Peng, Y. Qian, D. Yuan, M. A. Addicoat, T. Heine, Z. Hu, L. Tee, Z. Guo and D. Zhao, *Chem. Mater.* **28** (2016) 1277.
34. C. Tan, X. Cao, X. J. Wu, Q. He, J. Yang, X. Zhang, J. Chen, W. Zhao, S. Han, G. H. Nam, M. Sindoro and H. Zhang, *Chem. Rev.* **117** (2017) 6225.
35. X. Cai, Y. Luo, B. Liu and H. M. Cheng, *Chem. Soc. Rev.* **47** (2018) 6224.
36. M. Xu, T. Liang, M. Shi and H. Chen, *Chem. Rev.* **113** (2013) 3766.
37. J.-W. Lee, Z. Dai, T.-H. Han, C. Choi, S.-Y. Chang, S.-J. Lee, N. De Marco, H. Zhao, P. Sun, Y. Huang and Y. Yang, *Nat. Commun.* **9** (2018) 3021.
38. M. J. Allen, V. C. Tung and R. B. Kaner, *Chem. Rev.* **110** (2010) 132.
39. A. H. Castro Neto, F. Guinea, N. M. R. Peres, K. S. Novoselov and A. K. Geim, *Rev. Mod. Phys.* **81** (2009) 109.
40. J. Kim, L. J. Cote, F. Kim, W. Yuan, K. R. Shull and J. Huang, *J. Am. Chem. Soc.* **132** (2010) 8180.
41. I. K. Moon, J. Lee, R. S. Ruoff and H. Lee, *Nat. Commun.* **1** (2010) 73.
42. C. R. Dean, A. F. Young, I. Meric, C. Lee, L. Wang, S. Sorgenfrei, K. Watanabe, T. Taniguchi, P. Kim, K. L. Shepard and J. Hone, *Nat. Nanotechnol.* **5** (2010) 722.
43. X. F. Jiang, Q. Weng, X. B. Wang, X. Li, J. Zhang, D. Golberg and Y. Bando, *J. Mater. Sci. Technol.* **31** (2015) 589.
44. Y. H. Wang, K. J. Huang and X. Wu, *Biosens. Bioelectron.* **97** (2017) 305.
45. K. F. Mak, C. Lee, J. Hone, J. Shan and T. F. Heinz, *Phys. Rev. Lett.* **105** (2010) 136805.
46. B. Radisavljevic, A. Radenovic, J. Brivio, V. Giacometti and A. Kis, *Nat. Nanotechnol.* **6** (2011) 147.

47. L. Li, Y. Yu, G. J. Ye, Q. Ge, X. Ou, H. Wu, D. Feng, X. H. Chen and Y. Zhang, *Nat. Nanotechnol.* **9** (2014) 372.
48. A. J. M. Giesbers, U. Zeitler, S. Neubeck, F. Freitag, K. S. Novoselov and J. C. Maan, *Solid State Commun.* **147** (2008) 366.
49. C. Lee, X. Wei, J. W. Kysar and J. Hone, *Science* **321** (2008) 385.
50. S. V. Morozov, K. S. Novoselov, M. I. Katsnelson, F. Schedin, D. C. Elias, J. A. Jaszczak and A. K. Geim, *Phys. Rev. Lett.* **100** (2008) 016602.
51. R. R. Nair, P. Blake, A. N. Grigorenko, K. S. Novoselov, T. J. Booth, T. Stauber, N. M. R. Peres and A. K. Geim, *Science* **320** (2008) 1308.
52. A. A. Balandin, S. Ghosh, W. Bao, I. Calizo, D. Teweldebrhan, F. Miao and C. N. Lau, *Nano Lett.* **8** (2008) 902.
53. Y. Cao, V. Fatemi, S. Fang, K. Watanabe, T. Taniguchi, E. Kaxiras and P. Jarillo-Herrero, *Nature* **556** (2018) 43.
54. H. Li, S. Ruan and Y.-J. Zeng, *Adv. Mater.* **31** (2019) 1900065.
55. S. Jiang, L. Li, Z. Wang, K. F. Mak and J. Shan, *Nat. Nanotechnol.* **13** (2018) 549.
56. J. Yi, H. Zhuang, Q. Zou, Z. Wu, G. Cao, S. Tang, S. A. Calder, P. R. C. Kent, D. Mandrus and Z. Gai, *2D Mater.* **4** (2016) 011005.
57. Y. Deng, Y. Yu, Y. Song, J. Zhang, N. Z. Wang, Z. Sun, Y. Yi, Y. Z. Wu, S. Wu, J. Zhu, J. Wang, X. H. Chen and Y. Zhang, *Nature* **563** (2018) 94.
58. H. Fang, S. Chuang, T. C. Chang, K. Takei, T. Takahashi and A. Javey, *Nano Lett.* **12** (2012) 3788.
59. O. Lopez-Sanchez, D. Lembke, M. Kayci, A. Radenovic and A. Kis, *Nat. Nanotechnol.* **8** (2013) 497.
60. D. Deng, K. S. Novoselov, Q. Fu, N. Zheng, Z. Tian and X. Bao, *Nat. Nanotechnol.* **11** (2016) 218.
61. F. Malekian, H. Ghafourian, K. Zare, A. A. Sharif and Y. Zamani, *Eur. Phys. J. Plus* **134** (2019) 212.
62. L. Shi and T. Zhao, *J. Mater. Chem. A* **5** (2017) 3735.
63. G. Reina, J. M. González-Domínguez, A. Criado, E. Vázquez, A. Bianco and M. Prato, *Chem. Soc. Rev.* **46** (2017) 4400.
64. Z. Wei, B. Li, C. Xia, Y. Cui, J. He, J.-B. Xia and J. Li, *Small Methods* **2** (2018) 1800094.
65. P. Chen, T. T. Zhang, J. Zhang, J. Xiang, H. Yu, S. Wu, X. Lu, G. Wang, F. Wen, Z. Liu, R. Yang, D. Shi and G. Zhang, *Nanoscale* **8** (2016) 3254.
66. S. Cao and J. Yu, *J. Phys. Chem. Lett.* **5** (2014) 2101.
67. A. D. Dillon, M. J. Ghidiu, A. L. Krick, J. Griggs, S. J. May, Y. Gogotsi, M. W. Barsoum and A. T. Fafarman, *Adv. Funct. Mater.* **26** (2016) 4162.
68. J. W. Suk, R. D. Piner, J. An and R. S. Ruoff, *ACS Nano* **4** (2010) 6557.
69. L. Song, L. Ci, H. Lu, P. B. Sorokin, C. Jin, J. Ni, A. G. Kvashnin, D. G. Kvashnin, J. Lou, B. I. Yakobson and P. M. Ajayan, *Nano Lett.* **10** (2010) 3209.
70. S. Bertolazzi, J. Brivio and A. Kis, *ACS Nano* **5** (2011) 9703.
71. C. Si, Z. Sun and F. Liu, *Nanoscale* **8** (2016) 3207.
72. K. S. Novoselov, A. K. Geim, S. V. Morozov, D. Jiang, M. I. Katsnelson, I. V. Grigorieva, S. V. Dubonos and A. A. Firsov, *Nature* **438** (2005) 197.

73. N. O. Weiss, H. Zhou, L. Liao, Y. Liu, S. Jiang, Y. Huang and X. Duan, *Adv. Mater.* **24** (2012) 5782.
74. Y. Cai, G. Zhang and Y.-W. Zhang, *Sci. Rep.* **4** (2014) 6677.
75. Y. Takao, H. Asahina and A. Morita, *J. Phys. Soc. Jpn.* **50** (1981) 3362.
76. I. Meric, N. Baklitskaya, P. Kim and K. L. Shepard, RF performance of top-gated, zero-bandgap graphene field-effect transistors, in *Proc. 2008 IEEE Int. Electron Devices Meeting* (IEEE, 2008).
77. T. Tsuchiya, K. Terabe and M. Aono, *Adv. Mater.* **26** (2014) 1087.
78. M. Hadadian, E. K. Goharshadi and A. Youssefi, *J. Nanopart. Res.* **16** (2014) 2788.
79. S. B. Hubbard, T. J. Kershaw, A. Usher, A. K. Savchenko and A. Shytov, *Phys. Rev. B* **83** (2011) 035122.
80. H. Inokawa, M. Nagase, S. Hirono, T. Goto, H. Yamaguchi and K. Torimitsu, *Jpn. J. Appl. Phys.* **46** (2007) 2615.
81. H. Zhou, J. Zhu, Z. Liu, Z. Yan, X. Fan, J. Lin, G. Wang, Q. Yan, T. Yu, P. M. Ajayan and J. M. Tour, *Nano Res.* **7** (2014) 1232.
82. A. N. Gandi and U. Schwingenschlögl, *Europhys. Lett.* **113** (2016) 36002.
83. J. K. Ellis, M. J. Lucero and G. E. Scuseria, *Appl. Phys. Lett.* **99** (2011) 261908.
84. V. Varshney, S. S. Patnaik, C. Muratore, A. K. Roy, A. A. Voevodin and B. L. Farmer, *Comput. Mater. Sci.* **48** (2010) 101.
85. S. Larentis, B. Fallahazad and E. Tutuc, *Appl. Phys. Lett.* **101** (2012) 223104.
86. X. Gu and R. Yang, *Appl. Phys. Lett.* **105** (2014) 131903.
87. B. Chamlagain, Q. Li, N. J. Ghimire, H. J. Chuang, M. M. Perera, H. Tu, Y. Xu, M. Pan, D. Xaio, J. Yan, D. Mandrus and Z. Zhou, *ACS Nano* **8** (2014) 5079.
88. H. J. Chuang, X. Tan, N. J. Ghimire, M. M. Perera, B. Chamlagain, M. M. C. Cheng, J. Yan, D. Mandrus, D. Tománek and Z. Zhou, *Nano Lett.* **14** (2014) 3594.
89. D. Akinwande, N. Petrone and J. Hone, *Nat. Commun.* **5** (2014) 5678.
90. V. Podzorov, M. E. Gershenson, C. Kloc, R. Zeis and E. Bucher, *Appl. Phys. Lett.* **84** (2004) 3301.
91. H. O. H. Churchill and P. Jarillo-Herrero, *Nat. Nanotechnol.* **9** (2014) 330.
92. S. Watcharotone, D. A. Dikin, S. Stankovich, R. Piner, I. Jung, G. H. B. Dommett, G. Evmenenko, S. E. Wu, S. F. Chen, C. P. Liu, S. T. Nguyen and R. S. Ruoff, *Nano Lett.* **7** (2007) 1888.
93. A. Chaves, J. G. Azadani, H. Alsalman, D. R. da Costa, R. Frisenda, A. J. Chaves, S. H. Song, Y. D. Kim, D. He, J. Zhou, A. Castellanos-Gomez, F. M. Peeters, Z. Liu, C. L. Hinkle, S.-H. Oh, P. D. Ye, S. J. Koester, Y. H. Lee, P. Avouris, X. Wang and T. Low, *npj 2D Mater. Appl.* **4** (2020) 29.
94. Z. Lin, A. McCreary, N. Briggs, S. Subramanian, K. Zhang, Y. Sun, X. Li, N. J. Borys, H. Yuan, S. K. Fullerton-Shirey, A. Chernikov, H. Zhao, S. McDonnell, A. M. Lindenberg, K. Xiao, B. J. LeRoy, M. Drndić, J. C. M. Hwang, J. Park, M. Chhowalla, R. E. Schaak, A. Javey, M. C. Hersam, J. Robinson and M. Terrones, *2D Mater.* **3** (2016) 042001.

95. A. C. Ferrari, J. C. Meyer, V. Scardaci, C. Casiraghi, M. Lazzeri, F. Mauri, S. Piscanec, D. Jiang, K. S. Novoselov, S. Roth and A. K. Geim, *Phys. Rev. Lett.* **97** (2006) 187401.
96. L. Liang and V. Meunier, *Nanoscale* **6** (2014) 5394.
97. X. Zhang, W. P. Han, J. B. Wu, S. Milana, Y. Lu, Q. Q. Li, A. C. Ferrari and P. H. Tan, *Phys. Rev. B* **87** (2013) 115413.
98. X. Zhang, X.-F. Qiao, W. Shi, J.-B. Wu, D.-S. Jiang and P.-H. Tan, *Chem. Soc. Rev.* **44** (2015) 2757.
99. X. Gu and R. Yang, *Annu. Rev. Heat Transf.* **19** (2015) 1.
100. Z. Luo, J. Maassen, Y. Deng, Y. Du, R. P. Garrelts, M. S. Lundstrom, P. D. Ye and X. Xu, *Nat. Commun.* **6** (2015) 8572.
101. X. Ling, W. Fang, Y.-H. Lee, P. T. Araujo, X. Zhang, J. F. Rodriguez-Nieva, Y. Lin, J. Zhang, J. Kong and M. S. Dresselhaus, *Nano Lett.* **14** (2014) 3033.
102. L. M. Malard, M. A. Pimenta, G. Dresselhaus and M. S. Dresselhaus, *Phys. Rep.* **473** (2009) 51.
103. X. Song, Y. Wang, F. Zhao, Q. Li, H. Q. Ta, M. H. Rümmeli, C. G. Tully, Z. Li, W.-J. Yin, L. Yang, K.-B. Lee, J. Yang, I. Bozkurt, S. Liu, W. Zhang and M. Chhowalla, *ACS Nano* **13** (2019) 8312.
104. M. Gibertini, M. Koperski, A. F. Morpurgo and K. S. Novoselov, *Nat. Nanotechnol.* **14** (2019) 408.
105. J. L. Lado and J. Fernández-Rossier, *2D Mater.* **4** (2017) 035002.
106. W. Xing, Y. Chen, P. M. Odenthal, X. Zhang, W. Yuan, T. Su, Q. Song, T. Wang, J. Zhong, S. Jia, X. C. Xie, Y. Li and W. Han, *2D Mater.* **4** (2017) 024009.
107. Z. Fei, B. Huang, P. Malinowski, W. Wang, T. Song, J. Sanchez, W. Yao, D. Xiao, X. Zhu, A. F. May, W. Wu, D. H. Cobden, J.-H. Chu and X. Xu, *Nature Mater.* **17** (2018) 778.
108. L. Xie, L. Guo, W. Yu, T. Kang, R.-K. Zheng and K. Zhang, *Nanotechnology* **29** (2018) 464002.
109. C. Gong and X. Zhang, *Science* **363** (2019) 4450.
110. C. Gong, L. Li, Z. Li, H. Ji, A. Stern, Y. Xia, T. Cao, W. Bao, C. Wang, Y. Wang, Z. Q. Qiu, R. J. Cava, S. G. Louie, J. Xia and X. Zhang, *Nature* **546** (2017) 265.
111. B. Huang, G. Clark, E. Navarro-Moratalla, D. R. Klein, R. Cheng, K. L. Seyler, D. Zhong, E. Schmidgall, M. A. McGuire, D. H. Cobden, W. Yao, D. Xiao, P. Jarillo-Herrero and X. Xu, *Nature* **546** (2017) 270.
112. B. Huang, G. Clark, D. R. Klein, D. MacNeill, E. Navarro-Moratalla, K. L. Seyler, N. Wilson, M. A. McGuire, D. H. Cobden, D. Xiao, W. Yao, P. Jarillo-Herrero and X. Xu, *Nat. Nanotechnol.* **13** (2018) 544.
113. N. Sivadas, S. Okamoto, X. Xu, C. J. Fennie and D. Xiao, *Nano Lett.* **18** (2018) 7658.
114. Y. Zhao, J. Gu and Z. Chen, *Adv. Funct. Mater.* **29** (2019) 1904782.
115. Z. Cui, C. Xiao, Y. Lv, Q. Li, R. Sa and Z. Ma, *Appl. Surf. Sci.* **527** (2020) 146894.
116. G. Eda, H. Yamaguchi, D. Voiry, T. Fujita, M. Chen and M. Chhowalla, *Nano Lett.* **11** (2011) 5111.

117. K. S. Kim, Y. Zhao, H. Jang, S. Y. Lee, J. M. Kim, K. S. Kim, J.-H. Ahn, P. Kim, J.-Y. Choi and B. H. Hong, *Nature* **457** (2009) 706.
118. C. Muratore, A. A. Voevodin and N. R. Glavin, *Thin Solid Films* **688** (2019) 137500.
119. X. Liu and M. C. Hersam, *Adv. Mater.* **30** (2018) 1801586.
120. R. Muñoz and C. Gómez-Aleixandre, *Chem. Vap. Deposition* **19** (2013) 297.
121. H. Li, J. Wu, X. Huang, G. Lu, J. Yang, X. Lu, Q. Xiong and H. Zhang, *ACS Nano* **7** (2013) 10344.
122. H. Chen, M. B. Müller, K. J. Gilmore, G. G. Wallace and D. Li, *Adv. Mater.* **20** (2008) 3557.
123. H. Sun, Z. Xu and C. Gao, *Adv. Mater.* **25** (2013) 2554.
124. D.-M. Tang, D. G. Kvashnin, S. Najmaei, Y. Bando, K. Kimoto, P. Koskinen, P. M. Ajayan, B. I. Yakobson, P. B. Sorokin, J. Lou and D. Golberg, *Nat. Commun.* **5** (2014) 3631.
125. V. Nicolosi, M. Chhowalla, M. G. Kanatzidis, M. S. Strano and J. N. Coleman, *Science* **340** (2013) 1226419.
126. F. Bonaccorso, A. Bartolotta, J. N. Coleman and C. Backes, *Adv. Mater.* **28** (2016) 6136.
127. W. Lei, V. N. Mochalin, D. Liu, S. Qin, Y. Gogotsi and Y. Chen, *Nat. Commun.* **6** (2015) 8849.
128. W. S. Hummers and R. E. Offeman, *J. Am. Chem. Soc.* **80** (1958) 1339.
129. R. Ma and T. Sasaki, *Adv. Mater.* **22** (2010) 5082.
130. H. Zhou, W. J. Yu, L. Liu, R. Cheng, Y. Chen, X. Huang, Y. Liu, Y. Wang, Y. Huang and X. Duan, *Nat. Commun.* **4** (2013) 2096.
131. J. Sun, C. Lu, Y. Song, Q. Ji, X. Song, Q. Li, Y. Zhang, L. Zhang, J. Kong and Z. Liu, *Chem. Soc. Rev.* **47** (2018) 4242.
132. J. B. Smith, D. Hagaman and H.-F. Ji, *Nanotechnology* **27** (2016) 215602.
133. Y. Shi, H. Li and L. J. Li, *Chem. Soc. Rev.* **44** (2015) 2744.
134. Y. Zhang, Y. Yao, M. G. Sendeku, L. Yin, X. Zhan, F. Wang, Z. Wang and J. He, *Adv. Mater.* **31** (2019) 1901694.
135. J. Jiang, N. Li, J. Zou, X. Zhou, G. Eda, Q. Zhang, H. Zhang, L.-J. Li, T. Zhai and A. T. S. Wee, *Chem. Soc. Rev.* **48** (2019) 4639.
136. T. Zhang, J. Liu, C. Wang, X. Leng, Y. Xiao and L. Fu, *Biosens. Bioelectron.* **89** (2017) 28.
137. B. Deng, Z. Liu and H. Peng, *Adv. Mater.* **31** (2019) 1800996.
138. X. S. Li, W. W. Cai, J. H. An, S. Kim, J. Nah, D. X. Yang, R. Piner, A. Velamakanni, I. Jung, E. Tutuc, S. K. Banerjee, L. Colombo and R. S. Ruoff, *Science* **324** (2009) 1312.
139. K. K. Kim, A. Hsu, X. Jia, S. M. Kim, Y. Shi, M. Hofmann, D. Nezich, J. F. Rodriguez-Nieva, M. Dresselhaus, T. Palacios and J. Kong, *Nano Lett.* **12** (2012) 161.
140. Y. Shi, C. Hamsen, X. Jia, K. K. Kim, A. Reina, M. Hofmann, A. L. Hsu, K. Zhang, H. Li, Z.-Y. Juang, M. S. Dresselhaus, L.-J. Li and J. Kong, *Nano Lett.* **10** (2010) 4134.

141. S. M. Kim, A. Hsu, M. H. Park, S. H. Chae, S. J. Yun, J. S. Lee, D.-H. Cho, W. Fang, C. Lee, T. Palacios, M. Dresselhaus, K. K. Kim, Y. H. Lee and J. Kong, *Nat. Commun.* **6** (2015) 8662.
142. G. Lu, T. Wu, Q. Yuan, H. Wang, H. Wang, F. Ding, X. Xie and M. Jiang, *Nat. Commun.* **6** (2015) 6160.
143. L. Tan, J. Han, R. G. Mendes, M. H. Rümmeli, J. Liu, Q. Wu, X. Leng, T. Zhang, M. Zeng and L. Fu, *Adv. Electron. Mater.* **1** (2015) 1500223.
144. Y. Zhan, Z. Liu, S. Najmaei, P. M. Ajayan and J. Lou, *Small* **8** (2012) 966.
145. X. Sang, X. Li, A. A. Puretzky, D. B. Geohegan, K. Xiao and R. R. Unocic, *Adv. Funct. Mater.* **29** (2019) 1902149.
146. J.-K. Huang, J. Pu, C.-L. Hsu, M.-H. Chiu, Z.-Y. Juang, Y.-H. Chang, W.-H. Chang, Y. Iwasa, T. Takenobu and L.-J. Li, *ACS Nano* **8** (2014) 923.
147. J. Zhou, J. Lin, X. Huang, Y. Zhou, Y. Chen, J. Xia, H. Wang, Y. Xie, H. Yu, J. Lei, D. Wu, F. Liu, Q. Fu, Q. Zeng, C.-H. Hsu, C. Yang, L. Lu, T. Yu, Z. Shen, H. Lin, B. I. Yakobson, Q. Liu, K. Suenaga, G. Liu and Z. Liu, *Nature* **556** (2018) 355.
148. L. Qu, Y. Liu, J. B. Baek and L. Dai, *ACS Nano* **4** (2010) 1321.
149. K. Kang, S. Xie, L. Huang, Y. Han, P. Y. Huang, K. F. Mak, C.-J. Kim, D. Muller and J. Park, *Nature* **520** (2015) 656.
150. N. Woehrl, O. Ochedowski, S. Gottlieb, K. Shibasaki and S. Schulz, *AIP Adv.* **4** (2014) 047128.
151. M. I. Ionescu, X. Sun and B. Luan, *Can. J. Chem.* **93** (2015) 160.
152. P. Sutter, J. Lahiri, P. Zahl, B. Wang and E. Sutter, *Nano Lett.* **13** (2013) 276.
153. S. Wu, C. Huang, G. Aivazian, J. S. Ross, D. H. Cobden and X. Xu, *ACS Nano* **7** (2013) 2768.
154. C. Huang, S. Wu, A. M. Sanchez, J. J. P. Peters, R. Beanland, J. S. Ross, P. Rivera, W. Yao, D. H. Cobden and X. Xu, *Nature Mater.* **13** (2014) 1096.
155. A. J. Mannix, B. Kiraly, M. C. Hersam and N. P. Guisinger, *Nat. Rev. Chem.* **1** (2017) 0014.
156. Y. Liu, Y. Huang and X. Duan, *Nature* **567** (2019) 323.
157. T. Jurca, M. J. Moody, A. Henning, J. D. Emery, B. Wang, J. M. Tan, T. L. Lohr, L. J. Lauhon and T. J. Marks, *Angew. Chem., Int. Ed.* **56** (2017) 4991.
158. J. M. P. Alaboson, C.-H. Sham, S. Kewalramani, J. D. Emery, J. E. Johns, A. Deshpande, T. Chien, M. J. Bedzyk, J. W. Elam, M. J. Pellin and M. C. Hersam, *Nano Lett.* **13** (2013) 5763.
159. J. Ding, O. ur Rahman, W. Peng, H. Dou and H. Yu, *Appl. Surf. Sci.* **427** (2018) 981.
160. C. Chen, S. Qiu, M. Cui, S. Qin, G. Yan, H. Zhao, L. Wang and Q. Xue, *Carbon* **114** (2017) 356.
161. Y. He, C. Chen, G. Xiao, F. Zhong, Y. Wu and Z. He, *React. Funct. Polym.* **137** (2019) 104.
162. S. Qiu, G. Liu, W. Li, H. Zhao and L. Wang, *J. Alloys Compd.* **747** (2018) 60.
163. L. Gu, S. Liu, H. Zhao and H. Yu, *ACS Appl. Mater. Interfaces* **7** (2015) 17641.
164. R. Imani, F. Mohabatpour and F. Mostafavi, *Carbon* **140** (2018) 569.
165. M. Cui, S. Ren, J. Pu, Y. Wang, H. Zhao and L. Wang, *Corros. Sci.* **159** (2019) 108131.

166. L. Shen, Y. Li, W. Zhao, L. Miao, W. Xie, H. Lu and K. Wang, *ACS Appl. Nano Mater.* **2** (2019) 180.
167. C. A. J. Richards, C. F. Glover, G. Williams, H. N. McMurray and J. Baker, *Corros. Sci.* **136** (2018) 285.
168. A. Tozar and İ. H. Karahan, *Surf. Coat. Technol.* **340** (2018) 167.
169. M. Majumder and A. K. Thakur, Graphene and its modifications for supercapacitor applications, in *Surface Engineering of Graphene*, eds. S. Sahoo, S. K. Tiwari and G. C. Nayak (Springer International Publishing, Cham, 2019), pp. 113–138.
170. T. Mahmoudi, Y. Wang and Y. B. Hahn, *Nano Energy* **47** (2018) 51.
171. M. Donarelli and L. Ottaviano, *Sensors* **18** (2018) 3638.
172. S. Bertolazzi, P. Bondavalli, S. Roche, T. San, S.-Y. Choi, L. Colombo, F. Bonaccorso and P. Samorì, *Adv. Mater.* **31** (2019) 1806663.
173. I. Gablech, J. Pekárek, J. Klempa, V. Svatoš, A. Sajedi-Moghaddam, P. Neužil and M. Pumera, *TrAC - Trends Anal. Chem.* **105** (2018) 251.
174. H. Li, Z. Yin, Q. He, H. Li, X. Huang, G. Lu, D. W. H. Fam, A. I. Y. Tok, Q. Zhang and H. Zhang, *Small* **8** (2012) 63.
175. T. Kim, D. Kang, Y. Lee, S. Hong, H. G. Shin, H. Bae, Y. Yi, K. Kim and S. Im, *Adv. Funct. Mater.* **30** (2020) 2004140.
176. H. S. Lee, S.-W. Min, M. K. Park, Y. T. Lee, P. J. Jeon, J. H. Kim, S. Ryu and S. Im, *Small* **8** (2012) 3111.
177. Y. T. Lee, H. Kwon, J. S. Kim, H.-H. Kim, Y. J. Lee, J. A. Lim, Y.-W. Song, Y. Yi, W.-K. Choi, D. K. Hwang and S. Im, *ACS Nano* **9** (2015) 10394.
178. C. Ko, Y. Lee, Y. Chen, J. Suh, D. Fu, A. Suslu, S. Lee, J. D. Clarkson, H. S. Choe, S. Tongay, R. Ramesh and J. Wu, *Adv. Mater.* **28** (2016) 2923.
179. E. C. Ahn, H. S. P. Wong and E. Pop, *Nat. Rev. Mater.* **3** (2018) 18009.
180. M. Si, P.-Y. Liao, G. Qiu, Y. Duan and P. D. Ye, *ACS Nano* **12** (2018) 6700.
181. C. Liu, H. Chen, S. Wang, Q. Liu, Y.-G. Jiang, D. W. Zhang, M. Liu and P. Zhou, *Nat. Nanotechnol.* **15** (2020) 545.
182. P. Zhang, F. Zhu, F. Wang, J. Wang, R. Dong, X. Zhuang, O. G. Schmidt and X. Feng, *Adv. Mater.* **29** (2017) 1604491.
183. X.-H. Li and M. Antonietti, *Chem. Soc. Rev.* **42** (2013) 6593.
184. D. Zhang, H. Mou, F. Lu, C. Song and D. Wang, *Appl. Catal. B, Environ.* **254** (2019) 471.
185. Z. Wang, I. Gutiérrez-Lezama, N. Ubrig, M. Kroner, M. Gibertini, T. Taniguchi, K. Watanabe, A. Imamoğlu, E. Giannini and A. F. Morpurgo, *Nat. Commun.* **9** (2018) 2516.
186. Z. Wang, D. Sapkota, T. Taniguchi, K. Watanabe, D. Mandrus and A. F. Morpurgo, *Nano Lett.* **18** (2018) 4303.
187. S. Yang, X. Feng, X. Wang and K. Müllen, *Angew. Chem., Int. Ed. Engl.* **50** (2011) 5339.
188. S. Chen, J. Duan, M. Jaroniec and S.-Z. Qiao, *Adv. Mater.* **26** (2014) 2925.
189. X. Wang, K. Maeda, A. Thomas, K. Takanabe, G. Xin, J. M. Carlsson, K. Domen and M. Antonietti, *Nature Mater.* **8** (2009) 76.
190. F. Goettmann, A. Thomas and M. Antonietti, *Angew. Chem., Int. Ed. Engl.* **46** (2007) 2717.

191. J. Deng, P. Ren, D. Deng, L. Yu, F. Yang and X. Bao, *Energy Environ. Sci.* **7** (2014) 1919.
192. X. Peng, L. Peng, C. Wu and Y. Xie, *Chem. Soc. Rev.* **43** (2014) 3303.
193. W. K. Chee, H. N. Lim, Z. Zainal, N. M. Huang, I. Harrison and Y. Andou, *J. Phys. Chem. C* **120** (2016) 4153.
194. K. W. Shing, G. C. How, P. W. Kee, Y. A. Akimov and B. Ping, *IEEE J. Sel. Top. Quantum Electron.* **20** (2014) 36.
195. H. J. Kim, S. Y. Lee, L. H. Sinh, C. S. Yeo, Y. R. Son, K. R. Cho, Y. Song, S. Ju, M. K. Shin, S. J. Park and S. Park, *J. Power Sources* **346** (2017) 113.
196. C. Ogata, R. Kurogi, K. Awaya, K. Hatakeyama, T. Taniguchi, M. Koinuma and Y. Matsumoto, *ACS Appl. Mater. Interfaces* **9** (2017) 26151.
197. C. Ogata, R. Kurogi, K. Hatakeyama, T. Taniguchi, M. Koinuma and Y. Matsumoto, *Chem. Commun.* **52** (2016) 3919.
198. R. R. Nair, H. A. Wu, P. N. Jayaram, I. V. Grigorieva and A. K. Geim, *Science* **335** (2012) 442.
199. W. S. Hung, T. J. Lin, Y. H. Chiao, A. Sengupta, Y. C. Hsiao, S. R. Wickramasinghe, C. C. Hu, K. R. Lee and J. Y. Lai, *J. Mater. Chem. A* **6** (2018) 19445.
200. L. Chen, G. Shi, J. Shen, B. Peng, B. Zhang, Y. Wang, F. Bian, J. Wang, D. Li, Z. Qian, G. Xu, G. Liu, J. Zeng, L. Zhang, Y. Yang, G. Zhou, M. Wu, W. Jin, J. Li and H. Fang, *Nature* **550** (2017) 380.
201. D.-E. Jiang, V. R. Cooper and S. Dai, *Nano Lett.* **9** (2009) 4019.
202. C. Sun, B. Wen and B. Bai, *Sci. Bull.* **60** (2015) 1807.
203. P.-Y. Chen and A. Alù, *ACS Nano* **5** (2011) 5855.
204. P. Y. Chen, J. Soric, Y. R. Padooru, H. M. Bernety, A. B. Yakovlev and A. Alù, *New J. Phys.* **15** (2013) 123029.
205. O. Balci, E. O. Polat, N. Kakenov and C. Kocabas, *Nat. Commun.* **6** (2015) 6628.
206. L. Xiao, H. Ma, J. Liu, W. Zhao, Y. Jia, Q. Zhao, K. Liu, Y. Wu, Y. Wei, S. Fan and K. Jiang, *Nano Lett.* **15** (2015) 8365.
207. O. Salihoglu, H. B. Uzlu, O. Yakar, S. Aas, O. Balci, N. Kakenov, S. Balci, S. Olcum, S. Suzer and C. Kocabas, *Nano Lett.* **18** (2018) 4541.
208. L. Zhao, R. Zhang, C. Deng, Y. Peng and T. Jiang, *Nanomaterials (Basel)* **9** (2019) 1096.
209. L. Phan, W. G. Walkup IV, D. D. Ordinario, E. Karshalev, J.-M. Jocson, A. M. Burke and A. A. Gorodetsky, *Adv. Mater.* **25** (2013) 5621.
210. Y. H. Zhang, X. L. Cai, D. Y. Guo, H. J. Zhang, N. Zhou, S. M. Fang, J. L. Chen and H. L. Zhang, *J. Mater. Sci., Mater. Electron.* **30** (2019) 7182.
211. Y. J. Yuan, F. Wang, B. Hu, H. W. Lu, Z. T. Yu and Z. G. Zou, *Dalton Trans.* **44** (2015) 10997.
212. F. Ma, Y. Wu, Y. Shao, Y. Zhong, J. Lv and X. Hao, *Nano Energy* **27** (2016) 466.
213. Z. Chen, N. Zhang and Y. J. Xu, *CrystEngComm* **15** (2013) 3022.
214. J. Zhang, J. Yu, M. Jaroniec and J. R. Gong, *Nano Lett.* **12** (2012) 4584.
215. Y. Zhong, Y. Shao, B. Huang, X. Hao and Y. Wu, *New J. Chem.* **41** (2017) 12451.

216. K. A. Jackson, *Kinetic Processes: Crystal Growth, Diffusion, and Phase Transformations in Materials* (John Wiley & Sons, 2006).
217. B. Ramezanzadeh, S. Niroumandrad, A. Ahmadi, M. Mahdavian and M. H. M. Moghadam, *Corros. Sci.* **103** (2016) 283.
218. D. Prasai, J. C. Tuberquia, R. R. Harl, G. K. Jennings and K. I. Bolotin, *ACS Nano* **6** (2012) 1102.
219. B. Jin, D. B. Xiong, Z. Tan, G. Fan, Q. Guo, Y. Su, Z. Li and D. Zhang, *Carbon* **142** (2019) 482.
220. M. Cui, S. Ren, S. Qin, Q. Xue, H. Zhao and L. Wang, *Corros. Sci.* **131** (2018) 187.
221. X. Chen, S. Chen, L. Liang, H. Hong, Z. Zhang and B. Shen, *Procedia CIRP* **71** (2018) 322.
222. B. Weng, M. Y. Qi, C. Han, Z. R. Tang and Y. J. Xu, *ACS Catal.* **9** (2019) 4642.
223. F. Khosravi, S. Nouri Khorasani, E. Rezvani Ghomi, M. K. Kichi, H. Zilouei, M. Farhadian and R. Esmaeely Neisiany, *Mater. Res. Express* **6** (2019) 086470.
224. C. H. Chang, T. C. Huang, C. W. Peng, T. C. Yeh, H. I. Lu, W. I. Hung, C. J. Weng, T. I. Yang and J. M. Yeh, *Carbon* **50** (2012) 5044.
225. L. Camilli, F. Yu, A. Cassidy, L. Hornekær and P. Bøggild, *2D Mater.* **6** (2019) 022002.
226. C. Zhang, Y. He, Y. Zhan, L. Zhang, H. Shi and Z. Xu, *Polym. Adv. Technol.* **28** (2017) 214.
227. M. Galbiati, A. C. Stoot, D. M. A. Mackenzie, P. Bøggild and L. Camilli, *Sci. Rep.* **7** (2017) 39770.
228. F. Mahvash, S. Eissa, T. Bordjiba, A. C. Tavares, T. Szkopek and M. Siaj, *Sci. Rep.* **7** (2017) 42139.
229. M. Yi, Z. Shen, X. Zhao, S. Liang and L. Liu, *Appl. Phys. Lett.* **104** (2014) 143101.
230. M. Cui, S. Ren, J. Chen, S. Liu, G. Zhang, H. Zhao, L. Wang and Q. Xue, *Appl. Surf. Sci.* **397** (2017) 77.
231. E. Husain, T. N. Narayanan, J. J. Taha-Tijerina, S. Vinod, R. Vajtai and P. M. Ajayan, *ACS Appl. Mater. Interfaces* **5** (2013) 4129.
232. Y. Göncü, M. Gecgin, F. Bakan and N. Ay, *Mater. Sci. Eng. C* **79** (2017) 343.
233. C. Parra, F. Montero-Silva, R. Henríquez, M. Flores, C. Garín, C. Ramírez, M. Moreno, J. Correa, M. Seeger and P. Häberle, *ACS Appl. Mater. Interfaces* **7** (2015) 6430.
234. S. Al-Saadi, P. C. Banerjee, M. Anisur and R. Raman, *Metals* **7** (2017) 518.
235. K. Hantanasirisakul, M.-Q. Zhao, P. Urbankowski, J. Halim, B. Anasori, S. Kota, C. E. Ren, M. W. Barsoum and Y. Gogotsi, *Adv. Electron. Mater.* **2** (2016) 1600050.
236. C. Zhang, S. J. Kim, M. Ghidiu, M.-Q. Zhao, M. W. Barsoum, V. Nicolosi and Y. Gogotsi, *Adv. Funct. Mater.* **26** (2016) 4143.
237. X. Li, Y. Qian, T. Liu, F. Cao, Z. Zang, X. Sun, S. Sun, Q. Niu and J. Wu, *J. Mater. Sci.* **53** (2018) 11078.
238. Y. Zhu, M. Dong, X. Zhao, J. Li, K. Chang and L. Wang, *J. Am. Ceram. Soc.* **102** (2019) 7521.

239. H. Zhang, Z. Wang, Y. Shen, P. Mu, Q. Wang and J. Li, *J. Colloid Interface Sci.* **561** (2020) 861.
240. H. Yan, W. Li, H. Li, X. Fan and M. Zhu, *Prog. Org. Coat.* **135** (2019) 156.
241. M. Yazdimamaghani, M. Razavi, D. Vashaee, K. Moharamzadeh, A. R. Boccaccini and L. Tayebi, *Mater. Sci. Eng. C* **71** (2017) 1253.
242. A. Tozar and İ. H. Karahan, *Appl. Surf. Sci.* **452** (2018) 322.
243. H. R. Bakhsheshi-Rad, E. Hamzah, M. Kasiri-Asgarani, S. N. Saud, F. Yaghoubidoust and E. Akbari, *Vacuum* **131** (2016) 106.
244. A. Pruna, D. Pullini and A. Soanca, Graphene-based coatings for dental implant surface modification, in *Carbon-related Materials in Recognition of Nobel Lectures by Prof. Akira Suzuki in ICCE*, eds. S. Kaneko, P. Mele, T. Endo, T. Tsuchiya, K. Tanaka, M. Yoshimura and D. Hui (Springer International Publishing, Cham, 2017), pp. 103–116.
245. W. Zhang, S. Lee, K. L. McNear, T. F. Chung, S. Lee, K. Lee, S. A. Crist, T. L. Ratliff, Z. Zhong, Y. P. Chen and C. Yang, *Sci. Rep.* **4** (2014) 4097.

https://doi.org/10.1142/9789811247859_0002

Chapter 2

Introduction to Molecular Interface Engineering of Transition Metal Dichalcogenide-based Devices

Kyungjune Cho*,†, Jae-Keun Kim*, Stuart S. P. Parkin† and Takhee Lee*,‡

**Department of Physics and Astronomy, and Institute of Applied Physics, Seoul National University, Seoul 08826, Korea*
†Max Planck Institute of Microstructure Physics Halle 06120, Germany
‡tlee@snu.ac.kr

Interface engineering of two-dimensional (2D) transition metal dichalcogenides (TMDs) has been regarded as a promising strategy for modulating their outstanding electrical and optoelectronic properties due to their inherent 2D nature and large surface-to-volume ratio. In particular, introducing organic molecules and polymers directly onto the surface of TMDs has been explored to passivate surface defects or achieve better interfacial properties with neighboring surfaces efficiently, thus leading to great opportunity for the realization of high-performance TMD-based applications. This review describes recent progress in the interface engineering of TMDs with organic molecules and polymers, which involves the modulation of their electrical and optoelectronic characteristics. Depending on the interfaces between the surfaces of the TMDs and the dielectric, conductive contacts or ambient environment, we present various strategies for introducing organic interlayers. In addition, the role of native defects on the surface of TMDs, such as adatoms or vacancies, in determining their electrical characteristics is also discussed in detail. Finally, future challenges and the outlook associated with interface engineering are highlighted.

Keywords: Two-dimensional materials; transition metal dichalcogenides; surface engineering; interface engineering; contact engineering; molecular treatment; organic molecules; organic polymers; electric transport properties; charge injection properties.

1. Introduction

In recent decades, after the successful exfoliation of atomic monolayer graphene accomplished by Geim and Novoselov,[1] two-dimensional (2D) layered materials, which are held together by weak van der Waals (vdW) interactions, have received much attention due to their attractive properties and the possibility of making a variety of heterostructures without considering their lattice mismatch. The family of 2D vdW materials is broadly classified into several groups, including the single-atom graphene family (e.g., graphene, germanene, and silicene), the transition metal chalcogen (TMC) family, which includes transition metal dichalcogenides (TMDs) (e.g., NiS, CuS, MoS_2, $NbSe_2$, and WTe_2), and the III–VI family (e.g., InS, GaSe, and GaTe).[1–9] Many 2D vdW materials have been discovered, and their material characteristics are so diverse that they have been studied in various research fields. For example, metallic graphene, semiconducting molybdenum disulfide (MoS_2), or tungsten diselenide (WSe_2) have tunable bandgaps depending on the number of layers, and insulating h-BN. Bulk crystal MoS_2 exhibits an indirect bandgap of 1.2 eV, whereas monolayer MoS_2 has a direct bandgap of 1.9 eV and can be utilized as a channel layer in a variety of optoelectronics.[10,11] Moreover, monolayer MoS_2 has been widely investigated in the field of spintronics and valleytronics due to the strong intrinsic spin-orbit coupling that results when its inversion symmetry is broken.[12–14] In addition, due to recent progress in the deterministic dry transfer technique using a polydimethylsiloxane (PDMS) stamp, fabricating the heterostructure of 2D vdW materials with an atomically clean interface has become feasible and reliable.[15] The feasibility of fabricating heterostructures with a certain twisted angle between adjacent vdW layers by using a PDMS stamp method has opened the field of twistronics.[16] Moiré superlattices can introduce interesting properties or change the properties of vdW materials or

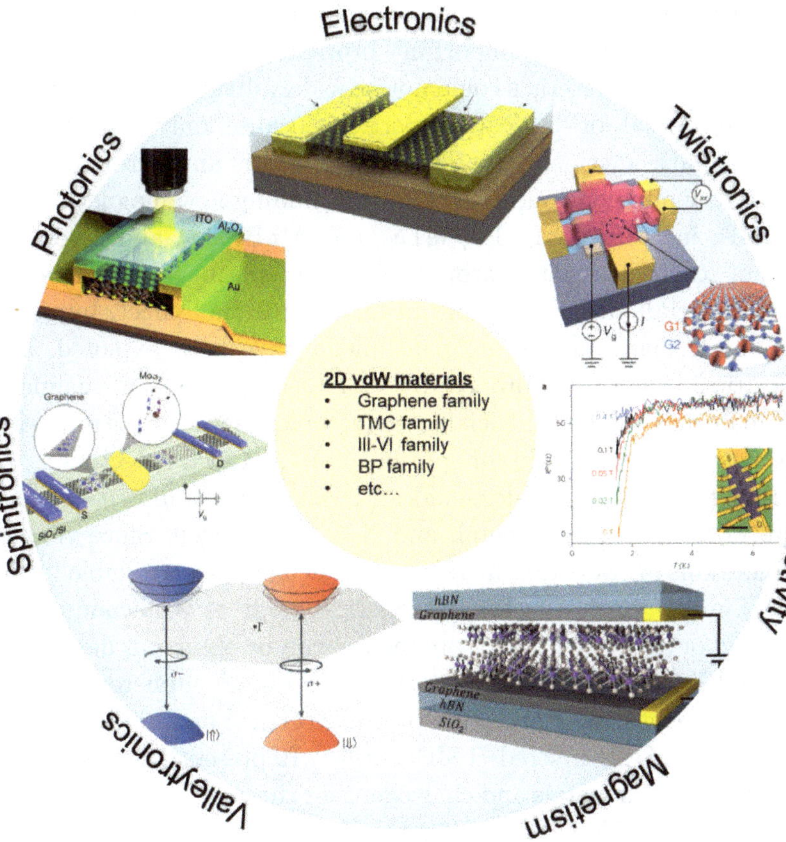

Fig. 1. Various research fields using 2D vdW materials. Images are from Refs. 4, 12, 13, 16, 25–27. Reproduced with permission from Ref. 4, Copyright 2011 Springer Nature Publishing; Ref. 12, Copyright 2017 Springer Nature Publishing; Ref. 13, Copyright 2016 Springer Nature Publishing; Ref. 16, Copyright 2018 Springer Nature Publishing; Ref. 25, Copyright 2012 American Chemical Society; Ref. 26, Copyright 2018 The American Association for the Advanced of Science; Ref. 27, Copyright 2016 Springer Nature Publishing.

heterostructures according to their size, as determined by the twisted angle.[16,17] Representatively, 1.1° twisted bilayer graphene, known as 'magic-angle' graphene, has shown superconductivity.[16] In addition to magic-angle graphene, intrinsic superconducting vdW materials such as $NbSe_2$, Bi_2Te_3, and FeSe exist.[5,18,19] It has been reported that even monolayer $NbSe_2$ shows superconductivity, and its critical temperature varies depending on its thickness. Although magnetism is also one of the most important properties of the material, the magnetism of 2D materials could not be widely investigated compared to the electrical or optical properties because only a few vdW materials are intrinsically ferromagnetic or antiferromagnetic. Recently, due to continued effort toward finding new magnetic vdW materials, including CrI_3, Fe_3GeTe_2, and $MnBi_2Te_4$, and investigations of their magnetic properties, such as magnetoresistance, anomalous Hall effect, or quantum anomalous Hall effect, by many researchers, many interesting phenomena have been explored, and applications have been proposed.[20–24] Recent progress in magnetic vdW materials will be beneficial for spintronics as well, since magnetic materials are usually used for spin injectors and spin detectors.

Transition metal dichalcogenides (TMDs) are composed of MX_2, where M is a transition metal, and X is a chalcogen. Since various combinations of transition metals and chalcogens are available in the TMD family, interesting characteristics, such as semiconducting, magnetic, or superconducting properties, can be observed, depending on M and X. As a result of having such advantages, numerous studies have been extensively conducted on the electrical and optoelectronic applications of 2D-layered TMDs.[25,30–36] Representative combinations of transition metals and chalcogens for the composition of TMDs are presented in Fig. 2(a). Furthermore, the large-area synthesis of layered TMDs by molecular beam epitaxy (MBE) or chemical vapor deposition (CVD) allows the realization of more practical TMD-based applications beyond a single device.[37–43] Interestingly, the electronic and optoelectronic properties of TMDs sensitively depend on the interface environment due to its 2D nature, such as the large surface-to-volume ratio. The interfaces of TMD-based semiconducting devices, for example, field-effect transistors (FETs), can be classified into three types: (1) the interface exists between the TMD and air (usually denoted by the surface, (2) the interface exists between the TMD and the top or bottom dielectric, and (3) the interface exists between the TMD and metal contacts, as denoted in Fig. 2(b).

Additionally, native defects, such as adatoms or vacancies on the surface of TMDs, play a pivotal role in the determination of the

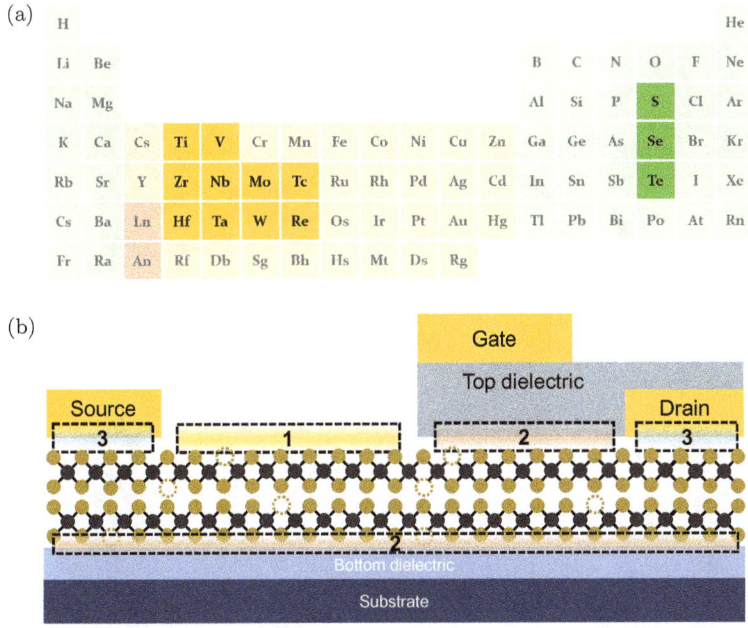

Fig. 2. (a) Representative constituent metal and chalcogen elements of transition metal dichalcogenides. (b) A schematic of the typical three-terminal FET structure with three representative interfaces: between the TMD semiconducting layer and (a) the ambient (air or vacuum), (2) the gate dielectric layer, and (3) metal S/D contacts. Reproduced with permission from Ref. 28, Copyright 2019 American Chemical Society.

material's electrical characteristics. It is well known that these interfacial properties can significantly affect the electrical and optoelectronic characteristics of TMD-based devices, such as the current level, subthreshold swing (*SS*) value, and carrier mobility. For example, as shown in Fig. 3(a), adsorbed water molecules or oxygen molecules on the surface of TMDs can capture charge carriers, resulting in degradation of the source-drain (S/D) current (I_{SD}) in FET applications.[44,45] In the channel layer, charged impurities, as well as the roughness of the dielectric layer, located between the dielectric layer and the TMDs, can behave as trap sites or scattering centers.[46–53] Additionally, even defects in the channel, such as a vacancy or an adatom, play an important role in their properties, including their electronic and optoelectronic properties.[54–58] In

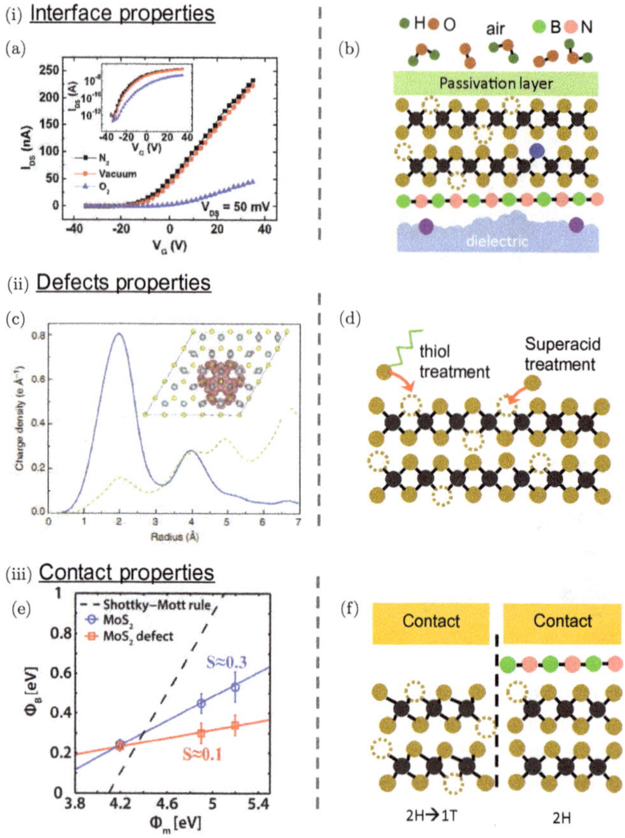

Fig. 3. (a) I_{DS}–V_{GS} characteristics of a nonpassivated MoS_2 FET in nitrogen, vacuum, and oxygen environments. The inset shows a semilogarithmic plot. Reproduced with permission from Ref. 45. Copyright 2013 IOP Publishing. (c) Blue solid line, radial distribution of the charge density for the localized midgap state in the defective MoS_2. The inset is the isosurface of the decomposed charge density Reproduced with permission from Ref. 56. Copyright 2013 Springer Nature Publishing. (e) The Schottky barrier heights for the pristine MoS_2 (blue) and the defects (red) for different work functions of the tip (ϕ_M). The pinning S-factor and the charge neutrality level (ϕ_{CNL}) were extracted. The dotted line is the standard Schottky–Mott rule. Reproduced with permission from Ref. 137. Copyright 2017 American Chemical Society. (b,d,f) Schematic images of interface engineering: (b) the introduction of passivation layers to prevent undesirable adsorption in the ambient and reduce charge scattering, (d) the introduction of organic molecule treatment for defect engineering and (f) changing the phase of TMDs from the semiconducting phase (2H) to the metallic phase (1T) or the introduction of a thin insulating layer to suppress Fermi-level pinning effects and reduce contact resistance.

addition, interfacial defects located at the contact regions can degrade the carrier injection properties *via* Fermi-level pinning phenomena, in such a way that they can hinder Schottky barrier height modulation by changing the work function of the metal electrodes (Fig. 3(c)).[59–61]

Therefore, a promising strategy for realizing further electrically improved TMD-based device applications is efficient interface engineering with consideration of the interfacial properties, such as an interfacial defect or coupling sites, as well as utilizing the intrinsic properties of TMDs, such as metallic (e.g., NbS_2), semimetallic (e.g., WTe_2), semiconducting (e.g., MoS_2 and WSe_2), and superconducting (e.g., TaS_2 and $NbSe_2$).[4,6,62–66] In this manner, there has been substantial effort toward introducing a surface passivation layer (Fig. 3(b)) or interlayer, e.g., *h*-BN, between the TMDs and the dielectric layer (Fig. 3(b)),[67] with phase engineering[68] and a thin insulating interlayer working as a tunneling layer between the metal contact and the TMD channel (Fig. 3(f)).[69] In particular, organic molecule and polymer treatments conducted directly on TMDs by using simple solution processing, such as immersion, spin-coating, or vapor-deposition processing, have attracted substantial attention as facile methodologies for modulating the electrical properties of TMD materials (Fig. 3(d)). For example, the electron density in TMD channel layers can be significantly increased *via* a simple surface treatment with benzyl viologen, which can serve as an *n*-type surface charge transfer dopant.[70] Additionally, a low-dielectric constant (low-k) polymer (e.g., CYTOP) interlayer, which is located between a high-k dielectric layer and a *p*-type TMD semiconductor, can enhance the electrical properties of TMD-based FETs, including the carrier mobility and *SS* value, by preventing unwanted effects from the scattering and dispersion of defects during the atomic layer deposition (ALD) process.[71] Furthermore, a self-assembled monolayer (SAM) can be introduced not only to passivate the surface of the dielectric layer or the structural vacancies of the TMDs but also to modulate the electrical properties of the TMDs *via* a change in the electric dipole moments.[72]

This book chapter covers recent progress in the interface engineering of TMDs with various organic molecules and polymers, for improving or modulating their performances, such as electrical and optical characteristics as well as magnetic characteristics, and how they can affect such characteristics of TMD-based devices from physical and chemical points of view. The key advantage of using organic molecules and polymers is to utilize the various combinations of organic materials that have different functionalities. Moreover, these materials can be directly applied to the different classes of interfaces to modulate the interfacial properties *via* relatively facile processing. Specifically, this chapter will introduce different engineering strategies and how they work depending on whether the interface is with air (or vacuum), the gate dielectric, or S/D contacts by using only organic molecules and polymers that enable tailoring of the electrical, optical and magnetic characteristics of TMDs with a high degree of design freedom. Therefore, this book chapter can be beneficial to the readers by reporting the current issues with the interfacial properties of TMDs and the efforts to address them *via* efficient interface engineering. Moreover, it will provide thorough insight into interface engineering, describing the significance, opportunities, and challenges of this research field.

2. Interface Engineering between TMDs and Air

2.1. *Background*

In the typically employed bottom-gated FET structure, the channel layer is inevitably exposed to ambient conditions. In particular, because TMDs exhibit a large surface-to-volume ratio, the electrical and optoelectronic characteristics are significantly affected by the surrounding atmosphere. For example, the adsorption of oxygen or water molecules onto the MoS_2 channel can affect the electrical characteristics of MoS_2 FETs by capturing electrons due to the large electronegativity of oxygen. Specifically, Late *et al.* reported a large hysteresis window of MoS_2 FETs during dual V_{GS} sweeps as the room humidity (RH) increased, which resulted in a reduced channel current.[44] Similar to this result, an oxygen-rich environment can also degrade the electrical performance, including the on/off ratio and carrier mobility.[45] Due to this environmental sensitivity, MoS_2 FETs could also be used as chemical sensors based on surface charge transfer by adsorbed NO_2 or NH_3.[73] Therefore, the introduction of functional molecules or polymers to passivate TMDs from air or eliminate natural defects on the surface has been widely investigated to address this issue. In this context, physical encapsulations have been introduced to protect channel TMDs against the adsorption of oxygen or water molecules from air. For example, Kufer *et al.* reported enhanced electronic and optoelectronic properties, such as a high on/off current ratio, narrow hysteresis window, high mobility, high sensitivity, and fast operation in monolayer and bilayer MoS_2 photodetectors by encapsulating them with atomic layer deposited hafnium oxide (HfO_2), which could isolate them from ambient conditions.[74] However, the processing for HfO_2 deposition can also create additional sulfur vacancies, which could degrade the long-term stability. Lee *et al.* reported highly stable MoS_2 FETs encapsulated by *h*-BN to improve the environmental stability and interfacial properties of the TMD channels;[75] however, achieving a well-aligned vdW heterostructure involves low-yield complicated processing that could be a bottleneck to realizing further practical applications. Others have utilized polymer thin films, such as CYTOP, poly(methyl methacrylate) (PMMA), and poly(4-vinylphenol) (PVP), as encapsulation layers to deposit from a processing perspective, which could have the attractive advantage of high-yield, low-cost, and facile implementation, although the robustness and cleanness of the interface between the organic molecules or polymers and TMDs would not be better than the interfaces of TMDs formed with h-BN.[45,71,76] In contrast to physical encapsulations, functionalized organic dopants, which can donate or withdraw charge carriers, facilitate surface charge transfer doping (SCTD), thus determining the carrier density. SCTD is a simple noninvasive and reversible strategy that is achieved without any physical changes in the TMD channel layer, such as lattice distortion. However, SCTD has difficulty changing the intrinsic states of the vacancies or adatoms due to the absence of chemical bonding. Therefore, physically adsorbed dopants could be easily removed under harsh environments, such as exposure to organic solvents or high temperature, compared to chemically adsorbed dopants. In this manner, although defect passivation *via* chemical bonding exhibits better environmental stability than SCTD, the required chemical reaction conditions, including overcoming the activation energy of the reaction, should be met. In addition, undesirable lattice distortion might be caused by steric effects from adsorbed molecules, and the doping concentration could be self-limited due to the limited number of bonding sites. Additionally, if the chemically adsorbed molecules have electric dipole moments, additional charge transfer will be feasible, similar to the same process of SCTD. Therefore, depending on the point of view or the circumstances, the advantages can become disadvantages, and vice versa. Therefore, it is important to choose which method is more appropriate while considering the purpose of the surface treatment. In this section, we will briefly describe the recent progress of surface charge transfer doping and defect passivation as examples in engineering the surface properties of TMDs.

2.2. *Surface charge transfer doping*

One of the important issues in the field of TMDs is the change in the charge concentration or the type of majority carrier that enables them to be utilized for realizing complementary applications. As a response to these demands, SCTD, which is a method of charge transfer between surface dopants and the semiconducting channel, is a widely used approach owing to its simple nondestructive method, which is achieved without any physical damage to the TMD channel layer and is even reversible by attaching and detaching the dopants, as well as substitutional doping and the work function engineering of electrodes. As shown in Fig. 4, we illustrated simple schematic diagrams to show how SCTD works in conventional semiconductor systems and 2D vdW materials. We assume the presence of a conventional semiconductor channel layer and a *p*-type surface dopant. When a *p*-type dopant that has a larger electron affinity than the neighboring semiconductor is doped, the energy level of the lowest unoccupied molecular orbital (LUMO) of the surface dopant will be placed below the valence band maximum of the semiconductor. When the surface

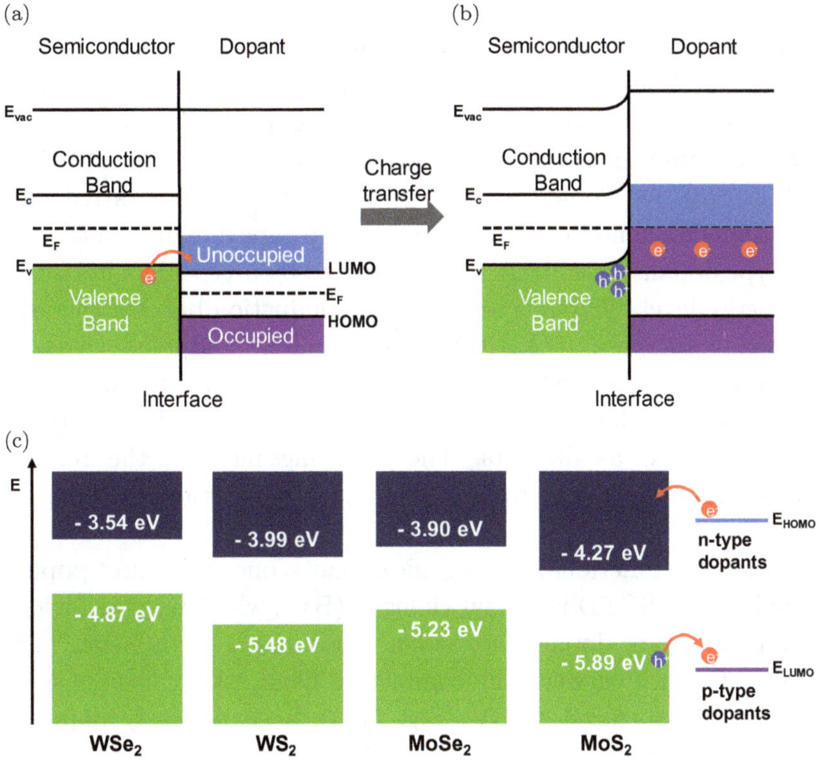

Fig. 4. Schematic illustration of energy diagrams for the conventional p-type SCTD process (a) before and (b) after charge transfer to the semiconductor by the organic dopant. (c) Schematic energy diagrams for SCTD processes of four representative TMDC semiconductors.

dopants are attached to the semiconductor layer, electrons of the semiconductor surface will transfer to the dopants, and holes will accumulate at the surface of the semiconductor layer until energy equilibrium is reached due to the difference in their energy levels. Therefore, the energy band structure of the semiconductor will be bent upward near its surface, resulting in the p-doped semiconductor layer. Similarly, when an n-type dopant is introduced, electron accumulation at the surface of the semiconductor will result in its energy band structure being bent downward near the semiconductor surface. It should be noted that this behavior is more effective in modulating the electrical or optoelectronic characteristics of 2D TMDs due to their large surface-to-volume ratio. Additionally, note that energy band bending is not in general considered in 2D vdW systems because the thickness of the 2D channel is frequently thinner than the band bending width considering their carrier concentrations. In Fig. 4(c), we depict the energy level of 4 representative vdW semiconductors and imaginary n-type and p-type dopants. In the case of n-type dopants, electrons will be donated by n-type dopants due to the energy level difference between the conduction band of the vdW semiconductors and the HOMO of dopants, and vice versa. In addition, many research groups have designed a variety of organic molecules with different energy levels (HOMO and LUMO levels) that can be utilized as dopants. The following includes the recently published representative results of SCTD with functionalized organic molecules.

Among the functionalized organic dopants, one of the most popular candidates for SCTD is benzyl viologen (BV), which has been widely used as an *n*-type dopant for carbon nanotubes and graphene.[70] Due to the difference between the conduction band energy level of MoS_2 and the HOMO level of BV (Fig. 5(b)), a significantly improved electrical performance in doped-MoS_2 FETs, especially in terms of the carrier concentration and contact resistance, was achieved by transferring electrons from BV to MoS_2 (Figs. 5(c) and 5(d)). In addition, because the neutral viologen molecule can spontaneously donate electrons to MoS_2 by a redox reaction, BV exhibited good stability as an *n*-type dopant in ambient air.[70] However, this processing results in a critical disadvantage in that the gate modulation ability was dramatically degraded because of the strong electron-donating behavior of BV. To address this issue, Kiriya *et al.* suggested an immersion process of doped MoS_2 in toluene to eliminate *n*-type dopants onto the MoS_2 surface. Depending on the immersion time, the ability to obtain current modulation *via* gating was facilely changed (Fig. 5(c)). Furthermore, significantly smaller contact resistance and improved *SS* values were extracted from the BV-doped MoS_2 FETs; therefore, this strategy enables us to operate MoS_2 FETs with a low operation voltage (Fig. 5(d)).

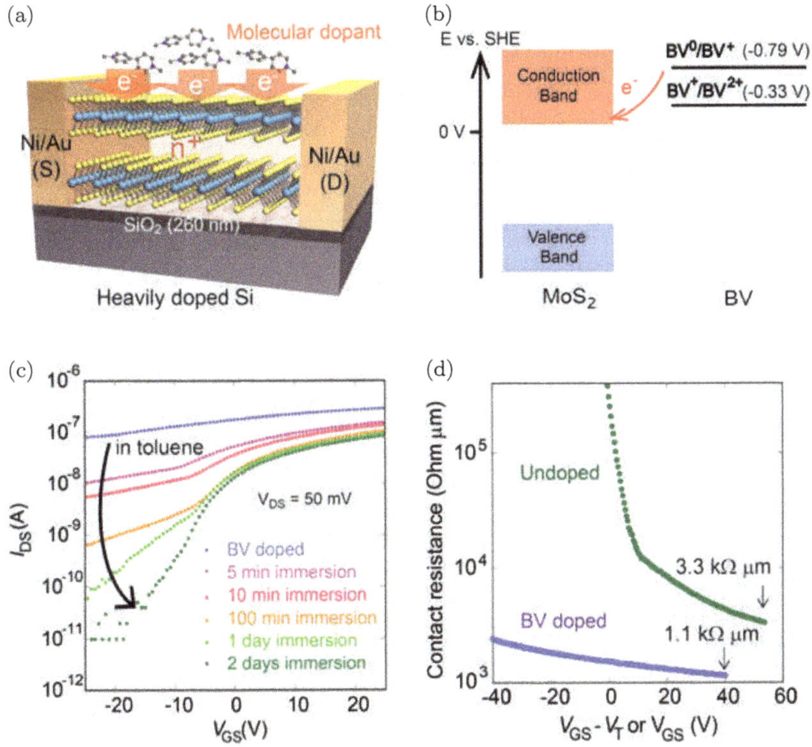

Fig. 5. (a) Schematic illustration of a back-gated MoS_2 FET to investigate the effect of surface charge transfer doping with BV. The phenyl rings are omitted in the schematics of BV molecules for simplification. (b) The energy band diagram of MoS_2 and BV redox states. The conduction band edge of MoS_2 is located at $\sim 0\,V$ vs standard hydrogen electrode (SHE), which is lower than the redox potentials of the BV molecules. This energy level offset results in electron donation from BV molecules to MoS_2. (c) Transfer characteristics of a BV-doped FET as a function of time of immersion in toluene, depicting the on-demand removal of BV surface charge transfer dopants. As the BV molecules were dissolved away in toluene, the $I_{DS} - V_{GS}$ curves were converted back to the original characteristic profile. (d) The gate-voltage dependence of the measured contact resistance before (green curve) and after (blue curve) BV doping. Reproduced with permission from Ref. 70. Copyright 2014 American Chemical Society.

SCTD can also be employed to change the optoelectronic properties of other TMDs. Kang *et al.* reported that octadecyltrichlorosilane (OTS), which has been widely used as a *p*-type dopant, could suppress the negative charge carrier density in the *p*-type WSe_2

channel layer.[77] By controlling the concentration of OTS, the doping level in the WSe_2 channel layer could be controlled from 2.1×10^{11} to $5.2 \times 10^{11}/cm^2$. As a result, substantially improved electrical performances were achieved, including a 10-fold higher I_{DS} with abrupt switching characteristics, good mobility, and a high on/off ratio as well as a threshold voltage (V_{TH}) shift in the positive direction due to the increase in the hole concentration (Fig. 6(b)). Furthermore, the ability to modulate the optoelectronic characteristics, such as responsivity and detectivity, of OTS-doped WSe_2 was shown by

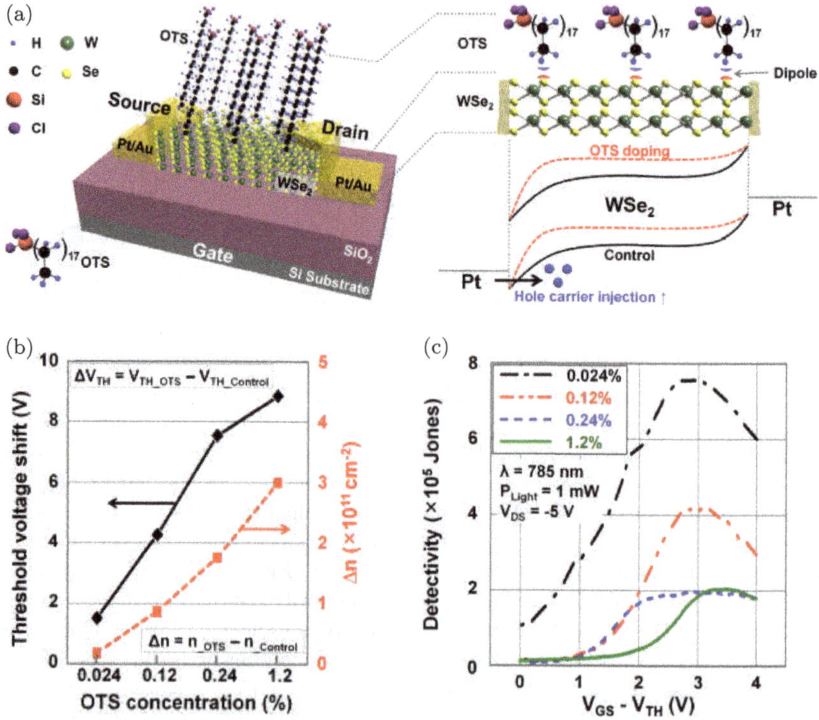

Fig. 6. (a) Schematic diagrams of a back-gated transistor fabricated on OTS-doped WSe_2 and the energy band diagrams of Pt–WSe_2–Pt junctions. (b) The threshold voltage shift ($\Delta V_{TH} = V_{TH-OTS} - V_{TH-Control}$) and carrier concentration increase ($\Delta n = n_{OTS} - n_{Control}$) as a function of OTS concentration (0.024%, 0.12%, 0.24%, and 1.2%). (c) Photodetectivity as a function of $V_{GS} - V_{TH}$ (between 0 and 4 V in the off-state) extracted from OTS-doped WSe_2 photodetectors ($V_{DS} = -5$ V). Reproduced with permission from Ref. 77. Copyright 2015 American Chemical Society.

changing the doping concentration (Fig. 6(c)). These doping effects were also revisable *via* additional postthermal treatments. The intrinsic dipole moment of the OTS molecules could cause charge transfer behavior, providing charges at the surface of WSe_2, as illustrated in Fig. 6(a).

For the SCTD methodology, vacuum processing has also been extensively used, where the doping effects are highly sensitive to the thickness of the organic doping layer. Recently, the phthalocyanine (Pc) molecule family, such as CuPc, CuTsPc, or FeTsPc, has shown attractive advantages as surface charge transfer dopants.[78] They are physically adsorbed on the surface of TMDs similar to other surface charge transfer dopants; therefore, they will not affect the crystalline lattice of the TMDs as noninvasive molecular dopants. Additionally, there are over 70 different types of molecules in the Pc compound family, which offers the opportunity to introduce a wide range of electrical characteristics compared to other conventional molecule dopants such as BV and OTS. In contrast to other organic polymers that exhibit a bulky nature, Pc molecules have a diameter that is compatible with emerging miniaturized devices. In this context, Pak *et al.* reported the modulation of the optoelectronic properties of MoS_2 phototransistors by depositing copper phthalocyanine (CuPc), a *p*-type organic semiconductor, onto the *n*-type MoS_2 channel layer using a thermal evaporator system.[79]

Figure 7(a) presents schematic images and the structure of the CuPc-deposited MoS_2 FETs. Due to the energy level difference between CuPc and MoS_2, as illustrated in Fig. 7(b), charge transfer occurred between CuPc and MoS_2, resulting in modulation of the carrier concentration in the MoS_2 channel layer. As expected, the threshold voltage of CuPC-doped MoS_2 FETs was shifted in the positive gate voltage direction (Fig. 7(c)). For CuPC thicknesses up to 10 nm, the electrical characteristics, including the threshold voltage, electron density, and channel current, were successfully modulated. In addition, better optoelectronic performances were exhibited by the CuPC-doped MoS_2 channel layer because the photogenerated electron-hole pairs could be easily separated at the junction between *n*-type MoS_2 and *p*-type CuPc due to the energy

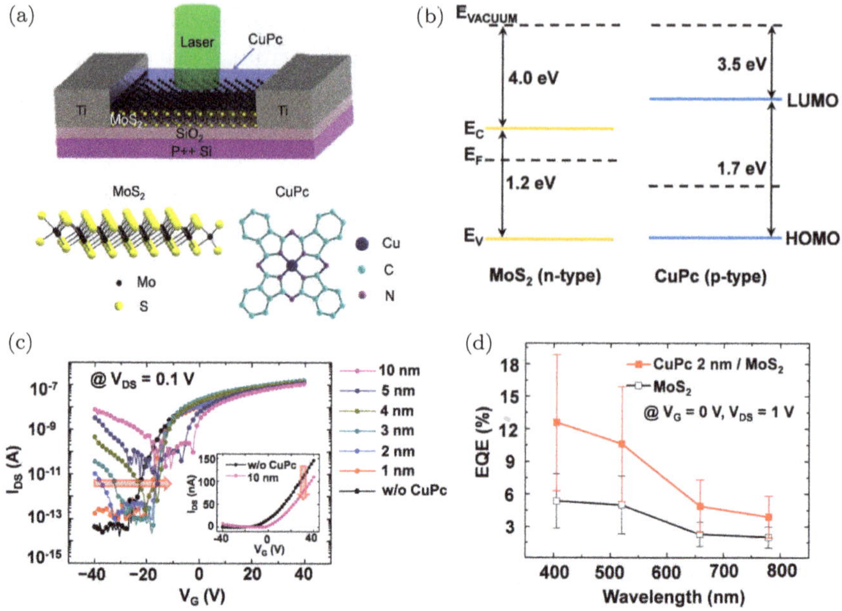

Fig. 7. (a) A schematic of a CuPc/MoS_2 device illuminated by a laser. The molecular structures of the MoS_2 and CuPc are also shown. (b) A schematic of the energy band alignment of CuPc/MoS_2 before stacking. (c) $I_{DS} - V_{GS}$ curves on a semilogarithmic scale of the MoS_2 FETs without (labeled with "*w/o* CuPc") and with a CuPc layer of various thicknesses (1, 2, 3, 4, 5, and 10 nm) measured at fixed $V_{DS} = 0.1$ V. The inset is a linear plot of the $I_{DS} - V_{GS}$ curves for a pristine MoS_2 FET without a CuPc layer and a 10 nm-thick CuPc/MoS_2 hybrid FET. The arrows indicate shifts in threshold voltages and currents. (d) The external quantum efficiency (EQE) of pristine MoS_2 and CuPc/MoS_2 photodetectors as a function of the wavelength of illumination. Reproduced with permission from Ref. 79. Copyright 2015 Royal Society of Chemistry.

level difference. Therefore, the external quantum efficiency (EQE), photoresponsivity, and detectivity of CuPC-doped MoS_2 were improved when CuPC with the optimum thickness was introduced (Fig. 7(d)).

The method of SCTD with organic dopants shows a promising possibility to facilely modulate the electrical and optoelectronic characteristics of TMD-based devices. As introduced above, because SCTD is enabled with customized organic molecules using low-temperature processing without any physical damage, many research groups have intensively studied this strategy to achieve desirable device performances. Recent progress in the SCTD of TMDs is summarized in Table 1.[70,72,77–90]

2.3. *Defect engineering with organic molecule treatment*

The electrical characteristics of TMDs are also sensitive to naturally or intentionally created surface defects. For example, sulfur vacancies or adatoms can affect the electrical characteristics, such as the carrier concentration and injection properties, between the MoS_2 channel layer and the metal contacts in FET applications. Specifically, it has been reported that the sulfur vacancies of MoS_2 could behave as electron donor sites or hopping sites, leading to a higher I_{DS} from the increased carrier concentration.[54–58] Therefore, we summarize some recent reports on the defect engineering of TMDs by employing organic molecules and polymers and its effects on the electrical and optoelectronic characteristics of TMD-based applications.

Regarding the effect of surface defects on TMDs, Qiu *et al.* reported theoretical analysis that the sulfur vacancies of MoS_2 could behave as electron donors, creating energy states in the bandgap of MoS_2.[56] Recent experimental evidence supports these theoretical predictions. Park *et al.* directly observed the electron density near the surface defects of CVD-grown monolayer MoS_2 to be higher than that of the surface without defects, especially near sulfur vacancies, as shown in the STM images (Fig. 8(a)) and cross-sectional profile (Fig. 8(b)).[55] Scanning tunneling spectroscopy (STS) (Fig. 8(e)) shows the local density of the states at two different positions: the black curve was measured far away from the surface defects, and the red curve was measured near the defect site. Interestingly, an almost metallic state located in the bandgap of MoS_2 was observed near the sulfur vacancies due to the highly concentrated electrons. By using spatially resolved STS, the band edge states were also investigated by changing the distance between the STM tip and the defect sites, as shown in Fig. 8(f). Note that the band edge states decreased noticeably as the STM tip was shifted away from the defect sites of MoS_2. For further investigation, many research groups have attempted to intentionally

Table 1. Summary of recent doping processes in TMDs with organic molecules. Reproduced with permission from Ref. 23, Copyright 2019 American Chemical Society.

Material	Dopant	Year	Dopant type	Method	Performance	Stability	Reference
MoS_2	F4-TCNQ, TCNQ	2013	p-type	Drop-casting	Increased n_h	—	[80]
	CuPc	2015	p-type	Thermal evaporation	$\Delta V_{th} = 12.4\,V$	Physisorption	[79]
	FeTsPc	2018	p-type	Soak	$\Delta V_{th} = 6.00\,V$	Physisorption, unstable	[78]
	CuTsPc	2018	p-type	Soak	$\Delta V_{th} = 18.0\,V$	Physisorption, stable	[78]
	APTMS	2013	n-type	Soak (substrate SAM)	$\Delta\mu_e = 1.41\,cm^2/V{\cdot}s$	—	[81]
	PEI	2013	n-type	Soak	$\Delta\mu_e = 12.3\,cm^2/V{\cdot}s$	Unstable	[82]
	TTF	2013	n-type	Soak	Increased n_e	—	[83]
	BV	2014	n-type	Soak	Degenerate doping	Chemisorption, stable	[70]
	APTMS	2015	n-type	Soak	$\Delta\mu_e = 113\,cm^2/V{\cdot}s$	Unstable	[84]
	MEA	2015	n-type	Soak	$\Delta n_e = 3.70 \cdot 10^{12}\,cm^{-2}$	Chemisorption, stable	[72]
	$(2\text{-Fc-DMBI})_2$	2015	n-type	Soak	$\Delta n_e = 6.30 \cdot 10^{12}\,cm^2$	—	[85]
	2-Fc-DMBI-H	2015	n-type	Soak	$\Delta n_e = 5.20 \cdot 10^{12}\,cm^{-2}$	—	[85]
	PTSA	2015	n-type	Soak	$\Delta\mu_e = 61.4\,cm^2/V{\cdot}s$, $\Delta n_e = 2.37 \cdot 10^{12}\,cm^{-2}$	Stable	[86]
	Alq_3 NPs	2017	n-type	Drop-casting	Increased n_e	—	[87]
	Na_2TsPc	2018	n-type	Soak	$\Delta V_{th} = -4\,V$	Physisorption, unstable	[78]
	PPh_3	2018	n-type	Spin-coating	$\Delta\mu_e = 229\,cm^2/V{\cdot}s$, $\Delta n_e = 8.19 \cdot 10^{12}\,cm^{-2}$	Physisorption, stable	[88]

WS_2	F4-TCNQ	2014	p-type	Drop-casting	$\Delta V_{th} = 50\,V$	—	[89]
$MoSe_2$	PDPP3T	2018	n-type	Spin-coating	$\Delta \mu_e = 72.2\,cm^2/V{\cdot}s$	—	[90]
WSe_2	OTS	2015	p-type	Soak	$\Delta \mu_h = 136\,cm^2/V{\cdot}s$	Unstable	[84]
	OTS	2015	p-type	Soak	$\Delta \mu_h = 162\,cm^2/V{\cdot}s$	Unstable	[77]
	FeTsPc	2018	p-type	Soak	$\Delta V_{th} = 27.0\,V$	Physisorption, stable	[78]
	PtTsPc	2018	p-type	Soak	$\Delta V_{th} = 33.0\,V$	Physisorption, unstable	[78]
	CuTsPc	2018	p-type	Soak	$\Delta V_{th} = 32.0\,V$	Physisorption, stable	[78]
	Na_2TsPc	2018	n-type	Soak	$\Delta V_{th} = 38.0\,V$	Physisorption, unstable	[78]

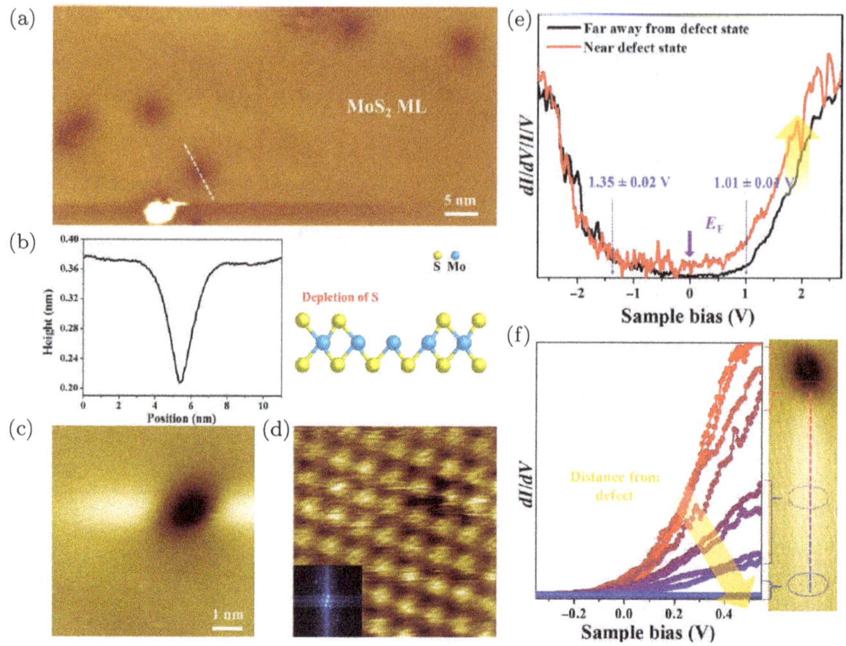

Fig. 8. (a) Large-area STM image showing the ML MoS_2 terrace ($V_s = 2\,V$, $I_t = 40\,pA$). (b) The line trace analysis of the white line in (a) and a schematic model of the defects in MoS_2. (c, d) Zoomed-in STM images of a single defect located on the ML MoS_2 in the empty ($V_s = 1\,V$) and filled ($V_s = -1\,V$) states ($I_t = 230\,pA$). Scale bar, 1 nm. (e) STS measured on the terrace of an ML MoS_2; the black spectra are measured far away from the defects, whereas the red spectra are measured near a defect. (f) Spatial STS near the CB edge as a function of the distance from the defect. Reproduced with permission from Ref. 55. Copyright 2017 American Association for the Advancement of Science.

create surface defects on TMDs by using mild plasma treatment, ion beam irradiation, or electron beam irradiation, and then, organic molecules and polymers have been introduced to passivate the created surface defects[91,93]

From a material perspective of passivation, the nonoxidizing acid poly(4-styrenesulfonate) (PSS) has been used to eliminate sulfur adatoms or passivate sulfur vacancies on MoS_2.[93] Zhang *et al.* reported that these defects on CVD-grown single-layer MoS_2 flakes could be successfully passivated by transferring the MoS_2 flakes onto a poly(3,4-ethylenedioxythiophene) poly(4-styrenesulfonate)

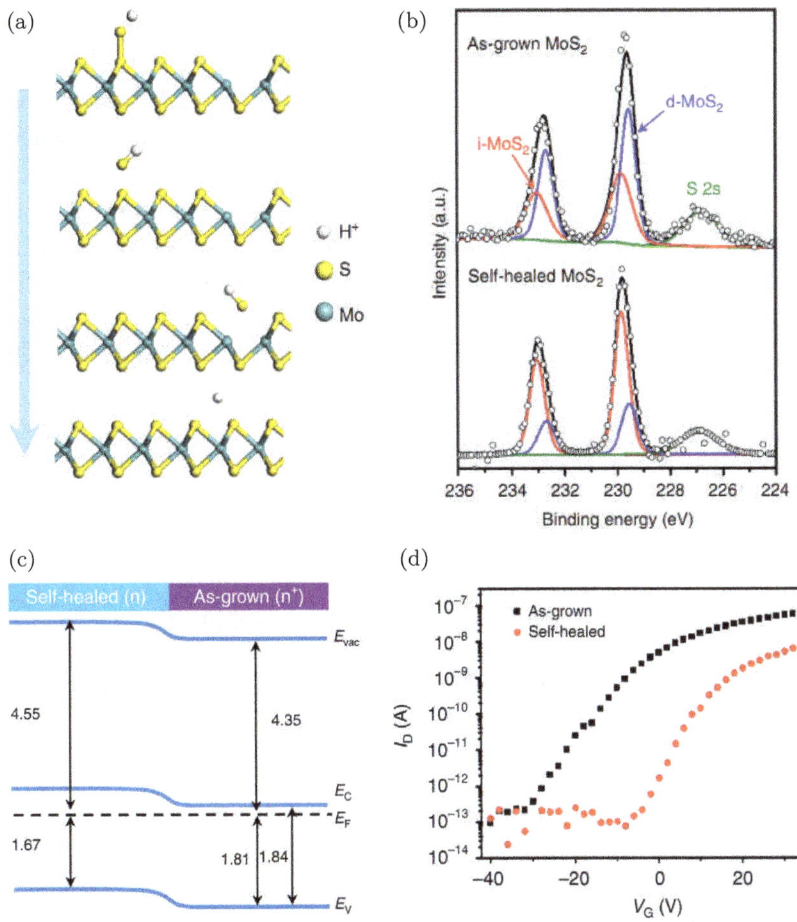

Fig. 9. (a) Change in the 2D chemical structure showing the PSS-induced sulfur vacancy self-healing (SVSH) effect. (b) High-resolution XPS of Mo 3d before (top) and after (bottom) the PSS treatment of MoS_2. Red and blue lines represent the intrinsic MoS_2 (i-MoS_2) and defective MoS_2 (d-MoS_2), respectively. (c) The band diagram of the monolayer MoS_2 homojunction obtained from UPS measurements. (d) Transfer characteristics of a monolayer MoS_2 transistor both before and after PSS-induced SVSH. Reproduced from Ref. 93.

(PEDOT:PSS) layer, which indicates a self-healing MoS_2 *via* PSS treatment.[93] Figure 9(b) shows the X-ray photoelectron spectroscopy (XPS) results of intrinsic CVD-grown MoS_2 and self-healed MoS_2 using the PSS treatment. The XPS results for CVD-grown MoS_2

with and without the PSS treatment support that the self-healing process was successfully conducted by revealing a change in the peaks for defective MoS_2 (d-MoS_2) and intrinsic MoS_2 (i-MoS_2). The authors also noted that these results could not be observed when pristine MoS_2 flakes were transferred onto a PEDOT layer without PSS.[93] From the ultraviolet photoelectron spectroscopy (UPS) measurements, an increased work function was observed for self-healed MoS_2; therefore, the difference in energy between the conduction band and Fermi level of the self-healed MoS_2 was larger than that of pristine MoS_2, which indicates that the concentration of electrons decreased, resulting in a V_{TH} shift in the positive gate voltage direction and reduced I_{DS} in FET applications (Fig. 9(d)). These results are evidence that the sulfur vacancies on MoS_2 work as electron donor states within the bandgap of MoS_2 as well as the defect sites. Therefore, through the PSS treatment on the MoS_2 surface, called the self-healing process, the electrical and optoelectronic characteristics could be changed by controlling the density of the surface defects.

Thiol-group-containing molecules can also be introduced to passivate the surface defects on MoS_2. Thiols and silanes are well-known materials for the formation of SAMs on the surface of metals and oxide layers, such as Au, Ag, Al_2O_3, and SiO_2.[94–98] The key strength of the thiol molecule is that the functional alkyl group or end group provides a variety of functionalities depending on the design of the molecule. Similar to PSS, the sulfur-containing group in the thiol molecules could passivate sulfur vacancies or eliminate sulfur adatoms on MoS_2.[54,72,92,99–107] Cho *et al.* investigated the effects on the electrical characteristics of MoS_2 FETs *via* surface engineering with thiol molecules.[101] After introducing thiol molecules onto the MoS_2 surface using a conventional immersion method, as shown in Fig. 10(a), I_{DS} was reduced and accompanied by a simultaneous threshold voltage shift to the positive gate voltage direction, which was similar to the results from the self-healing treatment with PSS (Figs. 10(b) and 10(c)). However, the *SS* value, primarily determined by the interfacial trap density, was not considerably changed (Fig. 10(d)). These results indicate that the adsorbed thiol molecules

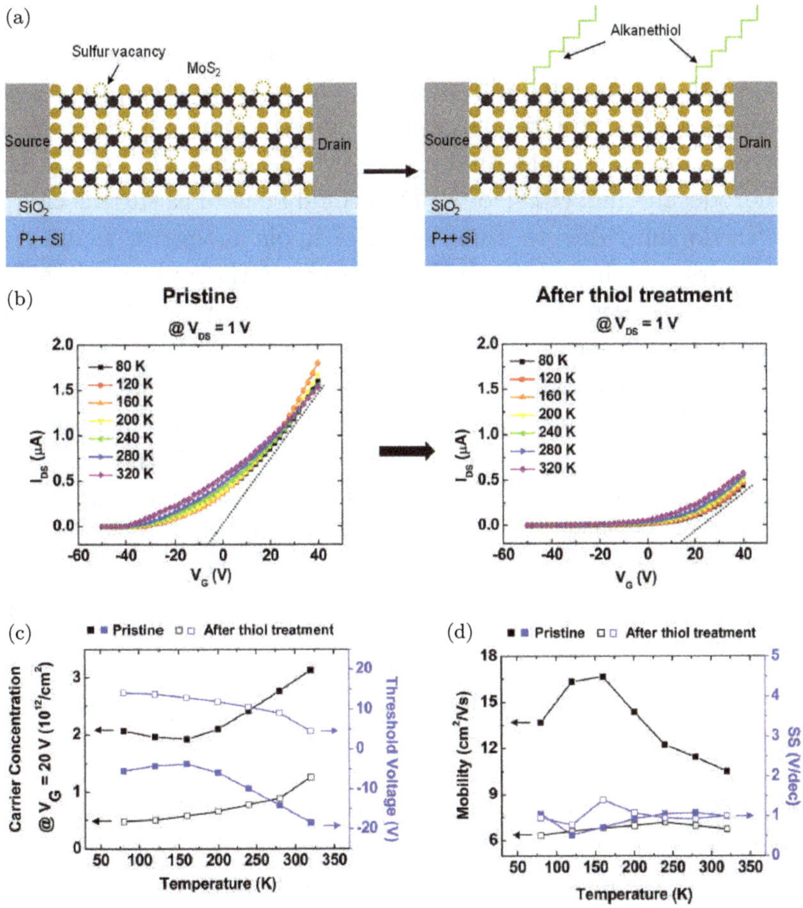

Fig. 10. (a) Schematic images of a MoS_2 FET before (left) and after (right) the alkanethiol molecule treatment. (b) I_{DS}–V_{GS} curves measured before and after hexadecanethiol treatment at varying temperatures from 80 to 320 K. (c) The carrier concentration and threshold voltage of a MoS_2 FET as a function of temperature acquired before and after the hexadecanethiol treatment. (d) The mobility and *SS* values of a MoS_2 FET as a function of temperature extracted before and after the hexadecanethiol treatment. Reproduced with permission from Ref. 101. Copyright 2015 American Chemical Society.

can decrease the carrier concentration in the MoS_2 channel by passivating the sulfur vacancies that might work as electron donor sites but not as additional trap sites. Furthermore, this approach provides insights into controlling the electrical properties of MoS_2 by introducing functional molecules that can be coupled with sulfur vacancies on the MoS_2 surface. This efficient method can also transfer charges from surface dopants to MoS_2 when the adsorbed functional molecules possess their own electric dipole moment. Sim *et al.* reported that mercaptoethylamine (MEA) and 1*H*,1*H*,2*H*,2*H*-perfluorodecanethiol (FDT or PFDT) could be used as surface charge transfer dopants due to their electric dipole moments as well as passivators of the MoS_2 defect sites.[72] NH_2-containing MEA molecules can donate electrons to MoS_2 because MEA molecules contain lone-pair electrons, while F-containing FDT molecules can withdraw electrons from MoS_2 due to the high electronegativity value of F; therefore, the two modifications can result in an increase and decrease in I_{DS} (Figs. 11(b) and 11(d)). For a further accurate investigation, an additional annealing process was conducted afterward to eliminate the adsorbed H_2O and O_2 molecules that can trap the electrons in the MoS_2 channel layer. After the annealing process, I_{DS} increased in MoS_2 layers treated with MEA or FDT due

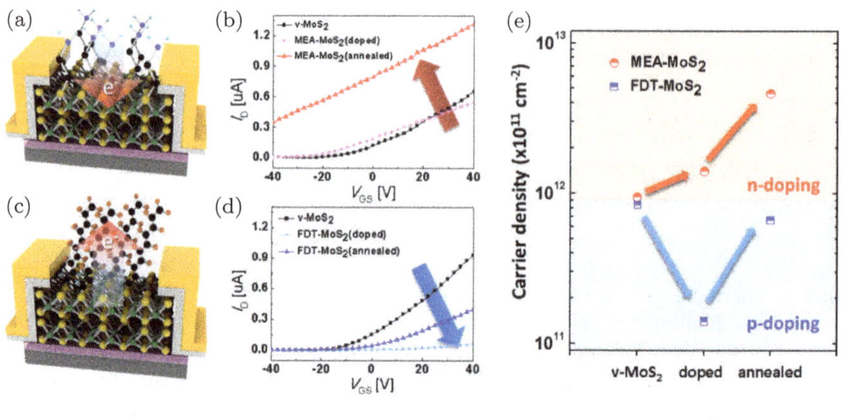

Fig. 11. (a, c) Schematic illustrations presenting the charge transfer doping of a back-gated MoS_2 FET with MEA and FDT molecules. (b, d) Transfer characteristics of MEA-MoS_2 and FDT-MoS_2 obtained with $V_{DS} = 0.5$ V; before doping (black line), after doping (blue line), and after doping and annealing (red line). (e) Changes in the carrier density before and after the doping process (measured at $V_{GS} = 0$ V). Reproduced with permission from Ref. 72. Copyright 2015 American Chemical Society.

to the detrapping of electrons from the adsorbed H_2O and O_2 molecules.

A near-unity photoluminescence quantum yield has also been successfully achieved in MoS_2 by using a nonoxidizing organic superacid, bis(trifluoromethane) sulfonamide (TFSI). According to Amani *et al.*, although the exact mechanism has not been fully understood, the PL in monolayer MoS_2 increased by approximately 190 times after organic superacid treatment, resulting in a quantum yield of more than 95% at low excitation densities, as shown in the PL spectra of a MoS_2 flake before and after treatment in Figs. 12(a) and 12(b).[108] There was no noticeable change in the relative intensity

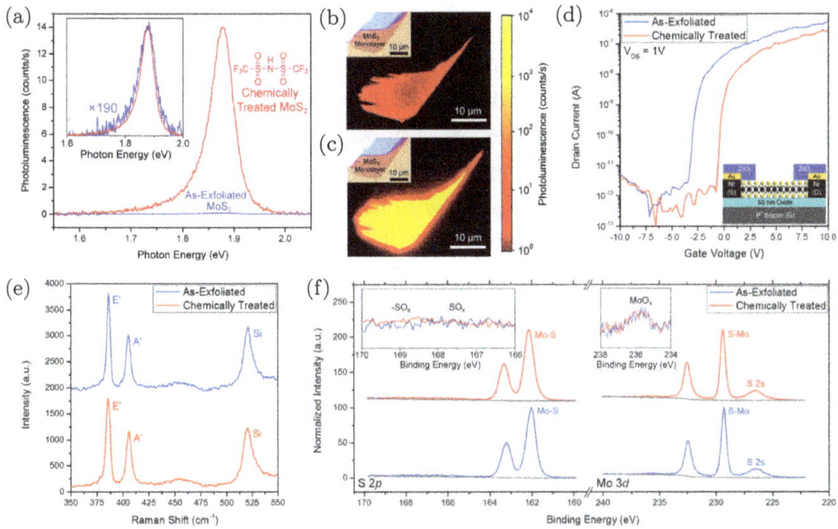

Fig. 12. (a) PL spectrum for both the as-exfoliated and TFSI-treated MoS_2 monolayers measured at an incident power of 1×10^{-2} W cm^{-2}. The inset shows normalized spectra. (b) and (c) PL images of a MoS_2 monolayer before (b) and after treatment (c). Insets show optical micrograph images. (d) Transfer characteristics of a monolayer MoS_2 transistor, both before and after treatment. (e) Raman spectrum of as-exfoliated and TFSI-treated MoS_2 samples. (f) XPS spectrum of the S 2p and Mo 3d core levels before and after treatment. The insets show that there is no appearance of SO_x or change in the MoO_x peak intensity after treatment. Reproduced with permission from Ref. 108. Copyright 2015 American Association for the Advancement of Science.

or peak position of the Raman spectra before and after treatment, which means that the structure of the MoS_2 was not changed, and the additional strain was not introduced by the treatment (Fig. 12(e)). In addition, from the electrical characterization shown in Fig. 12(d), the carrier concentration decreased, which resulted in a shift in the threshold voltage, and the subthreshold slope was improved. Such results indicate that the number of intrinsic n-type dopants was reduced by the treatment as well as the interface trap states, which contribute to the subthreshold swing value. By X-ray photoelectron spectroscopy measurements, they determined that although there were no observable changes in the Mo 3d and S 2p core levels (Fig. 12(f)), the ratio of bonded sulfur to molybdenum in MoS_2 increased on average from 1.84 ± 0.04 to 1.95 ± 0.05 after superacid treatment. Considering their calculation results, which were obtained by density functional theory, surface reconstruction due to superacid treatment could be one possibility. During this surface reconstruction process, the sulfur vacancies, which are usually considered to be n-type dopants in MoS_2, could be reduced by the rearrangement of sulfur adatoms on the surface, which can be facilitated by strong hydrogenation via superacid. This finding is similar to the results from both self-healing treatment with PSS and sulfur vacancy passivation with thiol molecules.[93,100,101] Defect-mediated nonradiative recombination is one of the important attributes in a low QY. Therefore, Amani *et al.* claimed that such a reduction in the number of defect states by organic superacids might be attributed to the near-unity QY.[108]

In addition to these experimental results on defect repair in TMDs, the reaction mechanism of thiol treatment has also been theoretically studied. From the DFT-PBE (Perdew-Burke-Ernzerhof) calculations, Förster *et al.* reported the procedure of the reaction between thiol-group-containing molecules and MoS_2.[54] Each of the following chemical reactions between thiol molecules (R–SH) and a defective MoS_2 monolayer, including sulfur monovacancies (V) or sulfur adatoms (A), was reported with experimental evidence.[99,103,105,107,109] The authors also reported the calculated results for the density of states (DOS) and energy levels of the reaction byproducts using *ab initio* DFT-PBE.

$$R{-}SH + MoS_2^V \rightarrow R{-}H + MoS_2 \quad (1)$$

$$R{-}SH + MoS_2^V \rightarrow 1/2H_2 + R{-}S{-}MoS_2^V \quad (2)$$

$$R{-}SH + MoS_2^A \rightarrow R{-}SS{-}H + MoS_2 \quad (3a)$$

$$R{-}SS{-}H + MoS_2^V \rightarrow R{-}SH + MoS_2 \quad (3b)$$

where R, V and A are alkyl chain, sulfur monovacancies, and sulfur adatoms, respectively. According to the calculations, the surface treatment with thiol molecules could also passivate the surface defects with that have low activation energy. This report strongly supports that strategies using organic functional molecules are highly efficient for tailoring the electrical or optoelectronic characteristics, especially the carrier concentration, in TMD-based applications.

3. Interface Engineering between TMDs and Dielectrics

3.1. *Background*

When TMDs are used for a semiconducting channel layer in FET applications, the interfacial properties with the gate dielectric and contacts are also important for the electrical characteristics. In particular, defects on the interface between TMDs and the gate dielectric layer can play a key role as trap sites, resulting in reduced channel current and poor reliability of the TMD FETs.[44,45,53,110–115] Therefore, for better electrical and optoelectronic characteristics, many strategies for improving the quality of the interfaces have been reported. For example, inserting *h*-BN has attracted attention for preventing undesirable effects from charge trapping sites on the oxide layer, such as impurity atoms, potassium or sodium, and natural defects.[116–118] However, the realization of a high-quality heterostructure with *h*-BN and TMDs still requires high cost, time consumption and low-yield processing, and as a result, this approach might not be suitable for practical applications beyond laboratory-scale applications. To address this issue, one of the promising approaches is the introduction of functional molecules that can be chemically coupled to the gate dielectric and contacts. Organic molecules can not only eliminate the defects located between the dielectric layer and TMDs but also lower the contact barriers, resulting in better carrier injection properties.[101] Among the representative organic material candidates, self-assembled monolayers (SAMs) have been widely investigated to eliminate interfacial defects and are a representative example of interface engineering using organic materials. Typically, molecules of SAM consist of three parts: a linker group located at the bottom, backbone located at the middle, and end group located at the top. It has been reported that four factors of the SAM interlayer can affect the electrical and optoelectronic characteristics of TMDs, including charge transfer, built-in molecular polarities, modified substrate defect properties, and interfacial properties.[117] Therefore, many combinations, which provide a variety of functionalities depending on the target applications, are available. End-functionalized PS directly coupled with hydroxyl groups on the underlying oxide layers can efficiently deliver hydrophobic surfaces and eliminate charge trapping sites from hydroxyl groups. Although CYTOP is also extensively utilized for introducing an interlayer or encapsulation *via* simple solution processing, relatively thick and low-k interlayers reduce the unit capacitance of the gate dielectric, which directly affects the I_{DS}. Moreover, the difficulty of the surface energy engineering does not allow for surface modification with functional molecules. In this section, we will introduce recent efforts to improve interfacial properties by employing functionalized molecules between the TMD channel and dielectric.

3.2. *Device structure*

To implement more reliable FETs, optimized device structures as well as physical or chemical engineering at the interface of the TMD channel have been reported to prevent degradation of the electrical characteristics during operation. Radisavljevic *et al.* reported that further reliable electrical characteristics of MoS_2 FETs could be exhibited by employing a dual-gated structure, even at low temperature, because of reduced phonon scattering, Coulomb scattering and charged impurities, as shown in Fig. 13(b).[119] In addition, the gate dielectric layer can also affect the electrical characteristics of MoS_2 FETs. Yu *et al.* reported that the dielectric constant, which is an important parameter of the screening effect of the dielectric layer, determined the effective size of the charged impurities.[120] Because a high-k dielectric layer can reduce the effective size of charged impurities, metal-oxide dielectrics, such as HfO_2 and Al_2O_3, have been widely used in MoS_2 FET applications. Cui *et al.* reported consistent results in WS_2 FETs in which enhanced carrier mobility could be observed when a high-k dielectric was introduced, and the theoretical expectation that accounted for the effects of intrinsic scattering, surface optical phonon scattering and charged impurities also explained the effect of the dielectric constant (Figs. 13(c) and 13(d)).[121]

3.3. *Interface modification with molecular treatments*

From the perspective of interface properties, much effort to eliminate interfacial traps and scattering sites has been reported. One of the most efficient approaches has been the introduction of end-functionalized polymers between the dielectric and TMD channel layers. Jeong *et al.* reported that dimethylchlorosilane-terminated polystyrene (PS) could be chemically coupled with hydroxyl groups (-OH) that work as interfacial traps on oxide-based gate dielectric layers, such as SiO_2 or Al_2O_3.[122] Therefore, improved electrical performances were achieved in interface-engineered MoS_2 and $MoSe_2$ FETs with ultrathin ($\sim$8.6 nm) and uniformly deposited chlorosilane-terminated PS (Fig. 14(a)). Additionally, charged impurities and optical phonons could be eliminated by passivating hydroxyl groups that resulted in significantly increased carrier mobility, thereby suppressing the hysteresis window, as shown in Fig. 14(b). Note that this strategy has been widely reported to introduce well-crystalline semiconductors in organic FETs, resulting in a substantially improved electrical performance and reliability.[123,124] Organosilanes ($-SH_3$) can also be an interface engineering material used to reduce the

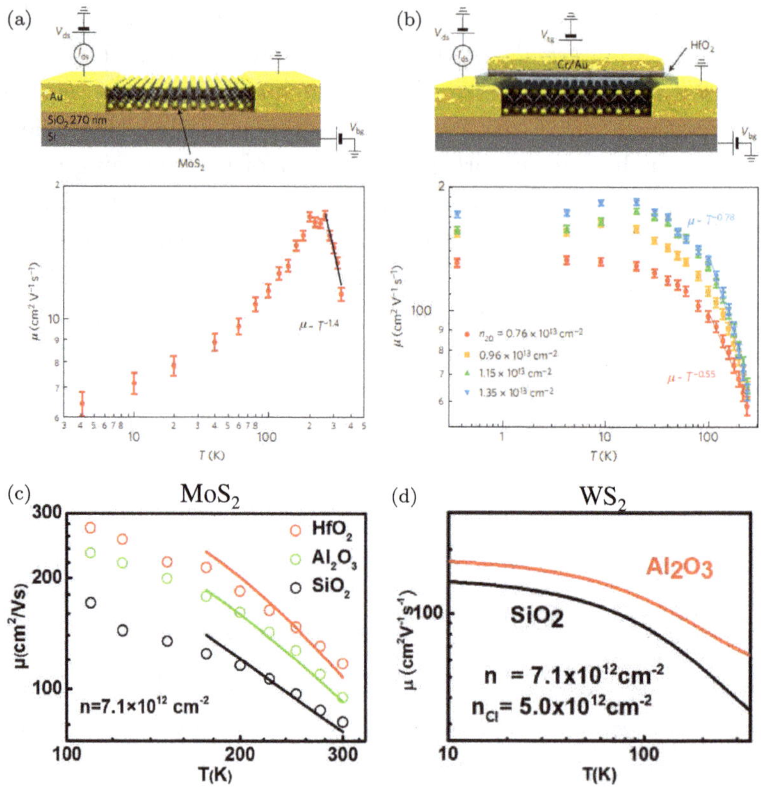

Fig. 13. (a) Cross-sectional views of a single-layer MoS_2 FET in a single-gate configuration. (top panel) The dependence of μ on T shows a pronounced low-temperature regime consistent with transport dominated by scattering from charged impurities. Above ~200 K, μ is limited by phonon scattering and follows a $\mu \sim T^{-1.4}$ dependence. (b) Cross-sectional views of a single-layer MoS_2 FET in a dual-gate configuration. (top panel) μ is practically independent of T under 30 K, indicating the screening of charged impurities due to the deposition of the top-gate dielectric. Above ~100 K, μ decreases owing to phonon scattering and follows a $T^{-\gamma}$ dependence with $\gamma = 0.55 - 0.78$. The strongly reduced value of the exponent γ with respect to the single-gated FET ($\gamma = 1.4$) is indicative of phonon mode quenching. Reproduced with permission from Ref. 119. Copyright 2013 Springer Nature Publishing. (c) Field-effect mobility as a function of temperature for three MoS_2 FETs on SiO_2 (black), Al_2O_3 (green), and HfO_2 (red) under $n = 7.1 \times 10^{12}\,\mathrm{cm}^{-2}$ (symbols). Solid lines are the modeling results at high temperature. Reproduced with permission from Ref. 120. Copyright 2016 John Wiley & Sons, Inc. (d) Theoretical charged impurities (CI)-limited mobility as a function of temperature for monolayer WS_2 on SiO_2 and Al_2O_3 at $n = 7.1 \times 10^{12}\,\mathrm{cm}^{-2}$ and $n_{CI} = 5.0 \times 10^{12}\,\mathrm{cm}^{-2}$. Reproduced with permission from Ref. 121. Copyright 2015 John Wiley & Sons, Inc.

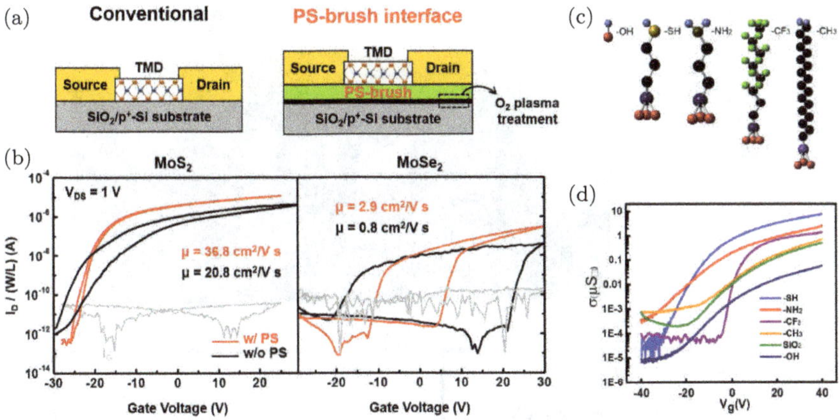

Fig. 14. (a) Cross-section schemes illustrating the conventional TMD FET structure and the structure employing a PS-brush layer. The thick black line indicates the O_2 plasma-cleaned SiO_2 surface before introducing a PS-brush layer. (b) $I_{DS} - V_{GS}$ curves of conventional (black line) and PS-brushed (red line) TMD transistors for two representative TMD nanosheets: n-MoS_2 and n-$MoSe_2$. The reduced hysteresis and higher mobility for the PS-brush treated FETs are noted. Reproduced with permission from Ref. 122. Copyright 2018 John Wiley & Sons, Inc. (c) The ball-and stick-models for SAMs used in the surface functionalization and their orientation relative to MoS_2 monolayers. (d) Transfer curves of representative measurements on a variety of modified and pristine substrates showing the ON-current modulation through interface engineering. Reproduced with permission from Ref. 125. Copyright 2014 American Chemical Society.

density of surface defects or to introduce functionalized dielectric surfaces (Fig. 14(c)). Najmaei *et al.* reported that by introducing a SAM formed with an organosilane on the SiO_2 gate dielectric, better electrical characteristics of TMD-based FETs could be achieved.[115] Because organosilanes can be designed with a variety of combinations, different functionalities, such as different surface energies or electric dipole moments, could be delivered to the surface, which resulted in facile modulation of the electrical characteristics, as shown in Fig. 14(d).[125]

3.4. *Organic polymer dielectric*

The introduction of thin low-k fluoropolymer layers is also an efficient approach to eliminating interfacial defects. Shokouh *et al.* reported

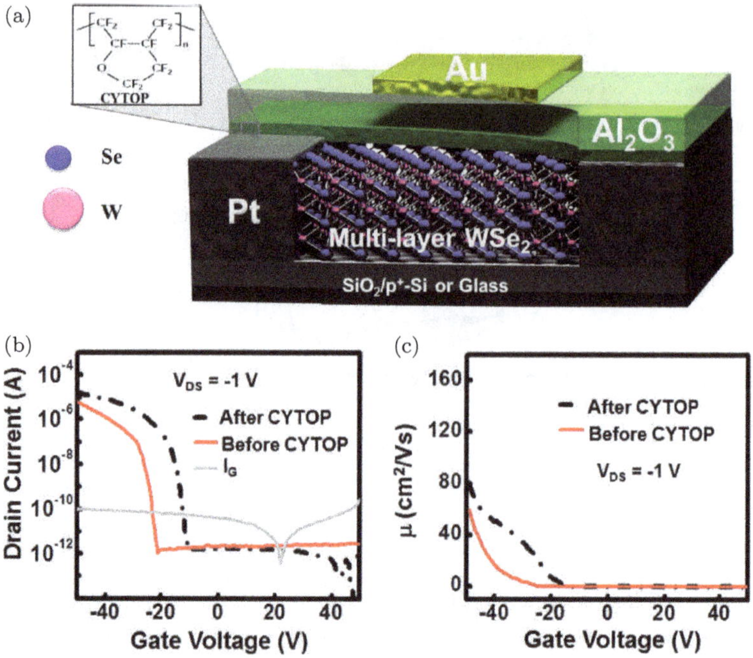

Fig. 15. (a) Schematic 3D view to show the inside of the top-gate FET with the bilayer CYTOP/Al_2O_3 gate dielectric. (b, c) Transfer and mobility curves of a bottom-gate FET based on a 4L–WSe_2 flake. Reproduced with permission from Ref. 71. Copyright 2015 John Wiley & Sons, Inc.

that CYTOP can be used to insert a protection buffer layer between the top Al_2O_3 dielectric layer and the WSe_2 channel layer to prevent channel charging (Fig. 15(a)).[71] It is well known that hydrogen diffusion from an Al_2O_3 dielectric to the underlying channel layer, which can degrade the *p*-type behavior, occurs during Al_2O_3 formation using atomic layer deposition (ALD) in the top-gate structure. Furthermore, because the effect of surface optical phonon scattering could be enhanced when only the high-k dielectric was introduced, Shokouh *et al.* claimed that such scattering could be suppressed by inserting the low-k CYTOP dielectric layer between the high-k dielectric and TMDs.[51] As shown in Figs. 15(b) and 15(c), the WSe_2 FETs with the CYTOP interlayer on the Al_2O_3 dielectric not only showed better electrical performance, especially the *SS* value, on/off ratio, and hole mobility (Figs. 15(b) and 15(c)), but also contributed to improving the device stability by passivating the WSe_2 surface.

4. Interface Engineering Between TMDs and *S*/*D* Contacts

4.1. *Background*

Interfacial properties, such as defects and energy barriers, between the *S*/*D* contacts and a TMD channel also dominantly determine the charge injection in FET applications.[30,60,126,127] In particular, the contact resistance also causes permanent failure *via* thermal stress at the contact region by Joule heating.[122] Therefore, many studies have been reported to reduce the contact resistance by using various contact engineering strategies, such as employing a low-work-function metal, phase engineering, laser annealing, a vdW heterostructure or 1D contacts with ultrathin graphene electrodes.[68,69,126,129–136] However, it remains a challenge to achieve ohmic contact due to a vdW gap between the metal electrodes and the TMD channel layer and interfacial states that can cause Fermi level pinning, as shown in Fig. 16.[127] Therefore, lowering the Schottky barrier height or changing the dominant mechanism of carrier injection is highly desirable to realizing high-performance TMD-based FET applications with low operation voltages. Usually, considering the relationship between the work function and the energy level of a semiconductor, the Schottky barrier height can be changed by using optimized metal electrodes. However, as mentioned above, due to Fermi level pinning by interfacial states, the Schottky-Mott rule is not vigorously followed. For example, a huge tunneling current through defects was observed in a three-dimensional tunneling current mapping image of MoS_2.[59] In addition, these metal-like defects of MoS_2 contributed to Fermi level pinning (Fig. 3(e)).[137] The extracted Schottky barrier height between MoS_2 and the metal tip of the STM depended only marginally on the work function of the metal tip, which indicates that MoS_2 barely follows the Schottky-Mott rule.[137] The pinning factor is the slope of the Schottky barrier height *versus* the work function of the metal tip. The pinning factor of the intrinsic MoS_2 surface was

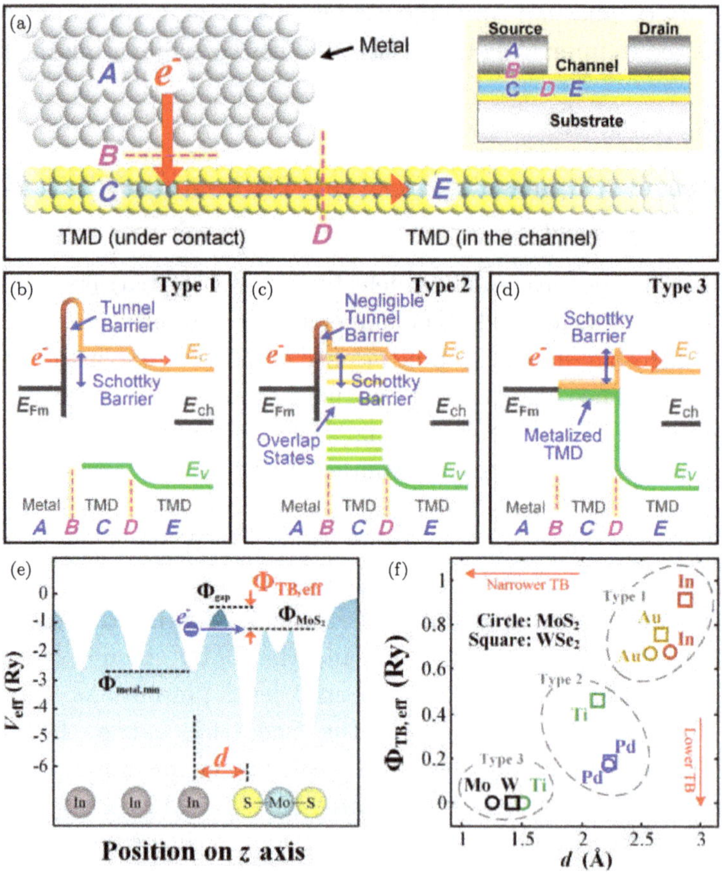

Fig. 16. (a) Schematic cross-sectional view of a typical metal–MoS_2 contact (n-type top contact). A, C, and E denote the three regions, while B and D are the two interfaces separating them. Red arrows show the pathway (A $\rightarrow$ B $\rightarrow$ C $\rightarrow$ D $\rightarrow$ E) of electron injection from the contact metal (A) to the MoS_2 channel (E). The inset shows the source and drain contacts and the channel region in a typical backgated FET. (b–d) The three possible band diagrams of (a): metal contacts with (b) very weak bonding (type 1), (c) medium bonding (type 2), and (d) strong bonding (type 3). E_c, E_v, E_{Fm}, and E_{ch} represent the conduction band edge, valence band edge, metal Fermi level, and channel potential, respectively. (e) The plot of minimum effective potential (V_{eff}) *versus* z position for In-MoS_2 top contact. Φ_{MoS2} is the V_{eff} of the Mo-S bond orbitals, and thereby, the effective tunnel barrier height ($\Phi_{TB;eff}$) is defined as the minimum barrier height that an electron from the metal has to overcome if it has the same potential energy as Φ_{MoS2}. (f) The $\Phi_{TB;eff}$ *versus* d plot depending on the top contact material. Reproduced from Ref. 127 (DOI: 10.1103/PhysRevX.4.031005).

determined to be ~0.3, which is larger than that of the defective MoS_2 surface (~0.1), indicating that the surface defects of MoS_2 mainly cause Fermi level pinning and influence the charge injection properties of MoS_2 FETs.

4.2. *Conventional contact engineering methods*

In this manner, many approaches have been proposed to achieve better charge injection properties with organic molecules. As a representative example, the aforementioned SCTD methods have also been employed to reduce the contact resistance of TMD-based FETs. Du *et al.* reported that chemical doping using organic molecules could reduce the contact resistance of multilayer MoS_2 FETs.[82] MoS_2 doped with polyethyleneimine (PEI), an amine-rich aliphatic polymer, exhibited reduced sheet resistance due to the strong electron-donating behavior of PEI. Similarly, Kiriya *et al.* also drastically reduced the contact resistance values of MoS_2 FETs by employing BV, a strong n-type dopant (Fig. 5(d));[70] however, these approaches left undesirable phenomena, such as a poor on/off ratio. Therefore, more efficient methods have been suggested to achieve better carrier injection properties, for example, phase engineering to form a metallic phase of the TMD (Fig. 17(a)), transferring the metal electrodes to prevent defects formed during high-energy deposition (Fig. 17(b)), depositing a thin In interlayer taking advantage of the soft nature of In (Fig. 17(c)) or inserting a thin tunneling layer to suppress the Fermi level pinning phenomenon (Fig. 17(d)).[68,131,139,140]

To understand the contact properties of TMDs, it is necessary to consider the formation of the vdW gap, which introduces a tunneling barrier between the TMDs and metal electrodes as well as a Schottky barrier. Phase engineering of MoS_2 has been used to reduce the contact resistance and enhance the electrical properties of the FETs by changing the phase of MoS_2 from the semiconducting 2H phase to the metallic 1T phase using n-butyl lithium.[68] Electron donation from organolithium stabilizes the metallic 1T phase during intercalation.[138] Kappera *et al.* treated n-butyl lithium on the contact region of MoS_2 only to make a metal-metal contact instead of a metal-semiconductor

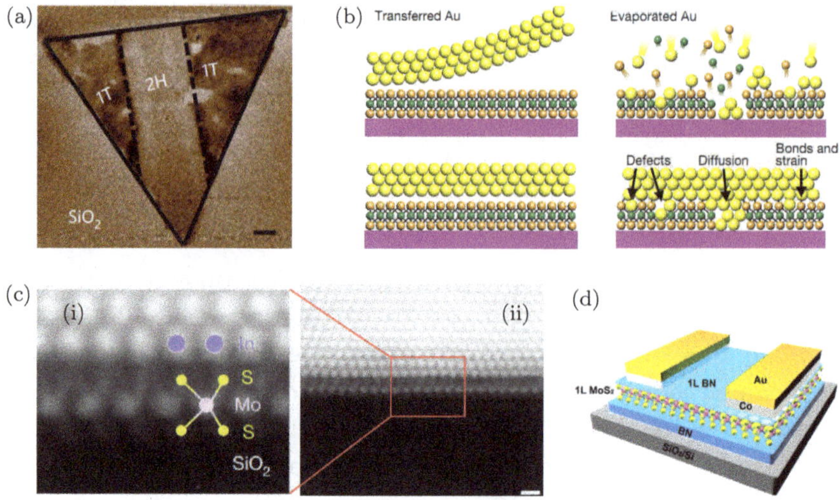

Fig. 17. (a) Electrostatic force microscopy phase image of a monolayer MoS_2 nanosheet showing the difference between locally patterned semiconducting 2H (bright colour) and metallic 1T (dark colour) phases. Reproduced with permission from Ref. 68. Copyright 2014 Springer Nature Publishing. (b) Cross-sectional schematics of the transferred Au electrode on top of MoS_2, with atomically sharp and clean metal-semiconductor interfaces (left panel) and conventional electron-beam-deposited Au electrodes on top of MoS_2, where the bombardment of the MoS_2 surface by high-energy Au atoms and clusters creates considerable damage to the MoS_2 surface, producing a glassy layer with apparent defects, interface diffusion, chemical bonding and atomic disorder. Reproduced with permission from Ref. 139. Copyright 2018 Springer Nature Publishing. (c) Atomic-resolution images of In/Au on monolayer MoS_2. (ii) Low-pass filtered ADF STEM image showing Mo, S and In/Au atoms as indicated by the enlarged image (i). Reproduced with permission from Ref. 140. Copyright 2019 Springer Nature Publishing. (d) Device schematic for Co/h-BN/1L MoS_2 contact. Reproduced with permission from Ref. 131. Copyright 2017 American Chemical Society.

contact.[68] They could change the phase of the MoS_2 contact region and successfully achieve low contact resistance. In addition, Liu *et al.* reported that the high-energy metal deposition process can create considerable damage to the MoS_2 channel layer, introducing additional interfacial states that cause the Fermi level pinning effect.[139] In this manner, they transferred predeposited metal electrodes to the target MoS_2 flake to prevent physical damage from the high-energy deposition process. From this transfer method, they observed nearly ideal Schottky-Mott behavior, which conventional deposition methods for electrode formation cannot achieve due to the high work function of Au. In addition, the creation of additional defects on the surface of MoS_2 was prevented by using the metal transfer method, although the contact resistance and Schottky barrier height of the Au electrode were found to be higher than the results in the molecule-treated TMD devices. On the other hand, Wang *et al.* reported the lowest contact resistance in single-layer MoS_2 FETs by delivering an ultraclean interface with indium (In) interlayers on MoS_2.[140] When they deposited a thin layer of In on MoS_2, no damage, including lattice distortion and defect creation, was observed during the deposition process. Consequently, they obtained the lowest contact resistance in CVD-grown NbS_2 FETs with In/Au electrodes. Additionally, the Schottky barrier height of MoS_2 FETs with In/Au electrodes was found to be approximately 110 meV. The activation energy of the In/Au electrode was found to be larger than that in the molecule-treated FETs (Fig. 17(c)).[140]

4.3. *Contact engineering method by using organic molecule treatment*

If there are many trap states located at the interface between the metal contacts and TMD channels, the Fermi level is pinned, and a Schottky barrier is formed regardless of the metals having different work functions. Therefore, it is difficult to achieve ohmic contact because of the Fermi level pinning effect. In this manner, eliminating interfacial states is considered to be a promising method for better charge injection with various contact engineering strategies. It has been theoretically predicted that the insertion of thin layers can reduce the interaction between the metal contacts and the TMD channel and remove the interfacial states that can be the origin of Fermi level pinning.[141,142] Additionally, Cui *et al.* reported that an inserted thin *h*-BN could improve the charge injection properties of TMD-based FETs by achieving ohmic contact behavior, even at low temperatures (Fig. 17(d)).[132] Instead of *h*-BN, similar effects could be delivered by inserting thiol molecules between the contact electrodes and MoS_2. Recently, simple contact engineering using the selective vapor deposition of thiol molecules was reported.[95] Thiol molecules tend to readily form covalent bonds with the sulfur vacancies of MoS_2, and as a result, thiol molecules selectively could be coupled only at the contact region (Fig. 16(a)). As a result, the effective activation energies of thiol-treated FETs showed significantly smaller values than those of their untreated counterparts, as shown in Fig. 18(b). Consequently, although the output curves of the treated and untreated MoS_2 devices are not presented in Fig. 18, the I_{DS}-V_{DS} curves even in the low-V_{DS} regime of the thiol-treated FETs exhibited good linearity without an S-shape, which is typically evidenced in FETs with ohmic contact, as reported in their paper.[101] The activation energy for the carrier injection can be dominantly determined by the characteristics of two barriers: the Schottky barrier and tunneling barrier. In particular, the tunneling behaviors are crucially affected by the tunneling barrier height and length. According to the recent work by Cho *et al.*, the activation energy for hexadecanethiol (HDT)-treated MoS_2 FETs was slightly higher than those of DT- or PFDT-treated MoS_2 FETs.[101] Because both HDT and DT have sulfur end groups, the tunneling barrier would not be much different. However, the longer carbon chain length of HDT increases the tunneling length, which results in slightly greater activation energy with lower tunneling probability. The inserted thiol molecules could eliminate the interfacial states that result in Fermi level pinning and create additional tunneling paths, which could tailor the dominant carrier injection mechanism from thermionic emission to field emission. In addition, Cho *et al.* treated one electrode among only source and drain electrodes in a MoS_2 FET with thiol molecules.[101] As shown in Fig. 18(c), the carrier injection through the treated contact was not affected by the temperature since the field emission was the dominant injection mechanism, while the injection through the untreated electrode showed temperature-dependent thermionic emission behavior.

The attractive advantage of the use of organic molecules compared to the other approaches is the wide versatility of modulation of the injection properties, including the barrier height and injection

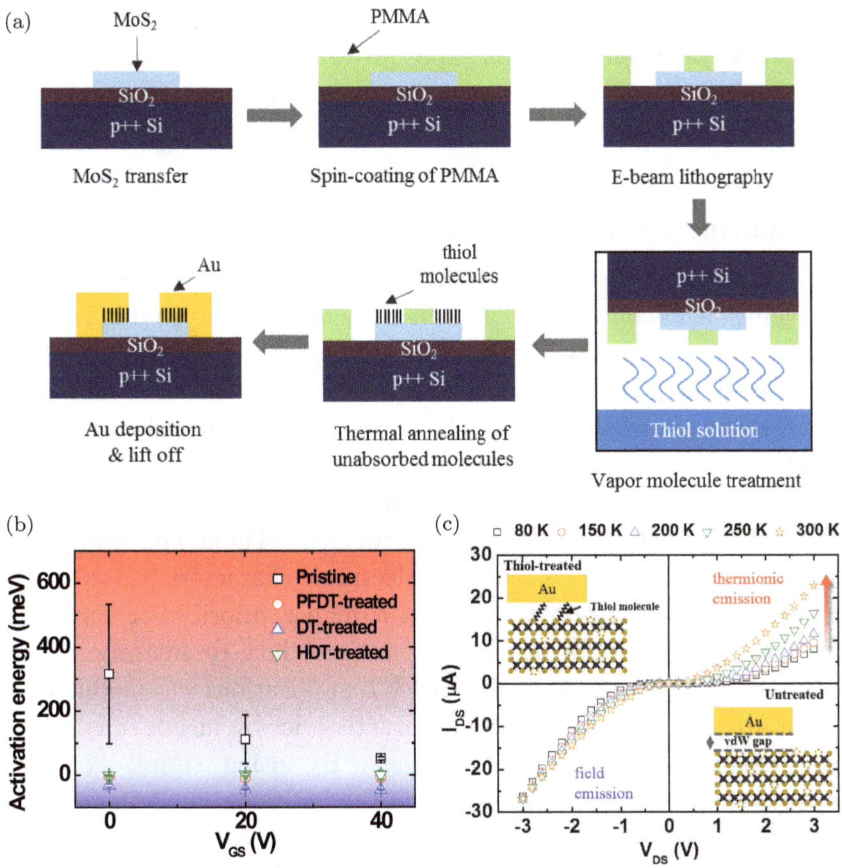

Fig. 18. (a) Schematic images of the FET fabrication process. The thiol molecules were deposited only at the contact region by using a vapor deposition method. (b) The extracted activation energies of untreated, DT-treated, PFDT-treated, and HDT-treated MoS_2 FETs under different gate bias conditions. The red region represents "thermionic emission dominant", and the blue region represents "field emission dominant". (c) Output curves at different temperatures at a fixed gate bias. The drain electrode is DT treated, and the source electrode is untreated. The insets are the schematic images of the thiol-treated and the untreated contacts. Reproduced with permission from Ref. 101. Copyright 2018 John Wiley & Sons, Inc.

mechanism, by introducing various molecular moieties. For example, the injection properties could be modulated by changing the molecular length, the conductivity of the molecular backbone and the electric dipole moment. However, according to Cho *et al.*,[101] the weak nature of organic molecules under harsh environmental or operating conditions, such as exposure to high temperature or high bias conditions, could be a disadvantage compared to previously reported results.

5. Conclusions and Outlook

Over the past decade, TMDs have been intensively explored as an emerging semiconducting material, which has provided solid potential for their use in the field of next-generation electronics. Due to their 2D nature, the implementation of ultrathin and transparent optoelectronics has been promising. Although significant effort has been devoted to realizing high-performance TMD-based FETs and sensor applications by producing high-quality TMDs or introducing vdW heterostructures, further reliable approaches to controlling their electrical and optoelectronic characteristics from the device perspective should be given a higher priority for practical applications. In this book chapter, we introduce recent progress in interface engineering of TMDs in FET applications *via* modulating the electrical and optoelectronic characteristics by means of employing organic molecules and polymers. By using simple solution processing, including immersion, vapor deposition, and spin coating, functionalized organic molecules and polymers can be adsorbed to coupling sites on TMDs, resulting in engineered interfacial properties. In addition, undesirable surface defects, such as adatoms or vacancies, that degrade the electrical performance can be efficiently eliminated.

Herein, three promising avenues that are dependent on the dominant interfaces were classified into TMDs and air, TMDs and the dielectric, and TMDs and contacts; these were systematically discussed. SCTD with organic molecules enables modulation of the electrical properties, such as the carrier concentration in TMD channels as well as optoelectronic characteristics, including photodetectivity and photoresponsivity. The ability to facilely reversibly de-dope TMDs provides a great opportunity to engineer the performance of the materials. Intrinsic defects located on the surface of TMDs can be passivated *via* simple treatments with polymers, such as PSS, or thiol-based molecules. It has been theoretically and experimentally determined that passivated TMDs exhibit a lower carrier concentration. At the interface between a TMD channel and dielectric layer, a self-assembled monolayer or grafted polymers, for example, end-functionalized PS on the gate dielectric can suppress charge trapping and scattering, leading to better electrical performance of TMD FETs, including the on/off ratio, a lower *SS* value and a smaller hysteresis window. At the interface between the TMD channel and the *S*/*D* contacts, the introduction of organic molecules can dramatically improve the charge injection properties by reducing the activation energy *via* additional tunneling paths. Furthermore, Fermi-level pinning can be successfully suppressed by eliminating the interfacial states, which promises reliable performance in FETs with metal-semiconductor interfaces. In addition, the ability to control the electrical and optoelectronic properties of TMDs without physical or chemical damage is a great advantage from the viewpoint of realizing structurally sensitive 2D device applications. In other words, surface engineering with organic molecules and polymers allows noninvasive processing onto TMDs directly as well as gradual dedoping *via* a simple rinsing process with solvents. This potential can be a key pathway for realizing high-performance device applications. For example, the output performance of TMD-based thermoelectrics could be improved by facilely modulating the carrier type, carrier concentration, electrical conductivity, and thermal conductivity, which are key parameters for determining the energy conversion performance of thermoelectrics. Additionally, it will attract much interest in the field of spintronics, especially in spin valve devices that require exquisitely optimized contact resistance values resulting in a maximized magneto resistance ratio. In this manner, interface engineering with various compatible organic molecules and polymers having different lengths and conductivities of the molecular backbone can be an attractive strategy for achieving the optimal characteristics of TMDs.

However, further technological advancements should be achieved to substitute mature semiconductors. For better electrical performance and low-voltage operation, high-quality high-k dielectrics, such as *h*-BN, must be used. Additionally, a critical bottleneck of organic materials, their weak nature under high temperature or vulnerability to moisture and oxygen, requires more reliable strategies in the interface engineering. From a manufacturing perspective, the scaling issue should be addressed with a high degree of design freedom, although many proposals have been examined, for example, the use of inkjet printing.[143] The ultimate goal of this research is the realization of ultrathin emerging optoelectronics based on TMD FETs and their utilization in transparent and flexible systems operating with low power consumption.

Acknowledgements

The authors appreciate the financial support of the National Research Foundation of Korea (NRF) grant (No. 2021R1A2C3004783 and 2021R1C1C2091728) funded by the Ministry of Science and ICT of Korea.

Vocabulary

self-assembled monolayer (SAM) — the monolayer which is spontaneously composed of ordered molecular assemblies on the surface

external quantum efficiency (EQE) — the ratio of the number of carriers collected by the devices to the number of photons illuminated on the device from the external light source

hopping transport — charge transport mechanism when the charges transport between localized sites *via* tunneling or overcoming potential barriers

phase engineering — the engineering of the material phase which is typically used to refer a physically homogenuous state of matter or crystal structure. For example, MoS_2 have two representative phases, semiconducting 2H phase (ABA stacked) and metallic 1T phase (ABC stacked)

the activation energy — the amount of energy needed to overcome energy barriers located between metal — semiconductor contact region for charge carrier injection.

References

1. K. S. Novoselov, A. K. Geim, S. V. Morozov, D. Jiang, Y. Zhang, S. V. Dubonos, I. V. Grigorieva and A. A. Firsov, *Science* **306** (2004) 666–669.
2. M. E. Dávila, L. Xian, S. Cahangirov, A. Rubio and G. Le Lay, *New J. Phys.* **16** (2014) 095002.
3. L. Tao, E. Cinquanta, D. Chiappe, C. Grazianetti, M. Fanciulli, M. Dubey, A. Molle and D. Akinwande, *Nat. Nanotechnol.* **10** (2015) 227–231.
4. B. Radisavljevic, A. Radenovic, J. Brivio, V. Giacometti and A. Kis, *Nat. Nanotechnol* **6** (2011) 147–150.
5. X. Xi, Z. Wang, W. Zhao, J.-H. Park, K. T. Law, H. Berger, L. Forró, J. Shan and K. F. Mak, *Nat. Phys.* **12** (2016) 139–143.
6. M. N. Ali, J. Xiong, S. Flynn, J. Tao, Q. D. Gibson, L. M. Schoop, T. Liang, N. Haldolaarachchige, M. Hirschberger, N. P. Ong and R. J. Cava, *Nature* **514** (2014) 205.
7. J. A. Hollingsworth, D. M. Poojary, A. Clearfield and W. E. Buhro, *J. Am. Chem. Soc.* **122** (2000) 3562–3563.
8. L. C. Gomes and A. Carvalho, *Phys. Rev. B* **92** (2015) 085406.
9. P. Zhang, F. Zhao, P. Long, Y. Wang, Y. Yue, X. Liu, Y. Feng, R. Li, W. Hu, Y. Li and W. Feng, *Nanoscale* **10** (2018) 15989–15997.
10. K. K. Kam and B. A. Parkinson, *J. Phys. Chem.* **86** (1982) 463–467.
11. K. F. Mak, C. Lee, J. Hone, J. Shan and T. F. Heinz, *Phys. Rev. Lett.* **105** (2010) 136805.
12. A. Dankert and S. P. Dash, *Nat. Commun.* **8** (2017) 16093.
13. J. R. Schaibley, H. Yu, G. Clark, P. Rivera, J. S. Ross, K. L. Seyler, W. Yao and X. Xu, *Nat. Rev. Mater.* **1** (2016) 16055.
14. K. F. Mak, K. L. McGill, J. Park and P. L. McEuen, *Science* **344** (2014) 1489–1492.
15. R. Frisenda, E. Navarro-Moratalla, P. Gant, D. Pérez De Lara, P. Jarillo-Herrero, R. V. Gorbachev and A. Castellanos-Gomez, *Chem. Soc. Rev.* **47** (2018) 53–68.
16. Y. Cao, V. Fatemi, S. Fang, K. Watanabe, T. Taniguchi, E. Kaxiras and P. Jarillo-Herrero, *Nature* **556** (2018) 43–50.
17. R. Ribeiro-Palau, C. Zhang, K. Watanabe, T. Taniguchi, J. Hone and C. R. Dean, *Science* **361** (2018) 690–693.
18. B. Sacépé, J. B. Oostinga, J. Li, A. Ubaldini, N. J. G. Couto, E. Giannini and A. F. Morpurgo, *Nat. Commun.* **2** (2011) 575.
19. J. Kang, R. M. Fernandes and A. Chubukov, *Phys. Rev. Lett.* **120** (2018) 267001.

20. B. Huang, G. Clark, D. R. Klein, D. MacNeill, E. Navarro-Moratalla, K. L. Seyler, N. Wilson, M. A. McGuire, D. H. Cobden, D. Xiao, W. Yao, P. Jarillo-Herrero and X. Xu, *Nat. Nanotechnol.* **13** (2018) 544–548.
21. X. Wang, J. Tang, X. Xia, C. He, J. Zhang, Y. Liu, C. Wan, C. Fang, C. Guo, W. Yang, Y. Guang, X. Zhang, H. Xu, J. Wei, M. Liao, X. Lu, J. Feng, X. Li, Y. Peng, H. Wei, R. Yang, D. Shi, X. Zhang, Z. Han, Z. Zhang, G. Zhang, G. Yu and X. Han, *Sci. Adv.* **5** (2019) eaaw8904.
22. K. Kim, J. Seo, E. Lee, K. T. Ko, B. S. Kim, B. G. Jang, J. M. Ok, J. Lee, Y. J. Jo, W. Kang, J. H. Shim, C. Kim, H. W. Yeom, B. Il Min, B.-J. Yang and J. S. Kim, *Nat. Mater.* **17** (2018) 794–799.
23. C. Hu, K. N. Gordon, P. Liu, J. Liu, X. Zhou, P. Hao, D. Narayan, E. Emmanouilidou, H. Sun, Y. Liu, H. Brawer, A. P. Ramirez, L. Ding, H. Cao, Q. Liu, D. Dessau and N. Ni, *Nat. Commun.* **11** (2020) 97.
24. S. Zhang, R. Wang, X. Wang, B. Wei, B. Chen, H. Wang, G. Shi, F. Wang, B. Jia, Y. Ouyang, F. Xie, F. Fei, M. Zhang, X. Wang, D. Wu, X. Wan, F. Song, H. Zhang and B. Wang, *Nano Lett.* **20** (2020) 709–714.
25. H. S. Lee, S.-W. Min, Y.-G. Chang, M. K. Park, T. Nam, H. Kim, J. H. Kim, S. Ryu and S. Im, *Nano Lett.* **12** (2012) 3695–3700.
26. T. Song, X. Cai, M. W.-Y. Tu, X. Zhang, B. Huang, N. P. Wilson, K. L. Seyler, L. Zhu, T. Taniguchi, K. Watanabe, M. A. McGuire, D. H. Cobden, D. Xiao, W. Yao and X. Xu, *Science* **360** (2018) 1214–1218.
27. D. Costanzo, S. Jo, H. Berger and A. F. Morpurgo, *Nat. Nanotechnol* **11** (2016) 339–344.
28. K. Cho, J. Pak, S. Chung and T. Lee, *ACS Nano* **13** (2019) 9713–9734.
29. S. Ahmed and J. Yi, *Nano-Micro Letters* **9** (2017) 50.
30. A. Allain, J. Kang, K. Banerjee and A. Kis, *Nat. Mater.* **14** (2015) 1195–1205.
31. S. Z. Butler, S. M. Hollen, L. Cao, Y. Cui, J. A. Gupta, H. R. Gutiérrez, T. F. Heinz, S. S. Hong, J. Huang, A. F. Ismach, E. Johnston-Halperin, M. Kuno, V. V. Plashnitsa, R. D. Robinson, R. S. Ruoff, S. Salahuddin, G. Shan, L. Shi, M. G. Spencer and M. Terrones, *ACS Nano* **7** (2013) 2898.
32. M. Chhowalla, D. Jena and H. Zhang, *Nat. Rev. Mater.* **1** (2016) 16052.
33. R. Ganatra and Q. Zhang, *ACS Nano* **8** (2014) 4074.
34. R. Mas-Balleste, C. Gomez-Navarro, J. Gomez-Herrero and F. Zamora, *Nanoscale* **3** (2011) 20.
35. Q. H. Wang, K. Kalantar-Zadeh, A. Kis, J. N. Coleman and M. S. Strano, *Nat. Nanotechnol.* **7** (2012) 699.
36. F. Xia, H. Wang, D. Xiao, M. Dubey and A. Ramasubramaniam, *Nat. Photonics* **8** (2014) 899.
37. S. M. Eichfeld, L. Hossain, Y. C. Lin, A. F. Piasecki, B. Kupp, A. G. Birdwell, R. A. Burke, N. Lu, X. Peng, J. Li, A. Azcatl, S. McDonnell, R. M. Wallace, M. J. Kim, T. S. Mayer, J. M. Redwing and J. A. Robinson, *ACS Nano* **9** (2015) 2080.
38. L. Jiao, H. J. Liu, J. L. Chen, Y. Yi, W. G. Chen, Y. Cai, J. N. Wang, X. Q. Dai, N. Wang and W. K. Ho, *New J. Phys.* **17** (2015) 053023.

39. Y. H. Lee, X. Q. Zhang, W. Zhang, M. T. Chang, C. T. Lin, K. D. Chang, Y. C. Yu, J. T. W. Wang, C. S. Chang, L. J. Li and T. W. Lin, *Adv. Mater.* **24** (2012) 2320.
40. C. H. Naylor, W. M. Parkin, J. Ping, Z. Gao, Y. R. Zhou, Y. Kim, F. Streller, R. W. Carpick, A. M. Rappe, M. Drndic, J. M. Kikkawa and A. T. C. Johnson, *Nano Lett.* **16** (2016) 4297.
41. H. Schmidt, S. Wang, L. Chu, M. Toh, R. Kumar, W. Zhao, A. Castro Neto, J. Martin, S. Adam, B. Özyilmaz and G. Eda, *Nano Lett.* **14** (2014) 1909.
42. X. Wang, H. Feng, Y. Wu and L. Jiao, *J. Am. Chem. Soc.* **135** (2013) 5304.
43. R. Yue, A. T. Barton, H. Zhu, A. Azcatl, L. F. Pena, J. Wang, X. Peng, N. Lu, L. Cheng, R. Addou, S. McDonnell, L. Colombo, J. W. P. Hsu, J. Kim, M. J. Kim, R. M. Wallace and C. L. Hinkle, *ACS Nano* **9** (2015) 474.
44. D. J. Late, B. Liu, H. S. S. R. Matte, V. P. Dravid and C. N. R. Rao, *ACS Nano* **6** (2012) 5635.
45. W. Park, J. Park, J. Jang, H. Lee, H. Jeong, K. Cho, S. Hong and T. Lee, *Nanotechnology* **24** (2013) 095202.
46. S. Yuan, R. Roldán, M. I. Katsnelson and F. Guinea, *Phys. Rev. B* **90** (2014) 041402.
47. K. Santosh, R. C. Longo, R. Addou, R. M. Wallace and K. Cho, *Nanotechnology* **25** (2014) 375703.
48. W. Bao, X. Cai, D. Kim, K. Sridhara and M. S. Fuhrer, *Appl. Phys. Lett.* **102** (2013) 042104.
49. J. Chen, B. Liu, Y. Liu, W. Tang, C. T. Nai, L. Li, J. Zheng, L. Gao, Y. Zheng, H. S. Shin, H. Y. Jeong and K. P. Loh, *Adv. Mater.* **27** (2015) 6722.
50. Y. Guo, X. Wei, J. Shu, B. Liu, J. Yin, C. Guan, Y. Han, S. Gao and Q. Chen, *Appl. Phys. Lett.* **106** (2015) 103109.
51. H. Ilatikhameneh, T. A. Ameen, G. Klimeck, J. Appenzeller and R. Rahman, *IEEE Electron Device Lett.* **36** (2015) 1097.
52. Y. Y. Illarionov, G. Rzepa, M. Waltl, T. Knobloch, A. Grill, M. M. Furchi, T. Mueller and T. Grasser, *2D Mater.* **3** (2016) 035004.
53. S. Borghardt, J. S. Tu, F. Winkler, J. Schubert, W. Zander, K. Leosson and B. E. Kardynał, *Phys. Rev. Mater.* **1** (2017) 054001.
54. A. Förster, S. Gemming, G. Seifert and D. Tománek, *ACS Nano* **11** (2017) 9989.
55. J. H. Park, A. Sanne, Y. Guo, M. Amani, K. Zhang, H. C. P. Movva, J. A. Robinson, A. Javey, J. Robertson, S. K. Banerjee and A. C. Kummel, *Sci. Adv.* **3** (2017) e1701661.
56. H. Qiu, T. Xu, Z. Wang, W. Ren, H. Nan, Z. Ni, Q. Chen, S. Yuan, F. Miao, F. Song, G. Long, Y. Shi, L. Sun, J. Wang and X. Wang, *Nat. Commun.* **4** (2013) 2642.
57. S. H. Song, M. K. Joo, M. Neumann, H. Kim and Y. H. Lee, *Nat. Commun.* **8** (2017) 2121.
58. W. Zhou, X. Zou, S. Najmaei, Z. Liu, Y. Shi, J. Kong, J. Lou, P. M. Ajayan, B. I. Yakobson and J. C. Idrobo, *Nano Lett.* **13** (2013) 2615.
59. S. McDonnell, R. Addou, C. Buie, R. M. Wallace and C. L. Hinkle, *ACS Nano* **8** (2014) 2880.

60. D. Liu, Y. Guo, L. Fang and J. Robertson, *Appl. Phys. Lett.* **103** (2013) 183113.
61. C. Gong, L. Colombo, R. M. Wallace and K. Cho, *Nano Lett.* **14** (2014) 1714.
62. Y. Zhou, Z. Wang, P. Yang, X. Zu, L. Yang, X. Sun and F. Gao, *ACS Nano* **6** (2012) 9727.
63. G. Seifert, H. Terrones, M. Terrones and T. Frauenheim, *Solid State Commun.* **115** (2000) 635.
64. B. E. Brown, *Acta Crystallogr.* **20** (1966) 268.
65. H. Fang, S. Chuang, T. C. Chang, K. Takei, T. Takahashi and A. Javey, *Nano Lett.* **12** (2012) 3788.
66. B. Sipos, A. F. Kusmartseva, A. Akrap, H. Berger, L. Forró and E. Tutiš, *Nat. Mater.* **7** (2008) 960.
67. G. H. Lee, Y. J. Yu, X. Cui, N. Petrone, C. H. Lee, M. S. Choi, D. Y. Lee, C. Lee, W. J. Yoo, K. Watanabe, T. Taniguchi, C. Nuckolls, P. Kim and J. Hone, *ACS Nano* **7** (2013) 7931.
68. R. Kappera, D. Voiry, S. E. Yalcin, B. Branch, G. Gupta, A. D. Mohite and M. Chhowalla, *Nat. Mater.* **13** (2014) 1128.
69. J. Wang, Q. Yao, C. W. Huang, X. Zou, L. Liao, S. Chen, Z. Fan, K. Zhang, W. Wu, X. Xiao, C. Jiang and W. W. Wu, *Adv. Mater.* **28** (2016) 8302.
70. D. Kiriya, M. Tosun, P. Zhao, J. S. Kang and A. Javey, *J. Am. Chem. Soc.* **136** (2014) 7853.
71. S. H. H. Shokouh, P. J. Jeon, A. Pezeshki, K. Choi, H. S. Lee, J. S. Kim, E. Y. Park and S. Im, *Adv. Funct. Mater.* **25** (2015) 7208.
72. D. M. Sim, M. Kim, S. Yim, M. J. Choi, J. Choi, S. Yoo and Y. S. Jung, *ACS Nano* **9** (2015) 12115.
73. B. Cho, M. G. Hahm, M. Choi, J. Yoon, A. R. Kim, Y. J. Lee, S. G. Park, J. D. Kwon, C. S. Kim, M. Song, Y. Jeong, K. S. Nam, S. Lee, T. J. Yoo, C. G. Kang, B. H. Lee, H. C. Ko, P. M. Ajayan and D. H. Kim, *Sci. Rep.* **5** (2015) 8052.
74. D. Kufer and G. Konstantatos, *Nano Lett.* **15** (2015) 7307.
75. G. H. Lee, X. Cui, Y. D. Kim, G. Arefe, X. Zhang, C. H. Lee, F. Ye, K. Watanabe, T. Taniguchi, P. Kim and J. Hone, *ACS Nano* **9** (2015) 7019.
76. J. Ma, K. Y. Choi, S. H. Kim, H. Lee and G. Yoo, *Appl. Phys. Lett.* **113** (2018) 013102.
77. D. H. Kang, J. Shim, S. K. Jang, J. Jeon, M. H. Jeon, G. Y. Yeom, W. S. Jung, Y. H. Jang, S. Lee and J. H. Park, *ACS Nano* **9** (2015) 1099.
78. C. J. Benjamin, S. Zhang and Z. Chen, *Nanoscale* **10** (2018) 5148.
79. J. Pak, J. Jang, K. Cho, T. Y. Kim, J. K. Kim, Y. Song, W. K. Hong, M. Min, H. Lee and T. Lee, *Nanoscale* **7** (2015) 18780.
80. S. Mouri, Y. Miyauchi and K. Matsuda, *Nano Lett.* **13** (2013) 5944.
81. Y. Li, C. Y. Xu, P. Hu and L. Zhen, *ACS Nano* **7** (2013) 7795.
82. Y. Du, H. Liu, A. T. Neal, M. Si and D. Y. Peide, *IEEE Electron Device Lett.* **34** (2013) 1328.
83. S. Dey, H. S. S. R. Matte, S. N. Shirodkar, U. V. Waghmare and C. N. R. Rao, *Chem. Asian J.* **8** (2013) 1780.
84. D. H. Kang, M. S. Kim, J. Shim, J. Jeon, H. Y. Park, W. S. Jung, H. Y. Yu, C. H. Pang, S. Lee and J. H. Park, *Adv. Funct. Mater.* **25** (2015) 4219.

85. A. Tarasov, S. Zhang, M. Y. Tsai, P. M. Campbell, S. Graham, S. Barlow, S. R. Marder and E. M. Vogel, *Adv. Mater.* **27** (2015) 1175.
86. S. Andleeb, A. Kumar Singh and J. Eom, *Sci. Technol. Adv. Mater.* **16** (2015) 035009.
87. G. Ghimire, K. P. Dhakal, G. P. Neupane, S. G. Jo, H. Kim, C. Seo, Y. H. Lee, J. Joo and J. Kim, *Nanotechnology* **28** (2017) 185702.
88. K. Heo, S. H. Jo, J. Shim, D. H. Kang, J. H. Kim and J. H. Park, *ACS Appl. Mater. Interfaces* **10** (2018) 32765.
89. N. Peimyoo, W. Yang, J. Shang, X. Shen, Y. Wang and T. Yu, *ACS Nano* **8** (2014) 11320.
90. H. Yoo, S. Hong, H. Moon, S. On, H. Ahn, H. K. Lee, S. Kim, Y. K. Hong and J. J. Kim, *Adv. Electron. Mater.* **4** (2018) 1700639.
91. M. Tosun, L. Chan, M. Amani, T. Roy, G. H. Ahn, P. Taheri, C. Carraro, J. W. Ager, R. Maboudian and A. Javey, *ACS Nano* **10** (2016) 6853.
92. S. Bertolazzi, S. Bonacchi, G. Nan, A. Pershin, D. Beljonne and P. Samorì, *Adv. Mater.* **29** (2017) 1606760.
93. X. Zhang, Q. Liao, S. Liu, Z. Kang, Z. Zhang, J. Du, F. Li, S. Zhang, J. Xiao, B. Liu, Y. Ou, X. Liu, L. Gu and Y. Zhang, *Nat. Commun.* **8** (2017) 15881.
94. R. G. Freeman, K. C. Grabar, K. J. Allison, R. M. Bright, J. A. Davis, A. P. Guthrie, M. B. Hommer, M. A. Jackson, P. C. Smith, D. G. Walter and M. J. Natan, *Science* **267** (1995) 1629.
95. J. C. Love, L. A. Estroff, J. K. Kriebel, R. G. Nuzzo and G. M. Whitesides, *Chem. Rev.* **105** (2005) 1103.
96. K. L. Prime and G. M. Whitesides, *Science* **252** (1991) 1164.
97. A. Ulman, *Chem. Rev.* **96** (1996) 1533.
98. C. Zhou, M. R. Deshpande, M. A. Reed, L. Jones and J. M. Tour, *Appl. Phys. Lett.* **71** (1997) 611.
99. X. Chen, N. C. Berner, C. Backes, G. S. Duesberg and A. R. McDonald, *Angew. Chem., Int. Ed.* **55** (2016) 5803.
100. K. Cho, M. Min, T. Y. Kim, H. Jeong, J. Pak, J. K. Kim, J. Jang, S. J. Yun, Y. H. Lee, W. K. Hong and T. Lee, *ACS Nano* **9** (2015) 8044.
101. K. Cho, J. Pak, J. K. Kim, K. Kang, T. Y. Kim, J. Shin, B. Y. Choi, S. Chung and T. Lee, *Adv. Mater.* **30** (2018) 1705540.
102. Q. Li, Y. Zhao, C. Ling, S. Yuan, Q. Chen and J. Wang, *Angew. Chem. Int. Ed.* **56** (2017) 10501.
103. M. Makarova, Y. Okawa and M. Aono, *J. Phys. Chem. C* **116** (2012) 22411.
104. E. P. Nguyen, B. J. Carey, J. Z. Ou, J. van Embden, E. D. Gaspera, A. F. Chrimes, M. J. S. Spencer, S. Zhuiykov, K. Kalantar-zadeh and T. Daeneke, *Adv. Mater.* **27** (2015) 6225.
105. S. L. Peterson and K. H. Schulz, *Langmuir* **12** (1996) 941.
106. B. Xu, Y. Su, L. Li, R. Liu and Y. Lv, *Sens. Actuators, B* **246** (2017) 380.
107. C. G. Wiegenstein and K. H. Schulz, *J. Phys. Chem. B* **103** (1999) 6913.
108. M. Amani, D.-H. Lien, D. Kiriya, J. Xiao, A. Azcatl, J. Noh, S. R. Madhvapathy, R. Addou, S. KC, M. Dubey, K. Cho, R. M. Wallace, S.-C. Lee, J.-H. He, J. W. Ager, X. Zhang, E. Yablonovitch and A. Javey, *Science* **350** (2015) 1065–1068.

109. Z. Yu, Y. Pan, Y. Shen, Z. Wang, Z. Y. Ong, T. Xu, R. Xin, L. Pan, B. Wang, L. Sun, J. Wang, G. Zhang, Y. W. Zhang, Y. Shi and X. Wang, *Nat. Commun.* **5** (2014) 5290.
110. J. M. Park, I. T. Cho, W. M. Kang, B. G. Park and J. H. Lee, *Appl. Phys. Lett.* **109** (2016) 053503.
111. K. Cho, W. Park, J. Park, H. Jeong, J. Jang, T. Y. Kim, W. K. Hong, S. Hong and T. Lee, *ACS Nano* **7** (2013) 7751.
112. H. J. Kwon, J. Jang, S. Kim, V. Subramanian and C. P. Grigoropoulos, *Appl. Phys. Lett.* **105** (2014) 152105.
113. L. Cheng, J. Lee, H. Zhu, A. V. Ravichandran, Q. Wang, A. T. Lucero, M. J. Kim, R. M. Wallace, L. Colombo and J. Kim, *ACS Nano* **11** (2017) 10243.
114. H. Y. Chang, S. Yang, J. Lee, L. Tao, W. S. Hwang, D. Jena, N. Lu and D. Akinwande, *ACS Nano* **7** (2013) 5446.
115. H. Ji, M. K. Joo, Y. Yun, J. H. Park, G. Lee, B. H. Moon, H. Yi, D. Suh and S. C. Lim, *ACS Appl. Mater. Interfaces* **8** (2016) 19092.
116. Z. Hu, Z. Wu, C. Han, J. He, Z. Ni and W. Chen, *Chem. Soc. Rev.* **47** (2018) 3100.
117. W. H. Lee and Y. D. Park, *Adv. Mater. Interfaces* **5** (2018) 1700316.
118. H. Wang, H. Yuan, S. S. Hong, Y. Li and Y. Cui, *Chem. Soc. Rev.* **44** (2015) 2664.
119. B. Radisavljevic and A. Kis, *Nat. Mater.* **12** (2013) 815.
120. Z. Yu, Z. Y. Ong, Y. Pan, Y. Cui, R. Xin, Y. Shi, B. Wang, Y. Wu, T. Chen, Y. W. Zhang, G. Zhang and X. Wang, *Adv. Mater.* **28** (2016) 547.
121. Y. Cui, R. Xin, Z. Yu, Y. Pan, Z. Y. Ong, X. Wei, J. Wang, H. Nan, Z. Ni, Y. Wu, T. Chen, Y. Shi, B. Wang, G. Zhang, Y. W. Zhang and X. Wang, *Adv. Mater.* **27** (2015) 5230.
122. Y. Jeong, J. H. Park, J. Ahn, J. Y. Lim, E. Kim and S. Im, *Adv. Mater. Interfaces* **5** (2018) 1800812.
123. S. Chung, M. Jang, S. B. Ji, H. Im, N. Seong, J. Ha, S. K. Kwon, Y. H. Kim, H. Yang and Y. Hong, *Adv. Mater.* **25** (2013) 4773.
124. J. Ha, S. Chung, M. Pei, K. Cho, H. Yang and Y. Hong, *ACS Appl. Mater. Interfaces* **9** (2017) 8819.
125. S. Najmaei, X. Zou, D. Er, J. Li, Z. Jin, W. Gao, Q. Zhang, S. Park, L. Ge, S. Lei, J. Kono, V. B. Shenoy, B. I. Yakobson, A. George, P. M. Ajayan and J. Lou, *Nano Lett.* **14** (2014) 1354.
126. S. L. Li, K. Komatsu, S. Nakaharai, Y. F. Lin, M. Yamamoto, X. Duan and K. Tsukagoshi, *ACS Nano* **8** (2014) 12836.
127. J. Kang, W. Liu, D. Sarkar, D. Jena and K. Banerjee, *Phys. Rev. X* **4** (2014) 031005.
128. J. Pak, Y. Jang, J. Byun, K. Cho, T. Y. Kim, J. K. Kim, B. Y. Choi, J. Shin, Y. Hong, S. Chung and T. Lee, *ACS Nano* **12** (2018) 7109.
129. S. Cho, S. Kim, J. H. Kim, J. Zhao, J. Seok, D. H. Keum, J. Baik, D. H. Choe, K. J. Chang, K. Suenaga, S. W. Kim, Y. H. Lee and H. Yang, *Science* **349** (2015) 625.

130. S. Chuang, C. Battaglia, A. Azcatl, S. McDonnell, J. S. Kang, X. Yin, M. Tosun, R. Kapadia, H. Fang, R. M. Wallace and A. Javey, *Nano Lett.* **14** (2014) 1337.
131. X. Cui, E. M. Shih, L. A. Jauregui, S. H. Chae, Y. D. Kim, B. Li, D. Seo, K. Pistunova, J. Yin, J. H. Park, H. J. Choi, Y. H. Lee, K. Watanabe, T. Taniguchi, P. Kim, C. R. Dean and J. C. Hone, *Nano Lett.* **17** (2017) 4781.
132. S. Das, H. Y. Chen, A. V. Penumatcha and J. Appenzeller, *Nano Lett.* **13** (2013) 100.
133. H. Kwon, W. Choi, D. Lee, Y. Lee, J. Kwon, B. Yoo, C. P. Grigoropoulos and S. Kim, *Nano Res.* **7** (2014) 1137.
134. W. S. Leong, X. Luo, Y. Li, K. H. Khoo, S. Y. Quek and J. T. L. Thong, *ACS Nano* **9** (2015) 869.
135. L. Wang, I. Meric, P. Y. Huang, Q. Gao, Y. Gao, H. Tran, T. Taniguchi, K. Watanabe, L. M. Campos, D. A. Muller, J. Guo, P. Kim, J. Hone, K. L. Shepard and C. R. Dean, *Science* **342** (2013) 614.
136. R. Yang, X. Zheng, Z. Wang, C. J. Miller and P. X. L. Feng, *J. Vac. Sci. Technol. B* **32** (2014) 061203.
137. P. Bampoulis, R. van Bremen, Q. Yao, B. Poelsema, H. J. W. Zandvliet and K. Sotthewes, *ACS Appl. Mater. Interfaces* **9** (2017) 19278.
138. G. Eda, H. Yamaguchi, D. Voiry, T. Fujita, M. Chen and M. Chhowalla, *Nano Lett.* **11** (2011) 5111–5116.
139. Y. Liu, J. Guo, E. Zhu, L. Liao, S. J. Lee, M. Ding, I. Shakir, V. Gambin, Y. Huang and X. Duan, *Nature* **557** (2018) 696.
140. Y. Wang, J. C. Kim, R. J. Wu, J. Martinez, X. Song, J. Yang, F. Zhao, A. Mkhoyan, H. Y. Jeong and M. Chhowalla, *Nature* **568** (2019) 70.
141. M. Farmanbar and G. Brocks, *Phys. Rev. B* **91** (2015) 161304.
142. M. Farmanbar and G. Brocks, *Adv. Electron. Mater.* **2** (2016) 1500405.
143. S. Chung, K. Cho and T. Lee, *Adv. Sci.* **6** (2019) 1801445.

https://doi.org/10.1142/9789811247859_0003

Chapter 3

On-Surface Molecular Reactions

Liangliang Cai and Wei Xu*

Interdisciplinary Materials Research Center,
Tongji-Aarhus Joint Research Center for
Nanostructures and Functional Nanomaterials,
College of Materials Science and Engineering,
Tongji University,
Shanghai 201804, P. R. China
**xuwei@tongji.edu.cn*

During the last decades, the bottom–fup strategy of on-surface molecular reactions has been extensively investigated in order to fulfill controllable fabrication of covalent interlinking nanostructures/nanomaterials at atomic scale. A variety of organic reactions have been introduced to substrates, such as Ullmann coupling, Glaser coupling, cyclodehydrogenation and so on. In this paper, these on-surface molecular reactions will be reviewed from three aspects: the precursor, surface and external stimuli. Finally, a summary of past achievements and an outlook of future scientific challenges will be discussed.

Keywords: On-surface molecular reactions; precursor; surface; external stimuli.

1. Introduction

Last decades have witnessed the prosperous development of two-dimensional (2D) materials including transition metal dichalcogenides (TMDs), black phosphorus and silicene, etc.[1–3] since the discovery of graphene in 2004.[4] These 2D materials display intriguing dimension-related properties, which promise great potential

*Corresponding author.

applications in molecular electronics, optoelectronic devices, sensors and so on. A basic principle shows that the property is determined by the structure. Multiple elements such as the scale, thickness and defects vary the properties of 2D materials dramatically. Taking graphene nanoribbons (GNRs) as an example, the band gap has been effectively regulated by different widths and edge structures.[5–7] Considering the synthetic method is the key factor to determine the structure, precise fabrication at atomic scale is of utmost importance. The traditional top–down approach following continuous miniaturization of structural patterns of 2D materials is intrinsically limited. Thus, the promising but challenging bottom–up strategy of on-surface molecular reactions has been extensively investigated in order to fulfill the atomically precise fabrication. Under ultrahigh vacuum (UHV) conditions, the development of scanning tunneling microscopy (STM) and noncontact atomic force microscopy (nc-AFM) allows high resolution of individual precursors, intermediates and products with submolecular level in real space.[8,9] X-ray photoelectron spectroscopy (XPS) makes it possible to monitor chemical states in real time during the reaction. In most cases, the underlying mechanisms of on-surface molecular reactions still remain unknown. With the combination of theoretical model especially the density functional theory (DFT) calculations, it is feasible to rationalize reaction pathways and unravel the mechanisms. All these advanced surface science techniques have paved the way for us to explore molecular reactions on surfaces by trial and error.

In recent years, controllable fabrication of covalent interlinking nanostructures/nanomaterials by on-surface molecular reactions has attracted great attention of researchers, including forming simple oligomers, 1D polymeric chains and 2D networks, etc. A variety of well-known organic reactions have been introduced to substrates, including the Ullmann coupling, alkane polymerization, Glaser coupling, Bergman cyclization, cyclodehydrogenation and so on. Particularly, on-surface reactions with hydrocarbons involved have aroused great attention owing to their captivating potential in constructing carbon-rich nanostructures and nanomaterials. Beyond forming carbon–carbon (C–C) bonds, heteroatom-containing connections have also been constructed on substrates, for example, by condensation reactions of boronic acids, imine formation, azide–alkyne click reactions, dimerization of N-heterocyclic carbenes, as well as cycloaddition reactions, etc.[10–13]

Owing to the irreversibility of the covalent bond formation, the initially formed defects cannot heal, which makes it a challenge to construct long-range ordered nanostructures. Therefore, the design of precursor and choice of surfaces are vital to optimize the precursor activation, diffusion, coupling, the surface reactivity, and to reduce the defects. As shown in Scheme 1, in this paper, some well-known on-surface molecular reactions will be reviewed in three aspects: the precursor, surface and external stimuli. The functional group and backbone of the precursor define the type and nature of a reaction. Different metal surfaces including surface materials and surface lattices could reduce the energy barrier thus rendering different reaction pathways, as well as steer morphologies of products and accomplish the selective formation to some extent. In addition to precursor designs and the choice of surfaces, ordered covalent interlinked nanostructures are achieved by external stimuli including thermal annealing, manipulation or irradiation. Finally, a summary of past achievements and an outlook of future scientific challenges will be discussed.

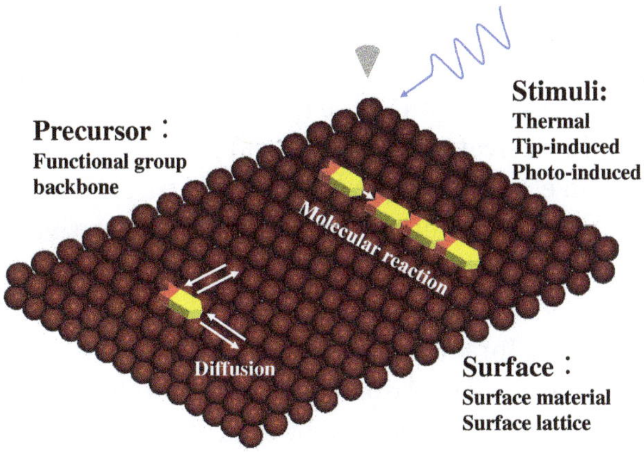

Scheme 1. (Color online) Key aspects of on-surface molecular reactions.

2. Precursor

2.1. *Functional group*

Trace back to 1904, Ullmann *et al.* discovered that the synthesis of biphenyl could be accomplished with a high purity by annealing the iodobenzene together with copper catalyst. Since then, this Ullmann reaction has become a textbook case and been intensively investigated in synthetic chemistry. Around a century later in 2000, Hla *et al.* demonstrated that the Ullmann coupling could be induced by the STM tip at atomic scale.[16] This process opens up a fascinating route for on-surface synthesis of individual molecular-electronic or molecular-mechanical devices. However, this approach is accompanied by the drawback of time-consuming and inefficiency, making it unsuitable for constructing desired architectures from a large number of building blocks. Later in 2007, Grill *et al.* employed thermal treatment to carbon–bromine (C–Br) activation,[14] which boomed the emerging field of on-surface synthesis. From then on, the Ullmann-type coupling has been widely studied by researchers across the whole world and heretofore still remains one of the most investigated on-surface molecular reactions. The authors chose a porphyrin molecular precursor with four phenyl legs as a building block. The legs were functionalized with 1, 2 and 4 bromine substituents, whose chemical structure is depicted in Figs. 1(a), 1(c) and 1(e), respectively. Two alternative methods for the activation of C–Br bond were employed. In method I, intact precursors are deposited on Au(111) surface, followed by thermal annealing of the precursor-covered surface, which finally led to the dissociation of the bromine substituent atoms. While in method II, the activation of C–Br bond has already occurred in the evaporator at an elevated evaporator temperature. Either method I or method II led to the similar results, that is, the activated building blocks were covalently interlinked on Au(111) surface. First, the BrTPP molecules were deposited on Au(111) kept at room temperature (RT). Intact BrTPP molecules were observed provided with 550 K or lower evaporator temperature. While the C–Br bond dissociation occurred in the evaporator at higher temperature of 610 K (method II). The activated TPP-type radicals were

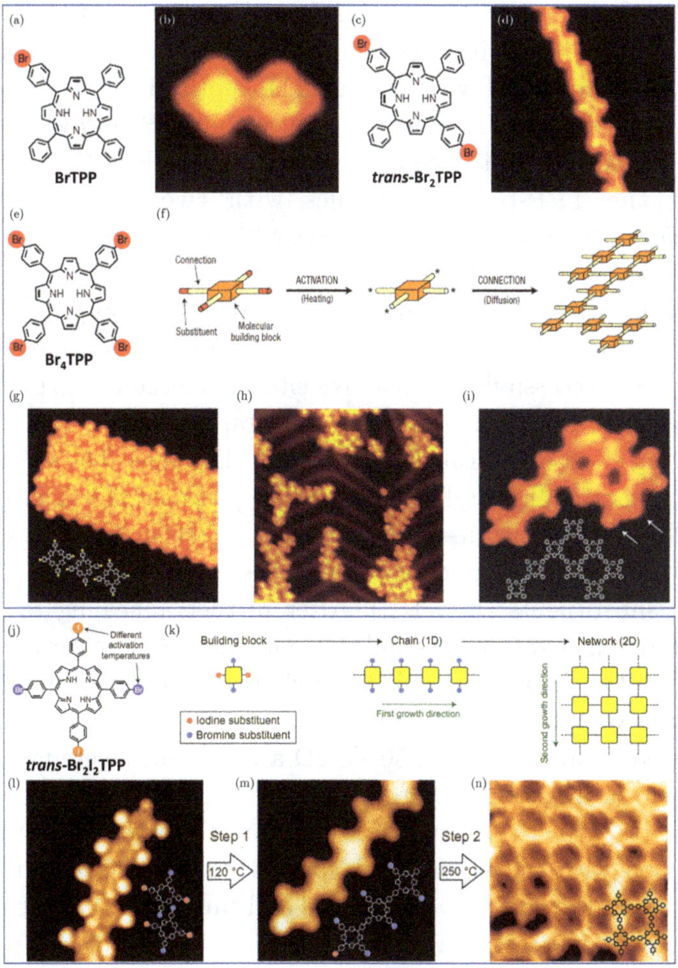

Fig. 1. (Color online) Functional groups of precursors regulate on-surface molecular reactions. Chemical structure of the (a) BrTPP, (c) trans-Br_2TPP, (e) Br_4TPP and (j) trans-Br_2I_2TPP molecule. STM image of the (b) dimer product and (d) 1D chain structure after activation and connection. (f) Scheme of the fabrication of covalently interlinked Br_4TPP molecule networks. (g) STM image of an island on Au(111) after deposition of the Br_4TPP molecule at 550 K. (h) STM image after deposition of the Br_4TPP molecule at 610 K, resulting in the connection of Br_4TPP molecules forming networks. (i) STM image of a network composed of eight molecular building blocks. (k) Scheme of the hierarchical fabrication by annealing. STM images of the trans-Br_2I_2TPP molecule on Au(111) (l) after deposition at 80 K, (m) after annealing to 120°C, (n) and after further annealing to 250°C.

Source: Panels (a)–(i) from Ref. 14; panels (j)–(n) from Ref. 15. Nature Publishing Group.

subsequently deposited onto Au(111), and then connected with each other to form new intermolecular chemical bonds by thermal diffusion, as shown in Fig. 1(b) (Au(111), kept at RT). Alternatively, further annealing to 610 K would induce the C–Br cleavage to aryl radicals, followed by the coupling reaction (method I). By further varying the TPP-based molecules with two or four bromine substituent atoms, the authors succeeded in the fabrication of linear chains (Fig. 1(d)) and 2D networks (Figs. 1(h) and 1(i)).

However, the networks constructed by this direct approach would encounter the limited extension, which could be ascribed to the uneasy steric accessibility of reactive sites. A method to get over this difficulty is to take full advantage of hierarchical reactions and the templating effect of intermediately formed 1D chains. The same group of Grill reported hierarchical on-surface polymerization in 2012.[15] After deposition of trans-Br_2I_2TPP (Fig. 1(j)) at low temperature (80 K), each molecule remained intact, i.e. contained both kinds of halogen substituents (Fig. 1(l)). After further annealing to 120°C, efficient polymerization was induced via the selective activation of iodine substituent atoms, and linear chains are created as shown in Fig. 1(m). Next, the remaining bromine substituent atoms were easily dissociated by annealing to 250°C, 2D architectures were fabricated, as shown in Fig. 1(n). Figure 1(k) schematically indicated the hierarchical growth following sequential thermal activation. In comparison with the 2D network formed by Br_4TPP through one-step reaction, the defects were reduced and thus the size of the regular network was increased. The reason lies in two adjacent polymer chains. Once a covalent bond between two adjacent polymer chains is formed, they are exactly suitable for the construction of the rest bonds between these two 1D chains, for all activated site are equally spaced along the chains. These results represent a versatile route for bottom–up construction of sophisticated electronic circuits and molecular machines based on individual functionalized molecular components. Moreover, the porphyrin core could act as a platform for the fabrication of highly functional nanostructures by coordinating with electronically, optically or magnetically active ions.

Since then a wide range of different functional groups regulating on-surface molecular reactions have been in blossoming development. In recent years, the group of Lackinger[17] and Beton[18] independently explored the synthesis of 2D nanostructures via on-surface Ullmann coupling of the TBPB precursor (Fig. 2(a)). After deposition of TBPB molecule precursors on Cu(111) substrate, C–Cu–C organometallic network was first formed. Subsequent annealing to 300°C led to the demetallation and 2D covalent organic frameworks (COFs) were built up. When the TBPB precursor molecules were deposited on Au(111) surface, no organometallic intermediates were observed, but only comparable 2D COFs were constructed, composed of polygons with 4, 5, 6, 7 and 8 edges as shown in Fig. 2(b), and could be used to control the organization of a sublimed layer of C_{60}.

If the bromine substituents of the TBPB precursor were replaced with terminal alkynyl groups, the Ext-TEB molecular, as shown in Fig. 2(c), was acquired. The homocoupling reaction of terminal alkynes via C_{sp}-H activation, i.e. Glaser coupling, is well established in solution. In 2012, Barth *et al.* made the first attempt to fabricate 2D porous networks by the surface-assisted Glaser coupling.[19] After annealing to 400 K, 2D frameworks with butadiyne connection moieties were constructed, as shown in Figs. 2(d) and 2(e). The covalent linking of acetylenes presented an important route for the fabrication of novel carbon-based scaffolds and 2D materials distinct from graphene. The formed 2D network is not ordered honeycomb structures owing to the inevitable and irreversible side reactions.

Dehalogenative homocouplings of aryl (sp^2-carbon), alkyl (sp^3-carbon)[23] and alkenyl (sp^2-carbon) halides[24] demonstrated that C–Br activation were always found to occur at lower temperatures in comparison with C–H activation. Therefore, further attempt of introducing the tBEP (Fig. 2(f)) molecules on Au(111) was worth mentioning. By delicately designing the molecular precursors with terminal alkynyl bromide, the Xu group demonstrated the dehalogenative homocoupling reaction of the alkynyl (sp^1-carbon) halide with high selectivity and fulfilled an effective approach to fabricate 2D nanostructures involving acetylenic scaffoldings.[20] After deposition of the tBEP precursor on Au(111) substrate at RT followed by slight

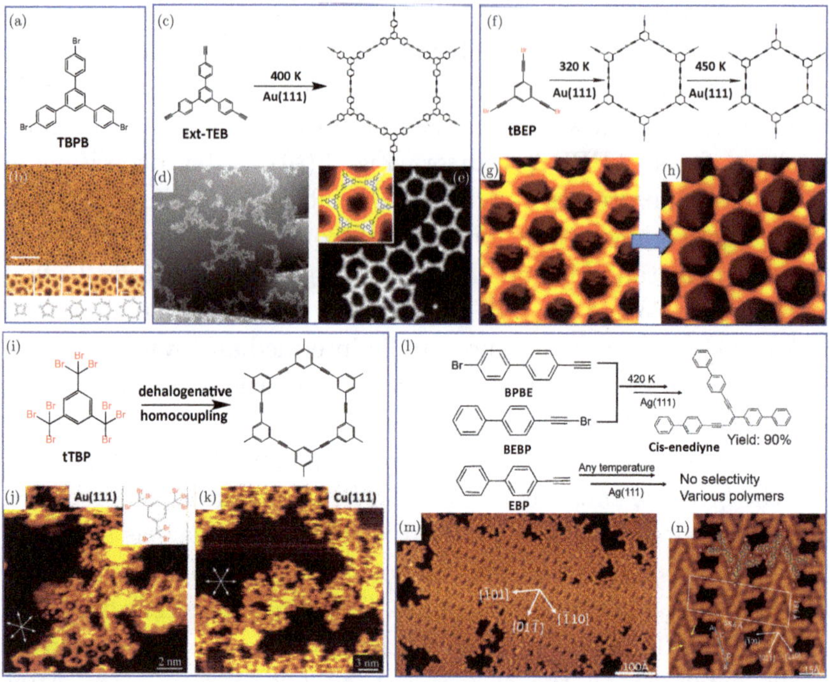

Fig. 2. (Color online) Functional groups of precursors regulate on-surface molecular reactions. (a) Chemical structure of the TBPB molecule. (b) STM images of porous covalent networks formed by deposition of the TBPB molecular on an annealed Au (111) surface, composed of polygons with 4, 5, 6, 7 and 8 edges. (c) Schematic illustration showing the dehydrogenative homocoupling of Ext-TEB on Ag(111). (d) Large-scale and (e) close-up STM images of covalent network through annealing to 400 K. (f) Schematic illustration showing the dehalogenative homocouplings of tBEP on Au(111). Close-up STM images of fabricating (g) the C–Au–C organometallic network after depositing tBEP molecules at RT followed by annealing to 320 K, and (h) the C–C coupled network with acetylenic scaffoldings after annealing to $\sim$ 450 K, respectively. (i) Scheme showing dehalogenative homocouplings of tTBP on noble metal surfaces. Large-scale STM images of the formed molecular network after deposition of tTBP on (j) Au(111) and (k) Cu(111) substrates at RT followed by annealing to $\sim$ 500 and $\sim$ 430 K, respectively. (l) Deposition of BPBE or BEBP on Ag(111) surface at 420 K leads to the fabrication of cis-enediyne with a yield of 90%, while, reaction of EBP on the Ag(111) surface shows no selectivity. (m) Large-scale STM image after deposition of BPBE on Ag(111) at 420 K. (n) Close-up STM image of the ordered island composed of cis-enediyne.

Source: Panels (a) and (b) from Ref. 18, Royal Society of Chemistry; panels (c)–(e) from Ref. 19, Nature Publishing Group; panels (f)–(h) from Ref. 20, American Chemical Society; panels (i)–(k) from Ref. 21, Wiley; panels (l)–(n) from Ref. 22, Wiley.

annealing to 320 K, C–Au–C organometallic network was first established, as shown in Fig. 2(g). Subsequent annealing to 450 K, the covalent interlinked honeycomb network with acetylenic scaffoldings was successfully achieved as depicted in Fig. 2(h).

Investigations of 2D nanostructures with acetylenic scaffoldings are thrilling in spite of the chemically vulnerable sp-hybridized carbon atoms. Thus, it is of utmost importance to find a functional group directly forming C–C triple bond on surfaces as a new strategy. The Xu group then designed precursors with three bromine substituent atoms attached at the same carbon location with the aim of directly forming C–C triple bonds on surfaces. As shown in Fig. 2(i), they have employed the tribromomethyl-substituted arenes (tTBP) on Au(111) and Cu(111) substrates, respectively.[21] After deposition of tTBP molecular precursors at RT followed by thermal annealing, the molecular networks were formed, as shown in Figs. 2(j) and 2(k). These networks cannot grow into larger and more ordered structures owing to the large amount of detached bromine atoms on the surfaces.

Except the contaminant, the detached bromine adatoms could also play an assistant role in on-surface synthesis.[25] Zhu *et al.* unambiguously confirmed the vital role of detached bromine adatoms in the selective synthesis of cis-enediyne[22] whose antibiotic and antitumor properties make itself potent antitumor antibiotics. As shown in Fig. 2(l), the authors investigated the BPBE, BEBP and EBP precursors, respectively. After deposition of the BPBE precursor on Ag(111) held at 420 K, ordered island composed of cis-enediyne products were formed (Figs. 2(m) and 2(n)). The research of BEBP precursor on Ag(111) also displayed a highly selective formation of cis-enediyne with a yield of 90%, while the reaction of EBP precursor showed no selectivity on the Ag(111) surface. It was also demonstrated by DFT calculations that this selective on-surface synthesis was implemented by the protection effect of the detached bromine adatoms. The bromine atoms lay between the formed cis-enediynes and acted as a "glue" to aggregate them into close-packed islands with Br···H bonds, which asked for a high steric barrier to further reactions of the cis-enediyne product.

2.2. *Backbone*

In addition to the functional groups, the backbone of precursors also plays an essential role in regulating on-surface synthesis of nanostructures. 2D networks are always constructed by activating multifunctional building blocks of three-fold, four-fold or six-fold symmetry, while ditopic monomers are used to forming chain structures. Lackinger *et al.* constructed 2D honeycomb networks by the hierarchical strategy, as shown in Fig. 3(a).[26] After deposition of the BIB precursor on Au(111) surface at RT, the carbon–iodine (C–I) bonds were first activated and the TBQ dimers were self-assembled to an ordered island (Fig. 3(b)). Subsequent thermal annealing to 250°C resulted in carbon–bromine activation, and finally the 2D honeycomb network is observed in Fig. 3(c).

Gottfried and co-workers designed a V-type molecule (DMTP as shown in Fig. 3(d)), and achieved the regulation between the zigzag chain and hyperbenzene molecules by steering the substrate temperature during deposition of the molecular precursor.[27] Depositing DMTP on Cu(111) surface held at 300 K resulted in the formation of a highly ordered zigzag-chain structure consisting of C–Cu–C organometallics (Fig. 3(e)). While deposition of DMTP on Cu(111) substrate held at 550 K led to the fabrication of hyperbenzene (Fig. 3(f)). They have also demonstrated the role of the substrate structure[28] and the pseudo-high-dilution method[29] in on-surface synthesis of organometallic and covalent oligophenylene chains.

Rosei *et al.* chose 1,4-diiodobenzene and 1,3-diiodobenzene as precursors to synthesize 1D polyphenylene wires by surface-confined polymerization.[30] Linear chains of poly(p-phenylene) (PPP in Fig. 3(g)) were successfully formed by the polymerization of 1,4-diiodobenzene, while zigzag chains of poly(*m*-phenylene) (PMP in Fig. 3(g)) and macrocyclic oligomers were simultaneously formed by annealing 1,3-diiodobenzene on Cu(110). The linear PPP chains, zigzag PMP chains and cyclic oligomers were resolved by STM, as shown in Figs. 3(h)–3(j). The formation of the latter two structures, i.e. zigzag PMP chains and cyclic oligomers was resembled to the case in Figs. 3(d)–3(f) for the reason of the resembled backbone.

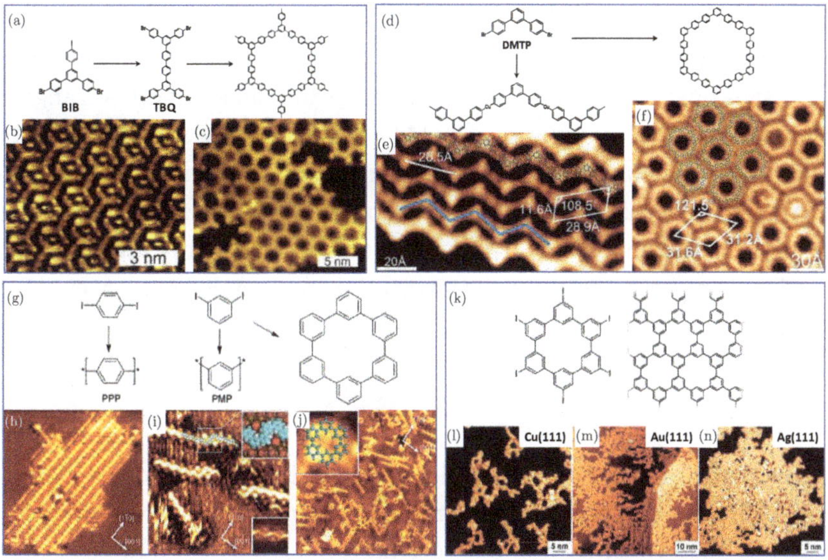

Fig. 3. (Color online) The backbone of precursors regulate on-surface molecular reactions. (a) Chemical structures of BIB and TBQ. (b) STM image of the self-assembled TBQ structure after deposition of BIB molecules on Au(111) at RT. (c) Annealing at 250°C, with the heating rate of 0.1°C/min. (d) Scheme indicates Ullmann reaction of the DMTP molecule. The constructed (e) zigzag chains at 300 K and (f) hyperbenzene at 550 K on Cu(111). (g) Ullmann coupling of diiodobenzene molecules. (h) STM image of linear PPP chains aligned on Cu(110) followed by dosing 1,4-diiodobenzene and annealing at 500 K. (i) STM image of zigzag PMP chains after deposition of 1,3-diiodobenzene on Cu(110) at 500 K. A model of PMP is superimposed. (j) STM image of PMP macrocyclic oligomers. (k) Chemical structure of hexaiodo-substituted CHP and a fraction of the polyphenylene network. STM images of polyphenylene networks on (l) Cu(111), (m) Au(111) and (n) Ag(111), respectively.

Source: Panels (a)–(c) from Ref. 26, American Chemical Society; panels (e)–(f) from Ref. 27, Wiley; panels (g)–(j) from Ref. 30, Wiley; panels (k)–(n) from Ref. 31, American Chemical Society.

The group of Fasel accomplished more uniform network structures by using building blocks with high conformational rigidity.[31,32] As shown in Fig. 3(k), the hexaiodo-substituted CHP was polymerized into polyphenylene networks. Figures 3(l)–3(n) display the fully polymerized polyphenylene networks supported by Cu(111), Au(111) and Ag(111) surfaces, respectively. On any substrate, the adsorption of the precursor led to a pathway with carbon–iodine bond dissociative, followed by intermolecular covalent bond formation between formed radicals through annealing. The fabricated networks were significantly different in morphology on the substrates, ranging from branched structures on Cu(111) to regular 2D networks on Ag (111). DFT calculations unraveled the balance between the diffusion and coupling steps: on Ag(111), diffusion prevailed over intermolecular coupling, resulting in an increased mobility of the radicals. A high mobility is a prerequisite for the fabrication of extended and ordered 2D polymer networks.

3. Surface

The substrates play a vital role in on-surface synthesis. Different metal surfaces could reduce the energy barrier thus rendering different reaction pathways, as well as steer morphologies of products and to some extent accomplish the selective formation. Taking the terminal alkynyl group, which is halogen free and without "capping" ions, as an example, a diversity of products such as phenyl groups, organometallic motifs, diynes and etc. could be formed on different substrates when triggering the reaction at higher temperatures.

Fuchs and co-workers studied on-surface Glaser coupling and found that Ag(111) surface outperforms the Au(111) surface for this particular on-surface molecular reaction.[37] They also discovered that the selectivity of on-surface Glaser coupling reaction could be dramatically increased by simple steric shielding of the active alkyne moieties.[33] As shown in Figs. 4(a)–4(c), they adopted the alkyne 1 as the precursor. It turned out that the dehydrogenative homocoupling, i.e. Glaser coupling was more efficient on Ag(111) surface than on Au (111). A step further, they decorated alkyl chains at the ortho-position of the reacting alkyne functionalities (alkyne 2 as shown in Fig. 4(d)), and this also led to a high selective of on-surface Glaser coupling and suppressed side reactions (Fig. 4(e)).

Gao *et al.* evaporated 4,4′-diethynyl-1,1′-biphenyl (DEBP) on Au (111) surface kept at RT and achieved cyclotrimerization with further annealing at 100°C as indicated by the scheme in Fig. 4(f).[34] After deposition of DEBP on Au(111) substrate kept at RT, a typical

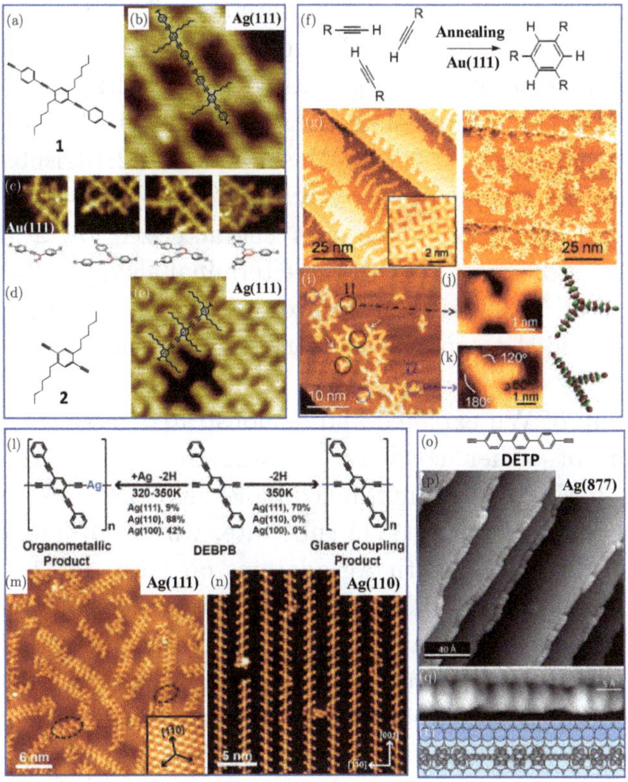

Fig. 4. (Color online) Substrates play a vital role in on-surface molecular reactions. Chemical structure of (a) alkyne 1 and (d) alkyne 2. STM image of (b) alkyne 1 and (e) alkyne 2 on Ag(111) after annealing. (c) STM images of alkyne 1 on Au(111) after on-surface oligomerization (annealing temperature 123–140°C) and the corresponding molecular structures. (f) Schematics of annealing causing cyclotrimerization of the alkyne molecules on Au(111). (g) STM images of DEBP molecules after deposition on Au(111). (h) STM image of networks generated by annealing at 100°C. (i) Magnified STM image showing two types of molecular interconnections, denoted as I1 and I2, respectively. (j, k) High-resolution STM images and molecular orbitals of I1 and I2. (l) Reaction pathway of DEBPB on Ag(111), Ag(110) and Ag(100). (m) Large-scale STM image of produced molecular chains on Ag(111) after annealing at $\sim 350\,K$. (n) Large-scale STM image of highly oriented organometallic chains on Ag(110) after annealing at $\sim 350\,K$. (o) Chemical structure of DETP. (p) Extended connected structures are constructed after annealing to 450 K. (q) High-resolution STM image of the terphenyl-butadiyne wire. (r) Model of the situation depicted in (q).

Source: Panels (a)–(e) from Ref. 33, Wiley; panels (f)–(k) from Ref. 34, American Chemical Society; panels (l)–(n) from Ref. 35, American Chemical Society; panels (o)–(r) from Ref. 36, American Chemical Society.

self-assembled structure via CH/π interaction[38,39] were formed as shown in Fig. 4(g). After annealing to 100°C, networks with branch and hexagon structures were generated (Figs. 4(h) and 4(i)). Two types of molecular interconnections denoted as I1 and I2 were assigned to symmetric 1,3,5- and asymmetric 1,2,4-trisubstitutions (Figs. 4(j) and 4(k)), respectively. Besides, Fasel *et al.* further explored the precursor which was characterized by a three-fold symmetrical aromatic core with three terminal alkynyl groups on Au (111), and cyclotrimerization proceeded readily, resulting in 2D covalently bonded polyphenylene honeycomb topology upon thermal activation.[40]

The group of Wu explored and demonstrated the lattice-directed fabrication of either covalent or organometallic wires on Ag substrates.[35] 1D covalent interlinked molecular wires were fabricated by Glaser coupling on Ag(111) substrate, while 1D organometallic wires were constructed with Ag adatoms on both Ag(110) and Ag (100) substrates (Figs. 4(l)–4(n)). Such lattice-dependent reactions were originated from the matching degree between the periodicities of the surface lattice structures and the fabricated molecular wires.

Barth and co-workers attempted to synthesize extended graphdiyne wires with lengths reaching 30 nm by vicinal surface templating.[36] On flat Ag(111), thermal activation triggered various side reactions leading to irregularly-branched networks. In order to increase the chemoselectivity and regioselectivity, the authors confined the reactants (DETP as depicted in Fig. 4(o)) to step edges of the vicinal Ag(877) template, and fortunately the construction of chemoselective butadiyne was preferred, as shown in Figs. 4(p)–4(r). The generated organic polymer chain was the first representative of the family of poly(oligophenylene-butadiynylene)s, distinct from the more common poly(phenylene-butadiynylene)s by the presence of several adjacent phenylene moieties in each monomer. The feature is important for tuning the energy gap and the band structure of the resulting π-conjugated polymer chain which represents a prospective 1D organic semiconductor. The ideal extended graphdiyne wire is characterized by DFT as a direct band gap semiconductor material with both highest occupied molecular orbital (HOMO) and lowest unoccupied molecular orbital (LUMO) derived bands promisingly isolated within the electronic structure, presenting significant promise for molecular-electronics and organic photovoltaic devices.

In addition to the terminal alkynyl, other functional groups such as the terminal methyl group and hydroxyl group followed different reaction pathways on different substrates as well. The group of Chi reported the highly selective C–H activation and subsequent dehydrogenative C–C coupling reaction of long-chain linear alkanes on Au(110), which resulted in the conversion of low to higher molecular-weight hydrocarbons. The reaction took place exclusively at terminal CH_3 or penultimate CH_2 groups in the chains at $420\,K \sim 470\,K$.[41] Xu *et al.* then explored the feasibility of dehydrogenative homocoupling of alkyl chains on a comparatively simpler surface Cu(110).[42] In comparison to the products constructed on Au (110), there was no preferential C site with respect to C–H activation of alkyl chains on Cu(110), which resulted in the formation of diverse coupling products. The growth of polymerized alkyl chains was confined in the (1×3) reconstruction grooves of Au(110), and this generated the different selectivity of products.

Chi and co-workers also steered the competition between dehydrogenation and deoxygenation of phenols on surfaces via introducing a directing group, that is, diselective and monoselective ortho C–H bond activation on Au(111) and Ag(111), respectively. By STM imaging, XPS analysis combined with DFT calculations, the authors clarified the reaction mechanisms. The activation for the first ortho-dehydrogenation process on both Au(111) and Ag(111) surfaces was lowered by the formation of keto-form intermediates. Next a catalytic reductive deoxygenation process was preferred over the second ortho-dehydrogenation on Ag(111), while on Au(111) the case was opposite. This resulted in the surface-controlled reaction selectivity of mono/diselective ortho C–H bond activation.[43]

The stereoselective synthesis of geometrically pure cis- or trans-isomers is challenging as well as promising in chemistry-related disciplines and applications. As shown in Fig. 5(a), Xu *et al.* successfully formed trans-diene moieties by dehydrogenative

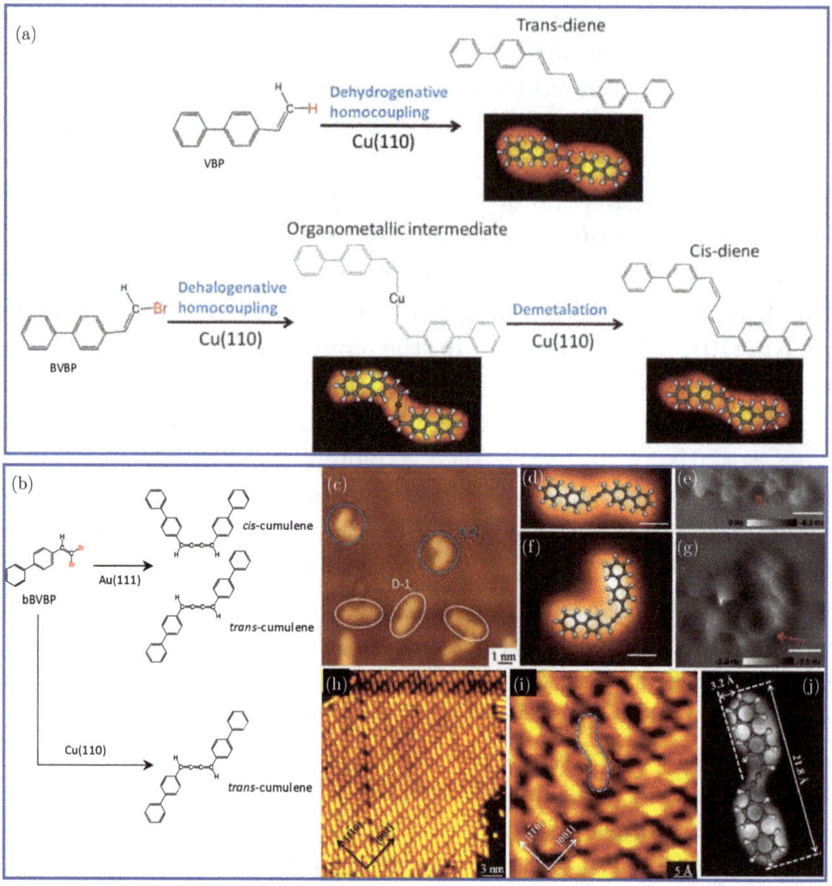

Fig. 5. (Color online) The stereoselective synthesis of trans- and cis-dimers through precursor design or choice of substrates. (a) Stereoselective synthesis of trans- and cis-diene products through different functional groups in on-surface synthesis. (b) Schematic illustration of both cis- and trans-cumulene through dehalogenative homocoupling of alkenyl gem-dibromide on Au(111), while on Cu(110) only the trans-cumulene moiety were observed. (c) STM image of dimer motifs after deposition of bBVBP on Au(111) at RT. (d, f) High-resolution STM images of cis- and trans-moieties. (e, g) Corresponding AFM images of D-1 and D-2. Red arrows indicate the cumulene group. (h) Large-scale and (i) close-up STM images showing the formation of an ordered island composed of trans-cumulene after annealing to $\sim 430\,K$ on Cu(110). (j) Simulated STM image with the scaled DFT relaxed trans-cumulene model superimposed.

Source: Panel (a) from Ref. 24, Royal Society of Chemistry; panels (b)–(g) from Ref. 44, Wiley; panels (h)–(j) from Ref. 45, Wiley.

homocoupling reaction of terminal alkenes (VBP) on Cu(110),[46] and the corresponding cis-diene moieties could be specifically formed by dehalogenative homocoupling reaction of terminal alkenyl bromides (BVBP) on the same surface.[24] The intermediate formed cis-organometallic should take charge of such a stereoselectivity revealed by DFT calculations.

In addition to functional groups, the choice of substrates could also regulate cis- and trans-isomerization of the reaction products and select the products to some degree. As shown in Fig. 5(b), Xu *et al.* studied the alkenyl gem-dibromide (bBVBP) on Au(111)[44] and Cu (110),[45] respectively. Both trans-(D-1) and cis-cumulene (D-2) products through dehalogenative C–C homocoupling of bBVBP on Au(111) were formed (Figs. 5(c)–5(g)), while only the trans-cumulene moiety were observed on Cu(110) (Figs. 5(h)–5(j)).

These researches provide a way to stereoselective synthesis of specific isomers, which may be of substantial utility in stereochemistry. But the underlying mechanisms still call for uncovering in order to further understanding and guidance, and it is also of general interest to extend such on-surface synthesis strategies to specific fabrication of cis- and trans-isomers involving more complicated nanostructures. The following section will focus on the external stimuli including thermal annealing, STM tip manipulation and irradiation to activate on-surface molecular reactions.

4. External Stimuli

4.1. *Thermal*

Since Grill constructed nanoarchitectures of molecular building blocks efficiently by thermal treatment in 2007, it has become the most frequently adopted approach in on-surface synthesis. In this section, thermal activation is needed for most of the on-surface molecular reactions presented. In order to make the section more readable, we subtitle this part as thermal external stimuli and take carbon-rich functional nanostructures as examples. This bottom–up strategy has demonstrated remarkable potential in the construction

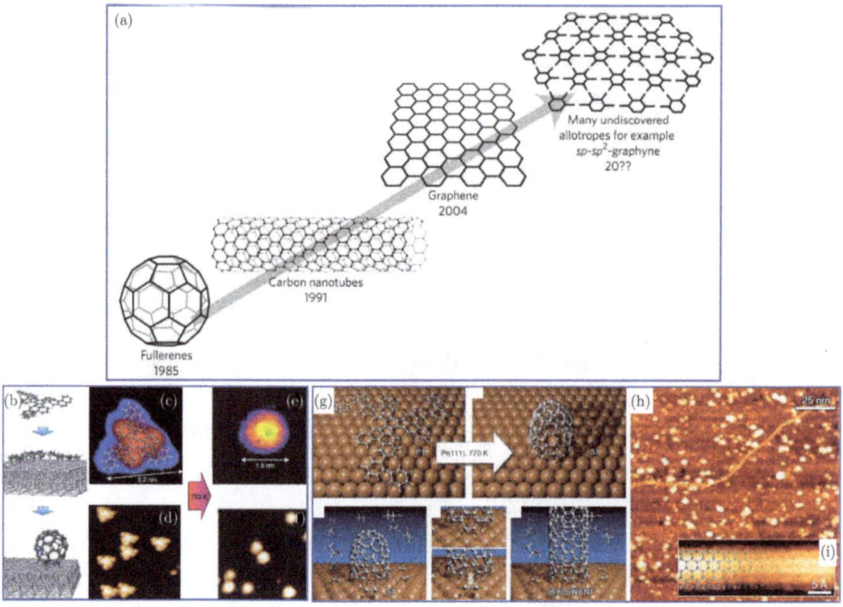

Fig. 6. (Color online) Investigating carbon allotropes by on-surface synthesis. (a) The carbon allotropes. (b) Optimized model of the cyclization process. (c, d) STM images after deposition of the $C_{57}H_{33}N_3$ molecule on Pt(111). With the calculated atomic model superimposed in (c). (e, f) STM images after annealing the molecular-covered sample at 750 K. (g) Formation and epitaxial elongation of (6,6) SWCNT seeds S1. (h) STM images of a long SWCNT lying on Pt(111). (i) The close-up STM image of the fabricated SWCNT with the molecular model superimposed.

Source: Panel (a) from Ref. 47; panels (b)–(f) from Ref. 48; panels (g)–(i) from Ref. 49. Nature Publishing Group.

of elusive carbon allotropes with atomic precision (Fig. 6(a)),[47] including fullerene, carbon nanotubes, graphene and GNRs.

The fullerenes represented the most intensively investigated class. Otero and co-workers designed polycyclic aromatic hydrocarbons (PAHs) $C_{57}H_{33}N_3$ (Fig. 6(b)). As shown in Figs. 6(c)–6(f), the authors found that after the deposition and thermal annealing of precursors at 750 K on Pt(111), an efficient surface-catalyzed cyclodehydrogenation process was performed, forming the triazafullerene products with about 100% yield.[48] The strong interaction between the precursor and the Pt(111) is considered to play an essential role. Different interactions between the precursors and surfaces could steer the competitive reaction pathways, i.e. cyclodehydrogenation and dehydrogenative polymerization. Martín-Gago *et al.* demonstrated that aromatic hydrocarbons tended to perform intramolecular cyclodehydrogenation on highly reactive surface such as Pt(111), while intermolecular dehydrogenative polymerization dominated on weakly reactive surface such as Au(111), Cu(110), or Cu(111).[50]

Single-walled carbon nanotubes (SWCNTs) have also received great attention for extraordinary properties promising for numerous applications, which depend sensitively on SWCNTs' structure. Thus the atomic precise and controlled synthesis of "single-chirality" SWCNTs has remained as a significant challenge. As shown in Fig. 6(g), Fasel and co-workers synthesized single-chirality SWCNTs with lengths up to a few hundred nanometers through a bottom–up strategy.[49] First, the (6,6) SWCNT seed S1 was constructed by surface-catalyzed cyclodehydrogenation of the precursor P1 on Pt(111). Then, epitaxial elongation of the seeds to the targeted nanotube was achieved by exposing to a carbon feedstock gas at 670–770 K.

GNRs were predicted to exhibit electronic properties which made them captivating for the construction of nanoscale electronic devices. Exploring the potential of GNRs had been hampered by their limited availability until Fasel and co-workers reported the atomically precise bottom–up method for the construction of GNRs with different topologies and widths.[5] As illustrated in Fig. 7(a), the authors designed the famous DBBA precursor. Thermal sublimation of DBBA molecules on Au(111) removed the halogen substituents, yielding surface-stabilized biradical species. In the first thermal activation step, these biradical species diffused across Au(111) and underwent radical addition reactions to constructed linear polymer chains. In the second thermal activation step, the prototypical armchair GNR (AGNR) of width $N = 7$ was obtained by the surface-assisted cyclodehydrogenation, as shown in Fig. 7(b). The width, topology and edge periphery of the formed GNRs are defined by the structure of delicately designed precursors, which could give access to a wide range of GNRs. Chevron-type GNRs with alternating widths of $N = 6$ and $N = 9$ were constructed by monomer 3, as shown in Figs. 7(c)

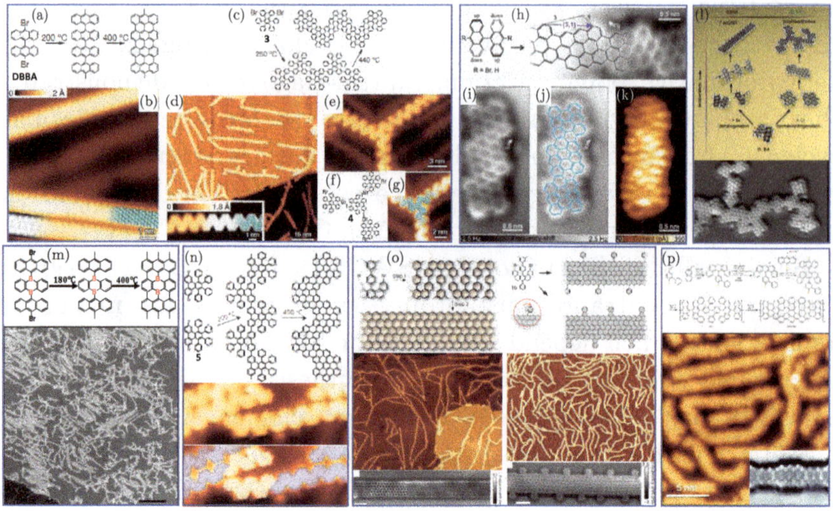

Fig. 7. (Color online) Investigating GNRs by on-surface synthesis. (a) Scheme from DBBA to $N = 7$ GNRs. (b) STM image with molecular model of the ribbon partly overlaid. (c) Scheme from monomer 3 to chevron-type GNRs. (d) STM image of chevron-type GNRs on Au(111). (e) STM image of three-fold GNR junction obtained from monomer 4 at the nodal point and monomer 3 for the ribbon arms after annealing to 440°C. (f) Chemical structure of the formed junction. (g) Model of the junction superimposed on the STM image. (h) Scheme for the construction of chiral (3,1)-GNRs. (i) nc-AFM image at constant height. (j) Image i with structural model superimposed. (k) Tunneling current map measured simultaneously with i. (l) Upper: Nanographene formation from DBBA and DCBA. Lower: AFM image of DCBA on Au(111) after annealing at 360°C. (m) Scheme showing the construction of boron-doped GNRs by on-surface molecular reactions. And STM image of B-GNR synthesized by annealing at 400°C. (n) Fabrication of N-GNRs from monomer 5. And STM image of p-N-GNR heterostructures. The identified p-GNR and N-GNR segments are highlighted in translucent gray and blue, respectively. (o) Left: Synthesis of GNRs with zigzag edges. Right: Synthesis of edge-modified 6-ZGNRs. (p) Synthesis of 7/9-AGNR. STM and bond-resolved STM images of the atomically precise 7/9-AGNR.

Source: Panels (a)–(g) from Ref. 5, Nature Publishing Group; panels (h)–(k) from Ref. 51, American Chemical Society; panel (l) from Ref. 52, Wiley; panels (m)–(p) from Refs. 7 and 53–55, Nature Publishing Group.

and 7(d). A three-fold GNR junction was obtained from monomer 4 at the nodal point and monomer 3 for the ribbon arms, as shown in Figs. 7(e)–7(g).

This recipe has been widely employed to fabricate a variety of GNRs with atomic precision ever since the seminal work. Substantial efforts have been devoted to the fabrication of GNRs with armchair edges including $N = 3, 5, 6, 7, 9, 13, 14$ and 21.[56–61] The group of Weiss and Fasel independently demonstrated that the coupling selectivity introduced by bromine substitutions could be overwhelmed by the surface structure and catalytic properties of Cu(111). After deposition of the DBBA molecular precursors on Cu(111),[51,62] identical chiral GNRs were obtained as shown in Figs. 7(h)–7(k). Swart and co-workers indicated that for the aryl chloride (DCBA), cyclodehydrogenation occurred before dehalogenation and polymerization.[52] In comparison to the staggered bianthryl radicals, the planar bisanthene radicals displayed a different coupling behavior, resulting in the construction of polybisanthenes as illustrated in Fig. 7(l). Besides, precursor design also provided routes to the substitutional doping of GNRs by replacing some carbon atoms with heteroatoms, such as boron (Fig. 7(m)), nitrogen (Fig. 7(n)) and sulfur.[53,54,63,64]

It is noteworthy that nanostructures with zigzag edges are expected to host spin-polarized electronic edge states, and could thus serve as key elements for spintronics. Fasel and co-workers further described the bottom–up synthesis of zigzag GNRs (ZGNRs) through surface-assisted polymerization and subsequent cyclodehydrogenation of specifically designed precursors to finally yield zigzag edges with atomic precision as indicated in Fig. 7(o).[7] In addition, as shown in Fig. 7(p), GNRs consisting of segments with alternating widths have also been fabricated on Au(111) by the bottom–up strategy.[55]

Such bottom–up strategy for atomically precise fabrication of nanostructures from specifically designed precursors by thermal activation has been intensively investigate and proved to be promising. In addition to aforementioned GNRs, other intriguing 1D conjugated polymer chains including PPV, PAEs and metalated carbyne were successfully constructed through this strategy as well. As indicated in Figs. 8(a) and 8(b), combined STM with nc-AFM, the group of Xu employed two and three bromine substituent atoms at the same location within the precursors, leading to on-surface

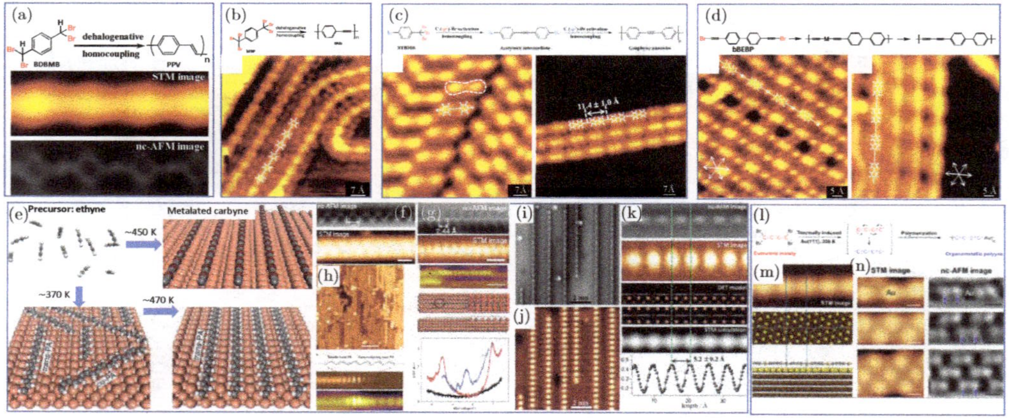

Fig. 8. (Color online) Investigating 1D chains involving vinylene or acetylenic linkages by on-surface synthesis. Dehalogenative homocoupling of the ditopic molecular precursors ((a) BDBMB; (b) bTBP) resulting in the fabrication of corresponding polymer chains ((a) PPV; (b) PAE). (c) Stepwise dehalogenative homocoupling reactions, leading to the fabrication of a graphyne nanowire. STM images of the acetylenic intermediate and the final graphyne nanowire. (d) Scheme of dehalogenative homocoupling of bBEBP, which results in the construction of the C–M–C organometallic intermediate and the final C–C coupled acetylenic moieties. (e) Illustration of the bottom–up synthesis of metalated carbyne and polyacetylene (PA) on Cu(110). nc-AFM and STM images of (f) the cis-PA segment and (g) the trans-PA segment. (h) The electronic properties of decoupled trans-PA. (i) nc-AFM image and (j) current image acquired simultaneously. (k) Equally scaled nc-AFM image, current image, DFT-optimized model, STM simulation and line-scan profile of a single metalated carbyne chain. (l) Scheme of the polymerization from a cumulene moiety to an organometallic polyyne on Au(111). (m) STM image and the corresponding DFT-optimized model of a single organometallic polyyne. (n) STM and nc-AFM images of the single, double and triple chains, respectively.

Source: Panel (a) from Ref. 65, American Chemical Society; panel (b) from Ref. 21, Wiley; panel (c) from Ref. 66, Royal Society of Chemistry; panel (d) from Ref. 20, American Chemical Society; panels (e)–(h) from Ref. 67, Nature Publishing Group; panels (i)–(k) from Ref. 68, American Chemical Society; panels (l)–(n) from Ref. 69, American Chemical Society.

polymerization via direct formation of double and triple C–C bonds, respectively.[21,65] Graphyne/graphdiyne nanowires, as shown in Figs. 8(c) and 8(d), were also well established on the isotropic Au (111) by thermal activation of delicately designed precursors.[20,66]

Recently, the clean and individual PA chain has been synthesized through polymerization of acetylene. The acetylenic group on surfaces faces various complications as mentioned above in Figs. 2(c) and 4. Although direct polymerization of acetylenic groups is energetically most favorable according to DFT calculations, such a reaction is always suppressed on various metal surfaces for the complications of adsorbate–substrate interactions. Xu *et al.* developed an on-surface synthesis approach to constructed PA chains by thermal treatment, especially the geometrically pure trans-PA.[67] As shown in Fig. 8(e), the anisotropic Cu(110) was chosen for its relatively high chemical activity and 1D templating along the $[1\bar{1}0]$ direction. The simple acetylene precursor was selected aiming at the formation of 1D polymer chains. After deposition of acetylene on Cu(110) at $\sim 220\,\mathrm{K}$ under UHV conditions, bright protrusions were observed and desorbed after gentle annealing to $\sim 330\,\mathrm{K}$, which agreed with a previous study revealing that acetylene could be cyclized into benzene on Cu(110).[70] After deposition of acetylene on Cu(110) at 300 K and followed by mild annealing to $\sim 370\,\mathrm{K}$, meandering 1D chain structures were constructed. From the STM image, bond-resolved nc-AFM image combined with XPS analysis and DFT calculations, the cis- and trans-PA chains were clearly assigned (Figs. 8(f) and 8(g)). Further annealing the sample to 470 K gave rise to a cis-to-trans isomerization, and well-aligned trans-PA chains along the $[1\bar{1}0]$ direction of Cu(110) were resolve by STM and nc-AFM. In order to suppress the strong charge transfer doping, the authors tried to electronically decouple trans-PA chains from the underlying metal surface by Cu oxide intercalation, and revealed a band gap of 2.4 eV for trans-PA on Cu(110)–(2×1)O, as illustrated in Fig. 8(h).

Next, the authors attempted to optimize the experimental conditions by regulating substrate temperature in order to facilitate the dehydrogenative coupling reaction.[68] After dosing acetylene on Cu(110) held at $\sim 450\,\mathrm{K}$, another kind of 1D chains consisting of periodic protrusions along the close-packed $[1\bar{1}0]$ direction were observed. The nc-AFM and current images acquired simultaneously as shown in Figs. 8(i) and 8(j) unambiguously confirmed the proposed metalated carbyne structure on Cu(110). From a comparison of the equally scaled nc-AFM image, STM image, DFT-optimized model, STM simulation and line-scan profile as shown in Fig. 8(k), the authors determined that the periodic protrusions of the 1D chain in the STM images should be assigned to the incorporated copper adatoms. While in the nc-AFM image, the copper adatoms were invisible. The C–C triple bonds were resolved instead. The substrate temperature played a vital role in steering on-surface synthesis including cyclization, direct polymerization and dehydrogenative organometallic coupling of acetylene molecules on Cu(110). The key for the direct polymerization was to dose acetylene molecules onto the Cu(110) surface held at relatively low temperatures considering the lower energy barrier needed. The relatively elevated temperatures ($\sim 450\,\text{K}$ under this circumstance) provided enough energy for dehydrogenative organometallic coupling of acetylene molecules on Cu(110), and abundant copper adatoms existed at such temperature could also facilitate the formation of metalated carbyne.

Two configurations of carbyne are proposed considering the carbon skeleton, i.e. polyyne with alternate single and triple bonds and cumulene with consecutive double bonds. Xu and co-workers recently demonstrated the carbon skeleton rearrangement from a cumulene moiety to a diyne one.[69] As shown in Figs. 8(l)–8(n), thermally-induced entire debromination led to the fabrication of 1D organometallic polyyne with semiconducting characteristic. Besides, STM tip manipulation was also employed to achieve the carbon skeleton rearrangement of the precursor from a cumulene moiety to a diyne one.

4.2. *Tip-induced*

STM tip manipulation provides a controllable and powerful tool for on-surface synthesis, even for those of extreme reactivity such as triangulene, acenes and cyclo[18]carbon. In 2000, Hla *et al.* pioneered

STM manipulation by introducing the well-known Ullmann reaction to Cu(111) substrate.[16] As shown in Fig. 9(a), the STM tip was utilized to inject electrons into the iodobenzene leading to the carbon–iodine bond cleavage. Further manipulation was performed to bring two phenyl radicals close to each other. Finally, by excitation of two phenyl radicals with inelastic tunneling, covalent coupling of the phenyl radicals was triggered, and the biphenyl products could be formed. Later, Abel *et al.* brought the STM tip close to the H-bonded monolayer of the boronic acid precursor and scanned at the position indicated by the arrow as shown in Fig. 9(b).[71] The self-condensation (dehydration) reaction was immediately initiated on the Ag(100), and slowly propagated until complete disappearance of the H-bonded phase.

Triangulene has been an enigmatic molecule ever since its existence was first hypothesized because of its extreme reactivity. The unique topology of such non-Kekulé hydrocarbons would result in open-shell π-conjugated graphene fragments and give rise to high-spin ground states, which is potentially useful in organic spintronic devices. Particularly, it will be of great interest to exhibit how the chemical environment affects the magnetic ground state of triangulene by magnetic atoms close by or adsorption on ferromagnetic surfaces. As illustrated in Fig. 9(c), Gross *et al.* have been successfully generated stable unsubstituted triangulene on NaCl, Xe and metal Cu (Fig. 9(d)) surfaces by the voltage pulse from the tip to dehydrogenate precursors.[72] STM images at voltages of positive ion resonance, the gap region and the negative ion are shown in Fig. 9(e). Among others, the STM image recorded in the gap region resembled the triangular shape.

Atomically precise fabrication of acenes and their derivatives is promising for applications in molecular electronics such as field-effect transistors, photovoltaic devices and light-emitting diodes for their attractive electronic properties including low band gap and high mobility. Moreover, acenes can be viewed as the narrowest members in the family of ZGNRs, which are predicted to host spin-polarized edge states and are postulated to be an important component in carbon-based spintronic devices. As shown in Fig. 9(f), Moresco and

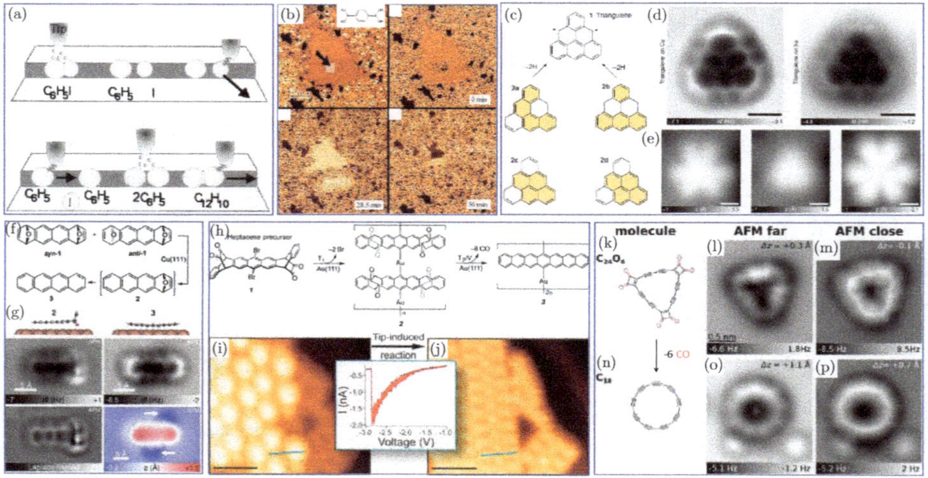

Fig. 9. (Color online) STM tip-induced reactions. (a) Schematic illustration presenting the STM tip-induced synthesis of a biphenyl molecule from the iodobenzene. (b) Tip-induced polymerization of BDBA. (c) Scheme illustrating the generation of triangulene by removal of two hydrogen atoms through atomic manipulation. (d) AFM images of triangulene on Cu and Xe. (e) STM images at voltages corresponding to the positive ion resonance, the gap region and the negative ion resonance from left to right, respectively. (f) Scheme showing on-surface reduction of diepoxytetracenes to give tetracene. (g) AFM and STM images of intermediate and the final tetracene. (h) On-surface synthesis of heptacene organometallic complexes. STM images of (i) Au-directed heptacene derivative organometallic complexes, and (j) Au-directed heptacene organometallic complexes after the tip-induced conversion. (k–p) The precursor and product generated by tip-induced decarbonylation. Chemical structures and AFM images of (k–m) the precursor, and (n–p) the cyclo[18]carbon.

Source: Panel (a) from Ref. 16, American Physical Society; panel (b) from Ref. 71, Royal Society of Chemistry; panels (c)–(e) from Ref. 72, Nature Publishing Group; panels (f) and (g) from Ref. 73, American Chemical Society; panels (h)–(j) from Ref. 74, American Chemical Society; panels (k)–(p) from Ref. 75, American Association for the Advancement of Science.

co-workers presented on-surface reduction of diepoxytetracenes to give tetracene on Cu(111) by STM tip manipulation.[73] The thermal activation could as well yield the formation of tetracene. The identification of the intermediate and the final tetracene was unambiguously demonstrated by nc-AFM (Fig. 9(g)). It is notable that acenes belong to the family of PAHs consisting of linearly fused benzene rings. However, the synthesis of acenes would become incrementally challenging with the length increases owing to their instabilities. As illustrated in Fig. 9(h), Fasel and co-workers constructed Au-directed heptacene organometallic complexes on Au (111) from an α-diketone protected heptacene precursor.[74] The authors demonstrated the tip-induced deprotection experiments in the complex formation as shown in Figs. 9(i) and 9(j). Additionally, Godlewski and co-workers achieved on-surface synthesis of nonacene by dehydrogenation of an air-stable partially saturated precursor on Au(111).[76]

Recently, Kaiser *et al.* generated the fascinating cyclo[18]carbon utilizing atom manipulation on bilayer NaCl on Cu(111) at 5 K by eliminating carbon monoxide from a cyclocarbon oxide molecular precursor $C_{24}O_6$.[75] Characterization of the oxide precursor and cyclo [18]carbon was revealed by high-resolution AFM, as shown in Figs. 9(k)–9(p).

4.3. *Photo-induced*

Light could be able to trigger on-surface molecular reactions as well. Photo-induced on-surface molecular reactions could effectively avoid some side reactions by reducing the activation temperature, thus selectively driving specific chemical reactions by regulating thermodynamic and kinetic aspects. Recently, Fasel *et al.* have demonstrated the on-surface synthesis of heptacene and nonacene via light-induced dissociation of α-bisdiketone protected precursors on Au(111) surface under UHV conditions, as indicated in Figs. 10(a) and 10(b).[77] The authors also provided insights into the electronic structure of the elusive products by STM, nc-AFM and DFT, and it presents an increasing open-shell nature with length increased.

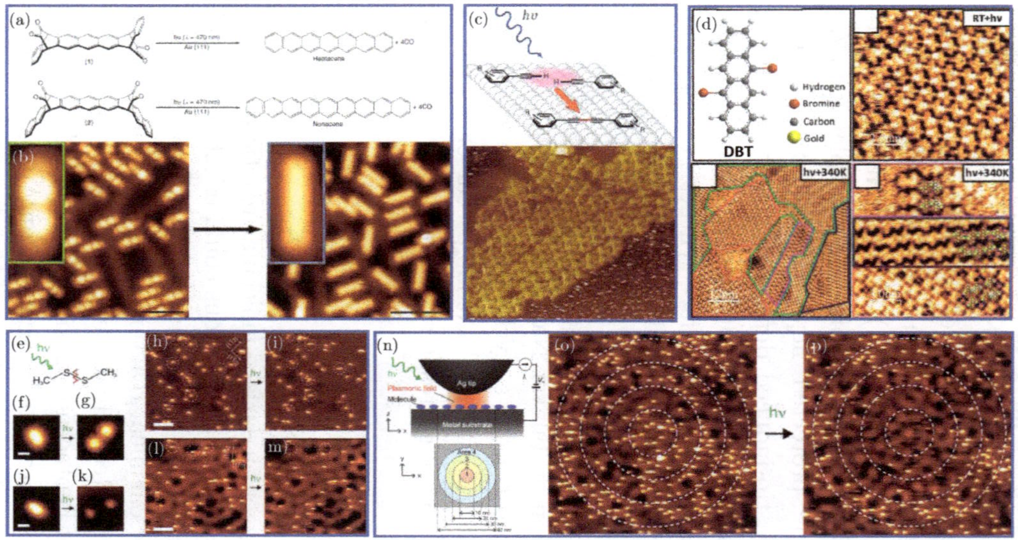

Fig. 10. (Color online) Photo-induced reactions. (a) Schematic representing the light-induced on-surface formation of heptacene and nonacene. (b) STM images showing the nonacene formation from the α-bisdiketone protected nonacene via photodissociation on Au(111). (c) Light-induced on-surface Glaser coupling. (d) Structure of DBT precursor. And STM images of the illuminated surface showing RT Ullmann coupling: on-surface photodissociation of C–Br bonds. (e) Chemical structure of $(CH_3S)_2$ molecule, illustrating the photodissociation of S–S bond. STM images of $(CH_3S)_2$ (f–i) on Cu(111) and (j–m) on Ag(111), (f, h, j, l) before and (g, i, k, m) after irradiation with 532 nm light. (n) Real-space investigation of the plasmon-induced chemical reaction. STM images of $(CH_3S)_2$ molecules on Ag(111) (o) before and (p) after irradiation.

Source: Panels (a) and (b) from Ref. 77, Nature Publishing Group; panel (c) from Ref. 78, American Chemical Society; panel (d) from Ref. 79, Royal Society of Chemistry; panels (e)–(m) from Ref. 80, American Chemical Society; panels (n)–(p) from Ref. 81, American Association for the Advancement of Science.

Fuchs *et al.* showed homocoupling of terminal alkynes on Ag(111) strongly accelerated upon UV irradiation at RT, as indicated in Fig. 10(c),[78] where highly efficient C–C bond formation occurred. Colazzo *et al.* also demonstrated a photo-induced homocoupling of terminal alkynes on highly oriented pyrolytic graphite.[82] Basagni *et al.* demonstrated that light stimulus could be applied to trigger the first step of the Ullmann reaction.[79] As indicated in Fig. 10(d), the 405 nm light irradiation led to C–Br bond dissociation at RT, which allows for obtaining well-ordered organometallic networks.

As shown in Figs. 10(e)–10(m), Kim *et al.* investigated the visible-light-induced photodissociation of S–S bond on Cu(111) and Ag(111) substrates by STM combined with DFT calculations on the projected density of states (PDOSs) and the spatial distribution of the frontier electronic states.[80] The hybridization between $(CH_3S)_2$ molecules and the underlying substrate not only reduced the optical energy gap, but also generated frontier states with less overlap with the substrate, opening a novel reaction pathway for the photodissociation by visible light of molecules adsorbed on metals, as well as resulting in longer excited-state lifetimes. Furthermore, Kim *et al.* achieved real-space and real-time observation of a plasmon-induced chemical reaction at the single-molecule level by STM together with theoretical calculations.[81] As illustrated in Figs. 10(n)–10(p), the $(CH_3S)_2$ molecule on Ag(111) and Cu(111) was dissociated by the optically excited plasmon at the STM junction, which could be explained by the direct intramolecular excitation mechanism.

Okawa and co-workers enabled topochemical polymerization of the diacetylene units after local initiation with the STM tip or ultraviolet light to give poly(diacetylene) wires, which were terminated by defects.[83,84] A decade later, the authors a step further utilized the same diacetylene monolayers together with phthalocyanine molecules. After tip-induced initiation along the aligned diacetylene rows, the reactive chain end was not terminated by defects but happened to be the phthalocyanine molecules, and covalently connected via C–H insertion presumably.[85] This so-called "chemical soldering" and controlled chain polymerization for single-molecule electronics provides perhaps the most impressive example thus far to combine top–down and bottom–up methods.[86]

5. Summary and Outlook

In this paper, pioneering and recent achievements in the young and emerging research area of on-surface molecular reactions have been reviewed from three aspects of precursors, surfaces and external stimuli. The atomically precise fabrication of covalent interlinked nanostructures through on-surface synthesis of deliberately designed precursors has proved to be powerful and promising. A diversity of functional groups have been introduced to surface-confined reactions, among which aryl halides and terminal alkynes were two of the most intensively investigated homocoupling reactions. Aryl halides yield the Ullmann coupling featuring an aryl–aryl bond in spite of the substrate. In contrast, terminal alkynes yield various products greatly depending on the choice of the surface. Bromine contamination often interferes with the reaction while dehydrogenative homocouplings are cleaner. Metal surfaces not only provide a 2D support but also catalyze surface-confined reactions by acting as electron reservoirs. In some cases, surface adatoms also contribute to the reactions and become essential part of the intermediates, i.e. C–M–C intermediates (M stands for metal).

With the continuous development of scanning probe microscopy (SPM) techniques performed in UHV, it allows for analysis in real space and detailed understanding of elementary chemical processes. A diversity of on-surface molecular reactions have been conducted in surface science community, nevertheless there still remain lots of problems and challenges. In contrast to the widely studied homocoupling reactions, cross-coupling reactions are still in infancy. Undoubtedly, the on-surface molecular reaction "toolbox" needs to be significantly extended to efficiently afford functional nanomaterials with diverse structures and properties. More reactions need to be introduced to the on-surface chemistry from the conventional solution chemistry, allowing to the greater diversity and flexibility in the design of targeted nanostructures. Besides, detailed insights into the chemical nature of the molecule–surface interaction are of urgent requirement to fully exploit and unravel mechanisms of on-surface synthesis, aiming to steer thermodynamic and kinetics of reaction pathways. Fabrication of well-defined π-conjugated molecular nanostructures with inviting properties including mechanical, electronic as well as optical properties would promise great potential applications in molecular-electronics devices, field-effect transistors, sensors and so on. The practical applications of the nanostructures further call for the large-scale construction of low-dimensional nanostructures, removal of underlying substrates and transfer to other substrates, which still faces challenges. Furthermore, some efforts are also urgent to exploit on-surface synthesis not only in UHV but also in ordinary vacuum and even at ambient pressure.

References

1. Y. L. Huang, Y. J. Zheng, Z. Song, D. Chi, A. T. S. Wee and S. Y. Quek, *Chem. Soc. Rev.* **47** (2018) 3241.
2. L. Li, Y. Yu, G. J. Ye, Q. Ge, X. Ou, H. Wu, D. Feng, X. H. Chen and Y. Zhang, *Nat. Nanotechnol.* **9** (2014) 372.
3. P. Vogt, P. De Padova, C. Quaresima, J. Avila, E. Frantzeskakis, M. C. Asensio, A. Resta, B. Ealet and G. Le Lay, *Phys. Rev. Lett.* **108** (2012) 155501.
4. K. S. Novoselov, A. K. Geim, S. V. Morozov, D. Jiang, Y. Zhang, S. V. Dubonos, I. V. Grigorieva and A. A. Firsov, *Science* **306** (2004) 666.
5. J. Cai, P. Ruffieux, R. Jaafar, M. Bieri, T. Braun, S. Blankenburg, M. Muoth, A. P. Seitsonen, M. Saleh, X. Feng, K. Müllen and R. Fasel, *Nature* **466** (2010) 470.
6. A. Kimouche, M. M. Ervasti, R. Drost, S. Halonen, A. Harju, P. M. Joensuu, J. Sainio and P. Liljeroth, *Nat. Commun.* **6** (2015) 10177.
7. P. Ruffieux, S. Wang, B. Yang, C. Sanchez-Sanchez, J. Liu, T. Dienel, L. Talirz, P. Shinde, C. A. Pignedoli, D. Passerone, T. Dumslaff, X. Feng, K. Müllen and R. Fasel, *Nature* **531** (2016) 489.
8. G. Binnig, H. Rohrer, C. Gerber and E. Weibel, *Appl. Phys. Lett.* **40** (1982) 178.
9. F. J. Giessibl, *Rev. Mod. Phys.* **75** (2003) 949.
10. S. Clair, M. Abel and L. Porte, *Chem. Commun.* **50** (2014) 9627.
11. A. Ciesielski, M. El Garah, S. Haar, P. Kovaricek, J. M. Lehn and P. Samori, *Nat. Chem.* **6** (2014) 1017.
12. F. Bebensee, C. Bombis, S. R. Vadapoo, J. R. Cramer, F. Besenbacher, K. V. Gothelf and T. R. Linderoth, *J. Am. Chem. Soc.* **135** (2013) 2136.
13. O. D. Arado, H. Mönig, H. Wagner, J. H. Franke, G. Langewisch, P. A. Held, A. Studer and H. Fuchs, *ACS Nano* **7** (2013) 8509.

14. L. Grill, M. Dyer, L. Lafferentz, M. Persson, M. V. Peters and S. Hecht, *Nat. Nanotechnol.* **2** (2007) 687.
15. L. Lafferentz, V. Eberhardt, C. Dri, C. Africh, G. Comelli, F. Esch, S. Hecht and L. Grill, *Nat. Chem.* **4** (2012) 215.
16. S. W. Hla, L. Bartels, G. Meyer and K. H. Rieder, *Phys. Rev. Lett.* **85** (2000) 2777.
17. R. Gutzler, H. Walch, G. Eder, S. Kloft, W. M. Heckl and M. Lackinger, *Chem. Commun.* **29** (2009) 4456.
18. M. O. Blunt, J. C. Russell, N. R. Champness and P. H. Beton, *Chem. Commun.* **46** (2010) 7157.
19. Y. Q. Zhang, N. Kepčija, M. Kleinschrodt, K. Diller, S. Fischer, A. C. Papageorgiou, F. Allegretti, J. Björk, S. Klyatskaya, F. Klappenberger, M. Ruben and J. V. Barth, *Nat. Commun.* **3** (2012) 1286.
20. Q. Sun, L. Cai, H. Ma, C. Yuan and W. Xu, *ACS Nano* **10** (2016) 7023.
21. Q. Sun, X. Yu, M. L. Bao, M. X. Liu, J. L. Pan, Z. Q. Zha, L. L. Cai, H. H. Ma, C. X. Yuan, X. H. Qiu and W. Xu, *Angew. Chem. Int. Ed.* **57** (2018) 4035.
22. T. Wang, H. Lv, Q. Fan, L. Feng, X. Wu and J. Zhu, *Angew. Chem. Int. Ed.* **56** (2017) 4762.
23. Q. Sun, L. Cai, Y. Ding, H. Ma, C. Yuan and W. Xu, *Phys. Chem. Chem. Phys.* **18** (2016) 2730.
24. Q. Sun, L. Cai, H. Ma, C. Yuan and W. Xu, *Chem. Commun.* **52** (2016) 6009.
25. J. Liu, Q. Chen, Q. He, Y. Zhang, X. Fu, Y. Wang, D. Zhao, W. Chen, G. Q. Xu and K. Wu, *Phys. Chem. Chem. Phys.* **20** (2018) 11081.
26. J. Eichhorn, D. Nieckarz, O. Ochs, D. Samanta, M. Schmittel, P. J. Szabelski and M. Lackinger, *ACS Nano* **8** (2014) 7880.
27. Q. Fan, C. Wang, Y. Han, J. Zhu, W. Hieringer, J. Kuttner, G. Hilt and J. M. Gottfried, *Angew. Chem. Int. Ed.* **52** (2013) 4668.
28. J. Dai, Q. Fan, T. Wang, J. Kuttner, G. Hilt, J. M. Gottfried and J. Zhu, *Phys. Chem. Chem. Phys.* **18** (2016) 20627.
29. Q. Fan, T. Wang, J. Dai, J. Kuttner, G. Hilt, J. M. Gottfried and J. Zhu, *ACS Nano* **11** (2017) 5070.
30. J. A. Lipton-Duffin, O. Ivasenko, D. F. Perepichka and F. Rosei, *Small* **5** (2009) 592.
31. M. Bieri, M. T. Nguyen, O. Gröning, J. M. Cai, M. Treier, K. Aït-Mansour, P. Ruffieux, C. A. Pignedoli, D. Passerone, M. Kastler, K. Müllen and R. Fasel, *J. Am. Chem. Soc.* **132** (2010) 16669.
32. M. Bieri, S. Blankenburg, M. Kivala, C. A. Pignedoli, P. Ruffieux, K. Müllen and R. Fasel, *Chem. Commun.* **47** (2011) 10239.
33. H. Y. Gao, H. Wagner, D. Zhong, J. H. Franke, A. Studer and H. Fuchs, *Angew. Chem. Int. Ed.* **52** (2013) 4024.
34. H. Zhou, J. Liu, S. Du, L. Zhang, G. Li, Y. Zhang, B. Z. Tang and H. J. Gao, *J. Am. Chem. Soc.* **136** (2014) 5567.
35. J. Liu, Q. Chen, L. Xiao, J. Shang, X. Zhou, Y. Zhang, Y. Wang, X. Shao, J. Li, W. Chen, G. Q. Xu, H. Tang, D. Zhao and K. Wu, *ACS Nano* **9** (2015) 6305.
36. B. Cirera, Y. Q. Zhang, J. Björk, S. Klyatskaya, Z. Chen, M. Ruben, J. V. Barth and F. Klappenberger, *Nano Lett.* **14** (2014) 1891.

37. H. Y. Gao, J. H. Franke, H. Wagner, D. Zhong, P. A. Held, A. Studer and H. Fuchs, *J. Phys. Chem. C* **117** (2013) 18595.
38. N. Kepčija, Y. Q. Zhang, M. Kleinschrodt, J. Björk, S. Klyatskaya, F. Klappenberger, M. Ruben and J. V. Barth, *J. Phys. Chem. C* **117** (2013) 3987.
39. Q. Li, C. Han, S. R. Horton, M. Fuentes-Cabrera, B. G. Sumpter, W. Lu, J. Bernholc, P. Maksymovych and M. Pan, *ACS Nano* **6** (2012) 566.
40. J. Liu, P. Ruffieux, X. Feng, K. Müllen and R. Fasel, *Chem. Commun.* **50** (2014) 11200.
41. D. Zhong, J. H. Franke, S. K. Podiyanachari, T. Blömker, H. Zhang, G. Kehr, G. Erker, H. Fuchs and L. Chi, *Science* **334** (2011) 213.
42. L. Cai, Q. Sun, C. Zhang, Y. Ding and W. Xu, *Chem. Eur. J.* **22** (2016) 1918.
43. Q. Li, B. Yang, H. Lin, N. Aghdassi, K. Miao, J. Zhang, H. Zhang, Y. Li, S. Duhm, J. Fan and L. Chi, *J. Am. Chem. Soc.* **138** (2016) 2809.
44. Q. Sun, B. V. Tran, L. Cai, H. Ma, X. Yu, C. Yuan, M. Stöhr and W. Xu, *Angew. Chem. Int. Ed.* **56** (2017) 12165.
45. L. Cai, F. Kang, Q. Sun, W. Gao, X. Yu, H. Ma, C. Yuan and W. Xu, *ChemCatChem* **11** (2019) 5417.
46. Q. Sun, L. Cai, Y. Ding, L. Xie, C. Zhang, Q. Tan and W. Xu, *Angew. Chem. Int. Ed.* **54** (2015) 4549.
47. A. Hirsch, *Nat. Mater.* **9** (2010) 868.
48. G. Otero, G. Biddau, C. Sánchez-Sánchez, R. Caillard, M. F. López, C. Rogero, F. J. Palomares, N. Cabello, M. A. Basanta, J. Ortega, J. Méndez, A. M. Echavarren, R. Pérez, B. Gómez-Lor and J. A. Martín-Gago, *Nature* **454** (2008) 865.
49. J. R. Sanchez-Valencia, T. Dienel, O. Gröning, I. Shorubalko, A. Mueller, M. Jansen, K. Amsharov, P. Ruffieux and R. Fasel, *Nature* **512** (2014) 61.
50. A. L. Pinardi, G. Otero-Irurueta, I. Palacio, J. I. Martinez, C. Sanchez-Sanchez, M. Tello, C. Rogero, A. Cossaro, A. Preobrajenski, B. Gómez-Lor, A. Jancarik, I. G. Stará, I. Starý, M. F. Lopez, J. Méndez and J. A. Martín-Gago, *ACS Nano* **7** (2013) 3676.
51. C. Sánchez-Sánchez, T. Dienel, O. Deniz, P. Ruffieux, R. Berger, X. Feng, K. Müllen and R. Fasel, *ACS Nano* **10** (2016) 8006.
52. P. H. Jacobse, A. van den Hoogenband, M. E. Moret, R. J. Klein Gebbink and I. Swart, *Angew. Chem. Int. Ed.* **55** (2016) 13052.
53. S. Kawai, S. Saito, S. Osumi, S. Yamaguchi, A. S. Foster, P. Spijker and E. Meyer, *Nat. Commun.* **6** (2015) 8098.
54. J. Cai, C. A. Pignedoli, L. Talirz, P. Ruffieux, H. Söde, L. Liang, V. Meunier, R. Berger, R. Li, X. Feng, K. Müllen and R. Fasel, *Nat. Nanotechnol.* **9** (2014) 896.
55. D. J. Rizzo, G. Veber, T. Cao, C. Bronner, T. Chen, F. Zhao, H. Rodriguez, S. G. Louie, M. F. Crommie and F. R. Fischer, *Nature* **560** (2018) 204.
56. W. Wang, X. Shi, S. Wang, M. A. Van Hove and N. Lin, *J. Am. Chem. Soc.* **133** (2011) 13264.
57. H. Zhang, H. Lin, K. Sun, L. Chen, Y. Zagranyarski, N. Aghdassi, S. Duhm, Q. Li, D. Zhong, Y. Li, K. Müllen, H. Fuchs and L. Chi, *J. Am. Chem. Soc.* **137** (2015) 4022.

58. A. Basagni, F. Sedona, C. A. Pignedoli, M. Cattelan, L. Nicolas, M. Casarin and M. Sambi, *J. Am. Chem. Soc.* **137** (2015) 1802.
59. L. Talirz, H. Sode, T. Dumslaff, S. Wang, J. R. Sanchez-Valencia, J. Liu, P. Shinde, C. A. Pignedoli, L. Liang, V. Meunier, N. C. Plumb, M. Shi, X. Feng, A. Narita, K. Müllen, R. Fasel and P. Ruffieux, *ACS Nano* **11** (2017) 1380.
60. Y. C. Chen, D. G. de Oteyza, Z. Pedramrazi, C. Chen, F. R. Fischer and M. F. Crommie, *ACS Nano* **7** (2013) 6123.
61. H. Huang, D. Wei, J. Sun, S. L. Wong, Y. P. Feng, A. H. Neto and A. T. Wee, *Sci. Rep.* **2** (2012) 983.
62. P. Han, K. Akagi, F. F. Canova, H. Mutoh, S. Shiraki, K. Iwaya, P. S. Weiss, N. Asao and T. Hitosugi, *ACS Nano* **8** (2014) 9181.
63. R. R. Cloke, T. Marangoni, G. D. Nguyen, T. Joshi, D. J. Rizzo, C. Bronner, T. Cao, S. G. Louie, M. F. Crommie and F. R. Fischer, *J. Am. Chem. Soc.* **137** (2015) 8872.
64. Y. Cao, J. Qi, Y.-F. Zhang, L. Huang, Q. Zheng, X. Lin, Z. Cheng, Y. Y. Zhang, X. Feng, S. Du, S. T. Pantelides and H. J. Gao, *Nano Res.* **11** (2018) 6190.
65. L. Cai, X. Yu, M. Liu, Q. Sun, M. Bao, Z. Zha, J. Pan, H. Ma, H. Ju, S. Hu, L. Xu, J. Zou, C. Yuan, T. Jacob, J. Björk, J. Zhu, X. Qiu and W. Xu, *ACS Nano* **12** (2018) 7959.
66. X. Yu, L. L. Cai, M. L. Bao, Q. Sun, H. H. Ma, C. X. Yuan and W. Xu, *Chem. Commun.* **56** (2020) 1685.
67. S. Wang, Q. Sun, O. Gröning, R. Widmer, C. A. Pignedoli, L. Cai, X. Yu, B. Yuan, C. Li, H. Ju, J. Zhu, P. Ruffieux, R. Fasel and W. Xu, *Nat. Chem.* **11** (2019) 924.
68. Q. Sun, L. Cai, S. Wang, R. Widmer, H. Ju, J. Zhu, L. Li, Y. He, P. Ruffieux, R. Fasel and W. Xu, *J. Am. Chem. Soc.* **138** (2016) 1106.
69. X. Yu, X. Li, H. Lin, M. Liu, L. Cai, X. Qiu, D. Yang, X. Fan, X. Qiu and W. Xu, *J. Am. Chem. Soc.* **142** (2020) 8085.
70. J. R. Lomas, C. J. Baddeley, M. S. Tikhov and R. M. Lambert, *Langmuir* **11** (1995) 3048.
71. S. Clair, O. Ourdjini, M. Abel and L. Porte, *Chem. Commun.* **47** (2011) 8028.
72. N. Pavliček, A. Mistry, Z. Majzik, N. Moll, G. Meyer, D. J. Fox and L. Gross, *Nat. Nanotechnol.* **12** (2017) 308.
73. J. Krüger, N. Pavliček, J. M. Alonso, D. Pérez, E. Guitián, T. Lehmann, G. Cuniberti, A. Gourdon, G. Meyer, L. Gross, F. Moresco and D. Peña, *ACS Nano* **10** (2016) 4538.
74. J. I. Urgel, H. Hayashi, M. Di Giovannantonio, C. A. Pignedoli, S. Mishra, O. Deniz, M. Yamashita, T. Dienel, P. Ruffieux, H. Yamada and R. Fasel, *J. Am. Chem. Soc.* **139** (2017) 11658.
75. K. Kaiser, L. M. Scriven, F. Schulz, P. Gawel, L. Gross and H. L. Anderson, *Science* **365** (2019) 1299.
76. R. Zuzak, R. Dorel, M. Krawiec, B. Such, M. Kolmer, M. Szymonski, A. M. Echavarren and S. Godlewski, *ACS Nano* **11** (2017) 9321.
77. J. I. Urgel, S. Mishra, H. Hayashi, J. Wilhelm, C. A. Pignedoli, M. D. Di Giovannantonio, R. Widmer, M. Yamashita, N. Hieda, P. Ruffieux, H. Yamada and R. Fasel, *Nat. Commun.* **10** (2019) 861.

78. H. Y. Gao, D. Zhong, H. Mönig, H. Wagner, P.-A. Held, A. Timmer, A. Studer and H. Fuchs, *J. Phys. Chem. C* **118** (2014) 6272.
79. A. Basagni, L. Ferrighi, M. Cattelan, L. Nicolas, K. Handrup, L. Vaghi, A. Papagni, F. Sedona, C. D. Valentin, S. Agnoli and M. Sambi, *Chem. Commun.* **51** (2015) 12593.
80. E. Kazuma, J. Jung, H. Ueba, M. Trenary and Y. Kim, *J. Am. Chem. Soc.* **139** (2017) 3115.
81. E. Kazuma, J. Jung, H. Ueba, M. Trenary and Y. Kim, *Science* **360** (2018) 521.
82. L. Colazzo, F. Sedona, A. Moretto, M. Casarin and M. Sambi, *J. Am. Chem. Soc.* **138** (2016) 10151.
83. Y. Okawa and M. Aono, *Nature* **409** (2001) 683.
84. Y. Okawa and M. Aono, *J. Chem. Phys.* **115** (2001) 2317.
85. Y. Okawa, S. K. Mandal, C. Hu, Y. Tateyama, S. Goedecker, S. Tsukamoto, T. Hasegawa, J. K. Gimzewski and M. Aono, *J. Am. Chem. Soc.* **133** (2011) 8227.
86. Y. Okawa, M. Akai-Kasaya, Y. Kuwahara, S. K. Mandal and M. Aono, *Nanoscale* **4** (2012) 3013.

https://doi.org/10.1142/9789811247859_0004

Chapter 4

Molecular Functionalization of 2D Materials

Miriam C. Rodríguez González*, Rahul Sasikumar and Steven De Feyter†

Department of Chemistry,
Division of Molecular Imaging and Photonics,
KU Leuven, Celestijnenlaan 200F,
B-3001 Leuven, Belgium
**miriamcandelaria.rodriguezgonzalez@kuleuven.be*
†steven.defeyter@kuleuven.be

In this chapter, we give an overview of different chemical modifications that can be done on the surface of two-dimensional (2D) layered materials. We place emphasis on the diversity of reactions that have been proposed and are now available to surface scientists working in 2D materials field. Using mainly, but not exclusively, MoS_2 as example, reactions involving covalent and non-covalent interactions are discussed.

Keywords: Chemical modification; 2D material; surface science MoS_2.

1. Introduction

With the isolation of graphene in 2004, a material theoretically predicted much earlier, materials chemistry has been dominated by two-dimensional (2D) materials in the 21st century.[1,2] A 2D material can be seen as a monolayer-thick inorganic crystal extending only in two dimensions. In the last decade, we have witnessed the emergence of new 2D layered materials from previously known 3D solid state structures based on stacked 2D layers. Some examples of these

*,†Corresponding authors.

materials are the transition-metal dichalcogenides (TMDCs),[3] such as molybdenum disulfide (MoS_2), hexagonal boron nitride (h-BN),[4] black phosphorus (BP)[5] or germanene.[6] A wide-ranging set of properties become accessible through these different materials. For example, in terms of electronic properties, h-BN is an insulator while TMDCs such as MoS_2 and $NbSe_2$ exhibit different behavior, being direct bandgap semiconductor and metallic, respectively. After the success in the isolation and processing of these 2D materials, the attention has been focused on the characterization of their special properties and the evaluation of its applicability.

Due to their high surface/volume ratio, 2D materials are sensitive to surface modification *via* chemical functionalization. Experimentally, different molecular chemistry approaches based on covalent or non-covalent interactions between pristine or defective TMDCs and organic moieties have been proposed. The simplest way for the functionalization is the physisorption of molecules *via* solution or gas phase in a non-ordered fashion. An additional level of control can be reached through the formation of ordered supramolecular assemblies. Non-covalent molecular functionalization of materials concerns with properties that emerge from electrostatic and dipole–dipole or induced–dipole interactions, H-bonds, or van der Waals interactions. These non-covalent approaches allow for surface modification in a straightforward manner and they do not require specific molecule-substrate bonds. Despite the feasibility of the non-covalent protocols, in the last few years, the interest in covalent functionalization approaches has increased. These covalent methodologies use specific chemical interactions with the substrate which can lead to the formation of strong and stable covalent bonds, with changes in the electronic structure of the material as a consequence. For example, the modification of the bandgap on monolayered MoS_2 materials can be achieved through these covalent interactions,[7] although some examples based on non-covalent interactions of strong organic dopants for this purpose can be found as well.[8] It has been theoretically predicted that the donor/acceptor nature of the substituents of the molecule and the nature of the bond on the surface can be systematically used to tune the band structure of the material on demand.[9] At first glance, covalent functionalization would seem to be superior to non-covalent methods due to the strength and stability of the interactions involved. But the large surface-to-volume ratios in 2D materials means that their properties are highly sensitive to external conditions, and covalent bonds could irreversibly deteriorate the interesting properties of the pristine material. For instance, strong chemical bonds have been shown to induce phase change from semiconducting 2H-MoS_2 into metallic 1T-MoS_2.[10] For these reasons, before proceeding to the chemical functionalization of a material, an evaluation of the properties required in the processed material and the advantages/drawbacks of the protocol chosen has to be carried out.

The importance of surface modification lies in the ease and selectivity to control the surface composition, structure, and properties of the material. 2D nanosheets of TMDCs have been extensively used in the last decade in multiple applications such as catalysis, energy storage, nanoelectronics, photonics or sensing.[11] With the chemical functionalization at the molecular level, it is possible to tune or engineer the electrical, optical, or thermal properties exhibited by these materials in a rational fashion, leading to promising opportunities for applications. Some of the advantages include the formation of hybrid structures or improved dispersibility[12] of the material for further processing.

In this work, we focus on the advances made in the functionalization of TMDCs, with special attention to the most intensively studied TMDC, MoS_2. Single sheets of TMDCs consist of an atomic layer of transition metals (M = Mo, W, Ti, etc.) between two atomic layers of chalcogens (X = S, Se or Te) in a sandwich-type structure. TMDCs exist in several structural phases depending on the stacking of the atomic planes of the material. Commonly, we find structural phases with a trigonal prismatic (2H) or octahedral (1T) coordination of metal atoms.[13] For MoS_2, several polymorphs can be found, the most important ones being the metallic 1T phase (1T-MoS_2) and the semiconducting 2H (2H-MoS_2) and 3R (3R-MoS_2) phases (Fig. 1). Although the 2H-MoS_2 phase is thermodynamically stable, the chemical exfoliation of bulk MoS_2 by lithium intercalation

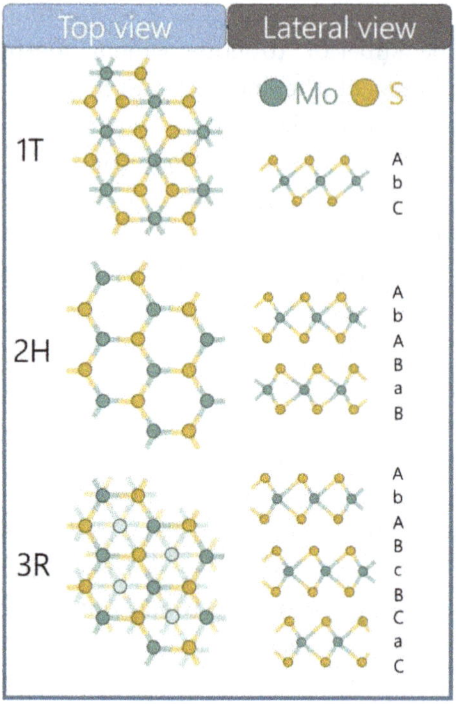

Fig. 1. (Color online) Scheme for the crystal structure of the different polytypes of MoS_2 (1T, 2H, and 3R). Adapted with permission from Ref. 13. Copyright 2015 The Royal Chemical Society.

(ceMoS_2) induces a phase transition from 2H-MoS_2 to the metallic 1T/1T′ polytype (≈65% of transformation).[14]

In terms of reactivity, the semiconducting phase (2H-MoS_2) cannot be easily functionalized, requiring in most of the cases the incorporation of defects or vacancies for its functionalization. In contrast, in the metallic phases (1T-MoS_2 or its distorted structures 1T′-MoS_2) strong bonds can be formed between molecules (especially electrophiles) and the surface due to the possibility of a direct molecule-substrate electron transfer. As commented before, these atomic-thick materials have interesting properties such as a direct bandgap or a strong spin–orbit coupling.

Here, we describe the experimental approaches for the molecular modification of 2D TMDCs that have been explored so far, in particular those applied for the model system of these materials, MoS_2. However, these modification strategies can also be applied to other TMDCs such as WS_2 or $MoSe_2$.

2. Non-Covalent Functionalization of 2D Materials

Non-covalent functionalization is desirable as it is a non-destructive and reversible process. Weak interactions do not necessarily mean that the resulting property modulations are insignificant. For instance, for a molecule adsorbed on a 2D material, when the energy of the highest occupied molecular orbital (HOMO) of the molecule is higher than the Fermi level of the material, charge transfer pathways where electron density is transferred between the molecules and the material can be opened. Such transfer of electron density can influence electrical properties in ultrathin materials. This effect is known as contact doping or charge transfer doping.[15]

The picture of a monolayer of few molecules influencing the properties of a material might come off as surprising. It is compelling to ask how such a minor influence of molecules can be as effective on ultrathin materials, as conventional doping is on traditional semiconductors. The answer lies in the size of the materials in our discussion. 2D materials are characteristically different from conventional materials as they lack a bulk. This low dimensionality leads, as pointed out before, to extremely high surface-to-volume ratios in these materials.

Apart from these, many of their optical properties are strongly influenced by external factors as their monolayer nature stabilizes various many-body exciton effects.[16] The extreme sensitivity of these materials is a dream come true for every researcher. Unfortunately, high sensitivity without selectivity is a double-edged sword. Due to the extent to which immediate surroundings of 2D materials influence their properties, significant predictable results are only obtained with a good control of the environment. For example, adsorbed water molecules at the surface can lead to charge trapping states which induce a hysteresis in current response of MoS_2 field-effect transistor (FET) devices.[17] The photosensitivity of MoS_2 multiplies this hysteresis on shining light, as more charges are generated and populate the trap states. Some reports claim that adsorption of water quenches the photoluminescence (PL) in single layer (SL-MoS_2),[18] although this effect remains under discussion. These results are an outcome of the polar nature of water. Water molecules adsorb at the interface orienting to form a template adlayer. This adlayer acts as an n-type dopant that quenches radiative recombination in MoS_2. Recent studies have shown that common organic solvents also have a similar influence on the electronic and optoelectronic properties of TMDCs. Samori and co-workers report doping in monolayer MoS_2 and WSe_2 by the controlled physisorption of benzene derivatives. They observe that the intensity and direction of doping depend not only on the functional group on the adsorbed molecule but also on the thermodynamic stability of the physisorption.[19] In another study, Tian and co-workers report that several common solvents could influence the electrical properties of TMDCs.[20] The doping effect is again influenced by different contributions: the charge donating characteristics and dipole interactions of the molecule, and the surface energy of the solvent. The effect of solvents could be eliminated from the surface by common washing procedures in most cases. But one particular solvent, N-methyl-2-pyrrolidone (NMP), could not be completely removed from the surface.

Beyond exploitable results, these experiments are more examples of the sensitivity of 2D materials than potential applications. These works succeed in showcasing the promises and pitfalls in the field of 2D materials functionalization. Any influence on electrical and optical properties of a 2D material must be viewed with scrutiny. Utmost caution and a good control of the environment is essential due to the extreme sensitivity of the systems in hand.

Our discussion on non-covalent functionalization of 2D materials is divided into two sections: (a) gas phase functionalization, and (b) condensed-phase functionalization.

2.1. *Gas phase functionalization*

Non-covalent gas-phase functionalization protocols are among the least explored areas of functionalizing 2D materials. Quite some literature predicts interesting properties of gas adsorbed 2D materials.

Experimental results have remained relatively scarce, some of which will be discussed here. One of the earlier reports of gas phase non-covalent functionalization of TMDCs is by Wu and co-workers. They reported an enhancement in the PL of annealed TMDCs materials through a charge transfer to physisorbed O_2 and H_2O molecules.[16] PL intensity of post-annealed MoS_2 is enhanced 10 times in the presence of H_2O, 35 times in the presence of O_2, and 100 times when they are present together. PL intensities reverted as the gases were removed, indicating that the change in the PL intensity is related to the non-covalent interactions of the gas molecules with the surface. These responses were also used to produce highly sensitive humidity sensors in the range of 0–35% humidity, beyond which the PL response saturates.[21] Monolayers of MoS_2 prepared by chemical vapor deposition (CVD) over oxide materials are unintentionally *n*-doped and have an excess of electrons. These excess electrons promote the formation of negative trions over excitons on irradiation. These decrease the PL intensity due to a competition between radiative and non-radiative recombination pathways in trions. With the annealing of the samples, a depletion of the excess electrons and improved PL of the material can be achieved. In the above study, MoS_2 samples annealed at 450°C for 40 min displayed the strongest PL enhancement among samples annealed for shorter and longer times. Lee and co-workers reported the work function modulation of MoS_2 by 0.4 eV (4.04–4.47 eV) through an oxygen-adsorption assisted pseudo *p*-*n* junction.[22] This was achieved from a relatively small surface coverage of oxygen molecules ($<10\%$). Their theoretical model suggests that the work function could be modulated up to a range of 1.2 eV by varying the surface oxygen coverage between 0% and 100%.

Another relevant work is from the group of Eom and co-workers where they report the functionalization of single, bi, and few layer MoS_2 devices with N_2 in presence UV light (Fig. 2).[23] N_2 doping shifts the threshold voltage of MoS_2 FETs towards negative gate voltages. It also improves the carrier density and charge-carrier mobility of *n*-type MoS_2 non-destructively. The prior electrical characteristics are completely recovered by treating the samples with O_2 under deep ultraviolet (DUV) light. First-principles calculations including van

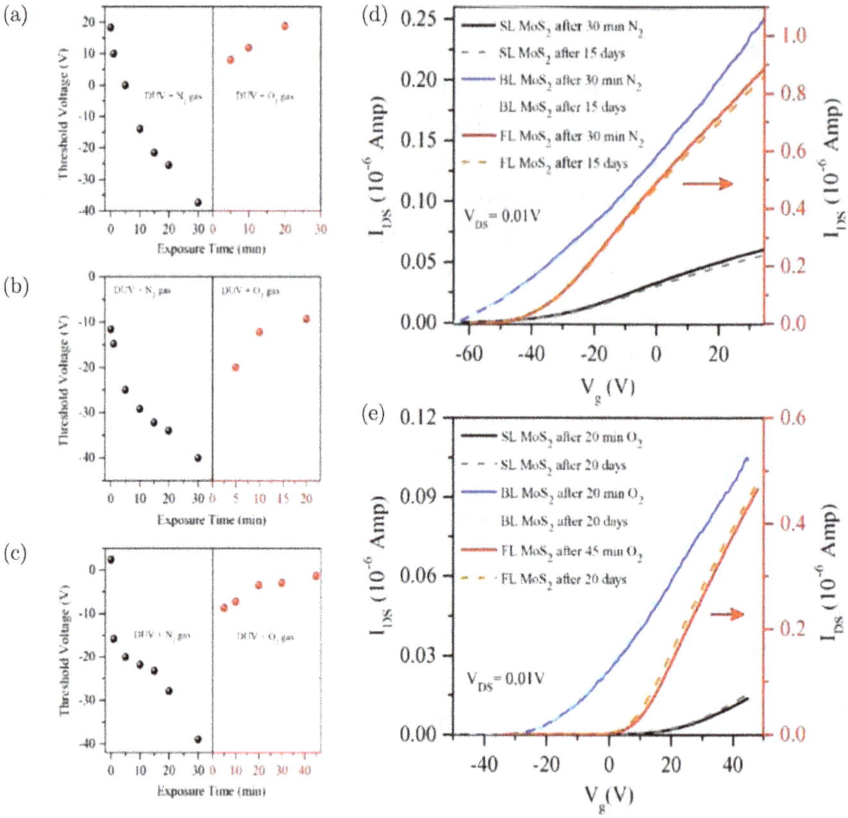

Fig. 2. (Color online) Dependence of the threshold voltage on time of exposure for N_2 and O_2 in presence of DUV light for (a) single, (b) bi, and (c) few layer MoS_2. (d) Stability of N_2 doping of SL (black), BL (blue), and FL (red) MoS_2 transistors as revealed by a I(DS)-V_g plot after exposure to N_2 molecules and DUV light, and the same plot after 15 days. (e) Same stability test but after recovery of the MoS_2 transistor with O_2 molecules and DUV light, and the same plot after 20 days. Reprinted with permission from Ref. 23. Copyright 2014 Wiley-VCH.

der Waals interactions have been carried out to predict adsorption energies of different gases.[24] The adsorption energy is different for each molecule, being in the range of -0.07–-0.15 eV for N_2 depending on the functional used in the calculations. Therefore, the improved properties are a result of the physisorption of N_2 at the interface. The doping is remarkably stable in ambient air and a non-significant loss of doping was reported even after 20 days of exposure.

Non-covalent functionalization in gas-phase has not attracted the same vigor of interest when compared to the condensed phase approaches. There is only limited room for tuning the properties of a molecule before it condenses to form a liquid or a solid. Hence, gas-phase strategy falls short in terms of degree of tunability when compared to the condensed phase techniques. This, in combination with the requirement of working in high vacuum deters the commercialization of these functionalization processes. Consequently, even if UHV conditions offer the better mechanistic understanding of the surface interactions, UHV is not suitable for most of the real-life applications of these materials.

2.2. *Condensed phase functionalization*

We have seen above the drawbacks of gas phase functionalization. Condensed phase reactions generally do not require an extreme control over external conditions. Additionally, there is a higher ceiling on tunability compared to gas phase techniques as there is an additional control in the form of solvent medium. The ease of performing reactions in liquid phase makes them instantly attractive. Our discussion here will focus on condensed phase techniques that are based on self-assembly of the interacting molecules. We will have a brief discussion on functionalization methods which are not based on self-assembly in the latter part of is section.

2.2.1. *Condensed phase techniques based on self-assembly*

The impact of an individual small molecule on the properties of a surface is in most of the cases not significant and/or detectable. However, a stable array of several packed and oriented molecules can be used to successfully tune the properties and behavior of these materials. Self-assembly is an association of molecules through weak and reversible interactions to form organized polymolecular systems. Such an ordered association allows a magnification of molecular properties by directional orientation and hence greater contact potential. Such control also leads to an improved understanding of the interface. Combined with the functional diversity of organic molecules, such self-assembled systems are ideal for tuning the properties of 2D materials. One should differentiate two self-assembly approaches. The first one is based on the formation of densely packed self-assembled monolayers (SAMs) where the molecules form a one-molecule thick layer and form a chemical bond with a solid support. The TMDC surface is then brought into physical, non-covalent contact with the other side of the monolayer. The second one is based on the formation of self-assembled molecular networks (SAMNs)[25] on the TMDC surface itself, where the molecules often form a crystalline film, typically again one molecule thick.

Some of the first works in modifying TMDCs' properties through SAMs proposed the use of functionalized silanes.[26] SAMs of alkyl silanes with functionalized chain-ends were formed on a SiO_2 substrate onto which MoS_2 flakes were transferred physically. Thus, in this setup, the MoS_2 films interact with the end-groups that stick out from the surface. As expected, the electrical properties of MoS_2 varied depending on the functional groups that interact at the interface. They also observed the effect silane monolayers have on a p-doped substrate (WSe_2) is opposite to that on an n-doped substrate (MoS_2).[27,28] SAMs of fluoroalkyltrichlorosilane and octyltrichlorosilane that have a positive dipole moment seem to accumulate electron carriers at the interface of TMDCs, shifting up the Fermi level in p-doped materials while shifting down the Fermi level in case of n-doped materials. This causes an inverse effect in threshold voltages of both types of materials. SAMs of 3-aminopropyltrimethoxysilane acted as n-doping agents by accumulating holes at the substrate interface, thus shifting down the Fermi level in p-doped materials and shifting it up in n-doped materials. These techniques also modulate the materials' work function as it directly depends on the Fermi level of the material. As previously seen in other systems, the magnitude of effects depends on the Fermi level of the material, the dipole moment of the molecule, and the strength of their interaction.

Another interesting approach of tuning properties of TMDCs has come *via* organic molecules that interact through heteroatoms (i.e. nitrogen) and pi clouds. Azobenzene derivative SAMs represent one such system explored so far.[29] Azobenzene molecules are interesting

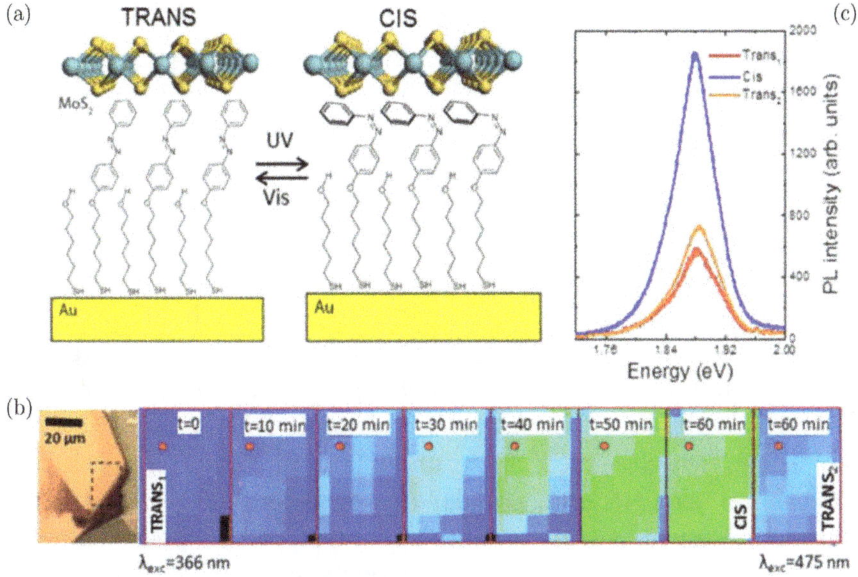

Fig. 3. (Color online) (a) Scheme of a monolayer MoS_2 sample gated in by trans- and cis-azobenzene. (b) PL intensity recorded for 2L-MoS2 on the SAM after UV exposure for different times at the position of the red spot. Last map shows the switching back to trans after 120 min of white light exposure. (c) PL spectra for 2L-MoS2 in trans (red), in cis (blue) and again in trans (orange). Reprinted with permission from Ref. 29. Copyright 2014 AIP Publishing.

due to their cis-trans photoisomerization (Fig. 3). The trans isomer can be converted into the cis form by UV radiation. The process is also photoreversible and turns from cis to trans under visible light, making it a photo-controllable system. Azobenzene is electropositive and the large contact potential in its trans state imparts an n-doping to the MoS_2. Among the isomers, the trans form has a larger contact potential on interacting with the MoS_2 surface than the cis form which leads to the higher doping impact of the former isomer.[30] The photoisomerization from trans to cis lowers the magnitude of doping and restores some of the intrinsic properties. It is shown that the magnitude of this doping can additionally be controlled by using mixed self-assembled monolayers (mSAMs) where the proportion of azobenzene units per area can be modulated. It is shown that the type of doping can also be controlled by varying other chemical species present in the SAM. A system populated with electronegative species such as chlorine can exert the opposite effect and act as a *p*-type dopant. This makes MoS_2 on azobenzene systems a conveniently tunable semiconductor with photo-switchable doping levels.

Other systems that interact with MoS_2 through heteroatoms and pi clouds are perylene derivatives, porphyrins, and phthalocyanine derivatives. Unlike silanes and azobenzene derivatives, these systems interact through SAMNs at MoS_2 surface rather than by means of SAMs on a substrate like in the former. Perylene-based dyes are excellent optoelectronic dopants due to their thermal and photo stability and have been used as *n*-type dopants in solar cells.[31] Porphyrins display excellent molar absorptivity the order of 10^6 $M^{-1}cm^{-1}$. Besides their high photo stability, porphyrins, and phthalocyanines (Pc) are popular due to their reduction potential that can be tuned by varying the metal centers and peripheric ligands. Metal Pc are also used in light harvesting systems where they act as both electron donors and electron acceptors to carbon nanotubes and graphene depending on their core metal.[32] These metal centers also provide the catalytic nature that is advantageous in various applications.

SAMNs of perylene *N*-alkyldiimides on MoS_2 are explored as well.[33] Deusberg and coworkers used a monolayer-thin film of perylene diimides (PDI) to facilitate atomic layer deposition (ALD) at TMDCs surfaces.[34] What makes it an important development is that ALD is a highly precise deposition technique that is used to print gate oxides and integrated circuits on semiconductor surfaces. But the common reaction for ALD needs surface hydroxy groups whose absence on TMD surface made it challenging to develop TMD-based electronic devices. They used PDI modified with carboxy alkanes to introduce hydroxy groups at the TMD surface that acted as a base for the ALD reaction without modifying its pristine properties. PDI and porphyrin have also been used as *n*-type dopants to improve the photocurrent and photoresponsivity of MoS_2 devices.[35] Additionally, the thermal stability and strong interactions of these molecules can withstand the 1T-2H phase transition in TMDCs. Thus, these molecules can even be used as non-covalent dopants in high-temperature applications.[36]

Pc are popular light-harvesting systems in solar cells and other optical applications due to their tunable HOMO–LUMO positions and bandgap that falls in the visible spectrum.[37] Most common approaches to tune the properties of these systems are either by replacing the metal cores or by substituents on the Pc. Recently, Samori and co-workers showed that the electronic nature of Pc can also be tuned by on-surface functionalization of the metal centers of Pc assembled on MoS_2.[38] The forced orientation and charge transfer in the SAMs of metallic phthalocyanines (MPc) at TMDC surface was taken advantage of to carry out efficient, unidirectional ligand functionalization by pyridine derivatives at the metal centers. This approach provided an additional tunability dimension to the properties of the molecular system. Pc are also interesting for photo-controllable doping. In the previous section, we discussed the photoswitchable doping in azobenzene functionalized MoS_2 devices. Pc offer a different mechanism of control. These species have dual emission; between HOMO and LUMO termed as S1, and between HOMO and LUMO+1 termed as S2 emission (Fig. 4). It happens that the conduction band minima of MoS_2 lay between the LUMO and LUMO+1 level in metal-free Pc. In this scenario, Pc acts as an n-type dopant or a p-type dopant depending on the excitation wavelength. But this possibility is not practically realized due to an unfavorable charge transfer from the dye's S1 state.[39] Nevertheless, Pc are important functionalizing systems for 2D materials with high potential in optoelectronic applications.

Functionalization and doping of MoS_2 is not limited to the above-mentioned systems. In general, dye materials have their HOMO–LUMO transitions in the visible range and hence absorb in this range, coinciding with the bandgaps of semiconducting TMDCs. Spatial interaction of these molecules with TMDCs allows charge transfer between these systems that leads to a manipulation of Fermi levels and subsequent outcomes on electronic properties. Due to this reason, various dyes like 2,3,5,6-tetrafluoro-7,7,8,8-tetracyanoquinodimethane (F4TCNQ), 7,7,8,8-tetracyanoquinodimethane (TCNQ), nicotinamide adenine dinucleotide (NADH),[40] benzyl viologen (BV),[41] methyl orange (MO), rhodamine 6G (R6G), and methylene

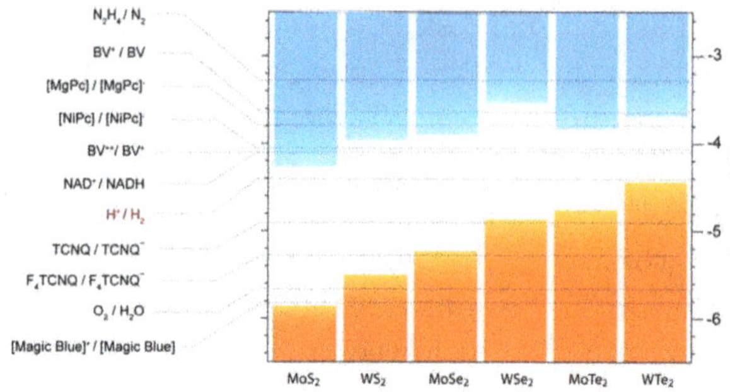

(a)

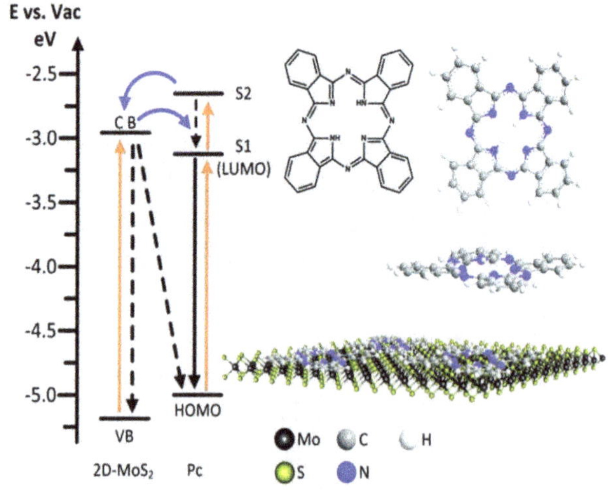

(b)

Fig. 4. (Color online) (a) Comparison between the electrochemical redox potentials of various molecular dopants and the valence and conduction band edges of group-6-TMDs. Reprinted with permission from Ref. 14. Copyright 2018 The Royal Society of Chemistry. (b) Energy diagram for MoS_2 functionalized with phthalocyanine. Orange (excitation), black (radiative recombination), and blue (charge-transfer). Dashed lines show non-radiative processes. Reprinted with permission from Ref. 39. Copyright 2016 The Royal Society of Chemistry.

blue (MB)[42] have been explored as dopants for TMDCs. This is not a comprehensive list, but an illustration of the array of possible dopants. Other functional molecules include triphenylphosphine, which was demonstrated to have air/water stable and reversible *n*-type doping–dedoping interactions with MoS_2,[43] and oleylamine for which a decrease in contact resistance was reported.[44]

Apart from small organic molecules, larger molecules could also be used to modulate TMDCs' properties through specific interactions. Polymers have been used for this purpose. Emrick and co-workers reported on polymers with tetrathiafulvalene (TTF) units in the sidechains as anchor group for non-covalent functionalization of MoS_2 (Fig. 5(a)).[45] TTF acts as an *n*-type dopant. The polymer chain increases the dispersibility of MoS_2 sheets in solvents. DNAs are another type of large molecules that have been used to functionalize TMDCs. Park and co-workers reported the self-assembly of DNA on MoS_2 and WSe_2 (Fig. 5(b)) where the phosphate groups in the backbone of the DNA act as anchoring groups as well as *n*-dopants due to their electron rich nature.[46] They also induce *p*-doping and tune the contact potential by means of DNA modified with metal ions (M-DNA). In this case, the metal ions withdraw the charge from the phosphate groups and reverse the effect. DNAs also present specific interactions that are useful in targeted sensing. Joo and co-workers reported the assembly of probe DNA (*p*-DNA) modified with a fluorescent dye, Cy3, at WSe_2 surface as *p*-type dopants.[47] They displayed the use of specific interactions of DNA as a tool for biosensing target DNA (*t*-DNA) indirectly through the PL intensity of WSe_2/Cy3/*p*-DNA system. The PL intensity of the system increases with the binding of *t*-DNA because of de-doping of WSe_2 as DNA forms double strands. Though still in its infancy, the DNA-based functionalization methods will become more important in the future. Techniques such as DNA origami that allow programmed folding and scaffolding of DNA can be used to design systems that allow non-arbitrary functionalization of MoS_2 surfaces, which could be switched between *n*-type, *p*-type, and dedoped selectively.

Self-assembly-based non-covalent functionalization of 2D materials is a rapidly growing field. We have summarized here various approaches that use self-assembly to functionalize TMD materials and tuning their semiconductor applications and optoelectrical properties. Although in this section we have focused on self-assembly

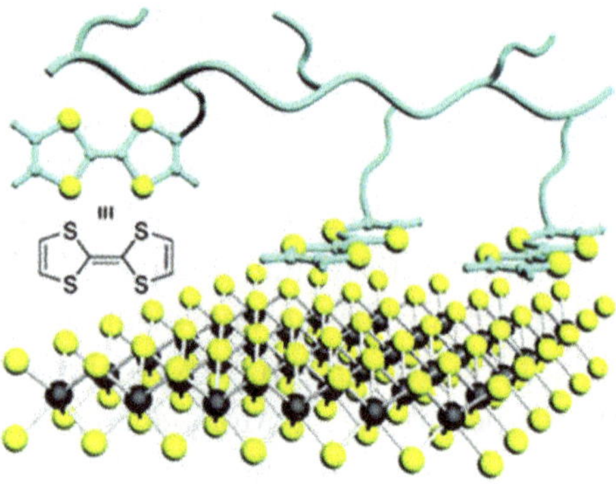

(a)

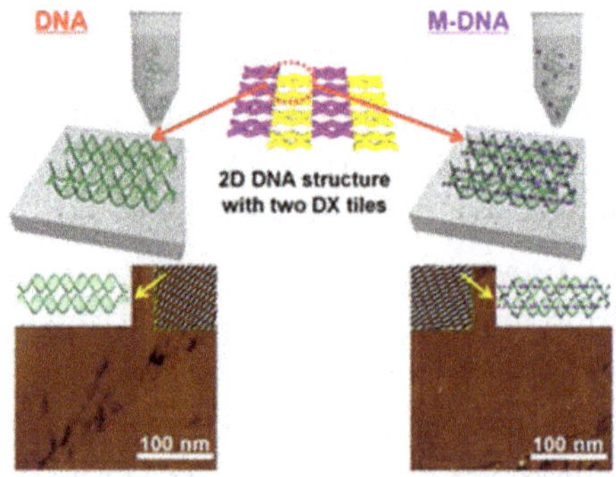

(b)

Fig. 5. (Color online) (a) Schematic illustration of the physisorption of TTF on MoS_2. Reprinted with permission from Ref. 45. Copyright 2016 Royal Society of Chemistry. (b) Schematic diagrams of the fabrication of DNA and M-DNA DX structures, AFM images of pristine DNA and M-DNA with 4 mM Cu^{2+}. Reprinted with permission from Ref. 46. Copyright 2014 American Chemical Society.

processes, other approaches for thin film formation on TMDCs have been reported as well.[8,48–50]

2.2.2. *Other condensed phase techniques*

In this section, we give an overview of other condensed phase molecular functionalization techniques for functionalizing TMDCs that do not fall under the category of self-assembly.

Polymers are a common dielectric material for all kinds of semiconductors and have also been popular for applications in TMDC devices. Polymers are also used to encapsulate functional molecules at the TMDCs surface. The charged nature of TMDCs in solution makes them suitable candidates for multilayer assembly through techniques like Langmuir–Blodgett (LB) films[51] and layer-by-layer (LbL) assembly.[52] Xie and coworkers prepared LbL films from a positively charged hybrid material composed by MoS_2 and poly(diallyl dimethyl ammonium chloride), and negatively charged poly(acrylic acid) (PAA). This material showed a good performance as functional separator in Li-S batteries.[53] MoS_2 and PAA in the membrane restrains polysulfide shuttling by impeding molecular diffusion and chemisorption. The membrane further improves cycle stability of Li-S batteries by regulating a homogenous Li-ion flux and hence suppressing Li dendrite formation. A similar approach was taken by Kim and co-workers who reported the LbL assembly of multilayered MoS_2 films spaced by polymer films (Figs. 6(a)–6(d)).[54] To this aim, the intrinsic negative charge of chemically exfoliated MoS_2 (ce-MoS_2) was used for the LbL deposition rather than introducing charge through a polymer hybrid. Polyallylamine hydrochloride (PAH), a positively charged polymer was used in the process as its counter-layer and a spacer for the next layer. The spacing could be increased by an interlayer of a negatively charged polymer, polystyrene sulfonate (PSS), resulting in a PAH/PSS/PAH spacer. Apart from acting as spacer, PAH also acts as a *p*-type dopant. They observed that films containing the three-layer spacer show improved PL intensity over a SL spacer as a result of better quenching of the interlayer coupling. Similar results were also obtained by Wang and co-workers where they observed a PL enhancement and preservation of direct bandgap in twisted bilayer MoS_2.[55] Residual poly(methyl methacrylate) (PMMA) present between layers from the transfer process here preserves the direct bandgap of MoS_2. MoS_2 encapsulated with a thin film of PMMA has been shown to act as ambipolar transistor.[56] A commercial polymer, cyclic transparent optical polymer, abbreviated as CYTOP, is an amorphous fluoropolymer with environmental stability and high

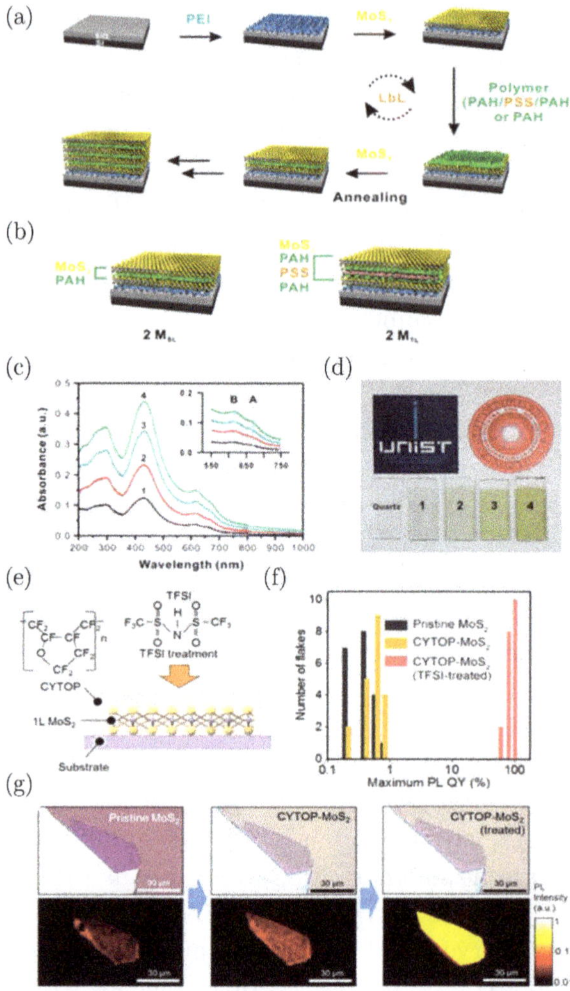

Fig. 6. (Color online) (a) Schematic representation of the layer-by-layer assembly of MoS_2 with polymer spacers. (b) Scheme showing the $2M_{SL}$ and $2M_{TL}$. (c) UV-Vis absorbance spectra of the growth of MoS_2 multilayers. (d) Photographs of MoS_2 multilayer films deposited on quartz slides with different number of layers. Reprinted with permission from Ref. 54. Copyright 2014 American Chemical Society. (e) Schematic of sample encapsulation/passivation showing a monolayer of MoS_2 coated by CYTOP and treated by TFSI. (f) Histogram showing the PL quantum yield of 20 flakes after exfoliation (black), coating with CYTOP (yellow) and treatment by TFSI. (g) PL images and corresponding optical micrographs for the three different cases. Reprinted with permission from Ref. 58. Copyright 2017 American Chemical Society.

optical transparency that is another interesting material. By virtue of its inertness, the CYTOP interlayer prevents unintentional doping of TMDCs by external species and substrate through Pauli blocking and carrier screening.[57] It provides an ideal platform to study the intrinsic electrical and optical properties of 2D materials. MoS_2 and WS_2 displayed highly stable, near unity PL quantum yield after encapsulation with CYTOP in combination with treatment of superacid bis(trifluoromethane)sulfonamide to suppress defects (Figs. 6(e)–6(g)).[58]

Other important and interesting contributions regarding the polymer-stabilized exfoliation of MoS_2 and other TMDCs using polyvinylpyrrolidone (PVP) have been reported.[59] Zhang and coworkers described for the first time the use of PVP as stabilizing agent.[60] They showed a facile method to prepare stable exfoliated few-layer MoS_2 in aqueous as well as many organic solvents. They expanded this approach to other TMDCs and achieved a narrow size distribution[61] improving their method by using the previously successful combination of grinding and ultrasonication.[62] Based on these results, Wang and coworkers obtained the highest yield of a polymer-based exfoliation using a coupled ultrasonication-milling process.[63]

Apart from PVP, biopolymers like proteins bovine serum albumin (BSA)[64,65] and gelatin[66] have also been used to stabilize exfoliated TMDCs in aqueous solutions. Hydrophobins, which are small proteins made of few hundreds of amino acids, were also used to exfoliate and stabilize TMDCs. These proteins interact through hydrophobic interactions between their domains and the surface. The hydrophilic part of these proteins forms an interacting link between the particles and aqueous solution.[67] Apart from stabilizing these colloids, the protein–TMDC interactions also improve the physiological stability and biocompatibility of these particles. The above properties combined with high absorption of TMDC particles in near-IR range have encouraged researchers to use functionalized TMDCs for various therapeutical applications including photothermal therapy of cancers and photoimaging of tumours.[65,68]

This hydrophobic interaction between TMDC and proteins was also exploited for biosensing of prostate specific antigen (PSA).

Anti-PSA antibodies were dispersed on the surface of an MoS_2 transistor sensor. When exposed to different concentrations of PSA, increasing antibody–antigen interaction emerges in the n-type device as a corresponding decrease in off-state current under negative bias.[69] The sensor could detect concentrations as low as 1 pg/mL, which is a few orders of magnitude below the clinical cut-off level for PSA.

Other studies of polymer–TMDC interactions report properties such as increased photoresponsivity,[70] and improved mechanical and thermal stability[71] of MoS_2 by polymer encapsulation.

For finishing this section, we should remark that even if we have focused the discussion on the relatively popular group of TMDCs, there is important literature concerning other 2D materials including hBN,[72,73] phosphorene,[74] or MXenes.[75] The next section will follow discussions on covalent functionalization protocols for 2D materials.

3. Covalent Functionalization of 2D Materials

The covalent functionalization of 2D materials has been deeply explored for graphene and related carbon-based materials with special focus on the use of highly reactive radicals (diazonium compounds, carbenes, etc.) or cycloaddition reactions.[76,77] Nevertheless, although some of these methodologies can be extended to the modification of other 2D materials, in some cases, their specific reactivities require new methods for their covalent functionalization.[77] Taking the MoS_2 case as example, the most obvious option for covalent functionalization would be the formation of a covalent bond between the molecular modifier and the sulfur atoms of the surface. Indeed, the first reported chemical modification using covalent chemistry on this material consisted of the ligand conjugation of a thiol moiety on defects formed during the chemical exfoliation of bulk MoS_2.[78] After the publication of this report, the chemical functionalization has been widely explored for the highly reactive metallic phase 1T-MoS_2 and MoS_2 with a considerable number of defects and sulfur vacancies. Nevertheless, the application of these approaches to functionalize pristine surfaces is not straightforward due to the inherent inertness of the basal plane of

MoS_2. Therefore, new strategies for the robust functionalization of the basal plane of 2H-MoS_2 have been proposed, for instance using electrophilic compounds such as maleimides or organohalides. In this section, we summarize the use of different precursors for the modification of defective/pristine TMDCs: (1) metal complexes to generate coordinate bonds, (2) sulfur-containing species, (3) organohalides, (4) radicals generated through diazonium chemistry, and (5) maleimides (Fig. 7).

3.1. *Metal complexes*

Metal complexes can be used as a bridge for the functionalization of layered TMDCs such as MoS_2. This strategy was previously shown to be successful for the modification of graphene oxide.[79] Later, it was extended to MoS_2 nanosheets in order to form MoS_2 paper cross-linked with polymeric ligands by adding divalent metal ions.[80] The basis of this methodology is the interaction of the divalent metal ion (Cu^{2+} ions) with the sulfur atoms on the surface and the simultaneous coordination of this ion to the carboxylate ($C=O$) groups of the

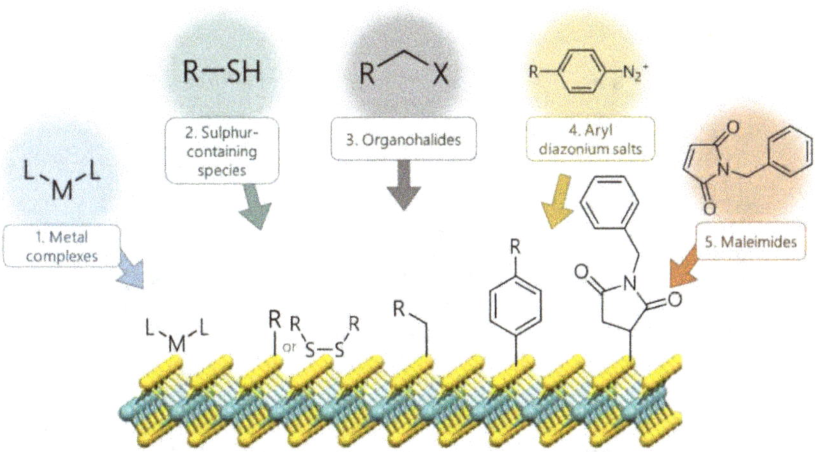

Fig. 7. (Color online) Schematic representation of the different molecular chemistry approaches that have been proposed in the last years for covalent functionalization of 2D TMDCs.

polymer ligand. The formation of this bond was confirmed by several techniques such as UV-Vis, FTIR or XPS. With this method, MoS_2/PMMA nanocomposites with enhanced mechanical properties were obtained.

A systematic study of the interaction of liquid-exfoliated MoS_2 sheets with acetate salts with the general formula $M(OAc)_2$ (M = Ni, Cu, Zn; OAc = acetate) was carried out by Backes and coworkers[81] (Fig. 8(a)). The sheets of MoS_2 were exfoliated using 2-propanol (IPA) as a solvent which stabilizes the 2H-MoS_2 phase. The formation of a coordination bond of the metal center with the sulfur could be confirmed by XPS measurements. The appearance of a second component in the S 2p core level spectra at higher energies in the functionalized materials is in line with a different chemical environment due to the coordination of the metal ion. The degree of

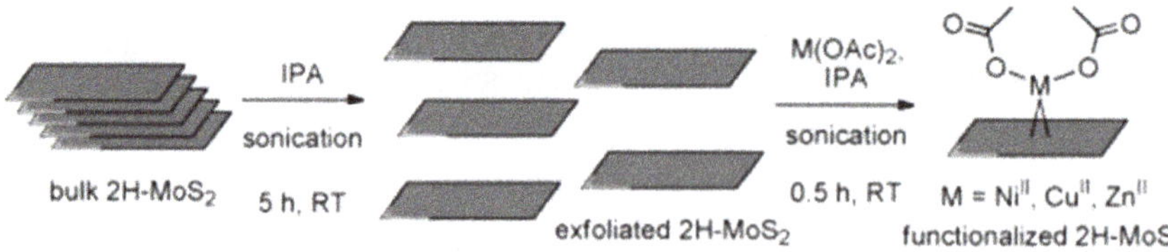

(a)

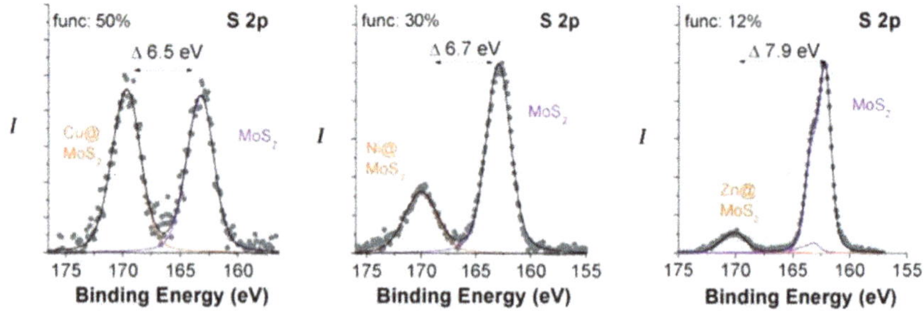

(b) (c) (d)

Fig. 8. (Color online) (a) Schematic representation of the exfoliation and functionalization of 2H-MoS_2 using metal acetate salts (IPA = 2 − *propanol*). (b)–(d) Fitted XPS high resolution spectra for the S2p core level spectra using different metals in the acetate salt showing the different reactivity found depending on the metal. Adapted with permission from Ref. 81. Copyright 2015 Wiley-VCH.

functionalization depends on the chemical nature of the metal ion due to the different affinity of the ions to bind the sulfur atom. The higher efficiency is obtained in the case of Cu (50% of the S of the surface for 2H-MoS_2-$Cu(OAc)_2$) when compared to Ni (30%) and Zn (10%) (Figs. 8(b)–8(d)). The main advantage of this functionalization method is the improvement in the dispersibility of the material in a wide range of solvents with different polarities.

3.2. *Sulphur-containing species*

The use of thiols to functionalize metal surfaces with SAMs has been thoroughly described in the literature.[82] This approach is based on the formation of a covalent bond between the sulphur atom and the metallic atom of the surface due to a specific molecule–substrate bond. The absence of this specific interaction impedes that thiols can be used in the covalent functionalization of 2D materials such as graphene, although some authors have suggested the formation of this covalent bond on graphite surfaces as well.[83] However, in the case of MoS_2, the functionalization can be achieved by conjugation of thiols at the sulphur vacancy sites or defects.

The early experiments using thiol chemistry for the modification of MoS_2 were carried out on ce-MoS_2 flakes.[78] The molecular functionalization of the surface was evaluated through the shift in the zeta (ζ)-potential. The same ligand with and without the thiol moieties shows different shifts in the ζ-potential, confirming the role of the thiol group. The material was further characterized by FTIR to check the presence of the functional groups on the surface. The electronic properties of the MoS_2 flakes functionalized with thiol molecules were explored as well.[84] Modified surfaces were characterized using XPS, photoelectron spectroscopy in air (PESA) and PL experiments. The results show a shift in the conduction and the valence bands for the functionalized materials, which was explained by considering the mild Lewis base character of the thiols, leading to an overall electron donating effect. PESA experiments showed a cathodic shift of both bands up to 500 meV, although no changes in the bandgap were observed. XPS confirmed changes in the Fermi level depending on the chemical nature of the thiol used (aromatic or aliphatic). Although the interaction between the thiol and the MoS_2 surface was assumed to be covalent, some doubts about the real nature of this bond emerged later.[85] In this case, the functionalization was targeted with the use of cysteine. The results showed that the oxidation of cysteine to the disulphide cysteine during the functionalization can take place on the MoS_2 surface. The presence of this oxidized molecule on the modified surface was confirmed through XPS and diffuse reflectance infrared spectroscopy (DRIFT) spectra. Both techniques showed the characteristic signals for the disulphide product. In a recent theoretical study, this discrepancy was explored in detail.[86] The calculations showed that the dimerization process which leads to the physisorbed material is thermodynamically favored when S adatoms are present on the surface. However, when there are sulphur vacancies, these disulphide species can be reduced back to the thiol form and anchor on the surface defect to repair these vacancies.

Recently, new insights into the functionalization of defective MoS_2 surfaces using organic thiols have been gained.[87] In this paper, the authors argued that the attachment of organic moieties such as thiophenols occurs on the defects of the surface forming a covalent bond. By tuning the functional group of the molecule, the functionalization density can be controlled. In this sense, the highest degree of functionalization is obtained for the NO_2-substituted molecules and the lower with the NH_2-sustitued ones. These results show the influence of the functional group on the efficiency of the modification and how the choice of the modifier can be used to control the number of vacancies or defects on the material. Interestingly, a spectroscopic indicator to monitor and quantify the degree of functionalization was discovered. The intensity of the 2LA(M) normalized peak to A1g peak correlates to the degree of functionalization and, therefore to the number of defects, being of special interest in the field of the molecular functionalization of MoS_2 (Fig. 9, left panel).

The effect of the incorporation of thiol moieties on the electrocatalytic performance of MoS_2 surfaces has been studied as

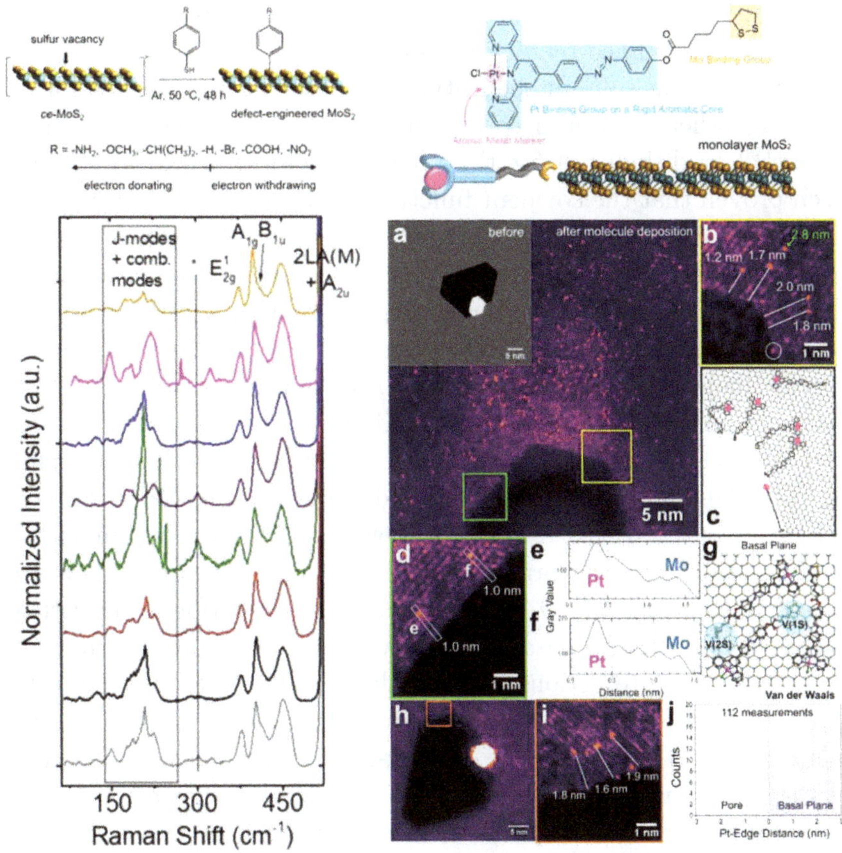

Fig. 9. (Color online) (left panel) Schematic illustration of the functionalization of MoS_2 sheets using thiophenols by using defect-engineering. Raman spectra corresponding to the functionalized material (gray = non-functionalized; black = NO_2; red = COOH; green = Br; purple = H; blue = Pro; magenta = OMe and orange = NH_2). Reprinted with permission from Ref. 87. Copyright 2020 Wiley-VCH. (right panel) ADF-STEM images of Pt-marked molecules and detailed study on how the molecule anchors to the surface. Reprinted with permission from Ref. 90. Copyright 2020 American Chemical Society.

well.[88] The results prove that the surface functionalization of the basal plane of monolayers of MoS_2 leads to a decrease in the number of active sites for the hydrogen evolution reaction (HER), showing a passivation of the material toward this reaction. In the same paper, the possibility of building heterostructures in a second step after the modification was demonstrated by incorporating PbSe quantum dots on a surface modified by dithiol molecules.

Regarding the sulphur chemistry, some studies show the possibility of using other molecules containing sulphur units such as dithiolanes[89,90] or dithiolenes for the functionalization of MoS_2.[91] It has been proven that the covalent functionalization using 1,2-dithiolanes derivatives occurs preferentially at the sheet edges of MoS_2.[89] Recent results using annular dark-field scanning transmission electron microscopy (ADF-STEM) have helped to gain insight into the attachment of these units to the surface.[90] For these experiments, a specific molecule containing a single transition metal atom (Pt) was designed in order to track the binding of the molecules on the material. The results confirmed that the covalent attachment occurs to the edges and the surface around pores on the MoS_2 surface. Importantly, with this technique, it was possible to determine the conformation and the orientation of the molecules once they are bound to the edge (Fig. 9, right panel). The interaction of the rest of the molecule with the basal plane of MoS_2 has been proved, showing van der Waals interactions between the aromatic core and the basal plane of the material. These results show the influence of the van der Waals interactions with the material after the covalent bond is established. The covalent attachment of dithiolanes has been used as a platform for further post-functionalization.[92] In this case, the surface of exfoliated MoS_2 and WS_2 is functionalized with 1,2-dithiolane-BOC followed by acidic deprotection of the covalently anchored organic moieties. After this second step, the surface contains charged ammonium units exposed to the solution which can interact with porphyrin molecules *via* electrostatic interactions due to the negative charge of the carboxylate-substituted porphyrin. Time-resolved PL experiments have shown that the transduction of energy from the photoexcited porphyrin to the exfoliated material takes place in a donor–acceptor process.

Thiol chemistry has been explored as an option to form robust interfaces between other 2D materials and complex molecules such as porphyrins, with a well-known activity toward electro- and photocatalysis. Thiolated tetraphenyl porphyrins have been attached covalently to the surface of chemically exfoliated $MoSe_2$ nanosheets.[93] The formation of a covalent bond was confirmed by using XPS, where the formation of a sulfo-selenide bridge can be detected. To fully understand the formation of this bond between the thiolated porphyrin and the $MoSe_2$ additional DFT calculations were carried out. In these calculations, only the adsorption of the thiolated alkyl chain (propanethiol) was considered given that ionic and van der Waals interactions are weak and the physisorbed material is removed after the washing protocol. Results show that the adsorption of the thiol on the basal plane of the material does not lead to an attractive interaction, which means no covalent bond is formed. Nevertheless, at the edge of the $MoSe_2$, the energy of adsorption is $-28.8\,\text{kcal mol}^{-1}$, a much favored interaction and, consequently, the formation of a covalent bond is possible. The functionalized nanosheets were further evaluated for their use in the electrocatalysis and photocatalysis of a model reaction such as the HER reaction. The hybrid materials showed an improvement in the photoelectrocatalytic activity with a shift in the onset potential for the HER reaction and a relative light enhancement (% of overpotential decrease due to the light illumination) four times higher than for unmodified $MoSe_2$.

3.3. *Organohalides*

The functionalization of monolayers of different TMDCs (MoS_2, WS_2, and $MoSe_2$) can be achieved with the use of organohalides.[10] This reaction can proceed spontaneously if the highly reactive 1T phase of these materials, obtained after their treatment with *n*-BuLi from the 1H phase, is used and it is based on the direct electron transfer between this 1T phase and the organohalide molecule. In this way, advantage was taken of the negative charge of the material after the chemical exfoliation with *n*-BuLi. The surface was treated with methyl iodide and 2-iodoacetimide. Chhowalla *et al.* indicated that the functionalization using this approach only occurs in the 1T phase of the material and with an extension of around 30 atomic percentage (at%) relative to the transition metal content. Interestingly, although scanning transmission electron microscopy (STEM) experiments show that the octahedral coordination of Mo atoms in the 1T phase was not altered, the electronic properties of this phase are profoundly affected from metallic to semiconducting. These changes in the optoelectronic properties are explained by an opening of the bandgap which gives rise to the tunable PL of the material. Authors show that this strategy is applicable to CVD monolayer MoS_2 with similar results. Raman and XPS results show that after the functionalization it is possible to recover the 1H phase with annealing at $T \geq 300°C$ while maintaining the functional group on the surface as revealed by ATR-FTIR measurements.

The use of different para-substituted iodobenzenes has been explored as well for the functionalization of MoS_2 and $MoSe_2$ nanosheets.[94] As mentioned before, the reaction takes place in the 1T metallic phase of the materials leading to the cleavage of the C–I bond and the formation of a C–S or C–Se bond. In this electron rich phase, the degree of functionalization can reach values around 24% for electron withdrawing substituents. Moreover, more complex molecules such pyrenes, coumarin or porphyrin derivatives were tested, showing the versatility of this protocol. The molecular functionalization of the 2H phase of both materials has been explored in a following step, showing a much lower reactivity and the need of using a Pd^0 catalyst to promote the reaction. DFT calculations shed light on the interaction of the molecule with the different phases. For all the functional groups analyzed (–OCH_3, –H, and –NO_2), the values for the binding energy in the 2H phase are smaller than for the 1T phase. These results confirm that the functionalization using organohalides is much more efficient and stronger in the 1T phase. The theoretical study of the band structure and the projected density of states show that after the functionalization, the valence band and conduction bands of the 1T phase are shifted and a gap opens between them. Moreover, it has been proven that the work function strongly changes for the 1T phase of MoS_2. The work function for the pristine MoS_2 surface (5.09 eV) is reduced when $R = -H$ (4.46 eV) and $R = -OCH_3$ (4.67 eV) are used. In contrast, the use of $R = -NO_2$ implies an increase in the value of the work function to 5.22 eV due to the electron-withdrawing nature of this functional group.

As mentioned earlier, the covalent functionalization of the 1T' phase of the MoS_2 and related materials can be achieved by using alkyl halides such as methyl iodide. However, the main disadvantage is the relatively low efficiency that is obtained (around the 25–30%). This low efficiency can be explained in terms of the dependence on the negative charge of the material for the direct electron transfer. To improve the grafting density using alkyl halides a new approach using the activation of the molecules by reducing species has been proposed.[95] The reducing agents employed consisted of one-electron metallocene (nickelocene, cobaltocene, etc.) which can react with the corresponding alkyl halide to produce a radical that binds to the surface. XPS and ATR-FTIR experiments confirm the presence of the alkyl chains in the modified material. The results of XPS can be used as an estimation of the efficiency of the modification and the surface coverage. Importantly, these data show that the reductant-activated functionalization is highly efficient when compared with the modification without reducing agent. In the best case, the coverage can increase up to 70% with a significant increase of 1.5–2 times compared to the one obtained previously (Fig. 10). The influence of the alkyl chain and the leaving group was evaluated as well. It has been shown that the best results are obtained for iodomethane. The grafting is not efficient when chloride is used as leaving group or when the molecule has a high steric hindrance such as for secondary and tertiary alkyl iodides. The use of reducing agents in combination with chemical exfoliation can be seen as a promising way of attaching alkyl chains by means of alkyl iodides to several TMDC materials.

3.4. *Aryl diazonium salts*

Aryl diazonium salts have emerged as the preferred reagents for the functionalization of a wide number of materials due to their high efficiency and versatility. The reaction consists of the reduction of the diazonium ions through a homolytic process leading to the formation of aryl radicals and the release of nitrogen. These aryl radicals are highly reactive and react with the surface forming a covalent bond

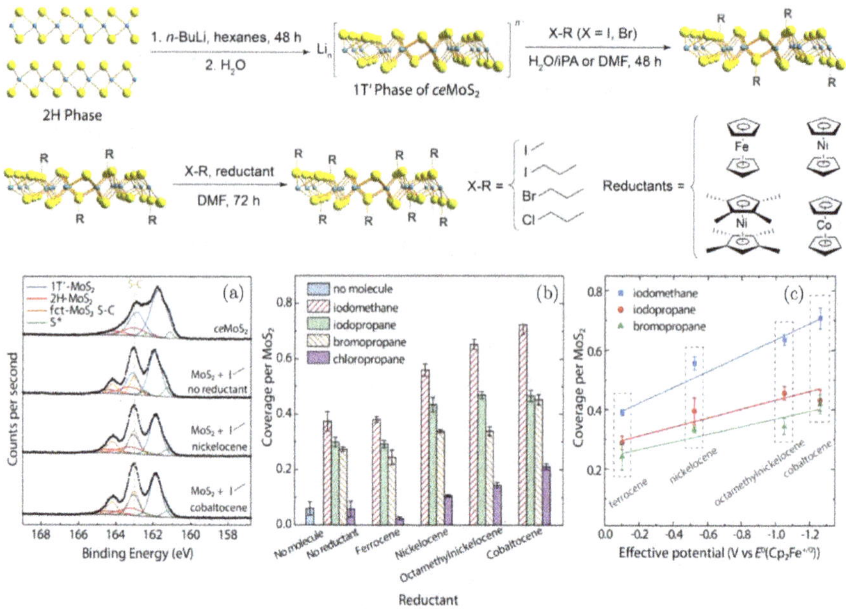

Fig. 10. (Color online) The functionalized 1T'-MoS_2 nanosheets are obtained by exfoliation in water and functionalized using alkyl halides (X-R). The functionalization can be promoted by adding one-electron metallocene reductants. (a) XPS high resolution spectra of the S2p region for the quantification of the coverage using the peak areas corresponding to the S atoms. (b) Calculated coverages for different alkyl halides using diverse reducing agents. (c) Coverage per MoS_2 unit depending on the effective potential of the reductants. Adapted with permission from Ref. 95. Copyright 2020 American Chemical Society.

between the modifier and the surface. Regarding the modification of 2D materials, diazonium ions have been shown to be very suitable for the covalent modification of both surface-supported graphene and graphene dispersions.[96–97] The functionalization of these materials has been carried out by spontaneous decomposition of the aryl diazonium salt or by electrochemical or chemical activation. Raman spectroscopy has been chosen in most of the cases as a sensitive technique to detect the functionalization of graphene due to the characteristic D band in the Raman spectra when the modification takes place. Given their success for the functionalization of graphene, the use of aryl diazonium salts has been extended for the functionalization of other 2D materials such as MoS_2, WS_2, $MoSe_2$, WSe_2,[98] black phosporous (BP)[99] or h-BN.[100]

The molecular functionalization of MoS_2 using diazonium salts emerges as an alternative that does not require the presence of defects on the surface and can be used, as a consequence, to modify pristine surfaces of MoS_2. Initial studies were done using ce-MoS_2 by intercalation of *n*-butyl lithium, dispersion in water and ultrasonication.[101] Once the negatively charged MoS_2 flakes are exfoliated, the aryl diazonium ions are incorporated and act as weak electrophiles. Interestingly, around 10–20% of the surface sulphur atoms are involved in the formation of a C–S bond. The covalent modification has been confirmed by TGA analysis, showing a mass loss in the temperature range of 220–450°C related to strongly bonded material. Indeed, it has been demonstrated that the annealing of the material at 350°C restores the pristine 2H-MoS_2. DFT calculations were performed to evaluate the C–S bond. The value for the reaction energy at low coverages ($E_{\text{react}} = 2.33\,\text{eV}$) and the analysis of the Bader charges showing a partial negative charge on the S atoms confirm that, although physisorption cannot be excluded, the covalent reaction is clearly favored.

The aforementioned functionalization protocol has been applied to pristine MoS_2 obtained by mechanical exfoliation from bulk MoS_2.[102] The coverage dependence on both numbers or layers and the defects on the surface were evaluated. The AFM results show that if a pristine surface is used, the surface coverage of the deposited aryl diazonium salts is higher in the case of a bilayer of MoS_2 when compared to the monolayer (Figs. 11(a)–11(d)). Some additional experiments in combination with DFT calculations showed that the S vacancies or defects can serve as nucleation centers in the initial stage of the functionalization. These calculations estimate that when an aryl group is attached on the surface, there is a charge reorganization and as a result, an increase in the reactivity of the surrounding area. It can be concluded that the growth of the layers is based on a cooperative surface reaction mechanism. Importantly, the semiconducting properties of the material are not disrupted as shown by the PL measurements. The reaction kinetics of this spontaneous functionalization of MoS_2 using aryl diazonium salts was explored in later experiments using 4-nitrobenzene diazonium (4-NBD) tetrafluoroborate.[103] Different adsorption isotherms were used to fit the dependence of the coverage on both time and concentration, extracted from AFM images (Figs. 11(e) and 11(f)). The results show that, studying the AFM images at different concentrations, the Freundlich and Temkin isotherms are the most suitable ones to describe the reaction kinetics of this system which includes both the inhomogeneity of the MoS_2 surface in terms of energy and the indirect adsorbate–adsorbate interactions with the already grafted nitrophenyl groups. Moreover, the dependence of the coverage with time shows that reaction kinetics can be described with a pseudo-second order model, confirming the interaction between 4-NBD moieties.

As commented earlier, the functionalization using spontaneous decomposition of aryl diazonium ions can be achieved. Nevertheless, although the efficiency of this process is high, it induces the growth of a polymer on the surface in an uncontrolled fashion. For conductive surfaces, the activation and control of the decomposition process can be achieved by means of electrochemical techniques. However, the semiconducting behavior of 2H-MoS_2 makes this approach extremely difficult due to the highly-impeded electron transfer through the material for the molecules to react. Recently, these difficulties could be overcome, allowing the electrochemical functionalization of MoS_2 surfaces with 4-bromobenzene diazonium tetrafluoroborate (4-BBDT) by means of cyclic voltammetry.[104] The deposition of a monolayer of 2H-MoS_2 on platinum allows the electron transfer to occur through an electron tunneling mechanism, promoting the reduction of the diazonium ions at the interface. Interestingly, after the formation of the C–S bond on the surface, the electronic structure of the material can be tuned. PL measurements show that the electrochemical process promotes the functionalization of the material but creates defects (sulphur vacancies) on the surface as well. The annealing of the functionalized samples up to 400°C shows the recovery of the PL intensity. A possible mechanism to explain these changes in the optoelectronic properties is shown in Figs. 11(g)–11(i).

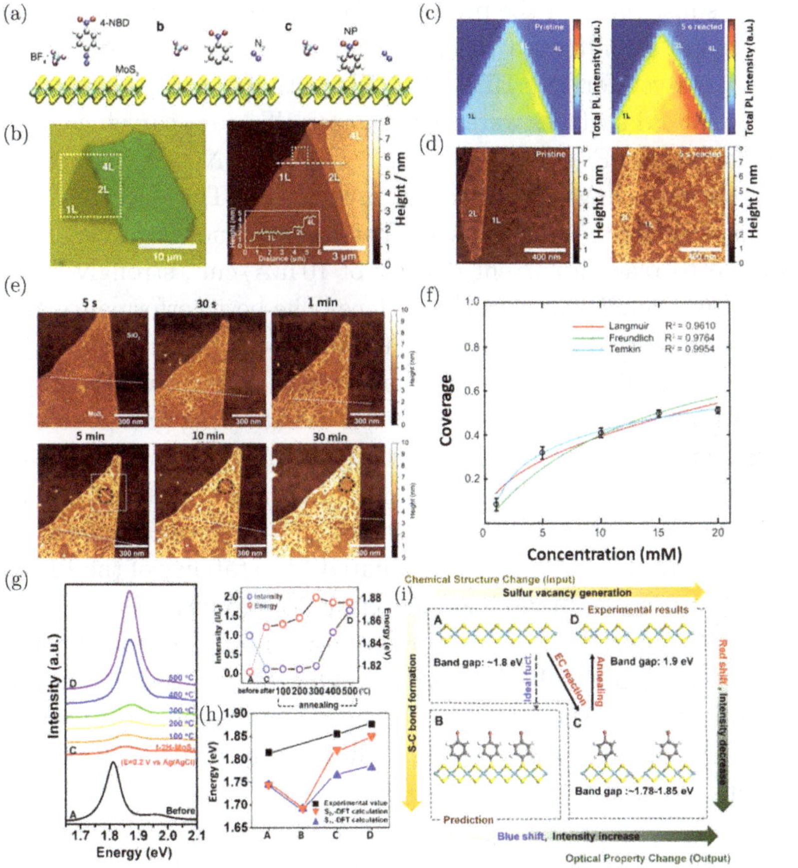

Fig. 11. (Color online) (a) Scheme of the functionalization of MoS_2 with NBD, (b) optical microscope images of mechanically exfoliated flakes and corresponding AFM image of the marked area, (c) PL maps before (left) and after (right) the functionalization process, (d) AFM images before (left) and after (right) the functionalization of the surface. Reproduced with permission from Ref. 102. Copyright 2018 American Chemical Society. (e) AFM images showing the evolution of the coverage on the surface with time for a 1 mM NBD solution, (f) Adsorption isotherms for the spontaneous adsorption of NBD on MoS_2. Reproduced with permission from Ref. 103. Copyright 2019 American Chemical Society. (g) PL spectra for functionalized 2H-MoS_2 after annealing at different temperatures and plot of the corresponding PL intensities and peak position, (h) Theoretical and experimental bandgap values for the different situations shown in scheme, (i) scheme of the mechanism showing the influence of the vacancies and the changes on the optical properties of the material. Reproduced with permission from Ref. 104. Copyright 2020 American Chemical Society.

The substituent in the precursor for the covalent modification has been found to influence the performance of the 1T-MoS_2 phase as electrocatalyst for the HER.[105] HER activity and stability of the functionalized surfaces were analyzed for 1T-MoS_2 carrying different functionalities (Et_2N^-, MeO–, Br^-, Cl_2– and NO_2–) with different electron donating/withdrawing properties. The electrocatalytic activity, evaluated in terms of Tafel slope and the overpotential required to reach a current density of $10\,mA/cm^2$, strongly depends on the functional group on the surface. The best performance among the functionalized substrates was found for the Et_2NPh-MoS_2 (Tafel slope: $75 \pm 3\,mV/dec$) case, showing a small decrease when compared to the bulk 1T-MoS_2 phase. Electron-withdrawing groups lead to a decrease in the electrocatalytic activity with a consequent increase of the Tafel slope (Tafel slope for Cl_2Ph-$MoS_2 = 213 \pm 33\,mV/dec$) and the overpotential. When the stability of the material under HER conditions of the material was evaluated, the stability of the 1T-MoS_2 phase in the Et_2NPh modified materials was confirmed.

These materials can be post-functionalized to form complex organic–inorganic hybrid interfaces. One of the easiest and more versatile ways is the use of click chemistry on the already grafted molecules on the surface. Aryl diazonium chemistry has been exploited for the modification of MoS_2 nanosheets with phenylazido moieties.[106] In a second step, through a copper-mediated azide-alkyne cycloaddition, single terminal acetylenes and defined binary acetylene mixtures are integrated to the surface. This strategy has been proved to be efficient to anchor an extensive number of functional groups on the surface and even more than one functional group when alkyne mixtures are used. A battery of techniques has been used for the characterization of the modified materials (SEM microscopy, Raman, XPS, UV-Vis, etc.). Interestingly, fluorescent labeling of the material has been achieved with this method which allows the tracking of the functionalization by fluorescence lifetime imaging (FLIM). The fluorescence images show a good distribution of the dye through the sample, confirming the efficiency of the grafting of the diazonium ions and the successful coupling of the alkyne cycloaddition. These modifications have been confirmed additionally by using XPS measurements where the specific signals for the functional groups introduced can be found. The post-synthetic functionalization of aryl diazonium-based platforms represent a simple and versatile method for the fabrication of multifunctional 2D inorganic materials.

Aryl diazonium salts have been employed for the functionalization of other TMDCs and pnictogen chalcogenides (PC).[98] The possibility of expanding the use of diazonium ions to a broad number of different materials has been proved. In all the cases, the functionalized materials have been characterized by spectroscopic and microscopic methods showing the efficiency of the molecular modification. FTIR and XPS confirms the presence of the molecule on the surface and the covalent attachment. Raman spectroscopy experiments have shown that the crystal structure of the material remains unaltered after the functionalization. It is remarkable that this chemistry works in both water and solution-based materials representing a valuable tool for the functionalization of different 2D materials. In fact, other popular 2D layered materials such as BP and h-BN have been modified using this approach. In the case of BP, this covalent functionalization allows the passivation of the material.[99] This passivation is of great importance due to the well-known chemical instability of BP. It avoids the surface degradation under ambient conditions. This can be analyzed by observing the morphological changes on the surface with time. It has been proved that after the modification, confirmed by the formation of P–C bonds in XPS, there is a clear difference in the evolution of the morphology of the surface with time, as confirmed by AFM images. When the surface is not modified, after 10 days of exposure of the samples to ambient conditions, the appearance of large morphological features appear on the surface. In contrast, when NBD groups are anchored on the surface, the sample morphology does not show differences when compared to the images for fresh samples. The surface roughness through AFM experiments for the non-modified and modified flakes confirm the observations. For the NBD modified samples, the invariance of the morphology is maintained for at least 25 days. The functionalization has been proved to produce significant changes in the electronic properties of the exfoliated BP. Especially at low degrees of functionalization, these changes result in an enhanced semiconductor behavior, with an improvement in the hole mobility and on/off current ratio in FETs devices.

3.5. *Maleimides*

Maleimides are electrophilic compounds with a well-known high selectivity toward thiols.[107] These chemical compounds can be conjugated to sulfhydryl groups, a reaction widely used for the covalent bonding of proteins. With this knowledge, it seems that these molecules are suitable for the functionalization of 2D materials containing sulphur in its structure. Recently, a mild chemistry based on the use of maleimides has been proposed for the covalent modification of 2H-MoS_2 and WS_2. This reaction is based on the soft nucleophilicity of the sulphur to react with a soft electrophile such as the maleimide compounds via a Michael addition.[108] The advantages of this method rely on the use of mild conditions, the possibility to functionalize pristine surfaces, the short times needed for the complete functionalization and its versatility. The surface coverage has been calculated using the S 2p (33%) and the Mo 3d (25%) regions in XPS. These values are in good agreement with those estimated using TGA measurements (21%) and are similar to the ones obtained for other TMDCs such as WS_2. As a consequence, with this protocol, it is possible to obtain a highly efficient covalent attachment without the use of harsh conditions or the transformation of the 2H-MoS_2 phase to the highly-reactive 1T-MoS_2 one. In fact, after the nanosheet functionalization, HR-TEM measurements reveal that the 2H-MoS_2 phase remains the dominant phase, proving this method as a suitable tool for the modification of the basal plane of TMDC.

After this first proof of concept, Pérez and coworkers evaluated the performance of different precursors based on this maleimide covalent chemistry.[109] They showed that the reaction can undergo two different pathways if a base such as Et_3N is used giving rise to polymeric growth (in presence of base) or limited growth to the monolayer (in absence of base) (Fig. 12). Additionally, the versatility of the reaction was tested by using different precursors demonstrating

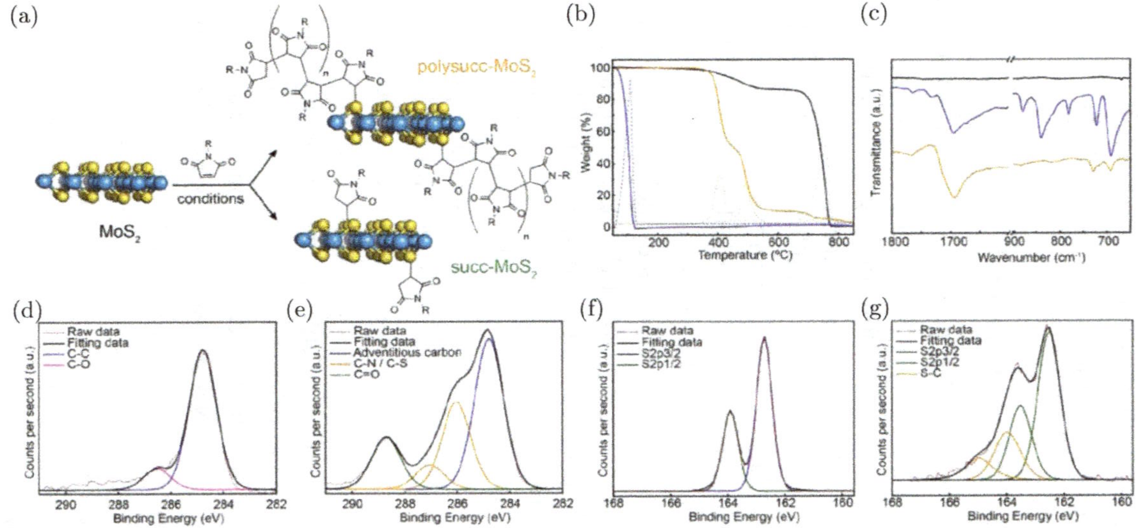

Fig. 12. (Color online) (a) Scheme of the functionalization of MoS_2 with maleimides through two different routes. (b) TGA of exfoliated flakes (black) and functionalized ones (Bn-poly-succ-MoS_2 = orange; Bn-mal = blue); (c) ATR-FTIR of the same materials. (d) and (e) XPS spectra for C core levels for bare and functionalized flakes, respectively, and (f) and (g) XPS spectra for the S core levels of the same samples. Reproduced with permission from Ref. 109. Copyright 2020 Wiley VCH.

that the maleimide chemistry is suitable to introduce different functional groups on the basal plane of 2H-TMDCs. Due to the commercially available reagents to produce protein–polymer conjugates based on maleimides, this chemistry is an interesting and promising tool for their incorporation into these TMDC materials.

4. Conclusions and Future Perspectives

Chemical functionalization of 2D materials is a field with plenty of possibilities and with a rapid growth in the last few years. When compared to the functionalization of graphene surfaces, the modification of these materials using molecules has been much less explored. Taking into account the availability of molecular building blocks that can be attached to TMDCs, there is plenty of room for the fundamental understanding of the chemical nature of these new interfaces and the changes in the properties of the functionalized materials. The future challenges in terms of molecular functionalization of 2D materials lie in the control of the structure and properties at the molecular level. The possibility of post-functionalization of the surfaces with covalently incorporated enzymes, antibodies or DNA can lead to the production of bio-functional surfaces with relevance in biomedicine. The use of MoS_2 in this field is especially promising due to its biocompatibility.[110,111] Additionally, control of the nature and the level of doping or bandgap through molecular modification can bring new functional materials in fields such as nanoelectronics or electrocatalysis. From the fundamental point of view, new insights in the functionalization of the 1T-MoS_2 phase are expected in the next years. To date, most studies have been focused on the thermodynamically stable 2H-MoS_2 phase. Nevertheless, regarding other prototypes of MoS_2, the 1T-MoS_2 has interesting properties as electrocatalyst and the 3R has exciting optical properties. Finally, in terms of covalent functionalization, there is a need of new characterization techniques which can be used as standard protocols for the quantification of the degree of functionalization, in the same way, Raman spectroscopy is employed for the characterization of modified graphene. It is therefore expected that research on the functionalization of 2D materials will keep growing and be extended to new materials, molecular building blocks and the use of novel techniques.

References

1. K. S. Novoselov, A. K. Geim, S. V. Morozov, D. Jiang, Y. Zhang, S. V. Dubonos, I. V. Grigorieva and A. A. Firsov, *Science* **306** (2004) 666.
2. K. Khan, A. K. Tareen, M. Aslam, R. Wang, Y. Zhang, A. Mahmood, Z. Ouyang, H. Zhang and Z. Guo, *J. Mater. Chem. C* **8** (2020) 387.
3. S. Manzeli, D. Ovchinnikov, D. Pasquier, O. V. Yazyev and A. Kis, *Nat. Rev. Mater.* **8** (2017) 1.
4. J. Wang, F. Ma and M. Sun, *RSC Adv.* **27** (2017) 16801.
5. F. Xia, H. Wang, J. C. M. Hwang, A. H. C. Neto and L. Yang, *Nat. Rev. Phys.* **1** (2019) 306.
6. A. Acun, L. Zhang, P. Bampoulis, M. Farmanbar, A. van Houselt, A. N. Rudenko, M. Lingenfelder, G. Brocks, B. Poelsema, M. I. Katsnelson and H. J. W. Zandvliet, *J. Phys. Condens. Matter.* **27** (2015) 443002.
7. Q. Tang and D. E. Jiang, *Chem. Mater.* **27**(10) (2015) 3743.
8. Z. Song, T. Schultz, Z. Ding, B. Lei, C. Han, P. Amsalem, T. Lin, D. Chi, S. L. Wong, Y. J. Zheng, M. Y. Li, L. J. Li, W. Chen, N. Koch, Y. L. Huang and A. T. S. Wee, *ACS Nano* **11** (2017) 9128.
9. L. O. Jones, M. A. Mosquera, M. A. Ratner and G. C. Schatz, *ACS Appl. Mater. Interf.* **12** (2020) 4607.
10. D. Voiry, A. Goswami, R. Kappera, C. D. C. C. E. Silva, D. Kaplan, T. Fujita, M. Chen, T. Asefa and M. Chhowalla, *Nat. Chem.* **7** (2015) 45.
11. X. Ding and J. Yi, *Synthesis and Applications of 2d Transition Metal Dichalcogenides, Functional Materials and Electronics* (Apple Academic Press, 2018), pp. 1–33.
12. M. J. Park, S. Gravelsins, J. Son, A. M. van der Zande and A. A. Dhirani, *ACS Nano* **13** (2019) 6469.
13. I. Song, C. Park and H. C. Choi, *RSC Adv.* **5** (2015) 7495.
14. S. Bertolazzi, M. Gobbi, Y. Zhao, C. Backes and P. Samorì, *Chem. Soc. Rev.* **47** (2018) 6845.
15. J. Ristein, *Science* **313** (2006) 1057.
16. S. Tongay, J. Zhou, C. Ataca, J. Liu, J. S. Kang, T. S. Matthews, L. You, J. Li, J. C. Grossman and J. Wu, *J. Nano Lett.* **13** (2013) 2831.
17. D. J. Late, B. Liu, H. S. S. R. Matte, V. P. Dravid and C. N. R. Rao, *ACS Nano* **6** (2012) 5635.
18. J. O. Varghese, P. Agbo, A. M. Sutherland, V. W. Brar, G. R. Rossman, H. B. Gray and J. R. Heath, *Adv. Mater.* **27** (2015) 2734.
19. Y. Wang, A. Slassi, M. A. Stoeckel, S. Bertolazzi, J. Cornil, D. Beljonne and P. Samorì, *P. J. Phys. Chem. Lett.* **10** (2019) 540.
20. X. Li, R. Sun, H. Guo, B. Su, D. Li, X. Yan, Z. Liu and J. Tian, *Adv. Electron. Mater.* **6** (2020) 1901230.

21. J. Zhao, N. Li, H. Yu, Z. Wei, M. Liao, P. Chen, S. Wang, D. Shi, Q. Sun and G. Zhang, *Adv. Mater.* **29** (2017) 1702076.
22. S. Y. Lee, U. J. Kim, J. Chung, H. Nam, H. Y. Jeong, G. H. Han, H. Kim, H. M. Oh, H. Lee, H. Kim, Y. G. Roh, J. Kim, S. W. Hwang, Y. Park and Y. H. Lee, *ACS Nano* **10** (2016) 6100.
23. A. K. Singh, S. Andleeb, J. Singh, H. T. Dung, Y. Seo and J. Eom, *Adv. Funct. Mater.* **24** (2014) 7125.
24. S. Zhao, J. Xue and W. Kang, *Chem. Phys. Lett.* **595** (2014) 35.
25. D. P. Goronzy, M. Ebrahimi, F. Rosei Arramel, Y. Fang, S. De Feyter, S. L. Tait, C. Wang, P. H. Beton, A. T. S. Wee, P. S. Weiss and D. F. Perepichka, *ACS Nano* **12** (2018) 7445.
26. Y. Li, C. Y. Xu, P. Hu and L. Zhen, *ACS Nano* **7** (2013) 7795.
27. D. H. Kang, M. S. Kim, J. Shim, J. Jeon, H. Y. Park, W. S. Jung, H. Y. Yu, C. H. Pang, S. Lee and J. H. Park, *Adv. Funct. Mater.* **25** (2015) 4219.
28. D. H. Kang, J. Shim, S. K. Jang, J. Jeon, M. H. Jeon, G. Y. Yeom, W. S. Jung, Y. H. Jang, S. Lee and J. H. Park, *ACS Nano* **9** (2015) 1099.
29. J. Li, J. Wierzbowski, Ö. Ceylan, J. Klein, F. Nisic, T. le Anh, F. Meggendorfer, C. A. Palma, C. Dragonetti, J. V. Barth, J. J. Finley and E. Margapoti, *Appl. Phys. Lett.* **105** (2014) 241116.
30. E. Margapoti, J. Li, Ö. Ceylan, M. Seifert, F. Nisic, T. le Anh, F. Meggendorfer, C. Dragonetti, C. A. Palma, J. V. Barth and J. J. Finley, *Adv. Mater.* **27** (2015) 1426.
31. D. Wöhrle, L. Kreienhoop, G. Schnurpfeil, J. Elbe, B. Tennigkeit, S. Hiller and D. Schlettwein, *J. Mater. Chem.* **5** (1995) 1819.
32. N. Karousis, J. Ortiz, K. Ohkubo, T. Hasobe, S. Fukuzumi, A. Sastre-Santos and N. Tagmatarchis, *J. Phys. Chem. C* **116** (2012) 20564.
33. Y. Kaneda, M. E. Stawasz, D. L. Sampson and B. A. Parkinson, *Langmuir* **17** (2001) 6185.
34. C. Wirtz, T. Hallam, C. P. Cullen, N. C. Berner, M. O'Brien, M. Marcia, A. Hirsch and G. S. Duesberg, *Chem. Comm.* **51** (2015) 16553.
35. A. J. Molina-Mendoza, L. Vaquero-Garzon, S. Leret, L. de Juan-Fernández, E. M. Pérez and A. Castellanos-Gomez, *Chem. Commun.* **52** (2016) 14365.
36. H. Zhang, J. Choi, A. Ramani, D. Voiry, S. N. Natoli, M. Chhowalla, D. R. McMillin and J. H. Choi, *Chem. Phys. Chem.* **17** (2016) 2854.
37. J. Choi, H. Zhang and J. H. Choi, *ACS Nano* **10** (2016) 1671.
38. Y. Wang, S. M. Gali, A. Slassi, D. Beljonne and P. Samorì, *Adv. Funct. Mater.* **30** (2020) 2002846.
39. E. P. Nguyen, B. J. Carey, C. J. Harrison, P. Atkin, K. J. Berean, E. della Gaspera, J. Z. Ou, R. B. Kaner, K. Kalantar-Zadeh and T. Daeneke, *Nanoscale* **8** (2016) 16276.
40. S. Mouri, Y. Miyauchi and K. Matsuda, *Nano Lett.* **13** (2013) 5944.
41. D. Kiriya, M. Tosun, P. Zhao, J. S. Kang and A. Javey, *J. Am. Chem. Soc.* **136** (2014) 7853.
42. Y. Huang, W. Zheng, Y. Qiu and P. Hu, *ACS Appl. Mat. Interf.* **8** (2016) 23362.
43. K. Heo, S. H. Jo, J. Shim, D. H. Kang, J. H. Kim and J. H. Park, *ACS Appl. Mat. Interf.* **10** (2018) 32765.

44. C. J. Lockhart De La Rosa, R. Phillipson, J. Teyssandier, J. Adisoejoso, Y. Balaji, C. Huyghebaert, I. Radu, M. Heyns, S. De Feyter and S. De Gendt, *App. Phys. Lett.* **109** (2016) 253112.
45. R. C. Selhorst, E. Puodziukynaite, J. A. Dewey, P. Wang, M. D. Barnes, A. Ramasubramaniam and T. Emrick, *Chem. Sci.* **7** (2016) 4698.
46. H. Y. Park, S. R. Dugasani, D. H. Kang, J. Jeon, S. K. Jang, S. Lee, Y. Roh, S. H. Park and J. H. Park, *ACS Nano* **8** (2014) 11603.
47. K. H. Han, J. Y. Kim, S. G. Jo, C. Seo, J. Kim and J. Joo, *Nanotechnol.* **28** (2017) 435501.
48. Y. J. Zheng, Y. L. Huang, Y. Chen, W. Zhao, G. Eda, C. D. Spataru, W. Zhang, Y. H. Chang, L. J. Li, D. Chi, S. Y. Quek and A. T. S. Wee, *ACS Nano* **10** (2016) 2476.
49. X. Lin, J. C. Lu, Y. Shao, Y. Y. Zhang, X. Wu, J. B. Pan, L. Gao, S. Y. Zhu, K. Qian, Y. F. Zhang, D. L. Bao, L. F. Li, Y. Q. Wang, Z. L. Liu, J. T. Sun, T. Lei, C. Liu, J. O. Wang, K. Ibrahim, D. N. Leonard, W. Zhou, H. M. Guo, Y. L. Wang, S. X. Du, S. T. Pantelides and H. J. Gao, *Nat. Mater.* **16** (2017) 717.
50. D. Jariwala, S. L. Howell, K. S. Chen, J. Kang, V. K. Sangwan, S. A. Filippone, R. Turrisi, T. J. Marks, L. J. Lauhon and M. C. Hersam, *Nano Lett.* **16** (2016) 497.
51. Y. Zhang, L. Xu, W. R. Walker, C. M. Tittle, C. J. Backhouse and M. A. Pope, *J. Mat. Chem. C* **5** (2017) 11275.
52. D. Li, W. Zhou, Q. Zhou, G. Ye, T. Wang, J. Wu, Y. Chang and J. Xu, *Nanotechnol.* **28** (2017) 395401.
53. J. Wu, H. Zeng, X. Li, X. Xiang, Y. Liao, Z. Xue, Y. Ye and X. Xie, *Adv. Energy Mater.* **8** (2018) 1802430.
54. P. Joo, K. Jo, G. Ahn, D. Voiry, H. Y. Jeong, S. Ryu, M. Chhowalla and B. S. Kim, *Nano Lett.* **14** (2014) 6456.
55. S. Zhang, W. Wang, L. Guan, X. Niu, J. Yao, X. Yan, B. Xing, Y. Yu, J. Sha and Y. Wang, *J. Lumin.* **213** (2019) 388.
56. W. Bao, X. Cai, D. Kim, K. Sridhara and M. S. Fuhrer, *Appl. Phys. Lett.* **102** (2013) 042104.
57. B. Liu, W. Zhao, Z. Ding, I. Verzhbitskiy, L. Li, J. Lu, J. Chen, G. Eda and K. P. Loh, *Adv. Mater.* **28** (2016) 6457.
58. H. Kim, D. H. Lien, M. Amani, J. W. Ager and A. Javey, *ACS Nano* **11** (2017) 5179.
59. J. Lu, M. Chen, L. Dong, L. Cai, M. Zhao, Q. Wang and J. Li, *Colloids Surf. B* **194** (2020) 111162.
60. J. Liu, Z. Zeng, X. Cao, G. Lu, L. H. Wang, Q. L. Fan, W. Huang and H. Zhang, *Small* **8** (2012) 3517.
61. X. Zhang, Z. Lai, Z. Liu, C. Tan, Y. Huang, B. Li, M. Zhao, L. Xie, W. Huang and H. Zhang, *Angew. Chem. Int. Ed.* **54** (2015) 5425.
62. Y. Yao, L. Tolentino, Z. Yang, X. Song, W. Zhang, Y. Chen and C. P. Wong, *Adv. Funct. Mater.* **23** (2013) 3577.
63. H. Dong, D. Chen, K. Wang and R. Zhang, *Nanoscale Res. Lett.* **11** (2016) 409.

64. G. Guan, S. Zhang, S. Liu, Y. Cai, M. Low, C. P. Teng, I. Y. Phang, Y. Cheng, K. L. Duei, B. M. Srinivasan, Y. Zheng, Y. W. Zhang and M. Y. Han, *J. Am. Chem. Soc.* **137** (2015) 6152.
65. Y. Yong, L. Zhou, Z. Gu, L. Yan, G. Tian, X. Zheng, X. Liu, X. Zhang, J. Shi, W. Cong, W. Yin and Y. Zhao, *Nanoscale* **6** (2014) 10394.
66. Y. Ge, J. Wang, Z. Shi and J. Yin, *J. Mat. Chem.* **22** (2012) 17619.
67. M. C. Demirel, M. Vural and M. Terrones, *Adv. Funct. Mater.* **28** (2018) 1704990.
68. H. Chen, T. Liu, Z. Su, L. Shang and G. Wei, *Nanoscale Horiz.* **3** (2018) 74.
69. J. Lee, P. Dak, Y. Lee, H. Park, W. Choi, M. A. Alam and S. Kim, *Sci. Rep.* **4** (2014) 1.
70. J. Ma, K. Y. Choi, S. H. Kim, H. Lee and G. Yoo, *Appl. Phys. Lett.* **113** (2018) 013102.
71. K. Zhou, J. Liu, P. Wen, Y. Hu and Z. Gui, *Appl. Surf. Sci.* **316** (2014) 237.
72. J. Biscarat, M. Bechelany, C. Pochat-Bohatier and P. Miele, *Nanoscale* **7** (2015) 613.
73. Y. Lin, T. V. Williams, T. B. Xu, W. Cao, H. E. Elsayed-Ali and J. W. Connell, *J. Phys. Chem. C* **115** (2011) 2679.
74. A. Ramasubramaniam, R. Selhorst, H. Alon, M. D. Barnes, T. Emrick and D. Naveh, *J. Mat. Chem. C* **5** (2017) 11158.
75. W. T. Cao, F. F. Chen, Y. J. Zhu, Y. G. Zhang, Y. Y. Jiang, M. G. Ma and F. Chen, *ACS Nano* **12** (2018) 4583.
76. A. Criado, M. Melchionna, S. Marchesan and M. Prato, *Angew. Chem. Int. Ed.* **54** (2015) 10734.
77. A. Hirsch and F. Hauke, *Angew. Chem. Int. Ed.* **57** (2018) 4338.
78. S. S. Chou, M. De, J. Kim, S. Byun, C. Dykstra, J. Yu, J. Huang and V. P. Dravid, *J. Am. Chem. Soc.* **135** (2013) 4584.
79. Y. T. Liu, Q. P. Feng, X. M. Xie and X. Y. Ye, *Carbon* **49** (2011) 3371.
80. Y. T. Liu, Z. Tan, X. M. Xie, Z. F. Wang and X. Y. Ye, *Chem. Asian J.* **8** (2013) 817.
81. C. Backes, N. C. Berner, X. Chen, P. Lafargue, P. LaPlace, M. Freeley, G. S. Duesberg, J. N. Coleman and A. R. McDonald, *Angew. Chem. Int. Ed.* **54** (2015) 2638.
82. C. Vericat, M. E. Vela, G. Benitez, P. Carro and R. C. Salvarezza, *Chem. Soc. Rev.* **39** (2010) 1805.
83. L. Soldi, R. J. Cullen, D. R. Jayasundara, E. M. Scanlan, S. Giordani and P. E. Colavita, *J. Phys. Chem. C* **115** (2011) 10196.
84. E. P. Nguyen, B. J. Carey, J. Z. Ou, J. van Embden, E. della Gaspera, A. F. Chrimes, M. J. Spencer, S. Zhuiykov, K. Kalantar-Zadeh and T. Daeneke, *Adv. Mater.* **27** (2015) 6225.
85. X. Chen, N. C. Berner, C. Backes, G. S. Duesberg and A. R. McDonald, *Angew. Chem. Int. Ed.* **55** (2016) 5803.
86. A. Förster, S. Gemming, G. Seifert and D. Tománek, *ACS Nano* **11** (2017) 9989.
87. X. Chen, P. Denninger, T. Stimpel-Lindner, E. Spiecker, G. S. Duesberg, C. Backes, K. C. Knirsch and A. Hirsch, *Chem. Eur. J.* **26** (2020) 6535.

88. Q. Ding, K. J. Czech, Y. Zhao, J. Zhai, R. J. Hamers, J. C. Wright and S. Jin, *ACS Appl. Mat. Interf.* **9** (2017) 12734.
89. R. Canton-Vitoria, Y. Sayed-Ahmad-Baraza, M. Pelaez-Fernandez, R. Arenal, C. Bittencourt, C. P. Ewels and N. Tagmatarchis, *NPJ 2D Mater. Appl.* **1** (2017) 13.
90. M. A. Gerkman, J. K. Lee, X. Li, Q. Zhang, M. Windley, M. V. Fonseca, Y. Lu, J. H. Warner and G. D. Han, *ACS Nano* **14** (2020) 153.
91. I. K. Sideri, R. Arenal and N. Tagmatarchis, *ACS Mater. Lett.* **2** (2020) 832.
92. R. Canton-Vitoria, C. Stangel and N. Tagmatarchis, *ACS Appl. Mat. Interf.* **10** (2018) 23476.
93. M. Blanco, M. Lunardon, M. Bortoli, D. Mosconi, L. Girardi, L. Orian, S. Agnoli and G. Granozzi, *J. Mater. Chem. A* **8** (2020) 11019.
94. P. Vishnoi, A. Sampath, U. V. Waghmare and C. N. R. Rao, *Chem. Eur. J.* **23** (2017) 886.
95. E. X. Yan, M. Cabán-Acevedo, K. M. Papadantonakis, B. S. Brunschwig and N. S. Lewis, *ACS Mater. Lett.* **2** (2020) 133.
96. G. Bottari, M. Ángeles Herranz, L. Wibmer, M. Volland, L. Rodríguez-Pérez, D. M. Guldi, A. Hirsch, N. Martín, F. D'Souza and T. Torres, *Chem. Soc. Rev.* **46** (2017) 4464.
97. J. M. Englert, C. Dotzer, G. Yang, M. Schmid, C. Papp, J. M. Gottfried, H. P. Steinrück, E. Spiecker, F. Hauke and A. Hirsch, *Nat. Chem.* **3** (2011) 279.
98. D. O. Li, M. S. Gilliam, X. S. Chu, A. Yousaf, Y. Guo, A. A. Green Bc and Q. H. Wang, *Mol. Syst. Des. Eng.* **4** (2019) 962.
99. C. R. Ryder, J. D. Wood, S. A. Wells, Y. Yang, D. Jariwala, T. J. Marks, G. C. Schatz and M. C. Hersam, *Nat. Chem.* **8** (2016) 597.
100. Z. Wang, Q. Li, Z. Chen, H. Li, S. Zheng and X. Tang, *Polym. Compos.* **40** (2019) 2346.
101. K. C. Knirsch, N. C. Berner, H. C. Nerl, C. S. Cucinotta, Z. Gholamvand, N. McEvoy, Z. Wang, I. Abramovic, P. Vecera, M. Halik, S. Sanvito, G. S. Duesberg, V. Nicolosi, F. Hauke, A. Hirsch, J. N. Coleman and C. Backes, *ACS Nano* **9** (2015) 6018.
102. X. S. Chu, A. Yousaf, D. O. Li, A. A. Tang, A. Debnath, D. Ma, A. A. Green, E. J. G. Santos and Q. H. Wang, *Chem. Mat.* **30** (2018) 2112.
103. D. O. Li, X. S. Chu and Q. H. Wang, *Langmuir* **35** (2019) 5693.
104. Y. Park, S. Shin, Y. An, J. G. Ahn, G. Shin, C. Ahn, J. Bang, J. Baik, Y. Kim, J. Jung and H. Lim, *ACS Appl. Mater. Interf.* **12** (2020) 40870.
105. E. E. Benson, H. Zhang, S. A. Schuman, S. U. Nanayakkara, N. D. Bronstein, S. Ferrere, J. L. Blackburn and E. M. Miller, *J. Am. Chem. Soc.* **140** (2018) 441.
106. G. Tuci, D. Mosconi, A. Rossin, L. Luconi, S. Agnoli, M. Righetto, C. Pham-Huu, H. Ba, S. Cicchi, G. Granozzi and G. Giambastiani, *Chem. Mat.* **30** (2018) 8257.
107. B. H. Northrop, S. H. Frayne and U. Choudhary, *Polym. Chem.* **6** (2015) 3415.
108. M. Vera-Hidalgo, E. Giovanelli, C. Navío and E. M. Pérez, *J. Am. Chem. Soc.* **141** (2019) 3767.

109. R. Quirós-Ovies, M. Vázquez Sulleiro, M. Vera-Hidalgo, J. Prieto, I. J. Gómez, V. Sebastián, J. Santamaría and E. M. Pérez, *Chem. Eur. J.* **26** (2020) 6629.
110. C. Martí, K. Kostarelos, M. Prato and A. Bianco, *Chem. Commun.* **55** (2019) 5540.
111. J. H. Appel, D. O. Li, J. D. Podlevsky, A. Debnath, A. A. Green, Q. H. Wang and J. Chae, *ACS Biomater. Sci. Eng.* **2** (2016) 361.

https://doi.org/10.1142/9789811247859_0005

Chapter 5

The Electronic Structure at Organic–2D Material Heterointerfaces

Yu Li Huang*,†,‡ and Andrew Thye Shen Wee†,§

**Joint School of National University of Singapore and Tianjin University, International Campus of Tianjin University, Binhai New City, Fuzhou, Fujian 350207, P. R. China*

†Department of Physics, Faculty of Science, National University of Singapore, 2 Science Drive 3, Singapore 117551, Singapore

‡hyl@tjufz.org.cn

§phyweets@nus.edu.sg

Organic–2D material heterostructures have attracted intensive research interest due to their intriguing properties, with a wide range of potential applications in multifunctional flexible electronic and optoelectronic devices. Central to the realization of such devices is a fundamental understanding of the electronic structures at organic–2D material heterointerfaces. The energy level alignment (ELA) at the interface is of paramount importance because it determines the charge transfer barriers between the two materials in contact. In this paper, we discuss the physical mechanisms determining the ELAs, with special attention on interfacial charge transfer at the heterostructures. We review the current understanding of electronic properties at the heterointerfaces formed by the integration of organics with graphene and 2D transition metal dichalcogenides (TMDs), and conclude with a perspective on the future development of organic–2D material heterostructures.

Keywords: Organic–2D material heterointerfaces; electronic structure; energy level alignment; scanning tunneling microscopy; photoemission spectroscopy.

‡Corresponding author.

1. Introduction

Recent progress in two-dimensional (2D) electronic materials has brought fascinating prospective applications in next-generation electronic and optoelectronic devices closer to realization. Marrying the fields of 2D materials and organics harnesses the advantages of both 2D materials and organic semiconductors (SCs) to create layered assemblies with unique properties for flexible electronic devices.[1–6] In particular, the deposition of organic molecular layers on 2D materials is a promising and controllable doping approach to avoid structural damage, thereby retaining the desired properties of the 2D materials, as such organic layers are mainly stabilized through noncovalent interactions.[7–10] Organics can also be used to enhance the carrier mobilities and optical response of 2D materials via noncovalent and covalent functionalization. Substantial research effort has been devoted to investigate the electronic and optical properties of organically functionalized 2D materials and their performance in devices, to obtain a comprehensive understanding of the organic–2D heterointerfaces at the atomic scale.

In organic–2D material nanodevices, the interface makes up most of the active layers for charge transfer and energy transport during the device operations. It is thereby important to have a clear understanding of the fundamental electronic properties at the interfaces. Energy level alignment (ELA), which describes how the energy levels of the different materials in contact are aligned relative to each other, plays a crucial role in determining the energy barriers for charge carriers to transfer across the interface.[11] These energy barriers, critical determinants of the tunneling rates across the interface and the interface resistance, are important figures of merit for device applications. In this paper, we discuss the current understanding of energy level alignments at the organic–2D material interfaces, and the physical processes determining the alignments. Graphene and 2D transition metal dichalcogenides (TMDs) are the two classes of materials we will focus on. We conclude with a future outlook on the scientific and device developments based on our understanding of organic–2D material heterointerfaces.

2. Energy Level Alignment

Energy level alignment refers to the relative positions of the quasiparticle energy levels of the materials at both sides of a heterointerface. Indeed, we need to distinguish the final ELA after two materials come into contact from the initial energy level lineup before contact, which can be quite different. In general, the initial energy level lineup describes the relative energy level positions of the different materials using the vacuum level as a reference, while the final ELA is established via subtle interfacial interactions, e.g. charge transfer, chemical bonding, interface dipole, and electronic screening, to equilibrate the Fermi level at both sides (Fig. 1).[12] In principle, the interface charge transfer and covalent chemical bonds both tend to induce interface dipoles and shift the energy levels upward or

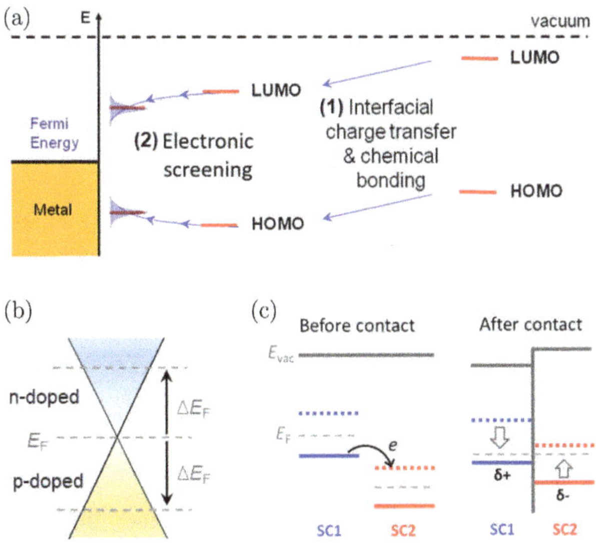

Fig. 1. Energy level alignments at heterointerfaces. (a) For a molecule physisorbed on a metal surface, (1) the frontier molecular levels shift upward or downward due to interfacial charge transfer and/or chemical bonds; and (2) the HOMO–LUMO gap reduces by electronic screening from the substrate. (b) Graphene can be p-doped or n-doped with the adsorption of selected organic dopant. (c) Charge transfer driven by the Fermi energy difference between SC1 and SC2. The schematic in panel (a) is adapted with permission from Ref. 13.

downward [Fig. 1(a)].[10] Interface dipoles can also be formed by dipolar molecules assembled into ordered organic dipole arrays. Electronic screening from the dielectric environment (e.g. substrates) is another important factor that affects the final ELA, which usually results in a noticeable reduction in the bandgap of a semiconductor. In this paper, we mainly focus on the energy level shifts induced by charge transfer, and a detailed discussion of the screening effects is presented elsewhere.

We consider heterostructures consisting of organic thin films physisorbed on 2D semiconducting TMD for simplification. In principle, charge transfer will occur at the interface to align the Fermi levels (E_F) at both sides when the two materials come into contact. A prototypical model is illustrated in Fig. 1(c), where SC1 and SC2 are inorganic or organic materials. Short solid lines represent the valence band maximum (VBM) for the inorganic SC or the highest occupied molecular orbitals (HOMOs) for the organic SC. Short dashed lines represent the conduction band minimum (CBM) or the lowest unoccupied molecular orbitals (LUMOs), respectively. Before contact, E_F lies midway between the VBM (HOMO) and CBM (LUMO) of the isolated SC1 and SC2 under thermodynamic equilibrium. To equilibrate the Fermi level, electrons flow from the highest filled states of SC1 to the lowest unfilled states of SC2. As thus, SC1 becomes positively charged and its energy levels shift downward, while SC2 is equally negatively charged and its energy levels shift upward. The Fermi levels finally align when the thermodynamic equilibrium is established.

The initial energy level lineup determines the direction and magnitude of the charge transfer, and the final ELA can be significantly different. The control of the ELA at organic–2D material heterointerfaces can be realized via a careful selection of the building blocks. ELA can be tuned by molecular orientation, substitutional groups, and molecular chain lengths, due to the formation of different interfacial dipoles. A few prototype studies will be discussed in the following section.

3. Organic–Graphene Heterointerfaces

3.1. *Tunability of the graphene band structure*

Surface adsorption of strong organic acceptors or donors is an effective method of tuning the energy level and carrier density of graphene.[7,14–16] When a molecular adsorbate possessing an electron affinity (EA) higher than or an ionization potential (IP) lower than the graphene work function (WF), spontaneous charge transfer occurs at the molecule–graphene interface. EA or IP is defined as the energy difference between the vacuum level and the LUMO or HOMO. Figure 2(a) shows a lineup of E_F (Dirac point) of neutral graphene with respect to the HOMOs and LUMOs of various organic molecules.[8,14] Since the LUMOs of fluorinated fullerene ($C_{60}F_{48}$) and tetrafluoro-tetracyanoquinodimethane (F_4-TCNQ) are below the Dirac point of graphene, these molecules will withdraw electrons from graphene to induce *p*-doping when brought into contact with graphene [Fig. 1(b)]. Other common organic dopants include *p*-type dopants such as tetracyanoquinodimethane (TCNQ)[17,18] and tetracyanoethylene (TCNE),[17] and *n*-type dopants such as 2-(2-methoxyphenyl)-1,3-dimethyl-2,3-dihydro-1H-benzoimidazole (o-MeO-DMBI),[19] benzyl viologen (BV),[20] and tetrathiafulvalene (TTF).[21]

Theoretical calculations based on density functional theory (DFT) have demonstrated that nondestructive *p*-type or *n*-type doping of graphene can be realized via interfacial charge transfer induced by organic decorations.[17,22] Figure 2(b) shows the calculated variation of the Dirac point relative to the Fermi level, ΔE_F, as a function of the change in WF (i.e. $\psi - \psi_g$) of graphene upon molecular deposition.[22] In the study, the molecule–graphene systems were modeled using organic monolayers, single components (TTF, TCNQ, and F_4-TCNQ) or binary complexes (TTF–TCNQ and TTF–F_4-TCNQ), adsorbed on a single-layer graphene. TTF is an *n*-type dopant, while TCNQ and F_4-TCNQ are *p*-type dopants. The doping effect can be tuned by the co-deposition of different molecules, e.g. TTF–TCNQ and TTF–F_4-TCNQ binary complexes. The linear dependence of ΔE_F on the WF changes suggests that the interaction between

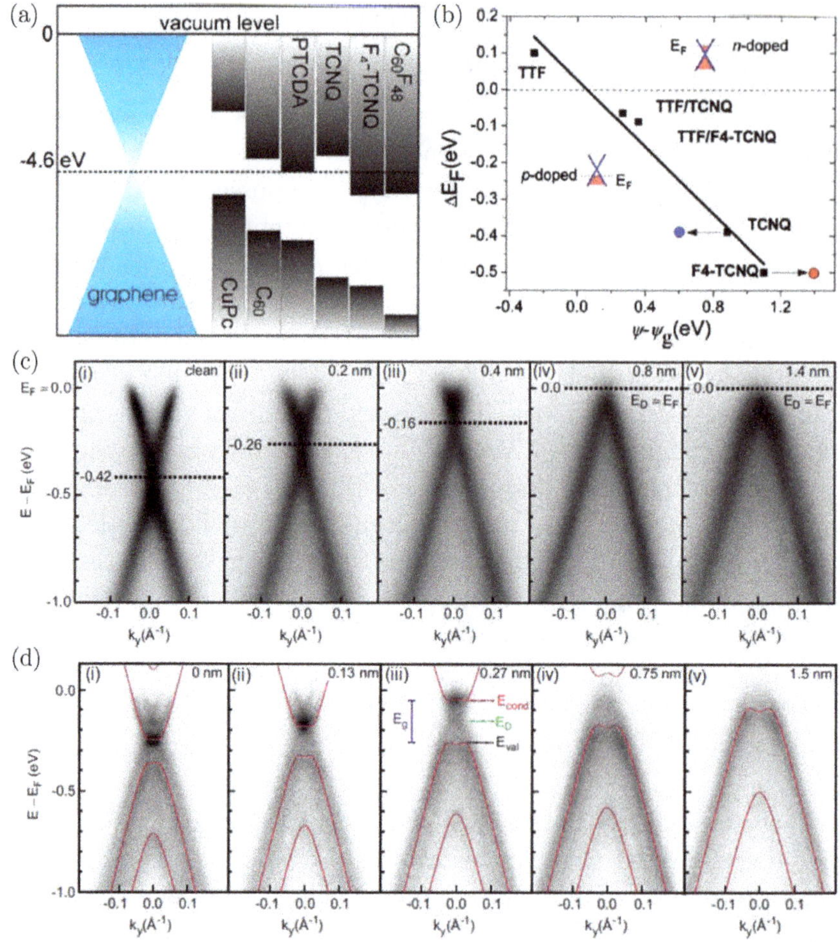

Fig. 2. The electronic structure of graphene tuning by molecular doping. (a) Initial energy level lineup of different organic molecules with graphene. (b) The calculated variation of the Dirac point relative to the Fermi level (ΔE_F) as a function of the potential change ($\psi - \psi_g$) in graphene upon the adsorption of a layer of different molecules or molecular complexes. Circles indicate experimental photoemission spectra (PES) data. The ARPES-measured band structures of (c) monolayer graphene and (d) bilayer graphene grown on SiC(0001) upon the adsorption of F_4-TCNQ with increasing thickness. Panel (a) adopted with permission from Ref. 8; panel (b) from Ref. 22; and panels (c) and (d) from Ref. 23.

adsorbed molecule and graphene is dominated by charge transfer at the interface.

Tuning the band structure of graphene via interfacial charge transfer doping can be directly measured by angular-resolved photoemission spectroscopy (ARPES).[23–25] An elegant example demonstrated in Figs. 2(c) and 2(d) is the epitaxy of graphene grown on SiC(0001) substrate doped by F_4-TCNQ.[22] The as-grown graphene on SiC is strongly negatively charged, revealed by the Dirac point located about 0.42 eV below the Fermi level in Fig. 2(c)(i). Upon the deposition of the strong electron acceptor, F_4-TCNQ, the Dirac point moves upward with increasing molecular coverage. Monolayer graphene finally reaches charge neutrality when the F_4-TCNQ thickness is higher than 0.8 nm, e.g. in panels (iv) and (v) of Fig. 2(c). A bandgap opening is observed in the bilayer graphene via F_4-TCNQ doping. As shown in Fig. 2(d), the size of the bandgap also increases with F_4-TCNQ thickness, i.e. increasing from 116 meV with a 0.13-nm thin layer, to 275 meV with a 1.5-nm thick layer. The opening of the bandgap is induced by the formation of interfacial dipole at the F_4-TCNQ–graphene–SiC heterostructure, and results from the electrostatic asymmetry between the top and bottom surfaces of the bilayer graphene. Such a bandgap opening has also been reported by dual molecular doping, where the exfoliated bilayer graphene is sandwiched between a bottom amine (NH_2)-functionalized self-assembled monolayer (SAM) and a top F_4-TCNQ film.[26] The NH_2-SAM donates electrons to the graphene while the F_4-TCNQ film withdraws electrons from it. As a result, a single-gate field-effect transistor (FET) based on organic-doped bilayer graphene with high current on/off ratios is successfully fabricated.

Figures 2(c) and 2(d) show that the epitaxial graphene grown on SiC is negatively charged with its Dirac points located below the Fermi level, also commonly observed for graphene grown on metal substrates (e.g. Cu, Au, Ni, and so on).[27–30] Such nonnegligible substrate effects could complicate the situation when graphene is covered with molecular dopants. For instance, graphene can be *p*-doped or *n*-doped with a strong molecular acceptor, e.g. 1, 3, 4, 5, 7, 8-hexafluorotetracyano-naphthoquinodimethane (F_6-TCNNQ),

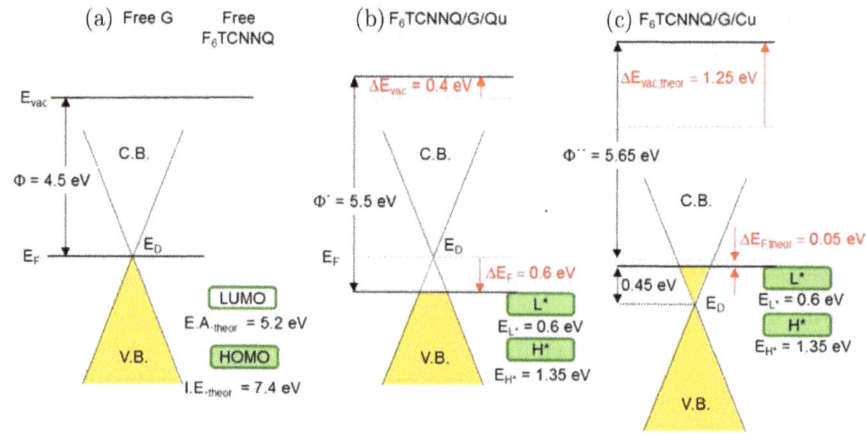

Fig. 3. Energy level diagrams for (a) free graphene (G) and free F_6-TCNNQ before contact, (b) F_6-TCNNQ–G–Qu (quartz) interface, and (c) F_6-TCNNQ–G–Cu interface. In panels (b) and (c), the energy position of the partially filled LUMO (L^*) is located at 0.6 eV (E_{L^*}), and that of the relaxed HOMO (H^*) at 1.35 eV (E_{H^*}). Φ corresponds to the work function; E_{vac} is the vacuum energy level. E_F is the Fermi level; and E_D is the Dirac point of graphene. Figures adopted with permission from Ref. 31.

depending on the underlying substrate.[31] As shown in Fig. 3, Christodoulou *et al.* compared the electronic structures of heterostructures comprising F_6-TCNNQ deposited on graphene layers supported by a quartz and a copper substrate. Graphene on quartz is neutral, and it exclusively donates electrons to F_6-TCNNQ molecules to partially fill the LUMO [Fig. 3(b)]. In contrast, graphene on Cu is n-doped and remains n-doped upon F_6-TCNNQ deposition, where the Cu electron reservoir provides an additional electron density flow to graphene upon the adsorption of the acceptor monolayer [Fig. 3(c)]. The different charge redistribution mechanisms of the two ternary systems contribute to the insulating and metallic natures of the substrates.

The doping level of graphene can be tuned by the substrate as well as intercalations of atoms at its interface.[27–30] For instance, epitaxial graphene on Ir(111) or SiC can be n-doped or p-doped by the intercalation of highly electropositive (e.g. Cs and Eu) or electronegative (e.g. O and F) elements.[32–34] Changes to the graphene doping level can consequently affect the binding energy of the organic adsorbates. The vdW binding energy of naphthalene to graphene increases by ~5%, when graphene on Ir(111) changes from *p*-doping with O intercalation to *n*-doping with Eu intercalation.[34] The frontier energy levels of organics can be changed with graphene doping level as well, and will be discussed below.

3.2. *Engineering of molecular frontier orbitals*

When organic molecules are combined with graphene and other materials, their properties could be significantly modified by the supporting environment. As has been illustrated in Fig. 1(a), interfacial charge transfer is a key factor determining the final positions of the molecular frontier energy levels. In addition to a shift of the energy levels, new electronic and magnetic states may be generated when an organic adsorbate strongly interacts with graphene, due to the charge transfer process.[10,35]

Figure 4 shows the site-dependent charge transfer of TCNQ molecules assembled on graphene–Ru(0001) surface at different coverages.[18,36] Unlike the blurred molecular features observed in TCNQ adsorbed on most metal substrates,[37] subtle intramolecular features corresponding to frontier orbitals are well resolved in the topographic scanning tunneling microscopy (STM) images, because the graphene interlayer prevents strong hybridization between the molecular orbitals and metallic electronic bands. At low coverage, TCNQ molecules demonstrate site-selective adsorption behavior on graphene–Ru(0001), when the lower parts of the graphene moiré pattern (fcc/hcp regions) are decorated first [Fig. 4(a)]. As a strong electron acceptor with low electron affinity, TCNQ tends to adsorb at the fcc/hcp regions, where the electron density is higher and the local work function is lower, to facilitate surface charge transfer. When the coverage increases to a full monolayer, the molecules assemble into well- ordered close-packed molecular arrays containing two distinct features, with the brighter features at top sites [Fig. 4(d)]. Detailed investigations confirmed that the TCNQ molecules adsorbed at top sites are neutral, whereas the ones at fcc/hcp sites are negatively

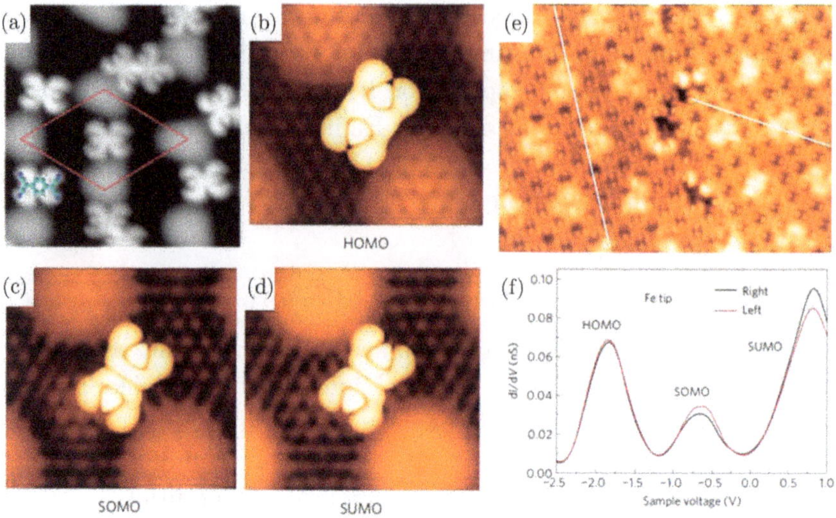

Fig. 4. Site-dependent charge transfer of TCNQ on graphene–Ru(0001). (a) STM image shows the selected adsorption of TCNQ molecules on graphene–Ru(0001) measured at 4.6 K. (b)–(d) Simulated dI/dV maps of HOMO, SOMO, and SUMO of individual TCNQ molecule. (e) TCNQ monolayer comprises two kinds of TCNQ molecules: neutral and charged. Spin-polarized dI/dV spectra recorded at the left and right domains of (e) reveal the long-range magnetic order in the organic layer on graphene. Figures adopted with permission from Ref. 18.

charged and give rise to Kondo resonance. In the negatively charged molecule, the LUMO is partially filled, and splits into a single occupied molecular orbital [SOMO, Fig. 4(c)] and a single unoccupied molecular orbital [SUMO, Fig. 4(d)], confirmed by the dI/dV spectra [Fig. 4(f)]. As a consequence, a sizeable magnetic moment of $0.4\,\mu_B$ develops in the single isolated TCNQ molecule, which decreases to a value of $0.18\,\mu_B$ after the formation of dimers or a full monolayer. Additionally, the spin-polarized dI/dV spectra taken at different molecular domains reveal the existence of magnetic contrast because of the formation of long-range magnetic order in the TCNQ monolayer, which might have potential applications as a spin filter or polarizer. In contrast, TCNQ adsorbed on graphene–Ir(111) shows negligible charge transfer and no magnetic moment.[36]

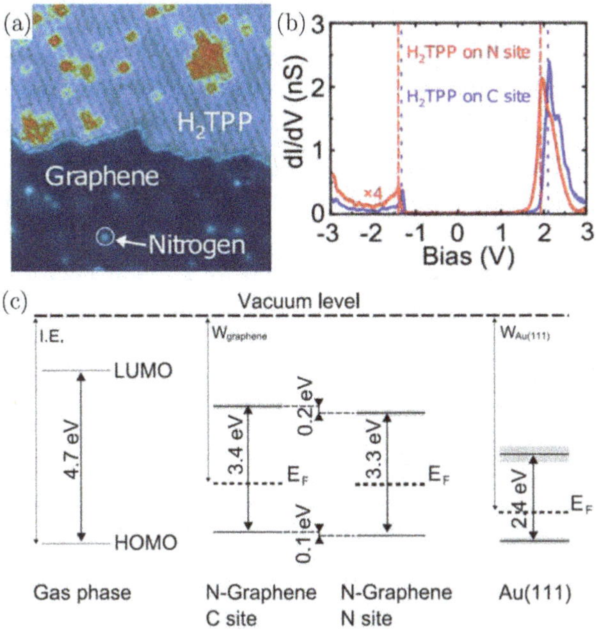

Fig. 5. (Color online) Organic molecules adsorbed on nitrogen-doped graphene. (a) STM image of H_2TPP molecules on the nitrogen-doped graphene on SiC(0001), where the brighter molecules adsorb on nitrogen sites. (b) The dI/dV spectra reveal that the molecular orbitals shift to low energy for H_2TPP molecules on nitrogen sites (red) as compared to those on carbon sites (blue). (c) ELA diagram of H_2TPP molecules from the gas phase to adsorption on N-doped graphene and Au(111) substrate. Figures adopted with permission from Ref. 38.

The interactions of graphene with molecular species can be steered by the nitrogen (N) doping, and thus the electronic properties. As shown in Fig. 5, STM and scanning tunneling spectroscopy (STS) studies were carried out to investigate the electronic interaction of tetraphenylporphyrin (H_2TPP) molecules with pristine and N-doped graphene on SiC.[38] In Fig. 5(a), the molecules with brighter contrast are adsorbed on N sites. Correspondingly, a local modification of the charge transfer around the N sites is evidenced as a downshift of both the HOMO and LUMO compared to the ones on the C sites [Fig. 5(b)]. A summary of the ELAs of H_2TPP molecules from the gas phase to adsorption on pristine and N-doped graphene, and on Au(111), is shown in Fig. 5(c). Clearly, the HOMO–LUMO gap of the molecule decreases from 4.7 eV in the gas phase to 3.4 eV on graphene, and finally to 2.4 eV on Au(111), due to screening from the dielectric substrate. Here, graphene reveals an electronic decoupling effect on the organic adsorbates. Similarly, the downshift of the LUMO state has been recently observed in TCNQ molecules adsorbed at the nitrogen site[39] DFT calculations revealed that a local charge transfer of 0.33 e occurs at the nitrogen site, as compared to 0.20e at carbon sites.[39]

In the case of magnetic molecules, e.g. iron (II) phthalocyanine (FePc), a spin transition can be induced by noncovalent interaction of the molecule with the N-doped graphene.[40] Analogous to H_2TTP and TCNQ molecules, FePc molecules adsorbed on N and C sites demonstrate different HOMO and LUMO positions in the dI/dV spectra due to their different electronic interactions [Fig. 6(b)]. The N-dopant can further affect the atomic orbital distribution.

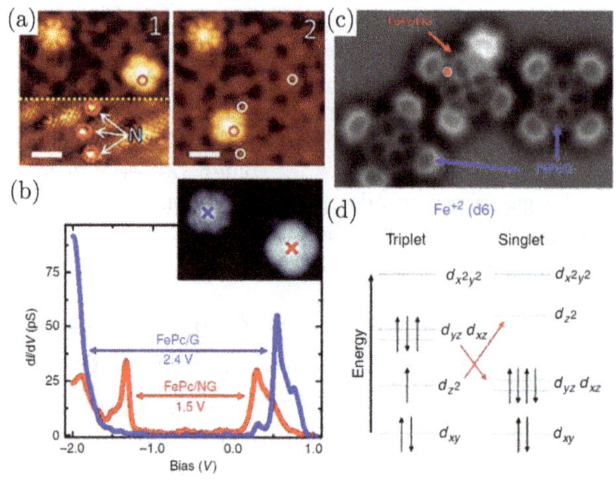

Fig. 6. (Color online) Control of electronic and spin states with N-doped graphene. (a) The controllable lateral manipulation of FePc on N-doped graphene, from the initial N site (1) to another N site (2), driven by the STM tip. (b) The dI/dV spectra acquired at FePc on pristine graphene (blue) and on an N-dopant (red). (c) Constant-height AFM image of three FePc molecules located on pure graphene (blue arrows) and N site (red arrow), revealing the different cross and square features of the molecular cores. (d) Calculated occupation of Fe^{+2} orbitals in the triplet and singlet states. Figures adopted with permission from Ref. 40.

As revealed by noncontact atomic force microscopy (nc-AFM), a cross-shaped feature is observed in the middle of both molecules adsorbed on pure graphene, whereas a square-like feature is clearly resolved for FePc adsorbed on the N-dopant [Fig. 6(c)]. This variation is attributed to the reordering of Fe^{2+} d-orbitals by weak intermixing between orbitals with the z-component (p_z) of the N-dopant and molecule (d_{xz}, d_{yz}, and d_z^2) [Fig. 6(d)], accompanied by an electron density redistribution within the molecule. These experiments show the unique capability of doped graphene to behave as a noncovalent tuner of the electronic and spin properties of molecules.

4. Organic–2D TMD Heterointerfaces

4.1. *Charge transfer at the interface*

Inspired by the success achieved in tuning graphene electronic properties by organic doping, researchers have also studied the interactions between molecular dopants and 2D TMDs. In organic–2D TMD systems, charge transfer depends on the electrochemical redox potential (closely relative to the HOMO and LUMO energies)[41] of the molecular dopant relative to the band edges (CBM and VBM) of 2D TMD semiconductors.[9] When the redox potential of a given molecule lies below the E_F of the TMD crystal, electrons will flow from the TMD to the molecule, resulting in p-type doping. Conversely, when the redox potential of the molecules is above the E_F of the TMD, electrons will transfer from the molecules to the TMD, inducing n-type doping. Similarly, TCNE, TCNQ, F_4-TCNQ, and $C_{60}F_{48}$ can be used as strong electron acceptors,[42–44] while TTF, BV, and dimethyl-p-phenylenediamine (DMPD) can act as effective donors.[42,45,46] Among them, $C_{60}F_{48}$ is a strong electron acceptor, which can withdraw about 0.29 e per molecule from a WSe_2 monolayer[44] BV is a strong electron donor, and is able to donate about 0.25 e per molecule to the semiconducting MoS_2.[45] Recently, Guo *et al.* reported a strong molecular organometallic donor, mesitylene pentamethylcyclopentadienyl ruthenium dimer ([RuCp* (mes)]$_2$), in which one [RuCp*(mes)] monomer can contribute 0.4 e to the underlying WSe_2 monolayer.[47] Such large charge transfer in turn brings significant changes in the ELA at the organic–2D TMD interfaces as well as the device transport properties.[46,47]

The differential charge density (DCD) obtained from DFT calculations, defined as the difference between the charge density for the combined system and that for the individual subsystems, i.e. $\Delta\rho = \rho_{\text{combined}} - \rho_{\text{TMD}} - \rho_{\text{organic}}$, provides a very useful way to analyze charge transfer effects. Figures 7(a), 7(c), and 7(g) show the DCD isosurfaces of different organic–2D TMD heterostructures, with different colors representing electron accumulation and depletion areas. Significant charge rearrangements occur at the interface. Clearly, TTF is an efficient electron donor with its HOMO close to the CBM of single-layered (SL) MoS_2, while TCNQ is an electron acceptor with the LUMO close to the VBM of MoS_2 [Fig. 7(b)].[42] Charge transfer also depends on the substrate phases as well as molecular orientations.[48,49] In the pentacene–MoS_2 interface, DFT calculations predict that charge transfer on metallic 1T-MoS_2 is more significant than that on semiconducting 2H-MoS_2.[48] For example, when pentacene is flat-lying on the TMD surface, it donates $\sim$0.437 e per molecule to 1T-MoS_2, but only $\sim$0.015 e per molecule to 2H-MoS_2. We need to also consider the orientation of the molecule, for example, pentacene oriented vertically on both 2H- and 1T-MoS_2 shows very little contribution to the charge redistribution.[48]

Analogous to organic–graphene heterointerfaces, the underlying substrate also plays a significant role in defining the electronic structure of the organic–2D TMD systems. For example, $C_{60}F_{48}$ is a strong hole dopant for both isolated SL-WSe_2 and isolated graphene, and is able to withdraw -0.29 e and -0.40 e from the supported substrates, respectively.[44] The situation becomes more complicated when the heterostructure comprises more than two stacked layers, e.g. $C_{60}F_{48}$–SL-WSe_2–graphene heterostructure as shown in Fig. 7(c). The amount of cumulative charges obtained by $C_{60}F_{48}$ from the SL-WSe_2–graphene substrate is as large as -0.38 e, with two-thirds of the negative charge from graphite and only one-third from the SL-WSe_2 interlayer [Fig. 7(d)]. Charge transfer mechanism between an acceptor molecule (F_6-TCNNQ) and

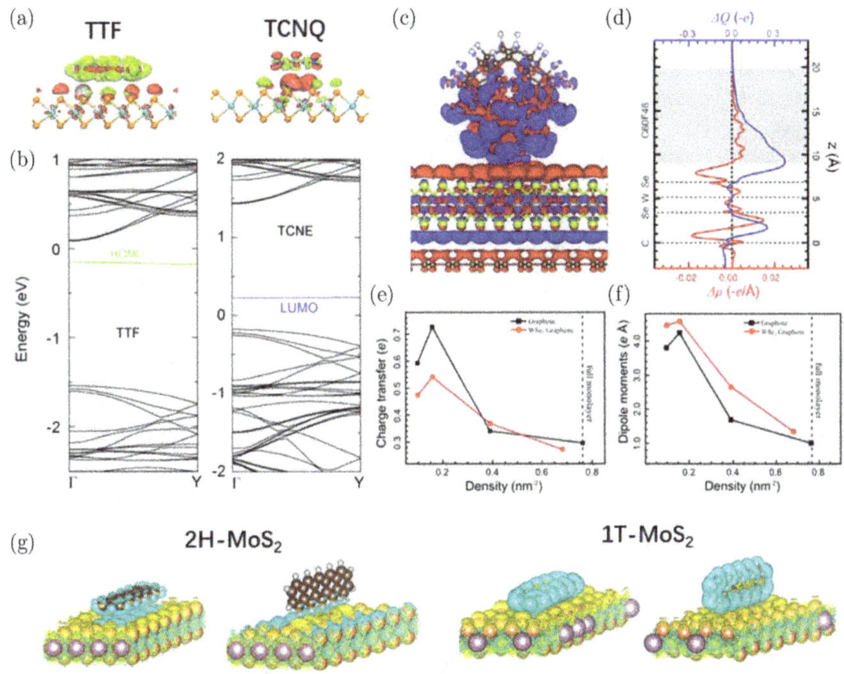

Fig. 7. (Color online) Charge transfer at organic–2D TMD interfaces. (a) Side views of the isosurfaces of the differential charge density $\Delta\rho$ for TTF and TCNQ on 2H-MoS_2 monolayer. Red (green) represents electron accumulation (depletion). (b) Band structures of TTF (left) and TCNQ (right) on SL-MoS_2. (c) Differential charge density isosurface for a $C_{60}F_{48}$ molecule adsorbed on SL-WSe_2 supported on graphite. Blue (red) represents electron accumulation (depletion). (d) The corresponding plot of the plane-averaged $\Delta\rho(z)$ (red) and $\Delta Q(z)$ (blue) for the $C_{60}F_{48}$–SL-WSe_2–graphene system. (e) The total amount of charge transfer and (f) the dipole moment per molecule as functions of increasing molecular density on both SL-WSe_2–graphene (red) and graphene (black) surfaces. (g) Differential charge density plots for pentacene on SL-MoS_2 with different molecular orientations and 2H/1T phases. Yellow (blue) represents electron accumulation (depletion). Panels (a) and (b) adopted with permission from Ref. 42; panels (c)–(f) from Ref. 44; and panel (g) from Ref. 48.

SL-MoS_2 depends on substrates.[50] For example, F_6-TCNNQ molecule withdraws electron from (i) MoS_2 gap states with insulating sapphire substrate, (ii) HOPG substrate with an electric field across the MoS_2 interlayer, and (iii) the combined SL-MoS_2–Au system.

Another factor that influences the amount of charge transfer is the molecular density, i.e. the number of molecules per area on the surface. In general, the amount of charge transferred per molecule decreases with increasing molecular density.[42,48,51] The plots in Figs. 7(e) and 7(f) reveal that the charge transfer as well as the interface dipole moment per molecule decrease with increasing molecular density in both the $C_{60}F_{48}$–graphene (black) and $C_{60}F_{48}$–SL-WSe_2–graphene (red) systems.[49] The molecular dipole per molecule and the charge transfer per molecule are larger at lower density, because each molecule possesses stronger electrostatic interactions with the substrate when it is more isolated. In addition, compared to the $C_{60}F_{48}$−graphite interface, $C_{60}F_{48}$−WSe_2−graphite has a relatively larger molecular dipole [Fig. 7(f)]. This enhancement can be explained by the increased molecule–graphite separation with the SL-WSe_2 interlayer, and thus the interfacial dipole is magnified as the electron accumulation and depletion regions are mainly located in the molecule and graphite, respectively.[44]

Experiments have also been carried out to understand the tunability of the WSe_2 energy levels upon $C_{60}F_{48}$ adsorption.[44] An intrinsic/*p*-doped homogeneous junction can be obtained in SL-WSe_2 by precise control of the molecular coverage [Figs. 8(a) and 8(b)]. From the STS spectra in Fig. 8(c), the VBM and CBM of SL-WSe_2 vary significantly as the STM tip approaches the $C_{60}F_{48}$ island. The schematic diagram in Fig. 8(d) clearly illustrates that VBM shifts gradually upward to the Fermi level with $\Delta U_{\mathrm{VBM}} = 0.62\,\mathrm{eV}$, and the CBM shifts upward with $\Delta U_{\mathrm{CBM}} = 0.38\,\mathrm{eV}$, implying a bandgap decrease of $\Delta E_g = 0.24\,\mathrm{eV}$. This suggests that the bare WSe_2 is intrinsic, and is *p*-doped when covered by $C_{60}F_{48}$, while a region of band bending forms between the two. The band bending over several nanometers observed at the lateral heterointerface can be explained by dielectric Thomas–Fermi screening.[52] In addition, the bandgap decrease is attributed to the Stark effect,[53,54] induced by a vertical electric field across the WSe_2 interlayer with electrons accumulating on the top Se interface with the molecule, and holes accumulating on the bottom Se interface with graphite [Fig. 7(c)].

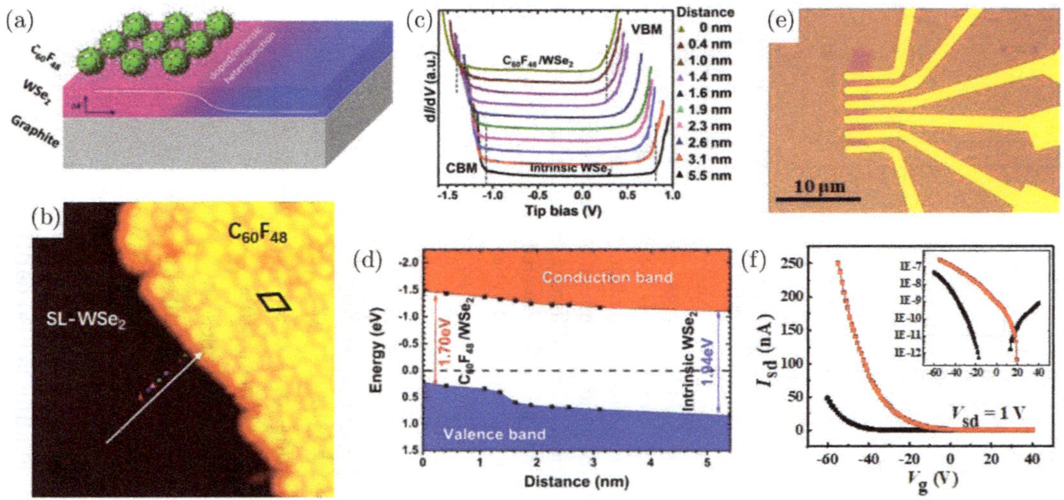

Fig. 8. (Color online) The 1D intrinsic/*p*-doped interface formed in $C_{60}F_{48}$–SL-WSe_2–graphite. (a) Schematic model of the lateral undoped/doped heterointerface. (b) An STM image shows a close-packed 0.5-ML $C_{60}F_{48}$ island on the SL-WSe_2 surface. (c) STS spectra recorded in the positions marked by multicolored triangles in panel (b), reveal that the WSe_2 bandgap changes with the distance to the $C_{60}F_{48}$ island; and panel (d) is the corresponding schematic diagram showing the upward band bending. (e) Optical image of an SL-WSe_2 FET device on a 300-nm SiO_2–Si substrate. (f) The transfer curves measured before (black) and after 20-nm $C_{60}F_{48}$ deposition (red) ($V_{sd} = 1$ V) reveal an efficient *p*-doped effect. Inset: Logarithmic plot of the transfer curve. Figures adopted with permission from Ref. 44.

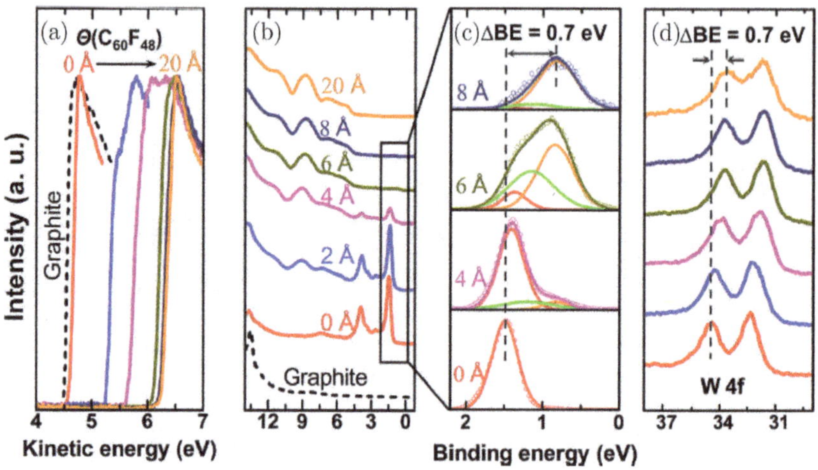

Fig. 9. (Color online) The PES of the $C_{60}F_{48}$–SL-WSe_2–graphite heterostructure. (a) WF increases with the increasing $C_{60}F_{48}$ coverage. (b) Valence band spectra show the sharp peaks at around 1.5 eV and 4 eV from SL-WSe_2 before molecular deposition (0 Å) and strongly attenuated for coverages higher than 4 Å. The dashed black line in panels (a) and (b) represents the clean graphite for comparison. (c) Zoomed-in spectra of the valence band region. Three different contributions can be clearly identified: the uncovered WSe_2 (red), the WSe_2 covered by $C_{60}F_{48}$ (orange), and the transition region between covered and uncovered WSe_2 (green). (d) W-4f core levels shift by 0.7 eV to lower binding energy as the $C_{60}F_{48}$ coverage increases. Figures adopted with permission from Ref. 44.

The tunability of the WSe_2 energy levels is further substantiated by PES measurements. Figure 9 demonstrates the coverage-dependent WF, valence bands, and core levels of the $C_{60}F_{48}$–WSe_2–graphite heterostructure.[44] Upon adsorption of a full $C_{60}F_{48}$ monolayer (8 Å), the WF increases by 1.56 eV indicating the formation of a significant interfacial dipole induced by the charge transfer across the heterostructure. Both WSe_2 valence band and W-4f core level shift to the Fermi level by ~ 0.7 eV, confirming that the WSe_2 interlayer is efficiently p-doped by the $C_{60}F_{48}$ overlayer. The transition area with band bending is also captured in the PES, giving rise to the transient peak (green) halfway between the pristine (red) and the $C_{60}F_{48}$-doped (orange) WSe_2 peak in Fig. 9(d). All these observations are consistent with the STM/STS results (Fig. 8) and theoretical modeling [Figs. 7(c)–7(f)].

Furthermore, the doping of WSe_2 with $C_{60}F_{48}$ is evaluated *in-situ* by an as-fabricated FET device on an insulating SiO_2–Si substrate. In Fig. 8(e), the intrinsic as-exfoliated SL-WSe_2 (black curve) demonstrates an ambipolar characteristic with hole-dominating transport behavior, i.e. a much faster current increase in the negative bias region compared to the positive side (inset). The threshold voltage V_{th} of the FET was determined to be −51.6 V for hole transport by linear extrapolation. V_{th} shifts to −36.6 V after the deposition of 20-nm $C_{60}F_{48}$ (red curve), indicating effective hole doping. The derived hole concentration is estimated to be $2.4 \times 10^{11}\ \text{cm}^{-2}$ for intrinsic SL-WSe_2, and increases to $1.2 \times 10^{12}\ \text{cm}^{-2}$ after $C_{60}F_{48}$ adsorption. These suggest that WSe_2 can be efficiently *p*-doped by the $C_{60}F_{48}$ layers under transport conditions, consistent with theoretical modeling.[44]

4.2. *Interaction with surface defects*

Compared to graphene, 2D TMDs are more susceptible to chemical interactions with the environment, and thus to defect formation during the growth and processing processes. The defects commonly observed in 2D TMDs, e.g. chalcogen vacancies and grain boundaries (GBs), are usually unfavorable for carrier/energy transport and optical applications.[55–58] However, they are often catalytically active due to their large density of states (DOS).[59–61] For instance, S vacancies and GBs in MoS_2 have demonstrated their potential in hydrogen evolution reactions.[62] Defect sites may also potentially be used as templates for the self-assembly of molecules. Site-selective adsorption of molecules on the defect sites via covalent or noncovalent bonding has been reported to be a promising method to functionalize 2D TMDs.[58,63–68]

Covalent functionalization can be achieved by grafting various thiol molecules with different functional groups onto the sulfur vacancy sites in MoS_2.[63,69,70,65] As shown in Fig. 10(a), a prototypical thiol molecule containing a carbon-bonded sulfhydryl is depicted by an *R*–SH formula, where *R* represents an alkyl or other organic substituent. Thiolation of the MoS_2 flakes is achieved as the molecular

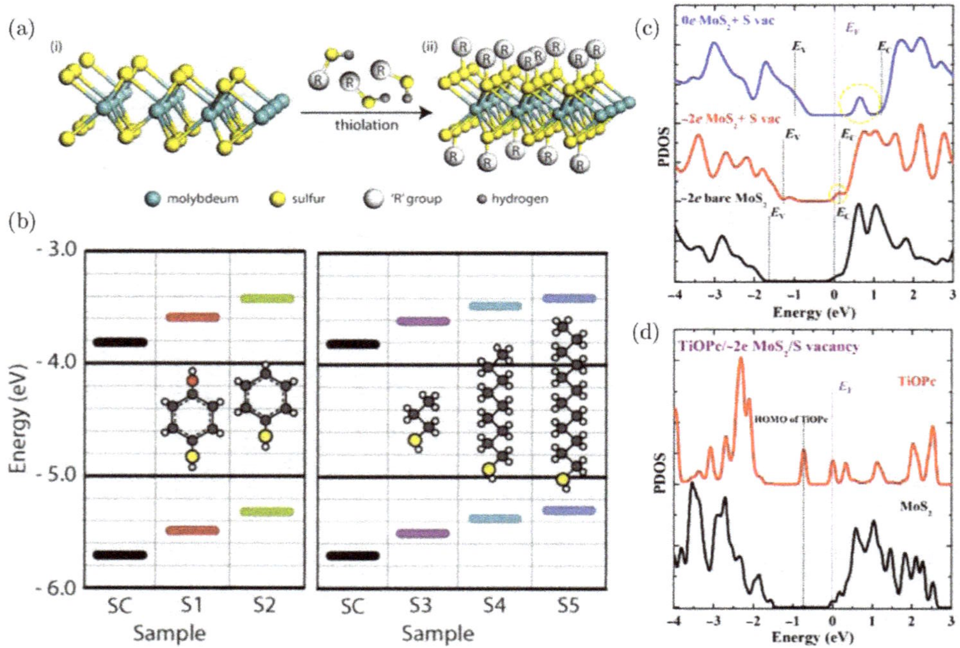

Fig. 10. Organics interacting with S vacancies. (a) Functionalization of MoS_2 monolayers by thiolation, where the thiol molecules with R group are bound to S vacancy sites. (b) Energy diagram shows that the MoS_2 energy levels change by different thiol molecules with various substituent groups and chain lengths, while the bandgap size remains almost constant. Here, SC represents the semiconducting MoS_2 without organic functionalization, and S1–S5 represent the organic molecules as depicted in the diagram. (c), (d) PDOS plots of neutral and charged MoS_2 with no defects and a sulfur vacancy, and TiOPc adsorbed on charged MoS_2 with a vacancy. Panels (a) and (b) are adopted with permission from Ref. 63; and panels (c) and (d) from Ref. 67.

ligand conjugated to the vacancy sites via the S atom with the dissociation of H atoms. The electronic band positions can be tuned by the different substituent *R* groups as well as the chain length through the formation of interfacial dipole, while the optical bandgap of such covalently modified MoS_2 flakes is often retained.[63,64] The energy diagram in Fig. 10(b) shows that the energy positions of the functionalized MoS_2 monolayer vary significantly with the molecular tail groups.[63] For example, the upward shift of the valence band position can reach as large as ~ 0.5 eV in MoS_2 flakes thiolated by the S2 and S5 molecules. In principle, the magnitude as well as the direction of the interfacial dipole vary with the substituent compound and the chain length, and therefore allow the tuning of the ELAs by careful selection of functional molecules.

The defect-induced trap states in 2D TMDs typically obscure material-intrinsic properties and hinder device performance, can be passivated by organic molecules via charge transfer at the organic–TMD interfaces.[67,68,71] Jiang *et al.* reported that the localized defect states in ReS_2 monolayer disappear upon selective decoration of protoporphyrin (H_2PP) molecules on the vacancy site, with a charge transfer of 0.65 *e* between ReS_2 and the molecule. As a result, the decay time of the ReS_2 device is shortened by several orders after the removal of deep traps via H_2PP decoration, and the specific detectivity of the phototransistor is also enhanced because of the reduced dark current. In another study, a strong negative charge transfer between SL-MoS_2 and a monolayer titanyl phthalocyanine (TiOPc) to remove the defect states was reported.[67] In Fig. 10(c), the projected density of states (PDOS) reveals that defect states near the conduction band (CB) can be observed in defectively neutral and charged MoS_2. Upon the adsorption of two TiOPc molecules on the MoS_2 monolayer with a single S vacancy, the defect states nearby the edges are suppressed, and the bandgap is restored to the original size in the intrinsic MoS_2 [Fig. 10(d)]. The formation of vdW organic–2D TMD interface results in an increase of I_{ON}/I_{OFF} ratio in backgated MoS_2 transistors by more than two orders in magnitude, as well as a great enhancement in PL signal.[67]

Grain boundaries are another class of common defects widely observed in 2D materials grown by bottom–up approaches such as chemical vapor deposition (CVD) and molecular beam epitaxy (MBE). Grain boundaries formed by irregular or regular lines containing four-, five-, seven-, and eight-membered dislocation cores have been reported in MoS_2, $MoSe_2$, WS_2, WSe_2, and so on, and usually result in deep gap states and/or bandgap tunability.[57,72–74] Figure 11(a) shows a dense mirror twin boundaries (MTBs) network formed in MBE-grown $MoSe_2$ on graphite surfaces.[58] MTBs in SL-$MoSe_2$ and other 2D TMDs have attracted increasing attention recently due to their metallic state with charge density wave transition and spin-charge separation property.[75–77] To explore their possible site-specific properties, He *et al.* deposited an amino derivative of phenazine, 2,3-diaminophenazine (DAP), onto the $MoSe_2$ surface.[58] The DAP molecule is essentially planar with two active amino groups at one end, which provide active sites that electrostatically interact with the metallic MTB [Figs. 11(e)–11(h)]. Figures 11(c) and 11(d) show the hexagonal DAP porous nanostructure formed which mimics the $MoSe_2$ wagon-wheel pattern substrate template. The self-assembled DAP molecules form preferential triangular-type configuration on the defect-free $MoSe_2$ domains and linear configuration on the MTBs, leaving the MTB intersections empty. Theoretical calculations reveal that the interaction between DAP and the MTB is enhanced, compared with DAP on pristine SL-$MoSe_2$, as evidenced by the shorter interfacial spacing, stronger charge redistribution, and larger adsorption energy [Figs. 11(e)–11(h)]. Furthermore, the charge redistribution is more pronounced at the DAP amino groups, resulting in a slightly bent DAP molecule with a smaller spacing between amino groups and $MoSe_2$. These results suggest that the MTB on $MoSe_2$ facilitates the template adsorption of DAP. Selective self-assembly of organic molecules offers a novel pathway to explore the local chemical properties of 2D defective surfaces, and create ordered nanopatterning with molecular precision.[78]

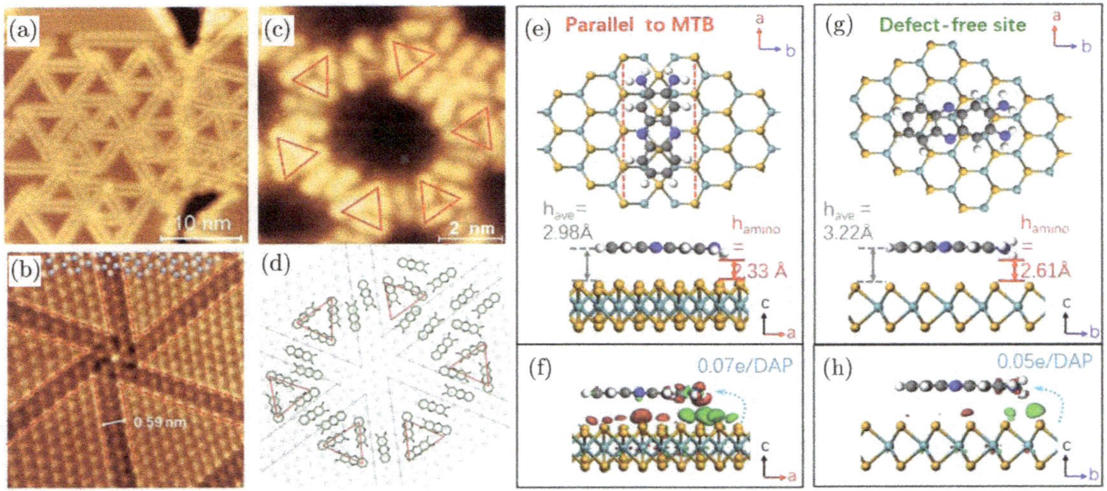

Fig. 11. Formation of DAP porous nanostructure on $MoSe_2$ MTB networks. (a) Topographic STM image shows the SL-$MoSe_2$ films on graphite substrate. (b) An atomic-resolution nc-AFM image shows a typical wagon-wheel pattern. (c) A close-up STM image of a prototypical DAP pore on the top of a wagon-wheel pattern; and (d) the corresponding atomic configuration. Optimized configurations of a DAP molecule adsorbed on (e) MTB region with parallel orientation and (g) defect-free region, and the corresponding charge redistributions at the interfaces are shown in panels (f) and (h), respectively. Figures adopted with permission from Ref. 58.

5. Conclusion and Outlook

In this paper, we present the current understanding of the fundamental electronic properties of organic–2D hybrid heterointerfaces. Special attention is focused on energy level alignments determined by the interfacial charge transfer at the hybrid interface. The physical and chemical interactions between organic molecules and 2D layers can be modified by the supported substrates as well as surface defects, such as atom dopants, vacancies, and grain boundaries. The interfacial integration of 2D materials with organic molecules is a promising and controllable approach to tune the electronic, optical, and even spin properties of the interface without structural damage. It brings the benefits not present in either material, hence enabling better functionalities for future flexible devices.

The field of organic–2D heterointerfaces and devices has been developing rapidly. Besides graphene and 2D TMDs, organic molecules have also been utilized to passivate, dope, and modify the electronic band structure of other 2D materials, e.g. boron nitride,[79] black phosphorus,[80–86] and X-enes (e.g. blue phosphorene[87] and borophene).[88] Recent research effort has also been devoted to the investigations of organic–2D hybrid structures with interesting optical,[89–91] magnetic,[92,93] and superconductivity.[94] properties, which is out of the scope of this paper. The field of organic–2D hybrid nanostructures and devices has immense potential for further developments in device applications.

In summary, the low dimensionality of organic and 2D materials brings new perspective to the field of interfacial physics and chemistry. A comprehensive understanding of ELA at the organic–2D hybrid interfaces, which is complicated by chemical bond formation and electronic screening, is still lacking at present. Furthermore, the theoretical studies on charge transfer at organic–2D interfaces have thus far focused on ordered organic submonolayer and monolayer, which may differ from their real geometries in devices. More intensive effort is required to investigate the effects of film thickness, packing structure,[95] orientation,[96] disorder, defects, and intermolecular interactions. Such studies on the atomic-scale interfacial properties will strengthen the link between fundamental scientific research and practical device applications.

References

1. D. Jariwala, T. J. Marks and M. C. Hersam, *Nature Mater.* **16** (2017) 170.
2. A. T. S. Wee, M. C. Hersam, M. Chhowalla and Y. Gogotsi, *ACS Nano* **10** (2016) 8121.
3. D. B. Velusamy, M. A. Haque, M. R. Parida, F. Zhang, T. Wu, O. F. Mohammed and H. N. Alshareef, *Adv. Funct. Mater.* **27** (2017) 1605554.
4. W. H. Lee, J. Park, S. H. Sim, S. Lim, K. S. Kim, B. H. Hong and K. Cho, *J. Am. Chem. Soc.* **133** (2011) 4447.
5. C.-H. Lee, T. Schiros, E. J. Santos, B. Kim, K. G. Yager, S. J. Kang, S. Lee, J. Yu, K. Watanabe, T. Taniguchi, J. Hone, E. Kaxiras, C. Nuckolls and P. Kim, *Adv. Mater.* **26** (2014) 2812.
6. S. Wieghold, J. Li, P. Simon, M. Krause, Y. Avlasevich, C. Li, J. A. Garrido, U. Heiz, P. Samori, K. Mullen, F. Esch, J. V. Barth and C. A. Palma, *Nat. Commun.* **7** (2016) 10700.
7. H. Y. Mao, Y. H. Lu, J. D. Lin, S. Zhong, A. T. S. Wee and W. Chen, *Prog. Surf. Sci.* **88** (2013) 132.
8. G. Hong, Q.-H. Wu, J. Ren, C. Wang, W. Zhang and S.-T. Lee, *Nano Today* **8** (2013) 388.
9. S. Bertolazzi, M. Gobbi, Y. Zhao, C. Backes and P. Samori, *Chem. Soc. Rev.* **47** (2018) 6845.
10. Y. L. Huang, Y. J. Zheng, Z. Song, D. Chi, A. T. S. Wee and S. Y. Quek, *Chem. Soc. Rev.* **47** (2018) 3241.
11. H. Kroemer, Quasi-electric fields and band offsets: Teaching electrons new tricks, in *Nobel Lectures in Physics 1996–2000* (World Scientific, 2000), pp. 449–469.
12. H. Ishii, K. Sugiyama, E. Ito and K. Seki, *Adv. Mater.* **11** (1999) 605.
13. S. Y. Quek and K. H. Khoo, *Acc. Chem. Res.* **47** (2014) 3250.
14. W. Chen, S. Chen, D. C. Qi, X. Y. Gao and A. T. S. Wee, *J. Am. Chem. Soc.* **129** (2007) 10418.
15. X. Zhang, Z. Shao, X. Zhang, Y. He and J. Jie, *Adv. Mater.* **28** (2016) 10409.
16. G. Bottari, M. A. Herranz, L. Wibmer, M. Volland, L. Rodriguez-Perez, D. M. Guldi, A. Hirsch, N. Martin, F. D'Souza and T. Torres, *Chem. Soc. Rev.* **46** (2017) 4464.
17. Y. H. Lu, W. Chen, Y. P. Feng and P. M. He, *J. Phys. Chem. B* **113** (2009) 2.
18. M. Garnica, D. Stradi, S. Barja, F. Calleja, C. Díaz, M. Alcamí, N. Martín, A. L. Vázquez de Parga, F. Martín and R. Miranda, *Nat. Phys.* **9** (2013) 368.
19. D. W. He, Y. M. Pan, H. Y. Nan, S. A. Gu, Z. Y. Yang, B. Wu, X. G. Luo, B. C. Xu, Y. H. Zhang, Y. Li, Z. H. Ni, B. G. Wang, J. Zhu, Y. Chai, Y. Shi and X. R. Wang, *Appl. Phys. Lett.* **107** (2015) 183103.
20. W. J. Yu, L. Liao, S. H. Chae, Y. H. Lee and X. Duan, *Nano Lett.* **11** (2011) 4759.

21. R. Voggu, B. Das, C. S. Rout and C. N. R. Rao, *J. Phys. Condens. Matter* **20** (2008) 472204.
22. J. T. Sun, Y. H. Lu, W. Chen, Y. P. Feng and A. T. S. Wee, *Phys. Rev. B* **81** (2010) 155403.
23. C. Coletti, C. Riedl, D. S. Lee, B. Krauss, L. Patthey, K. von Klitzing, J. H. Smet and U. Starke, *Phys. Rev. B* **81** (2010) 235401.
24. A. Grüneis and D. V. Vyalikh, *Phys. Rev. B* **77** (2008) 193401.
25. F. Joucken, L. Henrard and J. Lagoute, *Phys. Rev. Mater.* **3** (2019) 110301.
26. J. Park, S. B. Jo, Y. J. Yu, Y. Kim, J. W. Yang, W. H. Lee, H. H. Kim, B. H. Hong, P. Kim, K. Cho and K. S. Kim, *Adv. Mater.* **24** (2012) 407.
27. A. Varykhalov, M. R. Scholz, T. K. Kim and O. Rader, *Phys. Rev. B* **82** (2010) 121101.
28. S. Gottardi, K. Muller, L. Bignardi, J. C. Moreno-Lopez, T. A. Pham, O. Ivashenko, M. Yablonskikh, A. Barinov, J. Bjork, P. Rudolf and M. Stohr, *Nano Lett.* **15** (2015) 917.
29. Y. S. Park, J. H. Park, H. N. Hwang, T. S. Laishram, K. S. Kim, M. H. Kang and C. C. Hwang, *Phys. Rev. X* **4** (2014) 031016.
30. H. Peng, N. B. M. Schroter, J. Yin, H. Wang, T. F. Chung, H. Yang, S. Ekahana, Z. Liu, J. Jiang, L. Yang, T. Zhang, C. Chen, H. Ni, A. Barinov, Y. P. Chen, Z. Liu, H. Peng and Y. Chen, *Adv. Mater.* **29** (2017) 1606741.
31. C. Christodoulou, A. Giannakopoulos, G. Ligorio, M. Oehzelt, M. Timpel, J. Niederhausen, L. Pasquali, A. Giglia, K. Parvez, K. Mullen, D. Beljonne, N. Koch and M. V. Nardi, *ACS Appl. Mater. Interfaces* **7** (2015) 19134.
32. A. L. Walter, K.-J. Jeon, A. Bostwick, F. Speck, M. Ostler, T. Seyller, L. Moreschini, Y. S. Kim, Y. J. Chang, K. Horn and E. Rotenberg, *Appl. Phys. Lett.* **98** (2011) 184102.
33. S. Schumacher, T. O. Wehling, P. Lazic, S. Runte, D. F. Forster, C. Busse, M. Petrovic, M. Kralj, S. Blugel, N. Atodiresei, V. Caciuc and T. Michely, *Nano Lett.* **13** (2013) 5013.
34. F. Huttmann, A. J. Martinez-Galera, V. Caciuc, N. Atodiresei, S. Schumacher, S. Standop, I. Hamada, T. O. Wehling, S. Blugel and T. Michely, *Phys. Rev. Lett.* **115** (2015) 236101.
35. R. Otero, A. L. Vázquez de Parga and J. M. Gallego, *Surf. Sci. Rep.* **72** (2017) 105.
36. D. Maccariello, M. Garnica, M. A. Niño, C. Navío, P. Perna, S. Barja, A. L. Vázquez de Parga and R. Miranda, *Chem. Mater.* **26** (2014) 2883.
37. T. C. Tseng, C. Urban, Y. Wang, R. Otero, S. L. Tait, M. Alcami, D. Ecija, M. Trelka, J. M. Gallego, N. Lin, M. Konuma, U. Starke, A. Nefedov, A. Langner, C. Woll, M. A. Herranz, F. Martin, N. Martin, K. Kern and R. Miranda, *Nat. Chem.* **2** (2010) 374.
38. V. D. Pham, J. Lagoute, O. Mouhoub, F. Joucken, V. Repain, C. Chacon, A. Bellec, Y. Girard and S. Rousset, *ACS Nano* **8** (2014) 9403.
39. V. D. Pham, S. Ghosh, F. Joucken, M. Pelaez-Fernandez, V. Repain, C. Chacon, A. Bellec, Y. Girard, R. Sporken, S. Rousset, Y. J. Dappe, S. Narasimhan and J. Lagoute, *npj 2D Mater. Appl.* **3** (2019) 5.

40. B. de la Torre, M. Svec, P. Hapala, J. Redondo, O. Krejci, R. Lo, D. Manna, A. Sarmah, D. Nachtigallova, J. Tucek, P. Blonski, M. Otyepka, R. Zboril, P. Hobza and P. Jelinek, *Nat. Commun.* **9** (2018) 2831.
41. J. Conradie, *J. Phys. Conf. Ser.* **633** (2015) 012045.
42. Y. Cai, H. Zhou, G. Zhang and Y.-W. Zhang, *Chem. Mater.* **28** (2016) 8611.
43. S. Mouri, Y. Miyauchi and K. Matsuda, *Nano Lett.* **13** (2013) 5944.
44. Z. Song, T. Schultz, Z. Ding, B. Lei, C. Han, P. Amsalem, T. Lin, D. Chi, S. L. Wong, Y. J. Zheng, M. Y. Li, L. J. Li, W. Chen, N. Koch, Y. L. Huang and A. T. S. Wee, *ACS Nano* **11** (2017) 9128.
45. Y. Jing, X. Tan, Z. Zhou and P. W. Shen, *J. Mater. Chem. A* **2** (2014) 16892.
46. D. Kiriya, M. Tosun, P. D. Zhao, J. S. Kang and A. Javey, *J. Am. Chem. Soc.* **136** (2014) 7853.
47. R. Guo, Q. Li, Y. Zheng, B. Lei, H. Sun, Z. Hu, J. Zhang, L. Wang, E. Longhi, S. Barlow, S. R. Marder, J. Wang and W. Chen, *Mater. Today* **30** (2019) 26.
48. N. Shen and G. Tao, *Adv. Mater. Interfaces* **4** (2017) 1601083.
49. M. R. Habib, W. Wang, A. Khan, Y. Khan, S. M. Obaidulla, X. Pi and M. Xu, *Adv. Theory Simul.* **3** (2020) 2000045.
50. S. Park, T. Schultz, X. Xu, B. Wegner, A. Aljarb, A. Han, L.-J. Li, V. C. Tung, P. Amsalem and N. Koch, *Commun. Phys.* **2** (2019) 109.
51. Z. Song, Q. Wang, M.-Y. Li, L.-J. Li, Y. J. Zheng, Z. Wang, T. Lin, D. Chi, Z. Ding, Y. L. Huang and A. T. S. Wee, *Phys. Rev. B* **97** (2018) 134102.
52. R. Resta, *Phys. Rev. B* **16** (1977) 2717.
53. Y. Liu, Z. Qiu, A. Carvalho, Y. Bao, H. Xu, S. J. Tan, W. Liu, A. H. Castro Neto, K. P. Loh and J. Lu, *Nano Lett.* **17** (2017) 1970.
54. M. Kang, B. Kim, S. H. Ryu, S. W. Jung, J. Kim, L. Moreschini, C. Jozwiak, E. Rotenberg, A. Bostwick and K. S. Kim, *Nano Lett.* **17** (2017) 1610.
55. Y. L. Huang, Z. Ding, W. Zhang, Y. H. Chang, Y. Shi, L. J. Li, Z. Song, Y. J. Zheng, D. Chi, S. Y. Quek and A. T. Wee, *Nano Lett* **16** (2016) 3682.
56. Y. L. Huang, Y. Chen, W. Zhang, S. Y. Quek, C. H. Chen, L. J. Li, W. T. Hsu, W. H. Chang, Y. J. Zheng, W. Chen and A. T. Wee, *Nat. Commun.* **6** (2015) 6298.
57. W. Zhou, X. Zou, S. Najmaei, Z. Liu, Y. Shi, J. Kong, J. Lou, P. M. Ajayan, B. I. Yakobson and J. C. Idrobo, *Nano Lett.* **13** (2013) 2615.
58. X. He, L. Zhang, R. Chua, P. K. J. Wong, A. Arramel, Y. P. Feng, S. J. Wang, D. Chi, M. Yang, Y. L. Huang and A. T. S. Wee, *Nat. Commun.* **10** (2019) 2847.
59. H. Li, C. Tsai, A. L. Koh, L. Cai, A. W. Contryman, A. H. Fragapane, J. Zhao, H. S. Han, H. C. Manoharan, F. Abild-Pedersen, J. K. Norskov and X. Zheng, *Nature Mater.* **15** (2015) 364.
60. J. Peto, T. Ollar, P. Vancso, Z. I. Popov, G. Z. Magda, G. Dobrik, C. Hwang, P. B. Sorokin and L. Tapaszto, *Nat. Chem.* **10** (2018) 1246.
61. D. Voiry, R. Fullon, J. Yang, E. S. C. de Carvalho Castro, R. Kappera, I. Bozkurt, D. Kaplan, M. J. Lagos, P. E. Batson, G. Gupta, A. D. Mohite, L. Dong, D. Er, V. B. Shenoy, T. Asefa and M. Chhowalla, *Nature Mater.* **15** (2016) 1003.
62. Y. Ouyang, C. Ling, Q. Chen, Z. Wang, L. Shi and J. Wang, *Chem. Mater.* **28** (2016) 4390.

63. E. P. Nguyen, B. J. Carey, J. Z. Ou, J. van Embden, E. D. Gaspera, A. F. Chrimes, M. J. Spencer, S. Zhuiykov, K. Kalantar-Zadeh and T. Daeneke, *Adv. Mater.* **27** (2015) 6225.
64. D. M. Sim, M. Kim, S. Yim, M.-J. Choi, J. Choi, S. Yoo and Y. S. Jung, *ACS Nano* **9** (2015) 12115.
65. K. Cho, M. Min, T.-Y. Kim, H. Jeong, J. Pak, J.-K. Kim, J. Jang, S. J. Yun, Y. H. Lee, W.-K. Hong and T. Lee, *ACS Nano* **9** (2015) 8044.
66. Q. Ding, K. J. Czech, Y. Zhao, J. Zhai, R. J. Hamers, J. C. Wright and S. Jin, *ACS Appl. Mater. Interfaces* **9** (2017) 12734.
67. J. H. Park, A. Sanne, Y. Guo, M. Amani, K. Zhang, H. C. P. Movva, J. A. Robinson, A. Javey, J. Robertson, S. K. Banerjee and A. C. Kummel, *Sci. Adv.* **3** (2017) e1701661.
68. J. Jiang, C. Ling, T. Xu, W. Wang, X. Niu, A. Zafar, Z. Yan, X. Wang, Y. You, L. Sun, J. Lu, J. Wang and Z. Ni, *Adv. Mater.* **30** (2018) e1804332.
69. S. S. Chou, M. De, J. Kim, S. Byun, C. Dykstra, J. Yu, J. Huang and V. P. Dravid, *J. Am. Chem. Soc.* **135** (2013) 4584.
70. J. S. Kim, H. W. Yoo, H. O. Choi and H. T. Jung, *Nano Lett.* **14** (2014) 5941.
71. M. Amani, D.-H. Lien, D. Kiriya, J. Xiao, A. Azcatl, J. Noh, S. R. Madhvapathy, R. Addou, K. C. Santosh, M. Dubey, K. Cho, R. M. Wallace, S.-C. Lee, J.-H. He, J. W. Ager III, X. Zhang, E. Yablonovitch and A. Javey, *Science* **350** (2015) 1065.
72. X. Zhao, D. Fu, Z. Ding, Y. Y. Zhang, D. Wan, S. J. R. Tan, Z. Chen, K. Leng, J. Dan, W. Fu, D. Geng, P. Song, Y. Du, T. Venkatesan, S. T. Pantelides, S. J. Pennycook, W. Zhou and K. P. Loh, *Nano Lett.* **18** (2018) 482.
73. A. M. van der Zande, P. Y. Huang, D. A. Chenet, T. C. Berkelbach, Y. You, G. H. Lee, T. F. Heinz, D. R. Reichman, D. A. Muller and J. C. Hone, *Nature Mater.* **12** (2013) 554.
74. Z. Lin, B. R. Carvalho, E. Kahn, R. Lv, R. Rao, H. Terrones, M. A. Pimenta and M. Terrones, *2D Mater.* **3** (2016) 022002.
75. S. Barja, S. Wickenburg, Z.-F. Liu, Y. Zhang, H. Ryu, M. M. Ugeda, Z. Hussain, Z.-X. Shen, S.-K. Mo, E. Wong, M. B. Salmeron, F. Wang, M. F. Crommie, D. F. Ogletree, J. B. Neaton and A. Weber-Bargioni, *Nat. Phys.* **12** (2016) 751.
76. P. K. J. Wong, W. Zhang, F. Bussolotti, X. Yin, T. S. Herng, L. Zhang, Y. L. Huang, G. Vinai, S. Krishnamurthi, D. W. Bukhvalov, Y. J. Zheng, R. Chua, A. T. N'Diaye, S. A. Morton, C. Y. Yang, K. H. O. Yang, P. Torelli, W. Chen, K. E. J. Goh, J. Ding, M. T. Lin, G. Brocks, M. P. de Jong, A. H. Castro Neto and A. T. S. Wee, *Adv. Mater.* **31** (2019) e1901185.
77. H.-P. Komsa and A. V. Krasheninnikov, *Adv. Electron. Mater.* **3** (2017) 1600468.
78. M. A. Gerkman, J. K. Lee, X. Li, Q. Zhang, M. Windley, M. V. Fonseca, Y. Lu, J. H. Warner and G. G. D. Han, *ACS Nano* **14** (2020) 153.
79. D. Gunder, K. Watanabe, T. Taniguchi and G. Witte, *ACS Appl. Mater. Interfaces* **12** (2020) 38757.
80. C. R. Ryder, J. D. Wood, S. A. Wells, Y. Yang, D. Jariwala, T. J. Marks, G. C. Schatz and M. C. Hersam, *Nat. Chem.* **8** (2016) 597.

81. Z. Sofer, J. Luxa, D. Bousa, D. Sedmidubsky, P. Lazar, T. Hartman, H. Hardtdegen and M. Pumera, *Angew. Chem., Int. Ed.* **56** (2017) 9891.
82. Y. Cao, X. Tian, J. Gu, B. Liu, B. Zhang, S. Song, F. Fan and Y. Chen, *Angew. Chem., Int. Ed.* **57** (2018) 4543.
83. Y. He, F. Xia, Z. Shao, J. Zhao and J. Jie, *J. Phys. Chem. Lett.* **6** (2015) 4701.
84. C. Wang, D. Niu, B. Liu, S. Wang, X. Wei, Y. Liu, H. Xie and Y. Gao, *J. Phys. Chem. C* **121** (2017) 18084.
85. Y. Zhao, Q. Zhou, Q. Li, X. Yao and J. Wang, *Adv. Mater.* **29** (2017) 1603990.
86. R. Guo, Y. Zheng, Z. Hu, J. Zhang, C. Han, E. Longhi, S. Barlow, S. R. Marder and W. Chen, *Chem. Eur. J.* **26** (2020) 1.
87. D. Zhou, N. Si, Q. Tang, B. Jiang, X. Song, H. Huang, M. Zhou, Q. Ji and T. Niu, *J. Phys. Chem. C* **123** (2019) 12947.
88. X. Liu, Z. Wei, I. Balla, A. J. Mannix, N. P. Guisinger, E. Luijten and M. C. Hersam, *Sci. Adv.* **3** (2017) e1602356.
89. C. H. Cheng, D. Cordovilla Leon, Z. Li, E. Litvak and P. B. Deotare, *ACS Nano* **14** (2020) 10462.
90. C. Muccianti, S. L. Zachritz, A. Garlant, C. N. Eads, B. H. Badada, A. Alfrey, M. R. Koehler, D. G. Mandrus, R. Binder, B. J. LeRoy, O. L. A. Monti and J. R. Schaibley, *J. Phys. Chem. C* **124** (2020) 27637, doi/10.1021/acs.jpcc.0c06544.
91. S.-W. Shen, Y.-C. Wei, F.-Y. Fu, K.-J. Chou, S.-F. Wang, L.-W. Fu, W.-T. Chuang, M.-L. Tsai, M.-H. Chen, Y. Chi, P.-T. Chou and C.-I. Wu, *J. Phys. Chem. C* **124** (2020) 25538.
92. L. Zhang, T. Yang, W. Zhang, D. Qi, X. He, K. Xing, P. K. J. Wong, Y. P. Feng and A. T. S. Wee, *Appl. Mater. Today* **18** (2020) 100535.
93. R. Xu, F. Xuan and S. Y. Quek, *J. Phys. Chem. Lett.* **11** (2020) 9358.
94. F. Calavalle, P. Dreher, A. P. Surdendran, W. Wan, M. Timpel, R. Verucchi, C. Rogero, T. Bauch, F. Lombardi, F. Casanova, M. V. Nardi, M. M. Ugeda, L. E. Hueso and M. Gobbi, *Nano Lett.* (2020), doi/10.1021/acs.nanolett.0c03386.
95. M. Dreher, D. Günder, S. Zörb and G. Witte, *Chem. Mater.* **32** (2020) 9034.
96. J. Hagara, N. Mrkyvkova, P. Nadazdy, M. Hodas, M. Bodik, M. Jergel, E. Majkova, K. Tokar, P. Hutar, M. Sojkova, A. Chumakov, O. Konovalov, P. Pandit, S. Roth, A. Hinderhofer, M. Hulman, P. Siffalovic and F. Schreiber, *Phys. Chem. Chem. Phys.* **22** (2020) 3097.

https://doi.org/10.1142/9789811247859_0006

Chapter 6

Screening Effects at Organic–2D Material Heterointerfaces

Yu Jie Zheng*,‡ and Arramel†

*MOE Key Laboratory of Low-grade Energy Utilization Technologies and Systems, School of Energy and Power Engineering, Chongqing University, 174 Shazhengjie, Shapingba, Chongqing 400044, P. R. China

†Department of Physics, National University of Singapore, 2 Science Drive 3, Singapore 117551, Singapore

‡zhengyujie@cqu.edu.cn

Organic molecule–two-dimensional material (OM–2D material) heterostructures are relatively an emergent class of hybrid-based electronic devices and yet have gained enormous interests. These systems hold promise towards application in the realization of electronic, optoelectronic and energy conversion devices, due to their tunable thickness, rich raw materials, flexibility, and also their capability to merge the high light absorption efficiency of organic molecules and the high carrier mobility of 2D materials. Dielectric screening effects have a large influence on electronic structure and optical properties at the OM–2D material interface due to the low intrinsic dielectric polarizability of both organic molecules and 2D materials. Here, we will review current knowledge about the dielectric screening behaviors in 2D materials and the screening effect at the OM–2D material heterointerfaces, including the physical origin of the dielectric screening effect across such interfaces, how to investigate it, and the influence of the screening effect on OM–2D material interfaces. Engineering of the dielectric screening effects in OM–2D material systems is also discussed. We conclude the review by illustrating a brief introduction of other important factors, e.g. defects, charge transfer, and hybridization,

‡Corresponding author.

that will contribute to the screening effect and these features should be considered for the future scientific and device development of OM–2D material heterostructures.

Keywords: 2D materials; organic molecules; heterointerfaces; dielectric screening; electronic structure; exciton.

1. Introduction

Screening effects, also called dielectric screening, have a simplified interpretation that the screening of perturbations could be produced by the charged impurities and/or defects and various many-body interactions (i.e. electron–electron) in a solid.[1] For example, the Coulomb force between charged particles (e.g. two electrons) is considered as a long-range effect, and the flow of electrons might reduce the effective force between the contributing particles resulting in a short-range "screened" Coulomb interaction — this is called the screening effect. As shown in Fig. 1, the screening effect is described by the dielectric response function $\epsilon(r, r', t - t')$ which is the central concept of the electronic, transport, optical, and magnetic properties of solids.[1] In semiconductors and insulators, an incomplete screening is able to modify the Coulomb interactions between electronic carriers and therefore, it strongly influences the transport and optical properties. In the long-wavelength static limit, the dielectric response function of three-dimensional (3D) bulk insulators (semiconductors)

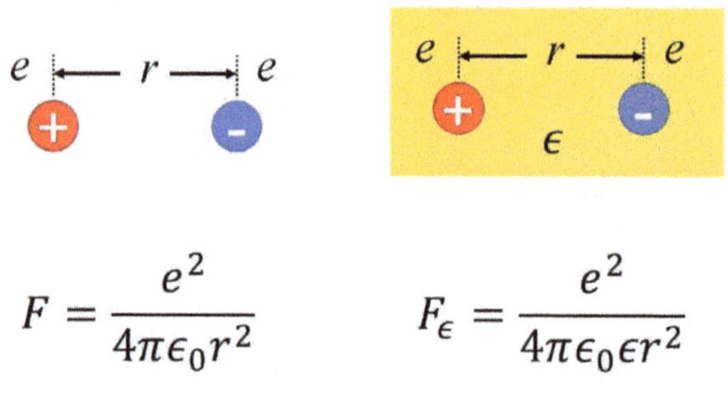

Fig. 1. The Coulomb forces between two charges with and without dielectric screening.

can be described as a constant ϵ. However, for two-dimensional (2D) materials, the electric field lines between charges extend significantly beyond the materials due to the reduction of dimension along the z (out-of-plane) direction. Hence, the dielectric screening of Coulomb interactions is reduced compared to that of their bulk crystals and thereby, 2D materials are more sensitive to the dielectric environment.

Heterojunction is an important basic component of electronic and optoelectronic devices. Logic chips containing billions of silicon-based metal–oxide field-effect transistors are commonly found in daily life. In recent trends, the rapid development of semiconductor electronic devices depends on the size or dimension reduction of field-effect transistors. However, with the high integration, miniaturization, and flexibility of devices, it is generally believed that silicon-based semiconductor heterojunctions have approached their electrical/optoelectronic performance limit. Therefore, seeking alternative solutions to maintain the continued development of the semiconductor industry is a major challenge facing the world. In 2004, the successful preparation of graphene and graphene-based field-effect transistors was expected to break through the "bottleneck" of silicon-based metal–oxide field-effect transistors, which stimulated people's exploration of new 2D materials and their heterojunctions.[2,3]

The 2D materials.[4–6] [such as graphene,[7] hexagonal boron nitride (hBN),[8] transition metal dichalcogenides (TMDs),[9] black phosphorene,[10] monochalcogenides,[11] chromium triiodide,[12] etc.] cover almost all material properties (from superconductors, conductors, semiconductors, to insulators), which can provide a wide range of choices for constructing 2D van der Waals heterostructures (vdWHs) with multifunctional properties. However, because they have only one or a few atomic layers, their light absorption capacities are poor.[13] Although few papers have reported the high yield with good quality of 2D materials,[14,15] it is still a challenging task to produce a multifunctional heterojunction solely based on 2D semiconductors.

Organic molecule–2D material (OM–2D material) heterostructures have received enormous attention in the last 10 years in both fundamental studies and emerging applications.[16–23] As shown in Fig. 2,

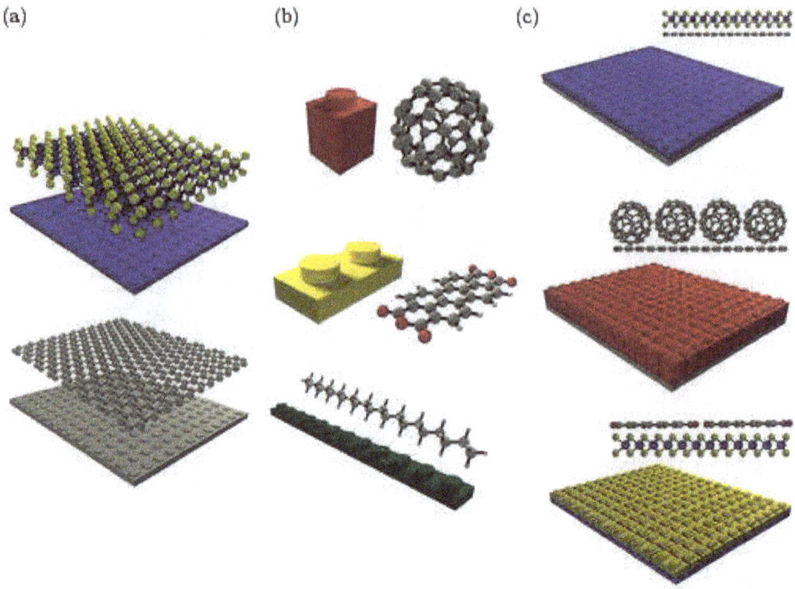

Fig. 2. Various van der Waals heterostructures with on-purpose properties that could be obtained by combining the 2D materials with organic molecules.
Source: Reproduced with permission from Ref. 24. Copyright 2018, Wiley-VCH.

the addition of organic molecules to a 2D heterojunction has great potential, because although nature provides a limited number of 2D materials, it is possible for people to design and synthesize an almost unlimited variety of organic molecules with predictable functions.[24] Most organic molecules have a large number of conjugated π-bonds which can form a 2D interface that has a similarity with respect to the inorganic counterparts. For example, vdWH interaction can be realized when organic molecules are deposited onto inorganic 2D materials. OM–2D material heterostructures are mainly comprised of single or few organic molecular layers that directly interface with 2D materials to form a hybrid heterojunction. OM–2D material vdWH provides a unique platform for combining the advantages of the organics and 2D materials and expands the horizon of 2D vdWH.[22,24] For example, organic molecules exhibit strong light absorption with a poor charge mobility, while 2D materials have a relatively low light absorption but a better in-plane charge transport performance. Hence, the high charge mobility and light absorption efficiency of OM–2D material heterojunctions can be obtained, which have shown promising direction for optoelectronic devices, such as photodectors,[25] photovoltaics,[26] and quantum information devices.[27] Moreover, OM–2D material vdWH also displays unprecedented interfacial phenomena, such as the two-dimensional growth mode of 2,3,5,6-Tetrafluoro-7,7,8,8-tetracyanoquinodimethane (F_4TCNQ) molecules on graphene–BN heterostructures[28] and the enhancement of magnetic moment of cobalt phthalocyanine (CoPc) molecules on vanadium diselenide,[29,30] which are completely different outcomes in comparison to the case of organic molecules adsorption on metal substrates.[31–33]

As discussed previously, the Coulomb interactions between charge carriers in organic molecules, 2D materials, and their combined heterostructures are enhanced due to quantum confinement and resulted in a reduction in their dielectric screening.[34] Therefore, the OM–2D material heterostructures are very sensitive to dielectric screening and this review focuses on the screening effect of 2D materials and its impact on the electronic, transport, optical, and magnetic properties of the OM–2D material interface.

This review is organized as follows. First, we briefly revisit the physical description of screening effect in the perspective of condensed matters and subsequently introduce how to investigate it both experimentally and theoretically. Then we introduce the intrinsic screening of 2D materials and the influences of screening effect on the electronic and optical properties of organic molecules and 2D materials at the heterointerface. Finally, engineering of the screening effects in OM–2D material systems and the potential unintended consequences of screening effect are highlighted. We will conclude the review by emphasizing that further study on the implication of other factors that could be merged with the screening effect is required.

2. Fundamental Concept of the Dielectric Screening Effects

2.1. *Dielectric function*

The dielectric response function of a material is used for describing the screening effect. Let us consider a weak external potential, represented by $\delta V_{\text{ext}}(r', t')$, being applied to a system of mobile charge carriers. Based on the linear response theory, we can assume that the change of total potential $\delta V(r, t)$ is linearly proportional to the week external potential derived from the following equation:

$$\delta V(r,t) = \int \epsilon^{-1}(r, r', t - t')\,\delta V_{\text{ext}}(r', t')dr'dt', \tag{1}$$

where ϵ^{-1} is called the inverse dielectric function.[1] In the periodic systems, the equation can be written in reciprocal and frequency space as follows:

$$\delta V_{\text{ext}}(q + G, \omega) = \sum_{G'} \epsilon(q + G, q + G', \omega)\delta V(q + G', \omega). \tag{2}$$

The dielectric function $\epsilon(q + G, q + G', \omega)$ describes the general responses of a system to an external potential perturbation. In general, ϵ is a complex quantity: $\epsilon = \epsilon_1 + i\epsilon_2$, where ϵ_1 and ϵ_2 satisfy the Kramers–Kronig relations.

Besides the dielectric function ϵ, polarizability χ is also widely used to describe the screening effect of solid systems. It describes the electron density response to a small external perturbation. Similar to ϵ, χ is defined by

$$\delta \rho_{\text{ind}}(r,t) = \int \chi(r, r', t - t')\delta V_{\text{ext}}(r', t')dr'dt', \tag{3}$$

where $\delta \rho_{\text{ind}}(r, t)$ is the induced charge density.

2.2. *Experimental observations and calculations of the dielectric screening effects*

The OM–2D material heterointerface consists of two components: organic molecules and 2D materials. Here, we may follow two approaches to investigate the heterostructure of either the organic molecule or the 2D material. Hence, conventional methods to characterize the screening effect at the interface can be divided into two categories. The first set of methods are used to determine the electronic properties of organic molecules, i.e. photoemission[35] combined with inverse photoemission[36] spectroscopy[37] and scanning tunneling microscopy/spectroscopy (STM/STS).[28,38,39] These methods can extract the outermost energies of the molecule, in the so-called highest occupied molecular orbital (HOMO), lowest unoccupied molecular orbital (LUMO), and HOMO–LUMO gaps. The degrees of orbital shifts and gap variation reflect the dielectric screening strength of the 2D substrates. For example, the removal of an electron from a molecule's HOMO results in a hole being left behind. As shown in Fig. 3(a), if the molecules are adsorbed on a substrate, the substrate electrons will screen the hole and make it easier to remove the electron from HOMO.[40] Therefore, the HOMO level shifts up. Similarly, the added electron into LUMO level is also stabilized by screening from the electrons, thus lowering the LUMO level. Alternatively, some other methods directly measure the interfacial structure of 2D materials by probing their surface potential as well as electronic and optical properties under different dielectric environments. These methods include electrostatic force microscopy (EFM)[41–44] and

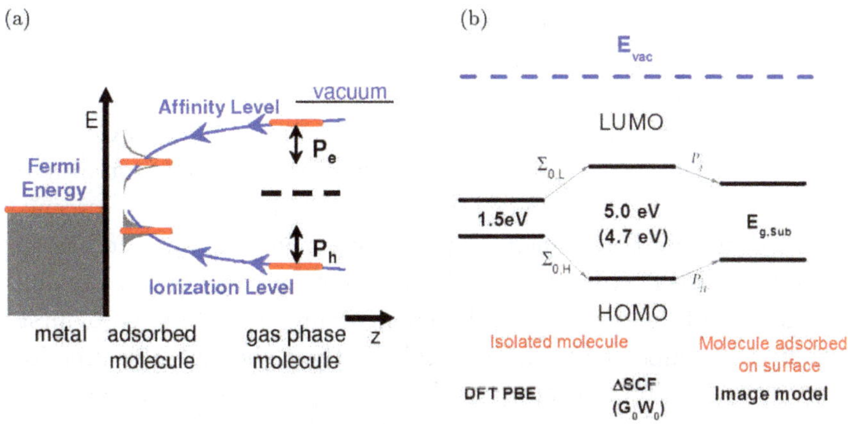

Fig. 3. Screening effects for a molecule on a substrate and the two-step approach. (a) Schematic energy-level diagram indicating polarization shifts in the frontier energy levels of a molecule upon adsorption on a metal surface. (b) Two-step approach for predicting the HOMO–LUMO gap of organic molecule on the substrate.

Source: Panel (a) reproduced with permission from Ref. 49. Copyright 2006, American Physical Society. Panel (b) reproduced with permission from Ref. 39. Copyright 2016, American Chemical Society.

Kelvin probe force microscopy (KPFM)[45,46] which measure the surface electrostatic potential, as well as reflectance contrast spectroscopy[34,47] and polarized magneto-absorption spectroscopy[48] related to the determination of their binding energies or the excited states of excitons.

The screening effect of materials is the nonlocal exchange–correlation effect. In principle, it should be calculated by first-principles within quasi-particle (GW) approximation[50,51] and this effort has been successfully implemented for the adsorbed benzene molecule on different substrates.[49,52] However, the existing GW calculations on the reported OM–2D interfaces are computationally expensive. Therefore, a two-step approach is widely used to consider the screening effect of 2D substrates and reproduce the HOMO–LUMO gap of organic molecule on 2D substrates.[39,53,54] As shown in Fig. 3(b), first, the corrections at the HOMO and LUMO levels of the isolated organic molecules are calculated within GW approximation. Then, the screening effects originating from the substrate are calculated separately, and subsequently, the corrections are adapted to the HOMO (P_H) and LUMO (P_L) positions. Some representative methods that take into account the screening effects of 2D substrates are image charge model,[49,55] GΔW method,[56] and XAF–GW method.[54]

Image charge model approximately replaces the screened Coulomb potential of the substrate with the electrostatic Coulomb potential of an image charge.[39,49,53,55] For GΔW method, the dynamical screened interaction ΔW is obtained using the quantum-electrostatic heterostructure model[57] (first, the dielectric functions of the freestanding versions of all the layers in the heterostructure are calculated by first-principles or imported from the database; then, the dielectric function of the whole heterostructure is obtained by solving a simple set of linear equations describing the electrostatic coupling between the layers). The XAF–GW method (X: eXpand-chi, A: Add-chi, and F: Full-wave function) contains three steps. First, the polarizabilities of the smallest cells of organic molecules and 2D materials are calculated, respectively, and then they are expanded to obtain the polarizability of the supercell which constitutes the heterostructure.

Then the total polarizability is obtained by adding the polarizabilities of the two supercells. Finally, the self-energy term with the calculated polarizability and full-wave function of the heterostructure is calculated.[54]

3. Intrinsic Dielectric Polarizability of 2D Materials

The intrinsic dielectric polarizability of 2D materials is an important parameter that plays a significant role in their electrical, optical, and magnetic properties compared to three-dimensional systems. This is mainly governed by the reduction of dimension and therefore, a decrease in screening along the z (out-of-plane) direction. The Coulomb interaction between charges in thin films increases strongly as the film thickness reduces. This interaction arises due to the generated fields which were produced by these charges in the vacuum surrounding the film being much larger than those in its homogeneous bulk.[58] Based on this outcome, one should consider that the interaction between charges in 2D materials should be much larger than that in their bulk counterparts.

3.1. *Nonlocal in-plane screening*

First, we discuss the characteristics of in-plane screening of 2D materials, which determine the scattering of the electrons and size of excitons. The nonlocal in-plane screening of the 2D materials has been rationalized using the first-principles calculations within the framework of GW approximation.[57,59–61] As shown in Fig. 4(a), the q-dependent dielectric functions show a logarithmic divergence for small distances and reach an unscreened Coulomb potential for large distances, which is considered as a nonlocal case. Moreover, this feature is a quite different behavior from 3D systems where macroscopic screening can be described by a static dielectric constant. However, unlike the isolated 2D materials, Latini *et al.*[60] reported that the screening characters of quasi-2D materials [Figs. 4(b) and 4(c)] such as supported layers or layers embedded in a vdWH resemble those of the bulks [Figs. 4(d) and 4(e)].

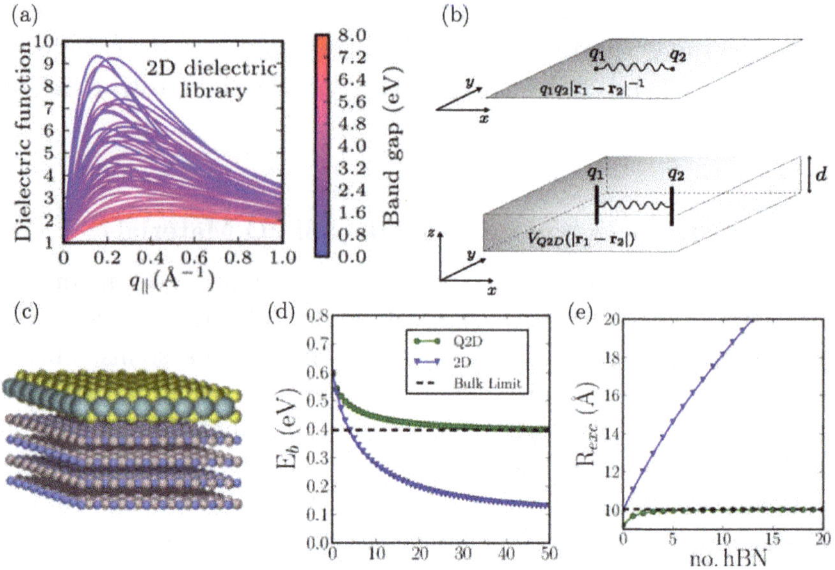

Fig. 4. (Color online) The q-dependent in-plane screening. (a) Static dielectric functions of 51 2D transition metal dichalcogenides and oxides with different sizes of the bandgap. (b) Sketch of the pure 2D and quasi-2D in-plane Coulomb interactions. (c) Atomic structure of MoS_2–hBN heterostructures. (d), (e) Energy and radius of the lowest bound exciton for MoS_2–hBN heterostructures as a function of the number of hBN layers obtained from the quasi-2D materials (green) and 2D (blue) approaches.

Source: Panel (a) reproduced with permission from Ref. 57. Copyright 2015, American Chemical Society. Panels (b)–(e) reproduced with permission from Ref. 60. Copyright 2015, American Physical Society.

3.2. *Thickness-dependent out-of-plane screening*

The 2D materials are ideal substrates that act as a convenient platform for the organic molecules since their deposition can be performed on their surface plane.[31,39] Hence, the vertical screening of 2D materials contributes mainly to the screening effect at the OM–2D material interface. Therefore, the characterization and investigation of vertical screening out of the plane are necessary. EFM has been widely used to investigate the screening effects of 2D materials, e.g. graphene,[62] 2D TMDs,[43,44] and hBN.[42] As shown in Fig. 5(a), Datta *et al.*[62] observed that the screening effects reduce with the decrease in number of layers for few-layer graphene and the dielectric constant is

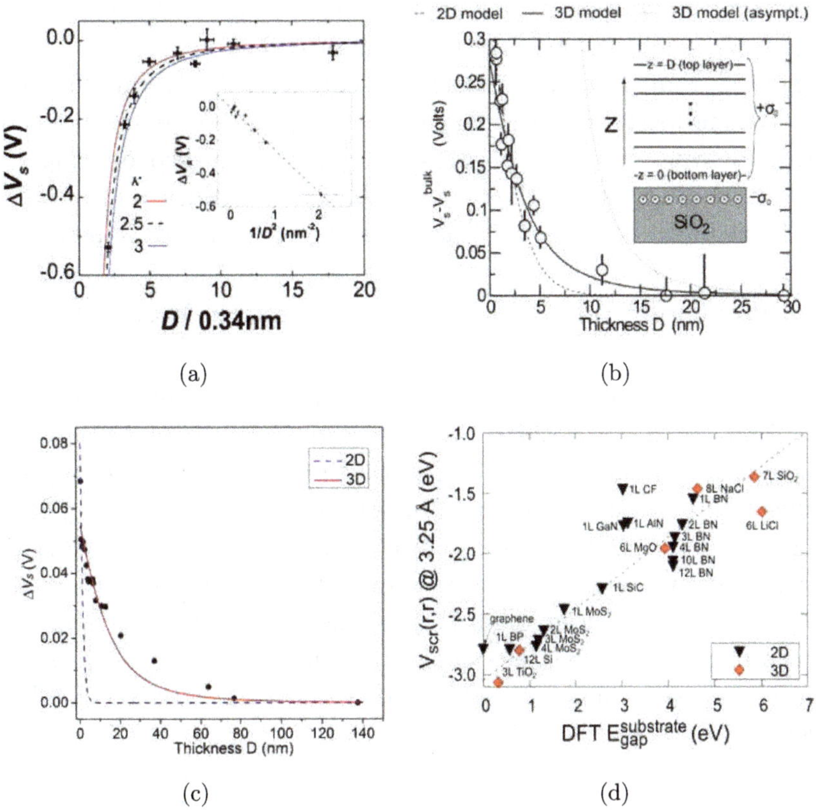

Fig. 5. Thickness-dependent out-of-plane screening. (a) Plot of the variation in surface potential ΔV_S ($\Delta V_S = V_S - V_S^{\max}$) versus few-layer graphene thickness D [single-layer (SL) graphene thickness of 0.34 nm]. (b), (c) Dependences of deviation from the bulk value of ΔV_S as a function of the flake thickness for MoS_2 and hBN. (d) Plot of the planar-average screening potential at 3.25 Å away from the substrate against the DFT Kohn–Sham gap of the substrate.

Source: Panel (a) reproduced with permission from Ref. 62. Copyright 2009, American Chemical Society. Panel (b) reproduced with permission from Ref. 43. Copyright 2013, Wiley-VCH. Panel (c) reproduced with permission from Ref. 42. Copyright 2015, American Chemical Society. Panel (d) reproduced with permission from Ref. 63. Copyright 2019, IOP Publishing, Ltd.

comparable to bulk graphite when the thickness is increased to five layers. Thickness-dependent screening is also observed for few-layer 2D TMDs[43,44] and hBN,[42] as shown in Figs. 5(b) and 5(c), respectively. However, the screening effects on other materials such as 2D TMDs and hBN are weaker by one order of magnitude than that of graphene case. Moreover, through systematic first-principles calculations within GW approximation, Noori *et al.*[63] suggested that this thickness-dependent screening is an intrinsic physical property of 2D materials. The proposed explanation is acquired from the variation of benzene gaps absorbed on different substrates [Fig. 5(d)], they found that the intrinsic out-of-plane screening effect is reduced as the bandgap of 2D materials increased.

4. Screening Effect at the OM–2D Inorganic Material Interface

Both the organic molecules and 2D materials are sensitive to dielectric environment due to their low intrinsic dielectric polarizability. Therefore, the screening effect at the interface will have a great influence on the electronic structures of the interface (i.e. energy-level alignment) and 2D materials. The energy-level alignment of the heterostructure determines the energy barrier of the electrons and holes transfer between organic molecules and 2D materials. The screening and shifting of electronic structure of the 2D materials will affect the scattering and mobility of the carriers and the binding energy of excitons.

4.1. *Screening effect of 2D materials on organic molecules*

As discussed in Sec. 3.2, the screening effects of 2D materials are lower than their bulk and therefore much lower than those of metallic substrates. Interestingly, a lower screening effect can also affect the growth mode of the organic molecule. Recently, as shown in Figs. 6(a) and 6(b), tightly packed 2D islands of F_4TCNQ assembled on graphene–BN are observed with STM and noncontact atomic force microscopy by Tsai *et al.*[28] This is in contrast with the F_4TCNQ on graphene supported by metal substrates where F_4TCNQ molecules are packed as 3D islands.[31,64] Through detailed comparison of the cohesion energy of the molecular crystal, molecular flat islands with and without graphene through first-principles calculations, they found that the stable 2D islands on graphene–hBN are observed due to the poor screening effect of graphene–hBN substrate

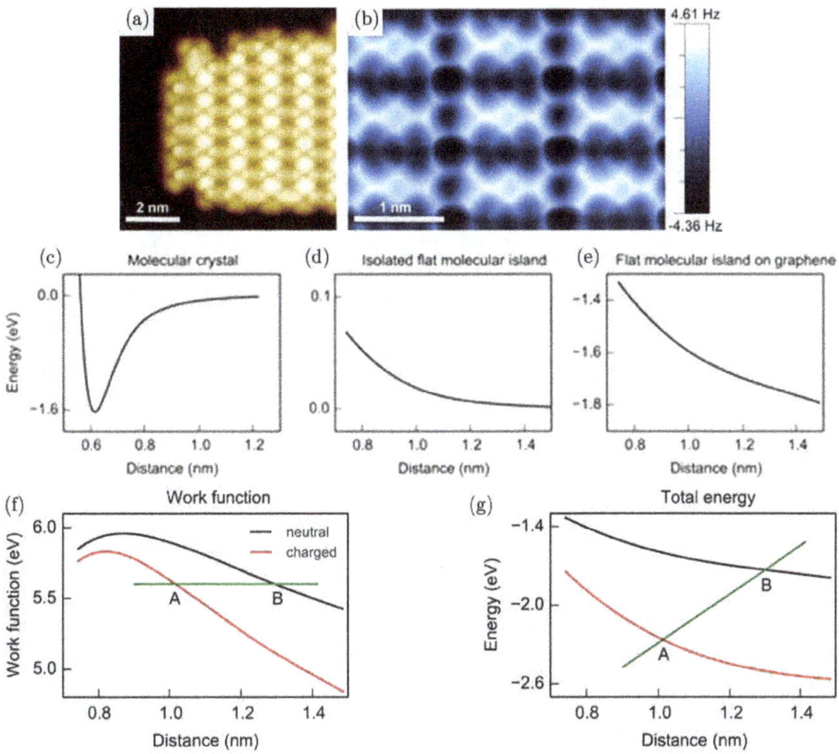

Fig. 6. (Color online) The 2D F_4TCNQ islands on graphene–hBN. (a) STM and (b) atomic force microscopy images of an F_4TCNQ 2D island on graphene–hBN. (c)–(e) Calculated total energy per F4TCNQ molecule as a function of the shortest distance between molecules for a bulk molecular crystal, an isolated flat 2D molecular island, and flat 2D molecular island on graphene substrate, respectively. (f), (g) Work function above an F_4TCNQ–graphene system and the total energy per F_4TCNQ molecule on graphene as a function of the shortest distance between molecules for both the neutral case (black line) and the electron-doped case (red line), respectively. Green lines show hypothetically constant electrochemical potential.

Source: Reproduced with permission from Ref. 28. Copyright 2015, American Chemical Society.

[Figs. 6(c)–6(f)]. Owing to the large binding energy (1.6 eV) of the 3D bulk [Fig. 6(c)] and negative energy-versus-distance slopes of 2D planar geometries [Figs. 6(d) and 6(e)], it is indicated that F_4TCNQ molecules preferably form 3D rather than 2D islands. However, the poor screening could promote the formation of 2D islands that can be understood as follows: F_4TCNQ molecules become negatively charged upon interfacing them in the F_4TCNQ–graphene system. This is attributed to the charge transfer that arises from graphene towards the F_4TCNQ molecular layers. When these negatively charged molecules come together to form an island, the local work function at the island increases [Fig. 6(f)], and subsequently attracts more electrons to flow into the island. Consequently, these extra electrons cause a decrease in the total energy of the F_4TCNQ–graphene complexes [Fig. 6(g)], resulting in the formation of 2D island cohesion. However, this mechanism is not applicable to the case of graphene adsorbates on metallic substrates. Comparing to graphene–hBN, graphene–metal substrates have higher screening effect due to their high density of states. Therefore, in the graphene–metal systems, the electric fields caused by charge transfer are screened, and charge transfer induced by work function differences does not lead to reduced total energy. These findings further indicate that 2D materials are suitable substrates to deposit organic molecules to form well-ordered functional 2D vdWHs.

In contrast to the geometry structure, the electronic structure of the organic molecules should be more sensitive to the screening effect. To understand the electronic screening effect of 2D materials on organic molecules, it is helpful to revisit the conceptual basis of HOMO and LUMO levels. The HOMO level represents the amount of energy required to remove an electron from the molecule, while the LUMO level represents the amount of energy needed to add an electron to the molecule. The removal of an electron from the molecule (e.g. by photoemission) results in a hole being left behind. As shown in Fig. 3(a), when a molecule is introduced on the 2D material substrate, the substrate electrons rearrange to screen the effect of the hole, thus stabilizing the hole, and therefore it is easier to remove the electron in HOMO. As a result, this event raises the HOMO level. Similarly, the added electron in the molecule (e.g. by inverse photoemission) is stabilized by screening through the electrons of the substrates, thus lowering the LUMO level. This reduction in the HOMO–LUMO gap can be as large as several eV, e.g. for benzene on metallic surfaces.[49,52] Recently, Zheng *et al.*[39] systematically

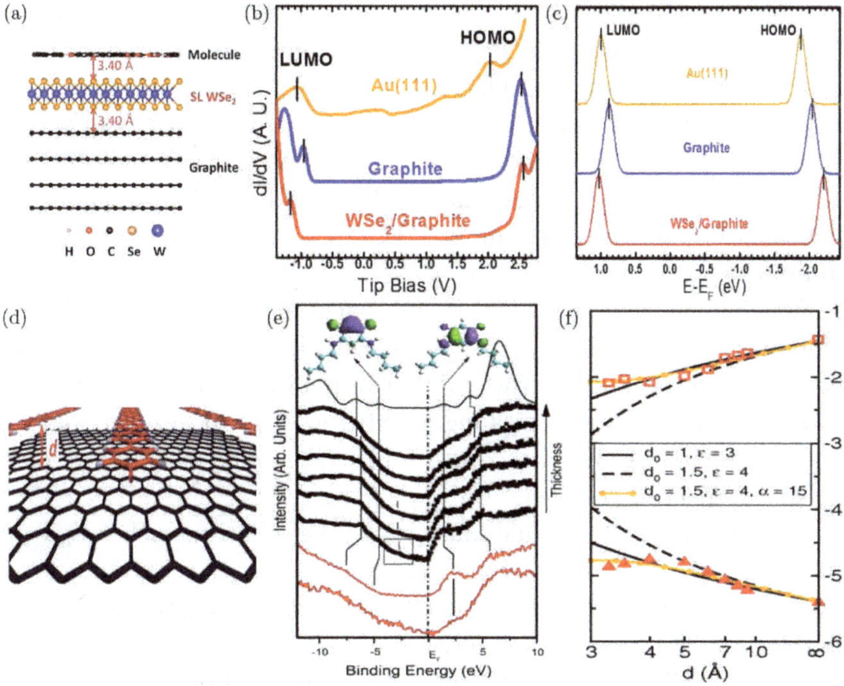

Fig. 7. (Color online) Renormalization of the HOMO–LUMO gap on 2D substrates. (a) Atomic structure of PTCDA on SLWSe_2 on graphite. (b) Structure of poly(para-phenylene) (red) adsorbed on graphene (black). (c), (d) STS measured and theoretically predicted HOMO and LUMO levels of monolayer PTCDA on different substrates. (e) Combined photoemission and inverse photoemission spectra (thick lines) of quinonoid zwitterion adsorbed on Au (black) and graphene on copper (red). (f) The G_0W_0 energies of the HOMO (filled triangles) and LUMO (open squares) of PPP as a function of the adsorption distance d.

Source: Panels (a)–(c) reproduced with permission from Ref. 39. Copyright 2016, American Chemical Society. Panels (d) and (f) reproduced with permission from Ref. 65. Copyright 2012, American Physical Society. Panel (e) reproduced with permission from Ref. 37. Copyright 2012, Elsevier.

compared the geometric and electronic structures of PTCDA molecules adsorbed on SL WSe_2–graphite [Fig. 7(a)], graphite, and Au(111) substrates by STM and STS. The cell shape and lattice parameters of the PTCDA monolayers are nearly the same on the three substrates because of the weak van der Waals forces between the molecules and the substrates. However, as shown in Fig. 7(b), the

HOMO–LUMO gaps of PTCDA are quite different, increasing from 3.10 eV, 3.49 eV to 3.73 eV on Au(111), graphite, and SLWSe_2–graphite, respectively. Through first-principles calculation and two-step approach, they found that the difference of HOMO–LUMO gaps is mainly due to the different screening effects of the substrates [Fig. 7(c)]. They also found that SL WSe_2 contributes to the main screening effect of the change of PTCDA's HOMO–LUMO gap by comparing the screening contribution with and without (the SLWSe_2 layer was substituted by vacuum thickness) SLWSe_2. Moreover, as shown in Fig. 7(e), the HOMO–LUMO gaps of quinonoid zwitterion measured by the combined photoemission and inverse photoemission spectroscopies increased from 5.9 ± 0.1 eV on gold to 7.4 ± 0.4 eV on graphene–copper substrates due to the weaker screening effect of graphene.[37]

Similar to small organic molecules, the quasi-particle gaps of polymers[65,66] also show obvious renormalization when adsorbed on 2D materials due to the screening effect. By comparing the calculated HOMO–LUMO gaps of polymer adsorbed on graphene [Fig. 7(d)] using both GW approximation and two-step approach with the classical image-potential model to model the screening effect of graphene, Puschnig *et al.*[65] found that the polarizability of the adsorbate must be taken into account for the adsorption distance lower than 4.5 Å [Fig. 7(f)]. The polarizability of the polymer will compensate the screening effect of the 2D substrate, i.e. the HOMO–LUMO gap will increase a bit for the adsorbed polymer than its monomer [Fig. 7(f)]. A similar phenomenon was also reported for the adsorbed PTCDA layer on the SL WSe_2–graphite substrate.[39] The calculated HOMO–LUMO gap of the isolated PTCDA monolayer within G_0W_0 is 4.1 eV, which is 0.6 eV smaller than that computed for the isolated PTCDA molecule. The inclusion of substrate image charge screening effects is proposed in the case of PTCDA adsorption via an embedding procedure[39] by means of the G_0W_0 calculations. Here, the metallic substrate is modeled as a semi-infinite electron gas and its screening properties are included via a renormalization of the bare Coulomb potentials. The predicted HOMO–LUMO gaps of the molecule with and without the polarizability of the PTCDA layer itself are found to be 2.9 eV and 2.4 eV, respectively. Hence, these results indicate that the polarizability of the organic molecule layer compensates for the screening effect of the 2D substrate.

However, a contrast behavior was observed in the very recent experiment results.[67] Pörtner *et al.*[67] employed STM to perform geometric manipulation of the F_{16}CoPc molecules [Fig. 8(a)] on hBN–Cu(111) substrate. They determined that the local molecular orbitals for the isolated F_{16}CoPc and segregated molecules are formed in different assemblies: a cluster of two molecules, a cluster of three molecules, and an island of molecules. It is obvious that the molecule orbital shifts up towards the Fermi level with the increase in the number of adjacent molecules [Figs. 8(b) and 8(c)] due to the increased screening by the neighboring molecules. To sum up the above studies, the interplay of screening of the 2D substrate and the molecule itself and/or induced by the presence of neighbor molecules is a complicated screening problem. Therefore, this subject requires further comprehensive and controllable studies, especially in the aspect of theoretical calculations.

Apart from the polarizability of molecules and their response to the adjacent molecules, the configuration of adsorbate and the charge transfer (between organic molecules and 2D materials) also play significant role towards the screening effect of 2D materials on the organic molecules at the interface.[65,67–69] Fu *et al.*[68] found that graphene has a notable influence on the optical excitations of azobenzene in both the *trans*- [Fig. 9(a)] and the *cis*-configurations [inset of Fig. 9(c)] due to its screening effect using GW calculations. The screening of graphene not only redshifts the intramolecular exciton peak and blueshifts the intermolecular exciton peak, but also modulates the intramolecular HOMO–LUMO transition (indicated by diamond) at 4.12 eV for unidirectional switching from *trans*-AB to *cis*-AB within the isolated monolayer [Figs. 9(b) and 9(c)].

As shown in Fig. 10, Tan *et al.*[69] compared the electronic structures of monolayer zinc phthalocyanine (ZnPc) and hexadecafluorinated zinc phthalocyanine (F_{16}ZnPc) self-assembled on the Si(111)–B surface by using STM and STS. Noticeable energy-level variations of an occupied molecular state (about −2.4–2 eV) across the F_{16}ZnPc

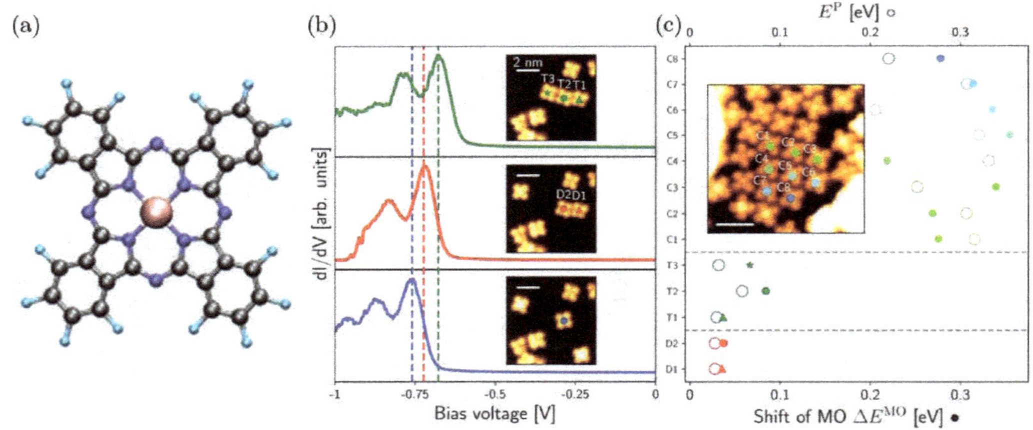

Fig. 8. Coupling with screening effect from the neighboring molecules. (a) Atomic structure of F_{16}CoPc. (b) Energy shift of the molecular orbital peaks of F_{16}CoPc as a function of the number of neighboring molecules. (c) Comparison of calculated polarization energies (empty circles) and the measured shifts of molecule orbital peak energies to an isolated F_{16}CoPc (filled markers) for molecules in different assemblies: a cluster of two molecules [$D1$ and $D2$ in (b)], a cluster of three molecules [$T1$–$T3$ in (b)], and an island of molecules (inset).

Source: Reproduced with permission from Ref. 67. Copyright 2020, Wiley-VCH.

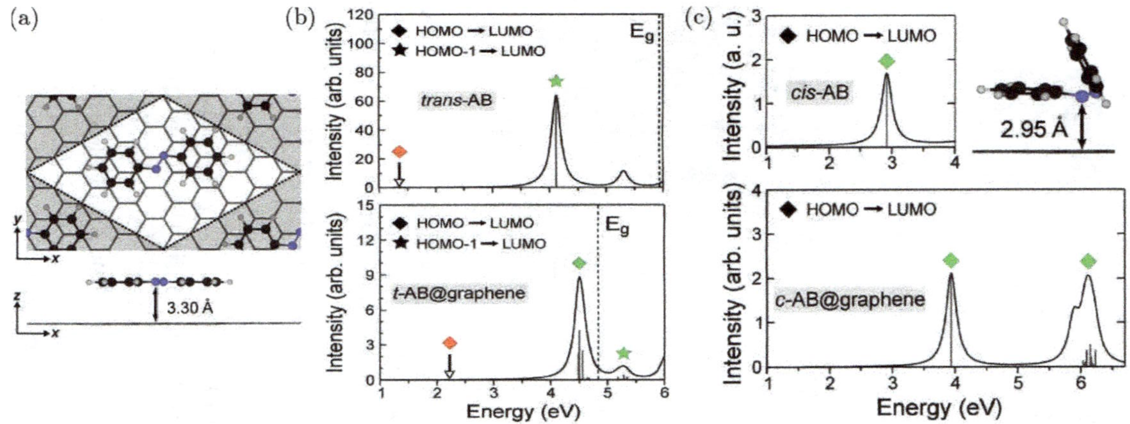

Fig. 9. Coupling with the configuration of the adsorbates. (a) The *trans*-AB monolayer adsorbed on graphene. (b) Optical spectra of the isolated *trans*-AB monolayer (top) and *trans*-AB on graphene (bottom) averaged over the in-plane diagonal components. (c) Optical spectra, averaged over the three Cartesian components, of the isolated *cis*-AB monolayer (top) and *cis*-AB on graphene (bottom).

Source: Reproduced with permission from Ref. 68. Copyright 2017, Royal Society of Chemistry.

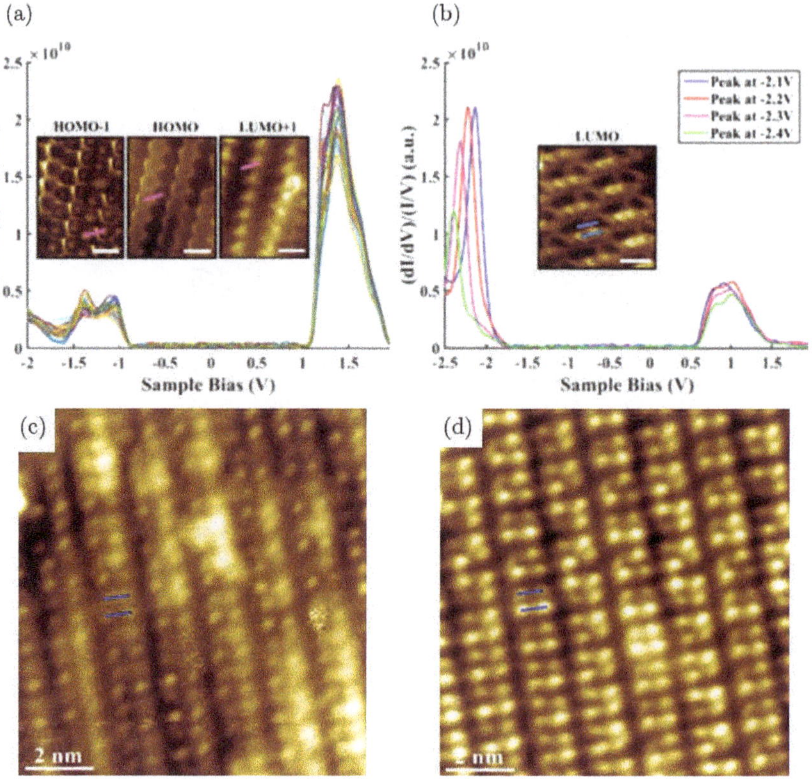

Fig. 10. Inhomogeneous electrostatic screening of the molecular orbitals. (a), (b) Typical STS spectra at multiple locations of ZnPc and F_{16}ZnPc overlayers, respectively. (c), (d) STM images of F_{16}ZnPc corresponding to HOMO and LUMO, respectively.

Source: Reproduced with permission from Ref. 69. Copyright 2017, American Physical Society.

assembly [Fig. 10(b)] were observed due to the substrate screening effect.[69] However, it is noted that both the unoccupied molecular orbital of F_{16}ZnPc and the molecular states of ZnPc [Fig. 10(a)] overlayers are constant. Hence, the influence of dielectric screening of the substrate on the occupied orbital of F_{16}ZnPc is different from the unoccupied molecular orbital of F_{16}ZnPc and the molecular states of ZnPc. They attribute this inhomogeneous electrostatic screening of intraorbital Coulomb interaction in the molecular adsorbates to the substrate boron distribution.[69] This event is mediated by charge transfer from substrate to the F_{16}ZnPc occupied molecular state, while no charge transfer occurs between the substrate and both the F_{16}ZnPc unoccupied orbital and molecular states of ZnPc. Moreover, a significant contrast variation is also observed in the STM image corresponding to the occupied orbital of F_{16}ZnPc [Fig. 10(c)] due to the inhomogeneity of the Si(111)–B substrate, while uniform features are observed in the STM image that corresponds to the unoccupied states of F_{16}ZnPc within the same area [Fig. 10(d)]. Based on this finding, the inhomogeneous electrostatic screening only impacts the molecular orbital that is directly involved in the OM–2D material interfacial charge-transfer process.[69]

4.2. *Screening effect of organic molecule on 2D materials by tuning the dielectric environment*

The integration of organic molecules, in particular, polymers, onto 2D materials has been widely adopted as a coating layer which is promising for flexible/wearable device applications.[70–72] The 2D materials especially semiconductors are very sensitive to dielectric environment due to their reduced screening of Coulomb interactions compared to their corresponding bulks. Hence, the screening of organic substrates and cover layers will have a great impact on the electronic and optical properties of 2D materials.[34,47,48]

Charge carriers and excitons in atomically thin OM–2D material heterostructure usually reside at or close to the interface, and therefore their properties are expected to be strongly influenced by the surrounding dielectric environment. By embedding them in different dielectric environments, the quasi-particle bandgap and exciton binding energies of 2D semiconductors can be tuned in the order of a few hundred meV.[47,48,57,73,74] Figure 11(a) outlines the schematic illustration of the relationship between exciton Rydberg states (exciton ground state $n = 1$ and excited states $n = 2, 3, 4, \ldots$) and quasi-particle bandgap of the optical response of an ideal 2D semiconductor.[47] The respective exciton binding energies can be defined as the energy difference between the quasi-particle bandgap E_B and the exciton resonance energies. Especially, the energy

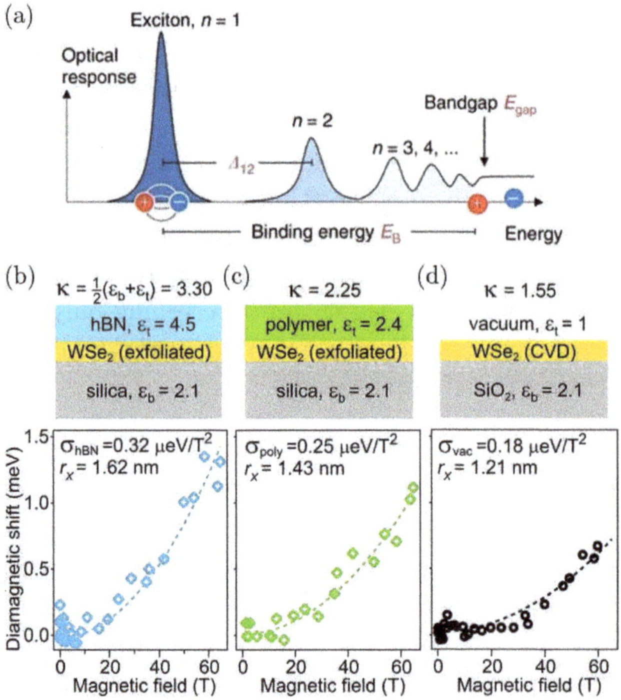

Fig. 11. The role of screening effect on the exciton properties of 2D materials. (a) Illustration of the optical response of an ideal 2D semiconductor, including the exciton ground- and excited-state resonances and the onset of (quasi-particle) bandgap. (b)–(d) The diamagnetic shifts measured for monolayer WSe_2 in three different dielectric environments.

Source: Panel (a) reproduced with permission from Ref. 47. Copyright 2017, Springer Nature. Panels (b)–(d) reproduced with permission from Ref. 48. Copyright 2016, American Chemical Society.

difference Δ_{12} between the first excited state ($n = 2$) and the exciton ground state ($n = 1$) scales with the ground-state exciton binding energy E_B. Hence, the quasi-particle gap can be determined via $E_{\text{gap}} = E_1 + E_B$, where E_1 is the transition energy of the exciton ground state. Using optical reflectance spectroscopy, Raja *et al.*[47] identified the energies of the excitonic resonances of monolayer 2D TMDs covered by graphene and hBN, and found that the quasi-particle bandgap and the exciton binding energy can be tuned up to hundreds of meV by embedding the excitons in different dielectric environments. By combining angle-resolved photoemission spectroscopy with micrometer spatial resolution (μ-ARPES) and optical spectroscopy, Waldecker *et al.*[73] further showed that the predominant effect of external dielectric screening is considered as a bandgap renormalization through a rigid shift of the occupied and unoccupied bands.

As pointed out previously, exciton's optical transition energy which is the most easily accessible exciton parameter, is insensitive to the surrounding medium due to the reduced exciton binding energy that is compensated by an equivalent reduction in the quasi-particle bandgap.[47,74,75] However, except for exciton binding energy, the size of exciton also depends on the dielectric environment. Using the polarized low-temperature magneto-absorption spectroscopy and tuning the dielectric environment of monolayer WSe_2 covered by transparent polymer and hBN, Stier *et al.*[48] analyzed in detail the diamagnetic shift that is dependent on the magnetic field. As depicted in Figs. 11(b)–11(d), they found that the size (1*s*-exciton radii) of the exciton increases as the dielectric screening of the environment increases.

The electromagnetically induced transparency (EIT) of 2D semiconductors can also be tuned by coating them with different polymers. Recently, Lin *et al.*[70] successfully realized the exciton energy and EIT tuning of single-layer WSe_2 by controlling its dielectric environment. Figure 12(a) shows the schematic of the structure comprising a monolayer WSe_2 encapsulated with hBN or various polymer overlayers; for example, poly(methyl methacrylate) (PMMA), polystyrene (PS), PMMA–PS copolymers, poly-(acrylic acid) (PAA), polyethylene (PE), and the transparent polymer Zeonex. It turns out that the energies of excitonic transition shifts can be correlated to the dielectric environment as observed in the photoluminescence (PL) [Figs. 12(b) and 12(c)] and second-harmonic generation (SHG) [Figs. 12(d) and 12(e)] spectra of monolayer WSe_2. Meanwhile, as shown in Fig. 12(f), the correlation coefficient r between them is as high as 0.96, highlighting that the EIT dip shifts linearly with the exciton energy in response to the dielectric environment.

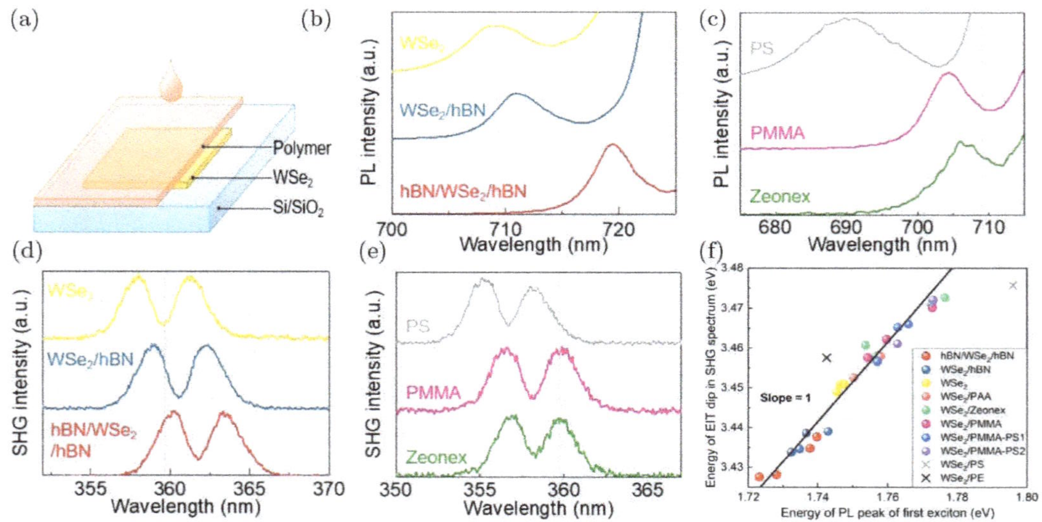

Fig. 12. Polymer coatings for tuning ETP in 2D semiconductors. (a) Schematic of tuning exciton energy and EIT of multilayer (ML) WSe_2 by depositing different polymer overlayers. (b), (c) and (d), (e) Plots of PL and SHG spectra showing the correlation of shifts with the dielectric environment. (f) Correlation between exciton energy and SHG–EIT.

Source: Reproduced with permission from Ref. 70. Copyright 2019, American Chemical Society.

5. Engineering of the Screening Effects on OM–2D Material Systems

5.1. *Tuning of electronic structure by depositing or inserting dipolar molecules*

In this subsection, we will introduce the potential design of OM–2D material heterostructure based on the dielectric screening engineering. The intrinsic dielectric polarizabilities of 2D materials are considerably low within the corresponding structural integrity; for example, Kraberger *et al.*[76] and Winkler *et al.*[77] proposed to tune the band structure of 2D materials or their vdWHs by depositing or inserting dipolar molecules onto the desired material surface. As shown in Fig. 13, the local shift of the density of states for both

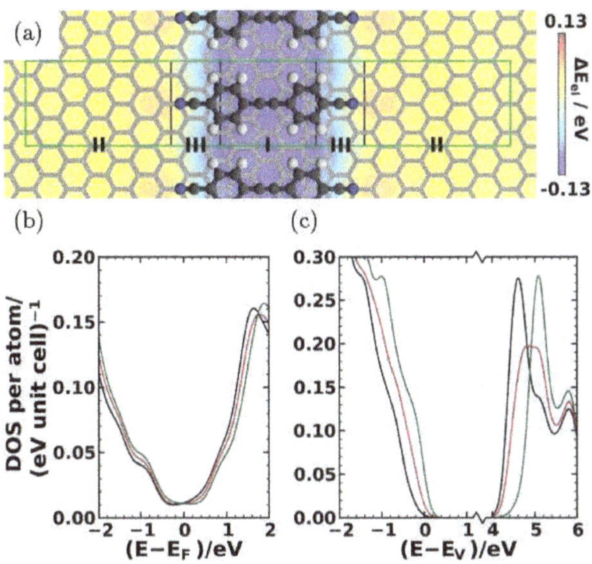

Fig. 13. (Color online) Tuning the band structure of 2D materials by depositing dipolar molecules. (a) Structure of a dicyano-substituted diphenylacetylene molecule adsorbed on graphene. The unit cell is shown as a green rectangle and divided into three regions I, II, and III. (b) Density of states projected onto the atoms in the regions I (black line), II (green line), and III (red line) of the graphene system. (c) Density of states of hBN in the same geometry as graphene with an adsorbed dicyano-substituted diphenylacetylene molecule.

Source: Reproduced with permission from Ref. 76. Copyright 2015, Wiley-VCH.

graphene and hBN can be achieved by the adsorbed polar dicyano-substituted diphenylacetylene molecules.[76] This type of OM–2D material heterostructure with periodic variation of electronic structure offers an atomically thin superlattice.

Moreover, the dipolar molecules can also be inserted in the interlayers of few-layer 2D materials and 2D vdWHs. As depicted in Fig. 14(a), by incorporating TiOPc between the MoS_2 bilayer (upper panel), type-II alignment is achieved (lower panel). In addition, this is a promising method to tune the energetic alignment of 2D vdWHs. Figures 14(b) and 14(c) show that after incorporating TiOPc between the NY_2–MoS_2 2D vdWHs, type-II alignment can be switched to type-I alignment.

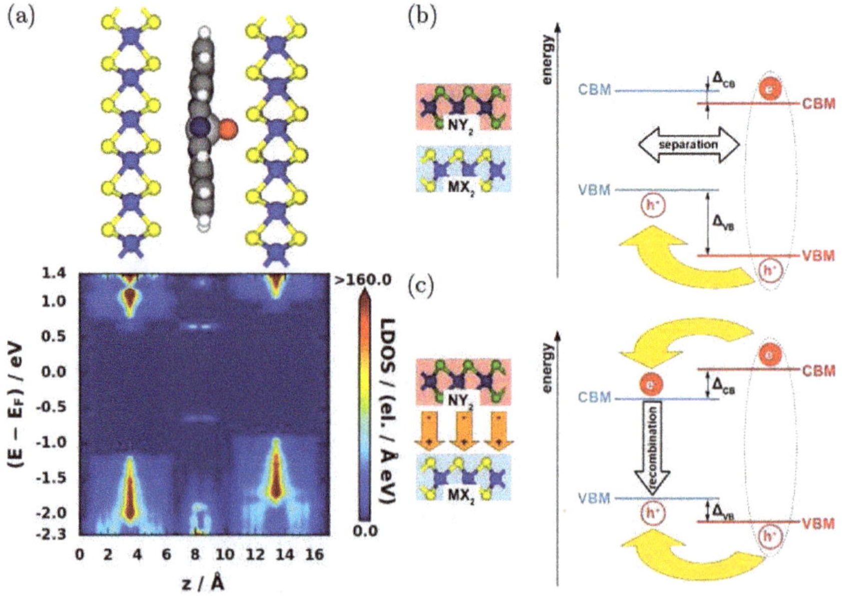

Fig. 14. Tuning the energetic alignment of 2D vdWHs. (a) Atomic structure (upper panel) and local density of states (lower panel) of the MoS_2–TiOPc–MoS_2 heterostructure spatially resolved with respect to the stacking direction. (b) Type-II band alignment conversion to (c) type I by applying the electrostatic response to the 2D vdWH.

Source: Reproduced with permission from Ref. 77. Copyright 2018, IOP Publishing, Ltd.

5.2. *Creating atomically thin lateral heterojunctions via structured dielectric environment*

Figure 15(a) shows one example to implement different structured lateral dielectric environments to create an atomically thin lateral heterojunctions based on 2D materials. A heterogeneous dielectric environment can be obtained by the controlled adsorption or deposition of different polarizable molecules on structured dielectric substrates. Through first-principles calculations, Rösner *et al.*[78] found that type-II heterojunction with a sharp bandgap crossover within less than five unit cells can be achieved [Fig. 15(b)].[78] Interestingly, this proof-of-concept device has been realized by Utama *et al.*[45] recently. As shown in Fig. 16(a), monolayer MoS_2 is placed on a substrate in which one segment is made up of an amorphous fluoropolymer (Cytop; low dielectric constant, $\varepsilon \approx 2$) and the other segment is made from hBN (high dielectric constant, $\varepsilon \approx 6.9$). As discussed previously, the degree of bandgap renormalization of 2D semiconductors is larger when the material is deposited onto the low-dielectric-constant substrates. Therefore, the bandgap of MoS_2 on Cytop substrate will be larger than that on the hBN substrate due to the different dielectric screening effects of the substrates, and the in-plane type-I alignment [Fig. 16(b)] is expected in monolayer MoS_2 when the two substrates are connected.

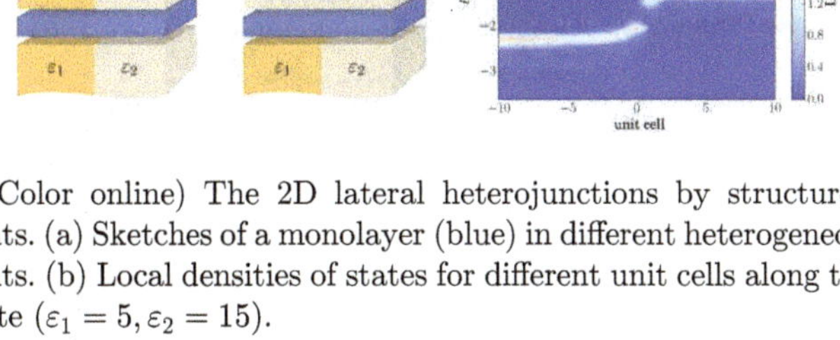

Fig. 15. (Color online) The 2D lateral heterojunctions by structured dielectric environments. (a) Sketches of a monolayer (blue) in different heterogeneous dielectric environments. (b) Local densities of states for different unit cells along the surface of the substrate ($\varepsilon_1 = 5, \varepsilon_2 = 15$).

Source: Reproduced with permission from Ref. 78. Copyright 2016, American Chemical Society.

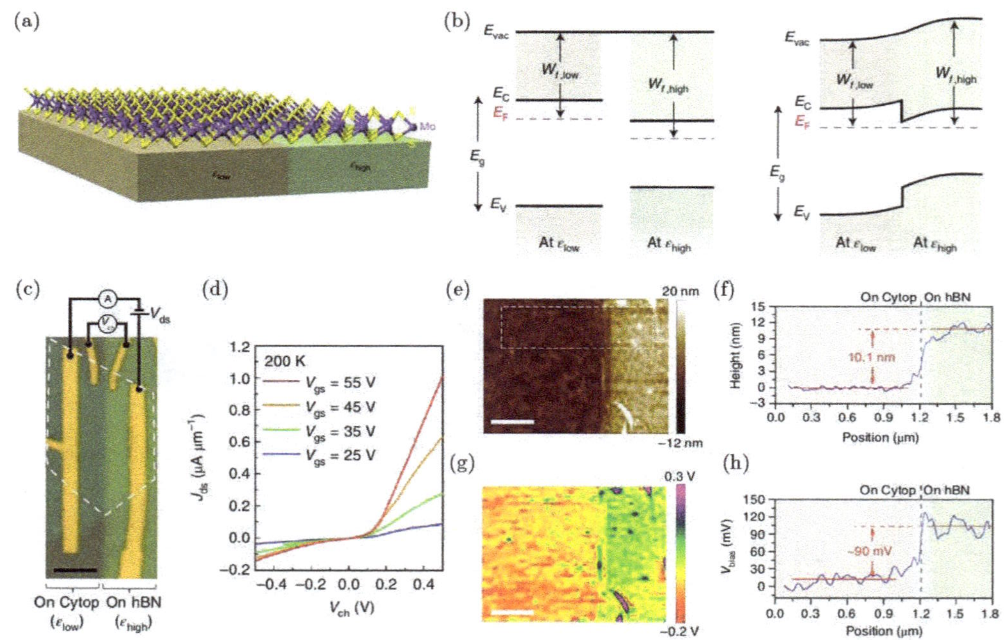

Fig. 16. Dielectric-engineered lateral heterojunction diodes. (a) Schematic illustration of an in-plane heterostructure of MoS_2 on the substrate connected by two substrates with different dielectric constants ($\epsilon_{low} < \epsilon_{high}$). (b) Type-I alignment form for MoS_2 segments on the two substrates. (c) Optical image of ML MoS_2 which is partially situated on Cytop and on hBN substrates. (d) The output characteristics of the device shown in (c). (e), (f) Atomic force microscopy image and height profile of the MoS_2 heterojunction. (g), (h) The spatially mapped V_{bias} and its profile from the same area in (e).

Source: Reproduced with permission from Ref. 45. Copyright 2019, Springer Nature.

The hBN was exfoliated on the spin-coated Cytop on the SiO_2–Si substrate and then the exfoliated monolayer MoS_2 was transferred on the flat edge of an hBN flake. The optical image of such a device is shown in Fig. 16(c). A strong asymmetric rectification behavior [Fig. 16(d)] with a turn-on voltage up to 90 meV at low temperature is observed using the current–voltage (I–V) measurement of the device with different back-gate voltages. This rectification behavior is consistent with the expected type-I heterojunction formation [Fig. 16(b)]. They also examined the local variation of surface potential across the heterojunction by using KPFM. As shown in Fig. 16(g), two distinct areal regions are observed in which two segments of MoS_2 on Cytop and hBN are shown in the topographic image [Fig. 16(e)] with the height difference is about 10.1nm [Fig. 16(f)]. The work function difference of MoS_2 on the Cytop and hBN substrates is about −90 meV [Fig. 16(h)] and therefore the calculated conduction-band offset and bandgap difference are about 90 meV and 180 meV, respectively. In addition to Cytop, many other low-dielectric-constant plastic substrates are available; for example, polyethylene naphthalate (PEN), polyethylene terephthalate (PET), and polyimide.[79] A combination of the low-dielectric-constant polymers with the high-dielectric-constant 2D materials is very appealing to develop high-yield flexible/wearable lateral heterojunction diodes.

6. Unintended Consequences and Perspective of the Screening Effect at the Interface

In this section, we will further introduce the unintended consequences due to screening effect and briefly discuss the perspective of screening effect induced by other contributions, e.g. defects, charge transfer, hybridization, etc.

6.1. *Dielectric disorder in 2D materials*

Disorder usually introduced by the local fluctuations of inherent material properties (e.g. chemical and structural composition, doping, or strain) is a very important aspect in nanotechnology and materials science. Recently, Raja *et al.*[34] found that the fluctuation induced by dielectric environment is a new source of disorder in 2D materials. Figure 17(a) schematically shows the influence of dielectric disorder caused by substrate roughness, impurity, and adsorbate on electron–electron and electron–hole interactions. The main influences of the Coulomb interaction on the electronic excitations are mainly originating from their binding energies and quasi-particle bandgap. In a similar approach as the exciton's Rydberg energy, the transition energies for the ground state ($n = 1$), first excited state ($n = 2$), and quasi-particle bandgap can be estimated as a function of the squared inverse of the external dielectric screening constant.[34] The consequences, i.e. the quasi-particle bandgap and exciton binding energy variations of the dielectric disorder, result in resonance broadening [Fig. 17(b)].

Since the exciton linewidths that are measured by linear optical reflectance spectroscopy contain the information of inhomogeneous broadening of the probed area, then the distinctive correlations between the ground- and excited-state linewidths can be used to clarify the underlying mechanism of dielectric disorder.[34] Therefore, Raja *et al.*[34] counted the linewidths of the ground and excited exciton states of exfoliated WS_2 and WSe_2 monolayer samples deposited on polydimethylsiloxane (PDMS) films, hBN, and SiO_2–Si. The theoretically predicted ratio of the relative shifts of $n = 1$ and $n = 2$ peaks depends on the external dielectric constant as shown in Fig. 17(c). The dielectric constants of PDMS and hBN are indicated by orange and purple rectangles, respectively. As depicted in Fig. 17(d), strong correlations of the $n = 2$ and $n = 1$ exciton peak broadening for both WS_2 and WSe_2 are observed experimentally. The linear fitted ratios of monolayer 2D TMDs on PDMS and hBN are 3.7 and 2.2, respectively. There is a consistent relationship between the experimental and theoretical predictions that exclude the potential interpretations of the excited-state broadening due to disorder (e.g. doping or strain fluctuations).[34]

Spatially resolved exciton spectroscopy images of an hBN-encapsulated WS_2 sample are also shown in Figs. 17(e)–17(g) obtained by using spatial reflectance mapping. Particularly, only

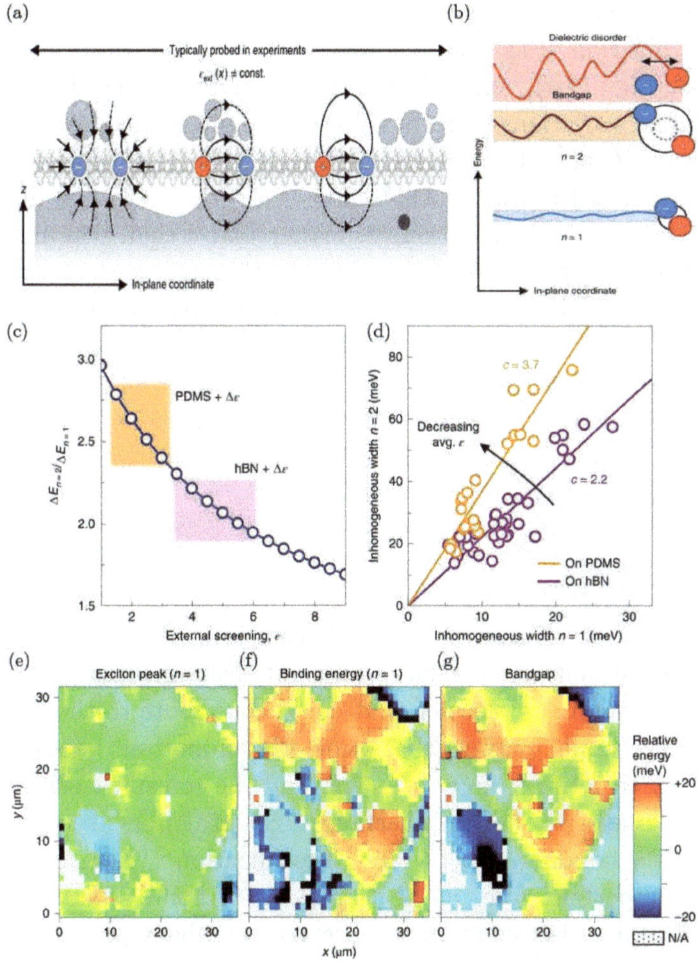

Fig. 17. Dielectric disorder in 2D materials. (a) Schematic illustration of the influence of dielectric disorder caused by substrate roughness, impurity, and adsorbate on electron–electron and electron–hole interactions. (b) Schematic diagram of resonance broadening caused by the fluctuation of bandgap and exciton energy due to dielectric disorder. (c) Ratio of the relative energy shifts of the $n = 2$ and $n = 1$ transitions for the exciton binding energies and bandgap as a function of external dielectric screening. (d) Correlation plot of $n = 2$ and $n = 1$ linewidths from 59 individual measurements on WS_2 samples on hBN and PDMS substrates. (e)–(g) Spatially resolved exciton spectroscopy images of the exciton $n = 1$ peak position, $n = 1$ exciton binding energy, and free-particle bandgap of an hBN-encapsulated WS_2 sample, respectively.

Source: Reproduced with permission from Ref. 34. Copyright 2019, Springer Nature.

small-to-moderate fluctuations in the image correspond to exciton $n = 1$ transition [Fig. 17(e)]. However, both the exciton binding energies [Fig. 17(f)] and quasi-particle bandgap [Fig. 17(g)] show much larger variations corresponding to the pronounced variations of the effective Coulomb interaction strength which is a characteristic of dielectric disorder from local changes of the external dielectric screening. Raja and co-workers[34] further found that even moderate fluctuations of the dielectric environment in the organic substrate layer will induce large variations of the bandgap and exciton binding energies (up to $\sim$100 meV) of 2D TMDs. Therefore, we should mitigate and avoid the introduction of dielectric disorder caused by the organic molecule or molecular adsorbate since it might hamper the optical and transport properties of 2D materials and the OM–2D material heterostructures based on them.

6.2. *Perspective*

In Sec. 4.1, we have introduced the interplay of adsorption configuration, charge transfer, and intrinsic dielectric screening strength of organic molecules that could contribute to the screening effects. One should consider that some other important factors should also be taken into account to understand the screening effect across OM–2D material heterointerfaces. Several features such as defects,[80,81] bonding type,[82] and hybridization[83,84] which usually play an important role in the formation of band structure of 2D materials, are also commonly observed at the OM–2D material interface. For instance, Singh *et al.*[85] found that the vertical dielectric screening of few-layered MoS_2 remarkably depends on the distribution of the vacancies that introduced out-of-plane polarization. Similar to band edges, the charge transition levels of defects have shown the presence of renormalization as they are affected by dielectric screening.[86] Meanwhile, to control the dielectric screening strength along the vertical direction, the fine-tuning of the nominal thickness and stacking sequence of the 2D materials is a reasonable approach to engineer the 2D vdWHs.[56,57,87] Moreover, Xu *et al.*[30] reported that the renormalization of the frontier orbitals of CoPc molecules on vanadium diselenide can only be corrected by considering both the screening and hybridization effects. Several findings that have been discussed here focused on how to couple the screening effects with the intrinsic defects of 2D materials, charge transfer, and hybridization that are essentially required to strengthen the link between fundamental scientific studies and device applications. Therefore, more extensive studies to investigate how these important factors impact the screening effect of the OM–2D material interface at the atomic scale are required.

In conclusion, the screening effects at the OM–2D material heterointerfaces form a very important knowledge that can be valuable for the development of fundamental science and technical applications. Here, we have reviewed several aspects of the screening effect within 2D inorganic materials and the OM–2D material interfaces that could open a new research line. Consequently, this effort might broaden their applicability as a means of tuning the dielectric-dependent electronic bandgap across diverse heterostructure systems.

Acknowledgments

This work was supported by the Natural Science Foundation of China (No. 12004057) and Key Laboratory of Low-grade Energy Utilization Technologies and Systems (No. LLEUTS-2020008). Arramel thank the Ministry of Education for supporting this work (Award No. MOE2018-T2-1-088).

References

1. M. L. Cohen and S. G. Louie, *Fundamentals of Condensed Matter Physics* (Cambridge University Press, Cambridge, 2016).
2. K. S. Novoselov, A. K. Geim, S. V. Morozov, D. Jiang, Y. Zhang, S. V. Dubonos, I. V. Grigorieva and A. A. Firsov, *Science* **306**(5696) (2004) 666.
3. K. S. Novoselov, D. Jiang, F. Schedin, T. J. Booth, V. V. Khotkevich, S. V. Morozov and A. K. Geim, *Proc. Natl. Acad. Sci. USA* **102**(30) (2005) 10451.
4. S. Z. Butler, S. M. Hollen, L. Y. Cao, Y. Cui, J. A. Gupta, H. R. Gutierrez, T. F. Heinz, S. S. Hong, J. X. Huang, A. F. Ismach, E. Johnston-Halperin, M. Kuno, V. V. Plashnitsa, R. D. Robinson, R. S. Ruoff, S. Salahuddin, J. Shan, L. Shi, M. G. Spencer, M. Terrones, W. Windl and J. E. Goldberger, *ACS Nano* **7**(4) (2013) 2898.
5. M. S. Xu, T. Liang, M. M. Shi and H. Z. Chen, *Chem. Rev.* **113**(5) (2013) 3766.

6. C. L. Tan, X. H. Cao, X. J. Wu, Q. Y. He, J. Yang, X. Zhang, J. Z. Chen, W. Zhao, S. K. Han, G. H. Nam, M. Sindoro and H. Zhang, *Chem. Rev.* **117**(9) (2017) 6225.
7. A. K. Geim and K. S. Novoselov, *Nature Mater.* **6**(3) (2007) 183.
8. C. H. Jin, F. Lin, K. Suenaga and S. Iijima, *Phys. Rev. Lett.* **102**(19) (2009) 195505.
9. Q. H. Wang, K. Kalantar-Zadeh, A. Kis, J. N. Coleman and M. S. Strano, *Nat. Nanotechnol.* **7**(11) (2012) 699.
10. H. Liu, A. T. Neal, Z. Zhu, Z. Luo, X. F. Xu, D. Tomanek and P. D. Ye, *ACS Nano* **8**(4) (2014) 4033.
11. R. X. Fei, W. B. Li, J. Li and L. Yang, *Appl. Phys. Lett.* **107**(17) (2015) 173104.
12. B. Huang, G. Clark, E. Navarro-Moratalla, D. R. Klein, R. Cheng, K. L. Seyler, D. Zhong, E. Schmidgall, M. A. McGuire, D. H. Cobden, W. Yao, D. Xiao, P. Jarillo-Herrero and X. D. Xu, *Nature* **546**(7657) (2017) 270.
13. M. Li, J.-S. Chen and M. Cotlet, *ACS Energy Lett.* **4**(9) (2019) 2323.
14. F. Liu, W. J. Wu, Y. S. Bai, S. H. Chae, Q. Y. Li, J. Wang, J. Hone and X. Y. Zhu, *Science* **367**(6480) (2020) 903.
15. S. Yang, P. P. Zhang, A. S. Nia and X. L. Feng, *Adv. Mater.* **32**(10) (2020) 1907857.
16. Q. H. Wang and M. C. Hersam, *Nat. Chem.* **1**(3) (2009) 206.
17. W. Chen, D. C. Qi, X. Y. Gao and A. T. S. Wee, *Prog. Surf. Sci.* **84**(9–10) (2009) 279.
18. D. Kiriya, M. Tosun, P. D. Zhao, J. S. Kang and A. Javey, *J. Am. Chem. Soc.* **136**(22) (2014) 7853.
19. D. Jariwala, S. L. Howell, K. S. Chen, J. M. Kang, V. K. Sangwan, S. A. Filippone, R. Turrisi, T. J. Marks, L. J. Lauhon and M. C. Hersam, *Nano Lett.* **16**(1) (2016) 497.
20. C. R. Ryder, J. D. Wood, S. A. Wells, Y. Yang, D. Jariwala, T. J. Marks, G. C. Schatz and M. C. Hersam, *Nat. Chem.* **8**(6) (2016) 597.
21. D. Jariwala, T. J. Marks and M. C. Hersam, *Nature Mater.* **16**(2) (2017) 170.
22. Y. L. Huang, Y. J. Zheng, Z. Song, D. Chi, A. T. S. Wee and S. Y. Quek, *Chem. Soc. Rev.* **47**(9) (2018) 3241.
23. S. Padgaonkar, J. N. Olding, L. J. Lauhon, M. C. Hersam and E. A. Weiss, *Acc. Chem. Res.* **53**(4) (2020) 763.
24. M. Gobbi, E. Orgiu and P. Samorì, *Adv. Mater.* **30**(18) (2018) 1706103.
25. S. H. Yu, Y. Lee, S. K. Jang, J. Kang, J. Jeon, C. Lee, J. Y. Lee, H. Kim, E. Hwang, S. Lee and J. H. Cho, *ACS Nano* **8**(8) (2014) 8285.
26. S. Bettis Homan, V. K. Sangwan, I. Balla, H. Bergeron, E. A. Weiss and M. C. Hersam, *Nano Lett.* **17**(1) (2017) 164.
27. H. Liu, M. Cui, C. Dang, W. Wen, X. Wang and L. Xie, *ACS Appl. Mater. Interfaces* **11**(37) (2019) 34424.
28. H.-Z. Tsai, A. A. Omrani, S. Coh, H. Oh, S. Wickenburg, Y.-W. Son, D. Wong, A. Riss, H. S. Jung, G. D. Nguyen, G. F. Rodgers, A. S. Aikawa, T. Taniguchi, K. Watanabe, A. Zettl, S. G. Louie, J. Lu, M. L. Cohen and M. F. Crommie, *ACS Nano* **9**(12) (2015) 12168.

29. L. Zhang, T. Yang, W. Zhang, D. Qi, X. He, K. Xing, P. K. J. Wong, Y. P. Feng and A. T. S. Wee, *Appl. Mater. Today* **18** (2020) 100535.
30. R. Xu, F. Xuan and S. Y. Quek, *J. Phys. Chem. Lett.* **11** (2020) 9358.
31. D. Stradi, M. Garnica, C. Díaz, F. Calleja, S. Barja, N. Martín, M. Alcamí, A. L. Vazquez de Parga, R. Miranda and F. Martín, *Nanoscale* **6**(24) (2014) 15271.
32. A. Zhao, Q. Li, L. Chen, H. Xiang, W. Wang, S. Pan, B. Wang, X. Xiao, J. Yang, J. G. Hou and Q. Zhu, *Science* **309**(5740) (2005) 1542.
33. X. Chen, Y.-S. Fu, S.-H. Ji, T. Zhang, P. Cheng, X.-C. Ma, X.-L. Zou, W.-H. Duan, J.-F. Jia and Q.-K. Xue, *Phys. Rev. Lett.* **101**(19) (2008) 197208.
34. A. Raja, L. Waldecker, J. Zipfel, Y. Cho, S. Brem, J. D. Ziegler, M. Kulig, T. Taniguchi, K. Watanabe, E. Malic, T. F. Heinz, T. C. Berkelbach and A. Chernikov, *Nat. Nanotechnol.* **14**(9) (2019) 832.
35. I. Lindau, P. Pianetta, S. Doniach and W. E. Spicer, *Nature* **250**(5463) (1974) 214.
36. F. J. Himpsel, *J. Phys. Chem. Solids* **49**(1) (1988) 3.
37. L. Kong, G. J. Perez Medina, J. A. Colón Santana, F. Wong, M. Bonilla, D. A. Colón Amill, L. G. Rosa, L. Routaboul, P. Braunstein, B. Doudin, C.-M. Lee, J. Choi, J. Xiao and P. A. Dowben, *Carbon* **50**(5) (2012) 1981.
38. G. Binning, H. Rohrer, C. Gerber and E. Weibel, *Phys. Rev. Lett.* **49**(1) (1982) 57.
39. Y. J. Zheng, Y. L. Huang, Y. Chen, W. Zhao, G. Eda, C. D. Spataru, W. Zhang, Y.-H. Chang, L.-J. Li, D. Chi, S. Y. Quek and A. T. S. Wee, *ACS Nano* **10**(2) (2016) 2476.
40. S. Y. Quek and K. H. Khoo, *Acc. Chem. Res.* **47**(11) (2014) 3250.
41. P. Girard, *Nanotechnology* **12**(4) (2001) 485.
42. L. H. Li, E. J. G. Santos, T. Xing, E. Cappelluti, R. Roldán, Y. Chen, K. Watanabe and T. Taniguchi, *Nano Lett.* **15**(1) (2015) 218.
43. A. Castellanos-Gomez, E. Cappelluti, R. Roldán, N. Agraït, F. Guinea and G. Rubio-Bollinger, *Adv. Mater.* **25**(6) (2013) 899.
44. D. Jeon, Y. Kang and T. Kim, *J. Phys. Chem. Lett.* **11**(16) (2020) 6684.
45. M. I. B. Utama, H. Kleemann, W. Zhao, C. S. Ong, F. H. da Jornada, D. Y. Qiu, H. Cai, H. Li, R. Kou, S. Zhao, S. Wang, K. Watanabe, T. Taniguchi, S. Tongay, A. Zettl, S. G. Louie and F. Wang, *Nat. Electron.* **2**(2) (2019) 60.
46. M. Nonnenmacher, M. P. Oboyle and H. K. Wickramasinghe, *Appl. Phys. Lett.* **58**(25) (1991) 2921.
47. A. Raja, A. Chaves, J. Yu, G. Arefe, H. M. Hill, A. F. Rigosi, T. C. Berkelbach, P. Nagler, C. Schüller, T. Korn, C. Nuckolls, J. Hone, L. E. Brus, T. F. Heinz, D. R. Reichman and A. Chernikov, *Nat. Commun.* **8**(1) (2017) 15251.
48. A. V. Stier, N. P. Wilson, G. Clark, X. Xu and S. A. Crooker, *Nano Lett.* **16**(11) (2016) 7054.
49. J. B. Neaton, M. S. Hybertsen and S. G. Louie, *Phys. Rev. Lett.* **97**(21) (2006) 216405.
50. L. Hedin, *Phys. Rev.* **139**(3A) (1965) A796.
51. M. S. Hybertsen and S. G. Louie, *Phys. Rev. B* **34**(8) (1986) 5390.
52. J. M. Garcia-Lastra, C. Rostgaard, A. Rubio and K. S. Thygesen, *Phys. Rev. B* **80**(24) (2009) 245427.

53. D. A. Egger, Z.-F. Liu, J. B. Neaton and L. Kronik, *Nano Lett.* **15**(4) (2015) 2448.
54. F. Xuan, Y. Chen and S. Y. Quek, *J. Chem. Theory Comput.* **15**(6) (2019) 3824.
55. S. C. Lam and R. J. Needs, *J. Phys., Condens. Matter* **5**(14) (1993) 2101.
56. K. T. Winther and K. S. Thygesen, *2D Mater.* **4**(2) (2017) 025059.
57. K. Andersen, S. Latini and K. S. Thygesen, *Nano Lett.* **15**(7) (2015) 4616.
58. L. V. Keldysh, *JETP Lett.* **29** (1978) 658.
59. P. Cudazzo, I. V. Tokatly and A. Rubio, *Phys. Rev. B* **84**(8) (2011) 085406.
60. S. Latini, T. Olsen and K. S. Thygesen, *Phys. Rev. B* **92**(24) (2015) 245123.
61. D. Y. Qiu, F. H. da Jornada and S. G. Louie, *Phys. Rev. B* **93**(23) (2016) 235435.
62. S. S. Datta, D. R. Strachan, E. J. Mele and A. T. C. Johnson, *Nano Lett.* **9**(1) (2009) 7.
63. K. Noori, N. L. Q. Cheng, F. Xuan and S. Y. Quek, *2D Mater.* **6**(3) (2019) 035036.
64. S. Barja, M. Garnica, J. J. Hinarejos, A. L. Vázquez de Parga, N. Martín and R. Miranda, *Chem. Commun.* **46**(43) (2010) 8198.
65. P. Puschnig, P. Amiri and C. Draxl, *Phys. Rev. B* **86**(8) (2012) 085107.
66. F. Marsusi, I. A. Fedorov and S. Gerivani, *J. Phys. Condens. Matter* **30**(3) (2018) 035002.
67. M. Pörtner, Y. Wei, A. Riss, K. Seufert, M. Garnica, J. V. Barth, A. P. Seitsonen, L. Diekhöner and W. Auwärter, *Adv. Mater. Interfaces* **7**(15) (2020) 2000080.
68. Q. Fu, C. Cocchi, D. Nabok, A. Gulans and C. Draxl, *Phys. Chem. Chem. Phys.* **19**(8) (2017) 6196.
69. A. Tan, S. R. Wagner and P. P. Zhang, *Phys. Rev. B* **96**(3) (2017) 035313.
70. K.-Q. Lin, R. Martin, S. Bange and J. M. Lupton, *ACS Photonics* **6**(12) (2019) 3115.
71. C. Zhong, V. K. Sangwan, C. Wang, H. Bergeron, M. C. Hersam and E. A. Weiss, *J. Phys. Chem. Lett.* **9**(10) (2018) 2484.
72. T. A. Shastry, I. Balla, H. Bergeron, S. H. Amsterdam, T. J. Marks and M. C. Hersam, *ACS Nano* **10**(11) (2016) 10573.
73. L. Waldecker, A. Raja, M. Rösner, C. Steinke, A. Bostwick, R. J. Koch, C. Jozwiak, T. Taniguchi, K. Watanabe, E. Rotenberg, T. O. Wehling and T. F. Heinz, *Phys. Rev. Lett.* **123**(20) (2019) 206403.
74. Y. Cho and T. C. Berkelbach, *Phys. Rev. B* **97**(4) (2018) 041409.
75. Z. Jiang, Z. Liu, Y. Li and W. Duan, *Phys. Rev. Lett.* **118**(26) (2017) 266401.
76. G. J. Kraberger, D. A. Egger and E. Zojer, *Adv. Mater. Interfaces* **2**(17) (2015) 1500323.
77. C. Winkler, S. S. Harivyasi and E. Zojer, *2D Mater.* **5**(3) (2018) 035019.
78. M. Rösner, C. Steinke, M. Lorke, C. Gies, F. Jahnke and T. O. Wehling, *Nano Lett.* **16**(4) (2016) 2322.
79. F. Giannazzo, *Nat. Electron.* **2**(2) (2019) 54.
80. Y. J. Zheng, Y. Chen, Y. L. Huang, P. K. Gogoi, M.-Y. Li, L.-J. Li, P. E. Trevisanutto, Q. Wang, S. J. Pennycook, A. T. S. Wee and S. Y. Quek, *ACS Nano* **13**(5) (2019) 6050.
81. Y. Cai, H. Zhou, G. Zhang and Y.-W. Zhang, *Chem. Mater.* **28**(23) (2016) 8611.

82. M. Amani, D.-H. Lien, D. Kiriya, J. Xiao, A. Azcatl, J. Noh, S. R. Madhvapathy, R. Addou, K. C. Santosh, M. Dubey, K. Cho, R. M. Wallace, S.-C. Lee, J.-H. He, J. W. Ager, X. Zhang, E. Yablonovitch and A. Javey, *Science* **350**(6264) (2015) 1065.
83. W. Chen, C. Tegenkamp, H. Pfnür and T. Bredow, *J. Chem. Phys.* **132**(21) (2010) 214706.
84. M. Wießner, J. Ziroff, F. Forster, M. Arita, K. Shimada, P. Puschnig, A. Schöll and F. Reinert, *Nat. Commun.* **4**(1) (2013) 1514.
85. A. Singh, S. Lee, H. Bae, J. Koo, L. Yang and H. Lee, *RSC Adv.* **9**(69) (2019) 40309.
86. M. H. Naik and M. Jain, *Phys. Rev. Mater.* **2**(8) (2018) 084002.
87. J. Koo, S. Gao, H. Lee and L. Yang, *Nanoscale* **9**(38) (2017) 14540.

https://doi.org/10.1142/9789811247859_0007

Chapter 7

Characterization of Electronic Structures at Organic–2D Materials Interfaces with Advanced Synchrotron-based Soft X-ray Spectroscopy

Liang Cao[*,§] and Dong-Chen Qi[†,‡,¶]

**Anhui Key Laboratory of Condensed Matter Physics at Extreme Conditions, High Magnetic Field Laboratory, HFIPS, Anhui, Chinese Academy of Sciences, Hefei 230031, P. R. China*

[†]School of Chemistry and Physics, Queensland University of Technology, Brisbane, Queensland 4001, Australia

[‡]Centre for Materials Science, Queensland University of Technology, Brisbane, Queensland 4001, Australia

[§]lcao@hmfl.ac.cn

[¶]dongchen.qi@qut.edu.au

The discovery of two-dimensional (2D) materials with single or a few-atom-thick layers has presented unprecedented opportunities to study, control and harness the properties of van der Waals (vdW) materials at their 2D limit. It also offers a platform that allows the arbitrary creation of heterostructures in a materials-by-design approach with properties tailored by the constituting components. Combining organic molecules with 2D materials to form hybrid heterointerfaces has recently emerged as a facile and versatile approach to engineer the electronic, optical, magnetic and chemical properties of 2D materials for new or optimized device applications. Underpinning the

[§,¶]Corresponding authors.

development of organic–2D materials heterostructures is the ability to interrogate the interfacial electronic structures and properties at multiple dimensions. This review provides a timely update on the application of synchrotron-based soft X-ray spectroscopies (SR-SXS) in the characterization of organic–2D materials interfaces. By harnessing the unparalleled high energy resolution, tunable energy, high brilliance and tunable polarization inherent to synchrotron radiation, electronic structures, charge transfer dynamics, molecular orientations and spin configurations at the hybrid interfaces can be examined, which helps us to formularize a coherent understanding of the organic–2D materials interfaces that will guide the design of new hybrid vdW structures and devices. We also offer our perspective on the future application of SR-SXS in the exploration of organic–2D materials heterostructures.

Keywords: Synchrotron radiation; photoemission spectroscopy; energy level alignment; NEXAFS; molecular orientation; core-hole clock spectroscopy; ultrafast charge transfer dynamics; XMCD; spin configuration.

1. Introduction

Since the discovery of graphene, tremendous efforts have been devoted to exploiting two-dimensional (2D) materials with extraordinary (opto)electrical, optical, magnetic, mechanical and many other emergent properties both for the interest of fundamental studies and transformative technologies. They have few surface dangling bonds and unique atomic-level uniformity which make them very appealing for developing optical, electronic and energy applications. Due to the weak van der Waals (vdW) interlayer interaction, 2D materials including graphene and its derivatives,[1] black phosphorus, transition metal dichalcogenides (TMDs), boron nitride (BN), and transition metal carbides/nitrides (MXene), can be mechanically exfoliated down to 2D limit of monolayer (ML). 2D materials also offer an unprecedented platform that allows the so-called vdW integrations by stacking different 2D materials in a rational way, bringing a totally new degree of freedom to nearly arbitrarily combine highly distinct materials at the atomic scale to create artificial superlattices and heterostructures.[2–5] Such a materials-by-design approach serves as an exciting frontier for exploring fundamental physics, chemistry, materials science and device engineering, opening up technological opportunities beyond traditional materials to allow advances across diverse areas from electronic, optoelectronics, energy storage, spintronics to bio-electronics.[6–12] The recent breakthrough in understanding and modulating exotic properties by precisely controlling the twist angle between layers of 2D materials brings a burgeoning new field of twistronics with relevance to high-temperature superconductivity, strongly correlated phases, and topological excitons.[13–16]

Another approach to build heterostructures with 2D materials is to incorporate organic molecules which themselves can serve as building blocks for functional devices with novel properties.[17] Organic semiconducting molecules incorporating π-conjugated molecular structures exhibit many advantages, including remarkable (opto) electronic properties, low cost, low temperature and solution processability, tremendous structural flexibility, and synthetic tailorability, which make them ideal candidates for next-generation printing and flexible electronics.[18,19] The development of organic–2D heterostructures not only optimizes the intrinsic properties of 2D materials,[20] but also creates new capabilities that are distinct from isolated components,[21] opening up a new route to achieve novel functionalities for various device applications.[17,22,23]

Underpinning the properties of organic–2D heterostructures is the interfacial electronic structure which is important for enabling a multitude of applications, including electronic devices, catalysts, and biosensors. It will be critical for establishing a fundamental understanding on the evolution of electronic structures at organic–2D heterojunctions for the rational design of organic–2D heterojunctions with tailored functionalities. However, characterization of interfacial electronic structures and processes often presents significant challenges to conventional experimental probes, as the interfaces are buried and many interfacial phenomena, such as ultrafast charge transfer dynamics, are elusive. Synchrotron-based soft X-ray spectroscopies (SR-SXS), enabled through high brilliance, polarized, and continuously tunable X-ray excitation source from a synchrotron facility, offer unique capacities and advantages in the quantitative characterization of interfacial properties, including electronic structures, spin configuration, energy level alignment, chemical composition, molecular orientation, and ultrafast charge transfer dynamics, with elemental, orbital and spin specificity and unprecedented resolution.[24] The purpose of this review is to summarise the application of SR-SXS on the characterization of electronic structures at organic–2D interfaces, which help to understand the implication of the interfacial properties to 2D materials based devices. For comprehensive reviews on the fabrication, properties and applications of organic–2D materials heterostructures, the readers are referred to recently published articles on these topics.[20,23]

This review is organized as follows: Following a brief introduction to SR-SXS (Sec. 2), in Sec. 3 SR-based photoemission spectroscopy (PES), and its application in determining interfacial energetics, and interfacial chemical interaction are introduced. The principles of near-edge X-ray absorption fine structures (NEXAFS) spectroscopy and its application in the determination of unoccupied electronic states and the determination of molecular orientation are presented in Sec. 4. The principle of the core-hole clock (CHC) spectroscopy and its application on quantifying charge transfer times at organic–2D interface are discussed in Sec. 5. The magnetic coupling at organic–2D interfaces investigated by X-ray magnetic circular dichroism (XMCD) is discussed in Sec. 6. Finally, a summary and outlook is presented in Sec. 7.

2. Synchrotron-Based Soft X-ray Spectroscopies

A synchrotron radiation (SR) source, as schematically depicted in Fig. 1(a), is essentially large-scale research facility that comprises a circular accelerator that can accelerate charged particles (e.g. electrons) nearly to the speed of light and maintain that speed in the storage ring. SR in the form of electromagnetic radiation with a wide energy spectrum is emitted whenever relativistic charged particles are deflected by a magnetic field, which is created either by the simple bending dipole magnet as in most 1st and 2nd generation SR facilities, or insertion devices including wigglers and undulators consisting of alternating magnet poles as in 3rd generation SR facilities with much higher photon flux. SR is highly brilliant with

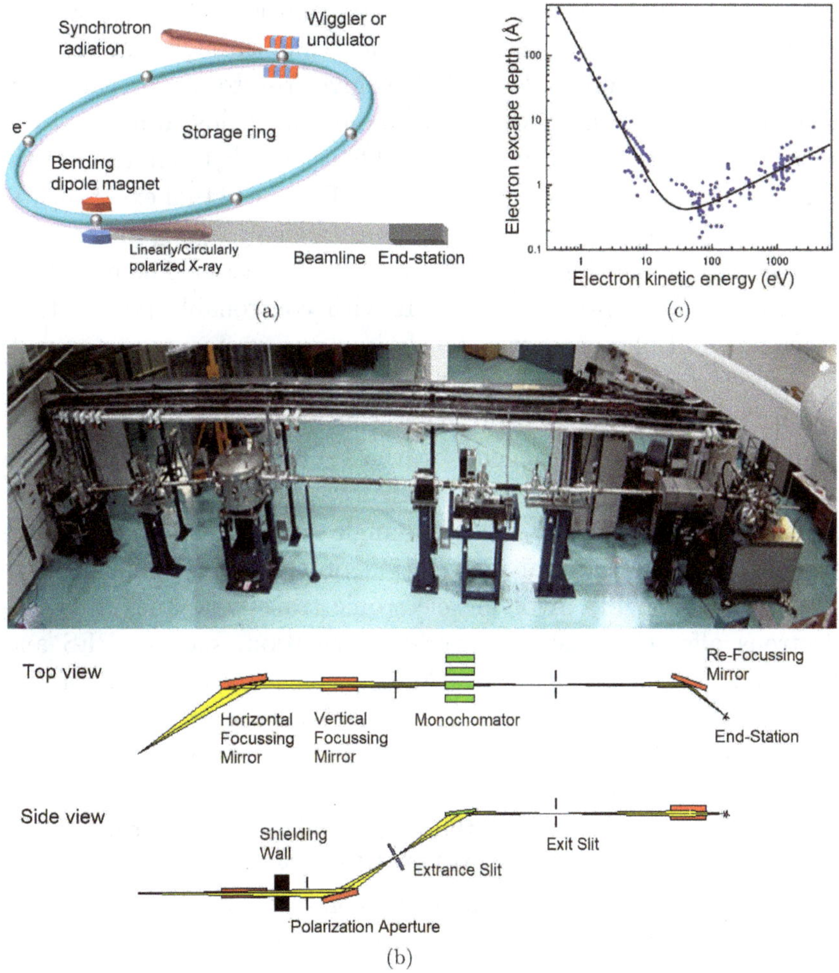

Fig. 1. (a) Schematic representation of SR source. (b) Surface, interface and nanostructure science (SINS) beamline of the Singapore Synchrotron Light Source (SSLS) for PES, NEXAFS and XMCD experiments. The available photon energy range is 50–1200 eV. (c) Universal curve of IMFP as a function of the kinetic energy of photoelectrons reproduced from Ref. 32.

a photon flux several orders of magnitude higher than conventional X-rays, highly coherent, and inherently polarized. The energy/wavelength of SR are widely tunable from far infrared-terahertz (THZ, below 1 meV) through the ultraviolet and into hard X-ray

(>5000 eV), offering unprecedented opportunities to gain fundamental insights into materials at atomic scale through light-matter interactions at a multitude level. This gives rise to a wide variety of applications in almost every technological fields ranging from materials science,[25] energy science,[26] life science,[27] to environmental science[28–30] and archaeological science.[31] For the actual experiments, the SR is guided through the beamlines tangential from the storage ring to end-stations for users to perform experiments. The beamlines are designed for optimizing the SR with controllable polarization, coherence and photon energy for desired measurements at the end-stations. Figure 1(b) shows an example for a typical SXS beamline.

The types of measurements that can be performed at different beamlines/end-stations depend on the type of interaction between X-rays and materials. Diffraction and/or scattering of X-rays, such as X-ray diffraction (XRD) and small-angle X-ray scattering (SAXS), provide atomic structural information of materials, while spectroscopic measurements involving electronic transitions and/or photoelectrons offer electronic structural information, such as PES and X-ray absorption spectroscopy (XAS), which are the focus of this review. When X-rays interact with atoms, a core-shell electron can be excited and emitted from the material by absorbing the energy of the photon provided that the photon energy is larger than the binding energy of the electron (plus the additional binding of the surface potential, i.e. work function (WF)). PES, also called photoelectron spectroscopy, exploits the photoemission process to provide rich information on the chemical composition and electronic structures of solids. Soft X-rays (100–5000 eV) generated by SR are particularly suitable for PES studies of organic materials since its energy range matches the binding energies of elements such as C, N, O and S, allowing much higher resolution and shorter data acquisition time in comparison to lab-based X-ray sources, as we later show in Sec. 3. Figure 1(c) shows the universal curve of inelastic mean free path (IMFP) as a function of the kinetic energy of photoelectrons.[32] A broad minimum of $\sim$1 nm is reached at kinetic energies of 50–100 eV, and several nanometer is obtained even for high electron energy of 2000 eV, reflecting the surface sensitivity of soft X-ray-based spectroscopy. While hard X-rays can be used for PES studies providing more bulk-sensitive information, the ionization cross-sections of common core-levels become extremely low placing stringent requirement on the brightness of light source. On the other hand, if the photon energy of the X-rays matches the energy difference between a core-level and an unoccupied energy state, the core-level electron will be resonantly excited into the unoccupied energy state by resonantly absorbing the X-ray photon and remains in the atom/molecule. The corresponding XAS essentially provides important information on the unoccupied density of states (DOS) of the system by scanning the photon energy across the absorption edge (hence SR must be used). The NEXAFS spectroscopy is the soft X-ray implementation of XAS, and is ideal for studying organic materials by providing both element and orbital specificity. The ability to tune the polarization of SR X-rays (e.g. linear and circular) allows for gaining further insight in the microscopic properties including molecular orientation (Sec. 4), and magnetic order (Sec. 6).[33–37] Lastly, the core-hole left behind by the X-ray absorption process has a well-defined life-time before various decay processes occur, and it can be used as a precise internal clock shielded from external environment allowing for quantitative determination of ultrafast charge transfer dynamics process, which is the focus of Sec. 5.

3. Synchrotron-Based PES

With the high energy resolution, brighter X-ray source and the tunability of the X-ray photon energy, SR-based PES offers a powerful multiscale probe allowing the study of a wide range of systems from molecules, nanoparticles to macroscopic solids. In the direct photoemission process as illustrated in Figs. 2(a) and 2(b), the binding energy (E_B) of photoelectron (referenced to the Fermi energy E_F) is related with the incident photon energy $h\nu$ and the kinetic energy of photoelectrons (referenced to the vacuum level) by $E_B = h\nu - E_K - \varphi$, in which φ is the WF of the sample. Each element has a distinctive set of binding energies for its core-levels, which serves

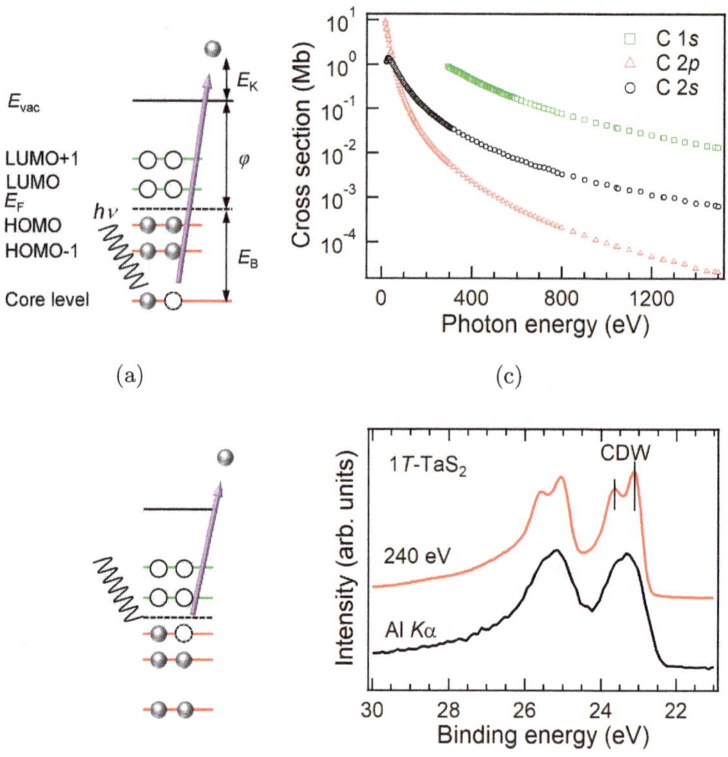

(a) (c)

(b) (d)

Fig. 2. Schematic view of photoemission from (a) core level and (b) valence state. (c) Carbon subshell photoionization cross section for 0–1500 eV. The values are extracted from Ref. 38. (d) Ta $4f$ core level spectra for $1T$-TaS_2 single crystal collected at room temperature by laboratory X-ray sources (Al-Kα) and synchrotron (National Synchrotron Radiation Laboratory (NSRL)) with photon energy of 240 eV.

as "finger prints" for elemental analysis. Moreover, the E_B of core levels is sensitive to the chemical environment surrounding the ionized atoms, resulting in a small deviation from that of free atoms. This chemical shift in the core-level PES peaks allows the chemical bonding environment of the probed elements and electronic structure of the materials to be determined. Consequently, PES is one of the most versatile and useful tools for the analysis of element/chemical composition of materials. It should be noted that an apparent shift in the binding energy of a core-level does not always entail a change in the chemical bonding environment; it could also directly reflect the shift of Fermi energy in a semiconductor due to doping, or other more intricate final-state effects.[37]

The tunability of photon energy has several important implications for SR-based PES. The ionization cross-sections of core-levels (and the corresponding peak intensity in PES) can be enhanced by several orders of magnitude by bringing the photon energy closer to the binding energy of the corresponding core-level (see Fig. 2(c) showing ionization cross-sections of carbon for example). It also alters the probing depth of the spectroscopy which depends sensitively on the kinetic energy of photoelectrons (see Fig. 1(c)).Consequently, by using a photon energy 30–50 eV above the binding energy of a particular core-level, both the ionization cross-section and surface sensitivity of measurements can be maximized simultaneously. This is particularly beneficial for studying interfaces with ML or sub-ML molecular coverage. In addition, the energy resolution of SR-based PES is usually superior to laboratory monochromatized X-ray sources, providing detailed information on chemical bonding and/or electronic structures. Figure 2(d) shows the Ta $4f$ core level spectra for $1T$-TaS_2 single crystal collected at room temperature by laboratory X-ray sources (Al-Kα) and synchrotron (NSRL) with photon energy of 240 eV. The splitting in Ta $4f$ spectra due to charge density wave (CDW) is clearly resolved resulting from the higher energy resolution of SR.

When the photon energy of soft X-rays is lowered down to less than 100 eV, the ionization cross-sections of valence bands of the solids (see Figs. 2(b) and 2(c)), or the frontier molecular orbitals of molecules (e.g. highest occupied molecular orbital (HOMO)), are greatly enhanced, and the SR-based PES can be implemented similar to ultraviolet photoelectron spectroscopy (UPS), providing important information on the valence band (VB) electronic structures of solids that dictate their (opto)electronic, magnetic, thermal and many other properties. Although VB features can also be measured using X-rays with higher photon energy, the spectral resolution and intensity can be greatly enhanced with SR at the VUV regime (10–100 eV). Altogether, these unique advantages afforded by SR-based PES allow a multiscale interrogation of molecule-2D material interfaces with unparalleled resolution, superior data quality and much shorter spectra acquisition time.

Because 2D materials are typically free of dangling bonds, organic molecules are expected to be predominantly physisorbed onto 2D materials with vdW-type interfacial interaction. However, in the presence of surface defects, such as vacancies, organic molecules can be covalently anchored to 2D materials and functionalize the surface. Consequently, the electronic structures and a variety of physical properties of 2D materials can be engineered by organic molecules in different ways depending on the nature of interfacial interactions.[39,40] As a consequence, interface engineering of 2D materials with organics, including interfacial energetic optimization, surface defect passivation and surface functionalization through chemical interaction, serves as a promising strategy to improve or tailor the properties of 2D materials toward device application. Underpinning this prospect is the ability to interrogate the organic–2D materials interface for which SR-based PES offers unique advantages for the reasons outlined above.

3.1. *Surface transfer doping: Establishing interfacial energetics in non-covalent interactions*

Although non-covalent functionalization of 2D materials with organic molecules does not directly alter the electronic structures of the underlying 2D materials through orbital hybridization, non-covalent interfacial interaction, in particular interfacial charge transfer, allows for the modulation of some of the intrinsic properties of 2D materials, such as electrical and optoelectronic properties. Surface charge transfer doping at organic–2D interface, which relies on the spontaneous interfacial charge transfer driven by band misalignment and intermediate electronic coupling (no orbital hybridization),[39] has recently emerged as a facile and non-destructive technique for the precise control of the carrier type and density of 2D materials in which conventional substitutional doping is ineffective and even damaging due to the atomically thin nature.[20,41,42] Figures 3(a) and 3(b) show

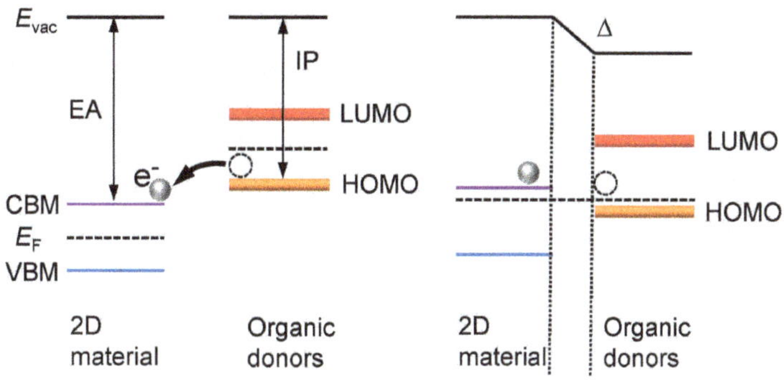

Fig. 3. Schematic energy level alignment diagrams at organic donors/semiconducting 2D material interface (a) before and (b) after charge transfer.

the schematic energy level alignment diagram to illustrate the n-type surface charge transfer doping process occurred at organic donors/semiconducting 2D material interfaces as an example. Before bringing them into contact (see Fig. 3(a)), the ionization potential (IP) of organic donors, which is the difference between HOMO and vacuum level E_{vac} is smaller than the electron affinity (EA) of 2D materials, which is the energy difference between conduction band minimum (CBM) of 2D materials and E_{vac}. Assuming a common vacuum level (Schottky–Mott limit) alignment, the HOMO of the organic donor would be placed directly above the CBM of 2D materials. Such an energy misalignment creates an energy offset which facilitates spontaneous electron transfer from the organic donor to the 2D material upon the formation of contact until thermodynamic equilibrium is reached (see Fig. 3(b)). The spatially separated electrons transferred to 2D materials and holes left behind in the organic donors give rise to a built-in electric field (or interface dipole, Δ) that has two implications: it confines the electrons to the surface of 2D materials while allowing them to move freely along the surface contributing to n-type electrical conduction; it also modifies the energy band alignment across the interface by pushing down the electron donating states of surface donors and self-limit the surface transfer doping process. p-type surface transfer doping occurs via a similar process, however between the VBM of 2D materials and lowest unoccupied molecular orbital (LUMO) of organic acceptors, leading to a hole conducting layer in the 2D materials.

The charge transfer process at the organic–2D interface can be monitored by using *in situ* SR-based PES. This enables us to establish the interfacial energy level alignment by following the evolution of interfacial electronic structures as a function of the *in situ* deposition of organic surface dopants. The schematic band alignment shown in Fig. 3 suggests several important spectroscopic features intimately associated with the surface transfer doping process. First, the shift of Fermi level in the 2D material, as a result of charge carrier accumulation, changes its energy distance to the corresponding core-levels of the 2D material; this entails a rigid shift of the core-level peak in the PES spectra toward lower/higher binding energy with the same offset amount of the Fermi energy. Secondly, as charge transfer progresses, the surface organic donors/acceptors become positively/negatively charged. Due to the weak intermolecular interaction, the excess charges tend to localize on individual molecules and modify the local electrostatic potential of the atomic core-level electrons, resulting in a chemical shift distinct form that of neutral molecules. However, these charged molecular species are mostly evident in spectroscopy at extremely small coverage where the deposited molecules are predominantly ionized.[43] Resolving these ionized species at very low coverages benefits from the use of SR-based PES in which one can dramatically enhance the photoionization cross-section of core-levels as well as surface sensitivity by tuning the photon energy close to the binding energies of core-levels of interest. The much higher energy resolution afforded by synchrotron X-rays also allows the ionized components to be more distinct from the neutral ones in core-level spectra. Finally, the interface dipole produced by interfacial charge separation, as discussed above, breaks the alignment of vacuum level at the interface (Schottky–Mott limit) and lifts the vacuum level. This is manifested as a rigid shift of the secondary electron cut-off (SECO) in UPS whose position indicates the position of vacuum level (hence WF), upon deposition of molecular dopants. Altogether, these spectroscopic characteristics unambiguously demonstrate surface transfer doping of 2D materials by molecular dopants, and allow us to form a detailed understanding of the interfacial electronic structures and energy level alignment at the 2D materials/organic molecules interface.

As an example, Fig. 4 shows the evolution of PES spectra for etrafluorotetracyanoquinodimethane (F_4TCNQ, with a molecular structure shown in Fig. 4(d) inset) on epitaxial graphene (EG) on silicon carbide (SiC).[44] No additional features are detected from core-levels spectra (Figs. 4(c) and 4(d)), ruling out the strong chemical interaction between F_4TCNQ and EG. The WF of 4.0 eV for EG

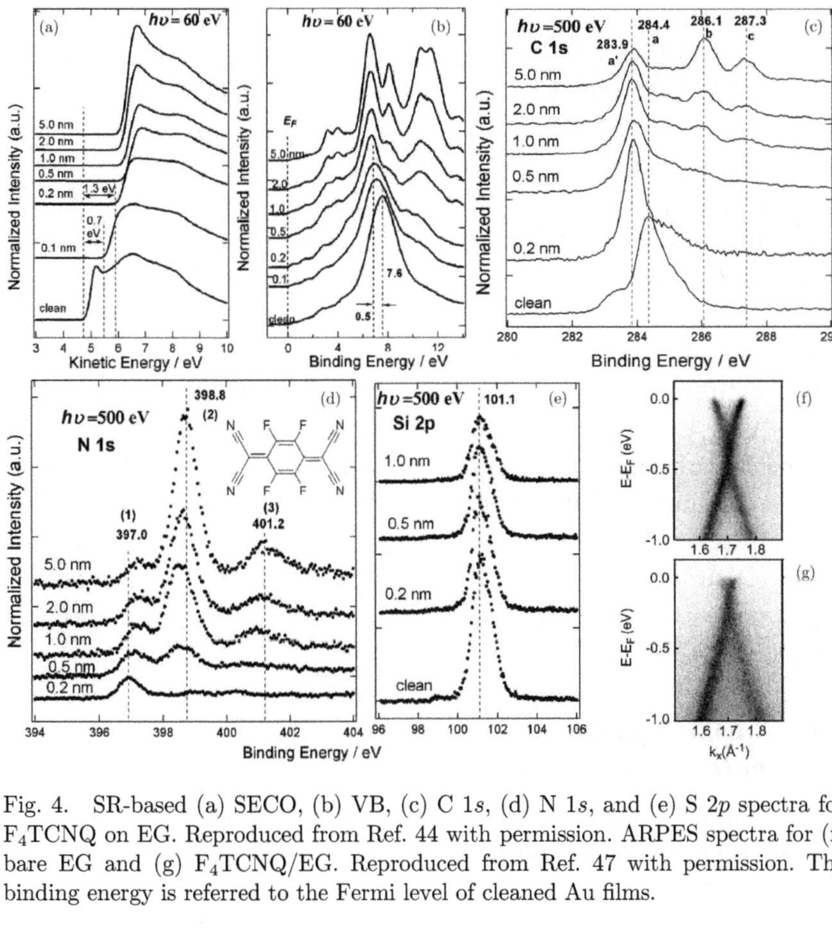

Fig. 4. SR-based (a) SECO, (b) VB, (c) C 1*s*, (d) N 1*s*, and (e) S 2*p* spectra for F_4TCNQ on EG. Reproduced from Ref. 44 with permission. ARPES spectra for (f) bare EG and (g) F_4TCNQ/EG. Reproduced from Ref. 47 with permission. The binding energy is referred to the Fermi level of cleaned Au films.

extracted from SECO (Fig. 4(a)) is smaller than the EA of ~5.24 eV for F_4TCNQ.[45,46] Upon the formation of contact, electrons are transferred from the Fermi level of EG to LUMO of F_4TCNQ molecules, leading to the formation of a built-in interface dipole which is accompanied by an upward shift vacuum level as revealed by the increase of WF, and the emergence of new features located at ~0.3 eV and ~1.3 eV below the Fermi level due to (partially) filled LUMO and the relaxed HOMO in the polaronic states of charged molecules. The dominant feature in the VB spectra (Fig. 4(b)) located at 7.6 eV and C 1*s* core level (Fig. 4(c)) at ~284.4 eV originates from the EG both shift 0.5 eV to the lower binding energy upon 0.2 nm F_4TCNQ deposition, and remain unchanged for further deposition, which is in line with the shift of E_F toward Dirac point of EG as revealed by angle-resolved PES (ARPES; see Figs. 4(f) and 4(g)) due to *p*-type doing of EG.[47] The induced holes are largely confined within the EG as the Si 2*p* peak in bulk SiC (Fig. 4(e)) remains at constant binding energy during the deposition of F_4TCNQ. N 1*s* core levels shown in Fig. 4(d) solely originate from the deposited F_4TCNQ film provide additional information on the charge transfer process. N 1*s* core level for negatively charged F_4TCNQ in direct contact with EG can be clearly identified at the lower binding energy of 397.0 eV at 0.2 nm thickness. The transferred electrons are localized at the charge withdrawing cyano groups (–C≡N) of F_4TCNQ causing the chemical shift. With the increase of molecular coverage, core-level features attributed to the neutral F_4TCNQ molecules at 398.8 and 401.2 eV become dominant. Altogether, the SR-based high-resolution PES clearly reveals the electron transfer from graphene to the adsorbed F_4TCNQ surface acceptors, leaving the underlying graphene *p*-type doped with its hole density controlled by the coverage of molecular acceptors.

3.2. *Covalent functionalization and passivation of 2D materials by organic molecules*

While non-covalent interaction with organic molecules provides a non-intrusive and facile route to modulate the electronic properties of underlying 2D materials, the absence of chemical grafting of molecules to the surface of 2D materials often lead to poor environmental and/or thermal stability. In contrast, covalent functionalization of 2D materials with organic molecules provides a much more robust interface against thermal desorption and harsh environment, and it has been thus adopted to modify the electronic band structure, and in turn magnetic, optical and mechanical properties of the 2D materials.[48–50] Although 2D materials present a vdW type surface without dangling bonds, the inevitable intrinsic defects, such as adatoms, vacancies, anti-sites, and grain boundaries on the surface of 2D materials, provide anchoring point to graft molecules via covalent interactions. On the other hand, these native defects in 2D materials,[51] acting as charge carrier trapping and recombination centers,[52,53] often play a critical role in determining the (opto) electrical characteristics of 2D materials.[54–56] The chemical passivation of 2D materials using organic molecules thus represents a powerful approach to tune the electronic and optical properties of 2D materials by defects engineering via the modulation of charge carrier mobility and conductivity,[57] or by phase engineering for device applications, such as polymorph phase transition between metallic T phase and semiconducting H phase.[58–60]

Fundamental insights of the bonding sites,[61] defect passivation,[62] and polymorph phase transition,[60] doping effect resulting from chemical interaction,[63,64] which is responsible for evolution of electronic structures, could be investigated using SR-based PES by taking advantage of the high energy resolution, enhanced surface sensitivity and tunable ionization cross-section.[50,62,65] Figure 5 shows the SR-based PES spectra and corresponding deconvoluted components to examine chemical interaction between thiophenol (Ph-SH, with a molecular structure shown in Fig. 5(e) inset) molecules and (hydrogen treated) graphene.[66] Four distinct components can be identified from the C 1s core level of pristine EG on SiC(0001) as shown in Fig. 5(a), and they are characteristic for graphene grown epitaxially on SiC surface. The dominant peak labeled as G is attributed to graphene. Component originated from SiC bulk substrate is located at lower binding energy side of dominant peak. Peaks C1 and C2 are assigned to C-atoms in $6\sqrt{3} \times 6\sqrt{3}$ buffer layer

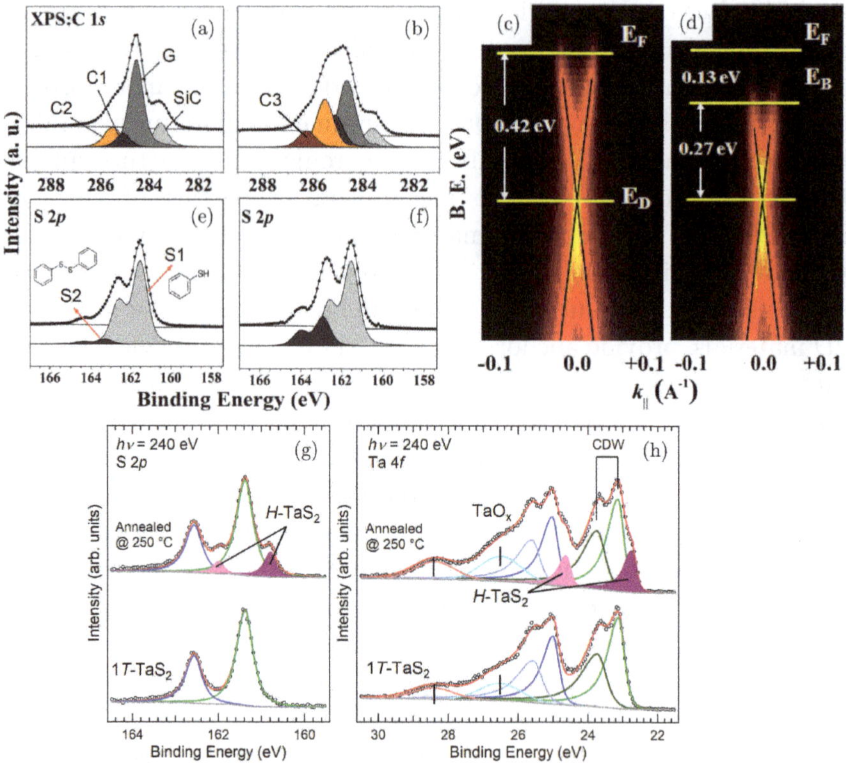

Fig. 5. SR-based (a) C 1*s* and (c) ARPES spectra for graphene grown epitaxially on SiC surface. (b) C 1*s* and (d) ARPES spectra for hydrogen treated graphene. S 2*p* spectra for Ph-SH on (e) graphene and (f) hydrogen treated graphene, respectively. Reproduced from Ref. 66 with permission. (g) S 2*p* and (h) Ta 4*f* spectra for 1*T*-TaS_2 and *H*-TaS_2 surface layer on 1*T*-TaS_2. Reproduced from Ref. 68 with permission.

interacted with dangling bonds of the underlying SiC and remaining sp^2 C atoms in the buffer layer, respectively.[66,67] The hydrogen treatment during graphene growth leads to an additional peak of C3 in the C 1*s* core level spectra (Fig. 5(b)) with a concomitant reduction in the G peak intensity, and a significant modification on the electronic band structure as revealed by the ARPES spectra (see Figs. 5(c) and 5(d)), suggesting the formation of defective structures due to hydrogen doping. The more defective structures for hydrogen treated graphene could provide more active sites to facilitate catalytic oxidation of Ph-SH molecules under 365-nm UV irradiation. This has been clearly shown by the comparison of S $2p$ core-level spectra in Figs. 5(e) and 5(f). Ph-SH molecules adsorbed on both graphene samples exhibit two distinct S species. The dominant component S1 was assigned to physisorbed unbound Ph-SH molecules, whereas the S2 to the disulfide state as a catalytic reaction product. The distinct fractions of two species indicates that hydrogen treated graphene enhance the catalytic reactivity of the surface.

The polymorph phase transition triggered by organic–2D chemical interaction can also be examined by chemical shift.[58] Generally, the binding energy of core-levels for metallic phase is smaller than that for semiconducting or insulating phase due to the different photo-hole screening effect induced by surrounding electrons.[37] For example, the Ta $4f$ and S $2p$ core levels for H-TaS_2 display the lower binding energy than those for T-TaS_2 (Figs. 5(g) and 5(h)), although the nominal valence states of Ta^{4+} and S^{2+} remain unchanged. In addition, the Ta 4f peaks for Ta^{4+} bonds to oxygen (Ta-O) exhibit highest binding energy due to insulating nature.[68]

4. Near-edge X-ray Absorption Fine Structure Spectroscopy

While PES provides access to core-levels and frontier molecular orbitals that are occupied, XAS and its variant NEXAFS monitor the resonant electronic transitions from a core level (e.g. $1s$, $2s$ or $2p$) of an atomic species to unoccupied states governed by the dipole selection rules ($\Delta l = \pm\ 1$ and $\Delta s = 0$), providing complementary electronic structure information. The absorption edges are named by the principal quantum number of excited electron: that is $n = 1, 2, 3,$ being related to K, L, M, respectively, and electronic orbital shells $p_{1/2}$, $p_{3/2}$, $d_{3/2}$, and $d_{5/2}$ refer to subscripts 2, 3, 4, and 5. Most organic molecules consist of light atoms of carbon, nitrogen and oxygen and hence K-edge is the most relevant absorption edge with an energy window located in the soft X-ray region. Figure 6 illustrates the X-ray absorption process using benzene molecules as an example. Resonances in C K-edge NEXAFS spectra are attributed to electronic

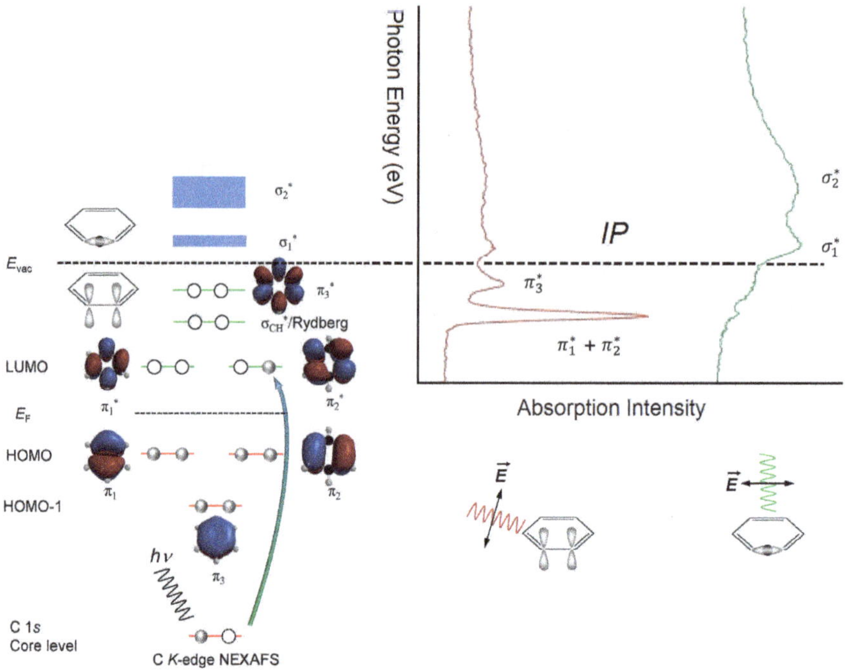

Fig. 6. Schematic view of X-ray absorption process in typical benzene molecular system and corresponding polarization dependence of C *K*-edge NEXAFS spectra for benzene on Ag(110) collected at grazing and normal incidence, respectively. The NEXAFS spectra are reproduced from Ref. 70 with permission.

transitions from a C 1*s* core level to unoccupied molecular orbitals with π^* or σ^* symmetry. The sharp π^* resonances (LUMO, LUMO +1,...) lie below the ionization potential (IP) of organic molecules, whereas the σ^* resonances reside above IP with a broad and asymmetric peak profile. The remaining C–H σ^*/Rydberg orbitals (σ^*_{CH}) are generally located between π^* orbitals and IP. The experimental C 1$s \rightarrow \pi^*$ and C 1$s \rightarrow \sigma^*_{\mathrm{CH}}$ resonances are typically sharper than C 1$s \rightarrow \sigma^*$ resonances due to the reduced lifetime of σ^* resonances resulting from σ^* orbital overlapping with continuum states.[69] The NEXAFS is surface sensitive, element specific, charging insensitive and local bonding environment sensitive, which makes it a valuable tool to investigate chemical interaction and electronic structures at organic–2D materials interfaces.

Compared with PES, NEXAFS is even more sensitive to the local bonding environment as it directly probes the unoccupied states which are hybridized with those of adjacent atoms through orbital overlapping. Consequently, NEXAFS is capable of providing the "finger print" of specific moieties in molecules (e.g. C–C, C=O, C≡N, etc.) through a building block approach.[71] Figures 7(a) and 7(b) show the C K-edge and N K-edge NEXAFS spectra for 3.5 Å azobenzene on HfS_2. The molecular structure of azobenzene is shown in Fig. 7(b) inset. The resonances below 289 eV (C K-edge) and 404 eV

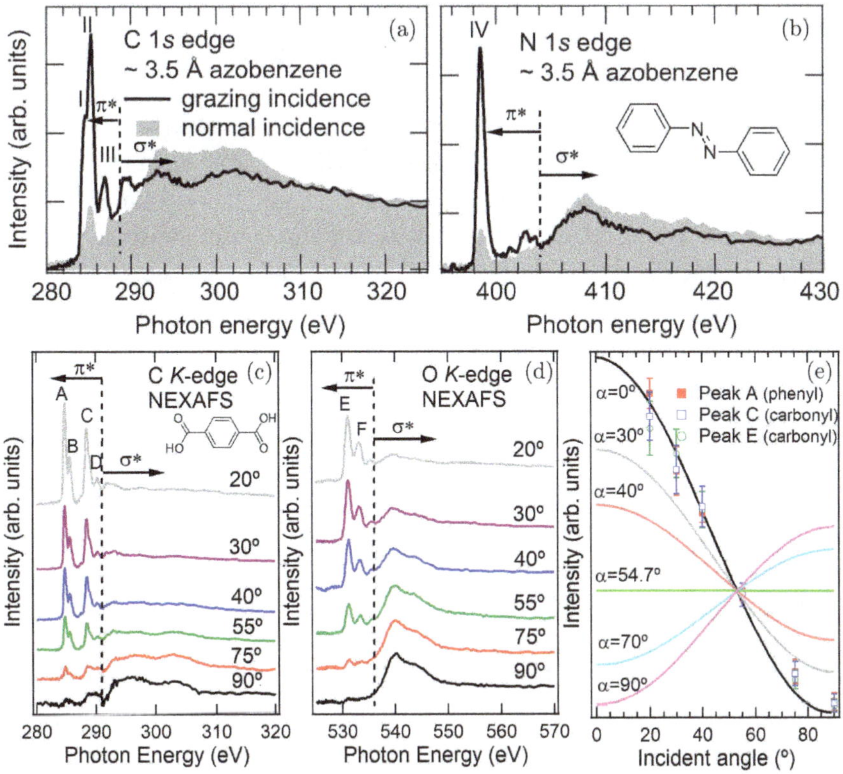

Fig. 7. Angular dependent C K-edge and N K-edge spectra for (a) 1ML (~3.5 Å) and (b) multilayer (~16 Å) azobenzene on HfS_2. Reproduced from Ref. 72 with permission. Angular dependent (c) C K-edge and (d) O K-edge NEXAFS spectra for 1ML TPA on graphene/Ni(111). (e) The intensity of π^* resonances as a function of incident angle θ and corresponding theoretical curve for various molecular tilt angles α. Reproduced from Ref. 73 with permission.

(N K-edge) are attributed to $1s \rightarrow \pi^*$ transitions, whereas the resonances at higher energy are due to $1s \rightarrow \sigma^*$ transitions. The resonances I and II in Fig. 7(a) correspond to C $1s$ to LUMO and LUMO+1 $\sim$ LUMO+3 transitions, respectively. The azo-bridge (–N=N–) dominates the LUMO density, which explains the lower intensity of resonance I. The resonance III ($\sim$285.3 eV) is attributed to higher unoccupied orbitals transition. A Rydberg resonance is observed at $\sim$286.9 eV. The π^* resonance (peak IV) detected from N K-edge NEXAFS spectra is assigned to N $1s$ to LUMO transition, suggesting that azo-bridge dominates the LUMO density, and higher lying LUMOs orbitals are mainly located at the phenyl rings. Strong chemical interaction can be ruled out due to absence of new features or broadening of π^* resonances.

In addition to probing unoccupied electronic states, molecular orientation can be determined from NEXAFS by harnessing the dependence of resonance intensity on the electric field polarization of incident X-rays. For planar π-conjugated molecules with a planar structure, such as benzene, π^* or σ^* orbitals are oriented essentially out-of-plane or in-plane, respectively. Resonance intensities depend on the orientation of the electric field vector **E** of the linearly polarized X-ray with respect to direction of the orbitals, exhibiting a distinct polarization dependence. For benzene on Ag(110) with a lying down geometry (Fig. 6), the π^* resonances are enhanced at grazing incidence with which the electric field vector is parallel to the orbital direction and suppressed at normal incidence, whereas σ^* resonances show an opposite polarization dependence. As a consequence, the spatial orientation of orbitals, and in turn molecular orientation, can be easily derived by the angular dependent intensities of π^* or σ^* resonances in its NEXAFS spectra.[36]

Polarization dependence of π^* resonances are commonly adopted in the derivation of molecular orientations, thanks to their sharp peak profiles in the NEXAFS spectra and flat background. For the substrate on which the molecules are adsorbed has at least threefold symmetry, π^* resonance intensities (I_{π^*}) can be expressed as[71]

$$I_{\pi}^* = CP(\sin^2\alpha\sin^2\theta + 2\cos^2\alpha\cos^2\theta) + C(1 - P)\sin^2\alpha, \qquad (1)$$

where the constant C describes the absolute angular integrated intensity, P denotes the linear polarization factor which can be controlled by beamlines, α is the average tilt angle of the molecular plane with respect to the substrate surface and θ is the incident angle. For the detailed derivation, readers are referred to Ref. 71.

Studying molecular orientation at the molecule/2D materials interface has enormous implications to the understanding and optimization of devices that harness the anisotropic transport characteristics inherent to molecular assemblies.[74] Additionally, unlike inorganic semiconductors, the IP (and EA) of an ordered molecular assembly on a substrate is known to depend on the molecular orientation.[60,75] Such orientation-dependent molecular energies inevitably affect the energy level alignment at device interfaces. Figures 7(c) and 7(d) show the angular dependent C K-edge and O K-edge spectra for one ML (~3.3Å) terephthalic acid (TPA) molecule on graphene.[73] As shown in Fig. 7(c), the resonances below 291 eV are attributed to C $1s \rightarrow \pi^*$ transitions, whereas the resonances at higher energy are due to C $1s \rightarrow \sigma^*$ transitions. The resonances A, B and C correspond to transitions from phenyl carbon core level to LUMOs orbitals, and the resonance D is due to carboxyl carbon core level to LUMOs transitions. The two sharp peaks of E and F in O K-edge spectra (Fig. 7(d)) below 536 eV are ascribed to the carbonyl oxygen core level to LUMOs transitions, and the O $1s \rightarrow \sigma^*$ transitions are located at higher energy.

The π^* resonances are enhanced at the grazing incidence and suppressed at the normal incidence, whereas the σ^* resonances follow the reverse trend. This strong angular-dependence indicates a lying-down orientation of TPA molecules on graphene to maximize the molecule-substrate π–π interactions. Figure 7(e) represents the evolution of intensity of resonances A, C and E as a function of incident angle together with theoretical curves for different tilt angles obtained using Eq. (1). The average tilt angle α is determined to be $0°$ indicating a perfect lying down geometry of molecules.

Often the tilt angles can be quantified quickly and accurately by using the resonance intensity ratio at two extreme X-ray incident angles (e.g. $\theta = 20^\circ$ and $\theta = 90^\circ$) with the following relationship.[71,76,77]:

$$\frac{I_\pi^*(\alpha, 90^\circ)}{I_\pi^*(\alpha, \theta)} = \frac{P(\sin^2\alpha\sin^2 90^\circ + 2\cos^2\alpha\cos^2 90^\circ) + (1-P)\sin^2\alpha}{P(\sin^2\alpha\sin^2\theta + 2\cos^2\alpha\cos^2\theta) + (1-P)\sin^2\alpha}. \tag{2}$$

Figures 7(a) and 7(b) also show the angular dependent C K-edge and N K-edge NEXAFS spectra for 1ML azobenzene on HfS_2 collected at grazing incidence of 30° and normal incidence of 90°.[72] The similar average tilt angles of $23 \pm 6^\circ$ are yielded from π^* resonance of both C 1s and N 1s transition, indicating that phenyl rings are co-planar with the –N=N– plane.

5. CHC Spectroscopy

5.1. *Basic principle*

CHC technique, using the intrinsic core-hole lifetime of several femtoseconds for low atomic number elements (6 fs for C 1s, 6 fs for N 1s, 4 fs for O 1s, 1.27 fs for S 1s and 0.5 fs for S 2s[78–80]) as an internal reference clock, can provide temporal information on femtosecond or even attosecond timescale. This superior temporal resolution makes CHC spectroscopy a promising tool to investigate the ultrafast interfacial charge transfer dynamics. Implementation of CHC spectroscopy requires soft X-rays with tunable photon energy, which necessitates the use of SR facilities.

Figure 8 summarizes photoexcitation and decay processes involved in the CHC technique. The first step involves the X-ray absorption process as in NEXAFS, in which a core-hole is created when a core-level electron is photoexcited to the LUMOs as shown in Fig. 8(a). Within the core-hole lifetime, the core-hole could decay through two competitive pathways: the autoionization process (i.e. spectator (SP) decay and participator decay) and interfacial charge transfer process; the relevant time constants determine which pathway dominates the decay process. Figure 8(b) illustrates the SP decay process. The photoexcited electron remains localized in the LUMOs as a

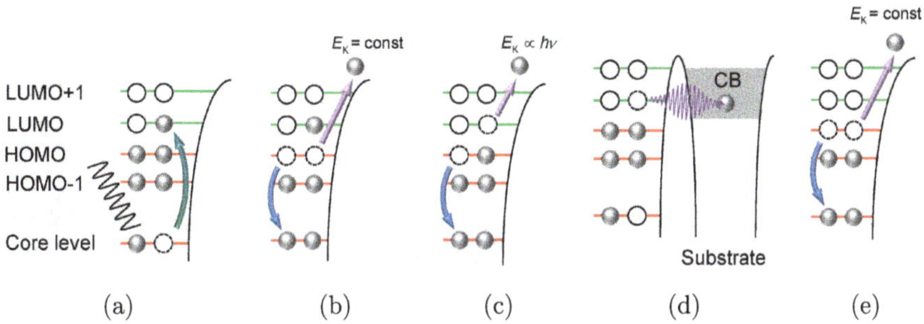

Fig. 8. Schematic overview of (a) resonant excitation of a core-electron to the LUMOs and subsequently core-hole decay processes of (b) SP decay (resonant Auger), (c) participator decay (resonant photoemission) and (d) charge transfer of excited electron to substrate conduction band. (e) NA decay process follows charge transfer process. Reproduced from Ref. 84 with permission.

"spectator" to "observe" the resonant Auger decay process, leaving two holes in the outer shell. Like a normal Auger (NA) process (Fig. 8(e)), the kinetic energy of resonant Auger electrons remains constant when the photon energy varies. However, the kinetic energy of resonant Auger electrons is larger than that of regular Auger electrons due to the core-hole screen of "spectator" electrons in the final state, which helps to distinguish between resonant Auger signals and NA signals. In the participator decay channel (Fig. 8(c)), HOMO electron fills the core-hole with the excessive energy promoting the photoexcited electron into vacuum. The resulting one-hole final state is energetically equivalent to direct valence state photoemission (Fig. 2(b)), but exhibit a resonant enhancement of the associated valence state features with a constant binding energy. Consequently, it is also known as resonant photoemission process. In contrast, if the photoexcited electrons transfer to the substrate conduction band within the core-hole lifetime (Fig. 8(d)), the autoionization decay process is greatly quenched with suppression of resonant Auger and resonant photoemission signals, whereas the NA signals become intense because the core-holes must decay via NA (non-resonant) process (Fig. 8(e)). For a comprehensive description of CHC approach, readers are recommended to Refs. 81–84.

For a quantitative evaluation, the charge transfer time τ_{CT} at the molecule/2D material interfaces can be obtained from

$$\tau_{CT} = \tau_{CH}\frac{I_{res}}{I_{Aug}}, \tag{3}$$

where τ_{CH} is the core-hole lifetime, I_{res} represents the intensity of autoionization component, and I_{Aug} the intensity of NA component. The two components are extracted from the measured CHC spectrum. For instance, a linear combination of NA spectrum conducted at non-resonant photon energy and resonant spectrum in isolated system with appropriate weigh factors is commonly employed to represent the CHC spectrum. This method is known as resonant Auger electron spectroscopy (RAES). For its qualitative application in self-assembled ML (SAM) systems, the readers are recommended to Ref. 85.

When the autoionization features and NA features are difficult to deconvolute, the charge transfer time can then be obtained from the relation

$$\tau_{CT} = \tau_{CH}\frac{I^{\text{iso}} - I^{\text{coup}}}{I^{\text{iso}}}, \tag{4}$$

where I^{iso} and I^{coup} represent the participator intensity for isolated and coupled system, respectively, and this method is named as resonant PES (RPES). For a detailed description and its application at organic/inorganic interfaces, the readers are recommended to Ref. 84. For a more detailed explanation and derivation of the equation described above, the readers are referred to the excellent review by Brühwiler *et al.*[81] and references therein.

It is worth noting that the CHC technique can yield the reliable charge transfer time in the range $0.1\tau_{CH} \leq \tau_{CT} \leq 10\tau_{CH}$, beyond which the extracted charge transfer times should be viewed in a more qualitative fashion.[82,86] CHC spectroscopy is elemental and orbital specific,[81] which allows the study of charge transfer dynamics at specific atomic sites (or functional groups) and unoccupied orbitals of organic molecules. In addition, spin resolution can be achieved using circularly polarized X-rays.[87] CHC technique thus offers an ideal pathway to study the charge transfer path and its dependence on molecular orientation, atomic sites, orbital symmetry, and spin. An example illustrating site and orbital dependent charge transfer dynamics at hybrid vdW heterojunction is the poly[3-hexylthiophene-2,5-diyl] (P3HT)/MoS_2 system.[88] Figure 9(a) shows the S K-edge NEXAFS spectrum and corresponding peak fitting curves for P3HT on ML MoS_2/SiO_2. The molecular structure of P3HT is shown in Fig. 9(a) inset. The first two resonances at 2470.5 and 2472.1 eV correspond to transition from the S $1s$ to $3p_{x,y}$ orbital parallel to the surface plane and $3p_z$ orbital perpendicular to the surface plane of MoS_2, respectively.[89,90] The higher energy resonances at 2472.9 and 2474.5 eV are associated with S $1s \rightarrow \pi^*$(C–C) and S $1s \rightarrow \sigma^*$(S–C) transitions of P3HT, respectively.[91] These distinct excitation sites enable a detailed investigation on the influence of species/atomic sites on interfacial charge transfer dynamics. Figure 9(b) shows a schematic energy level diagram at P3HT/MoS_2 interface at the core-excited state. The $3p_{x,y}$ electronic states lie within the band gap of P3HT along with the presence of a core-hole. The transfer of photoexcited electrons from $3p_{x,y}$ of MoS_2 to the conduction band of

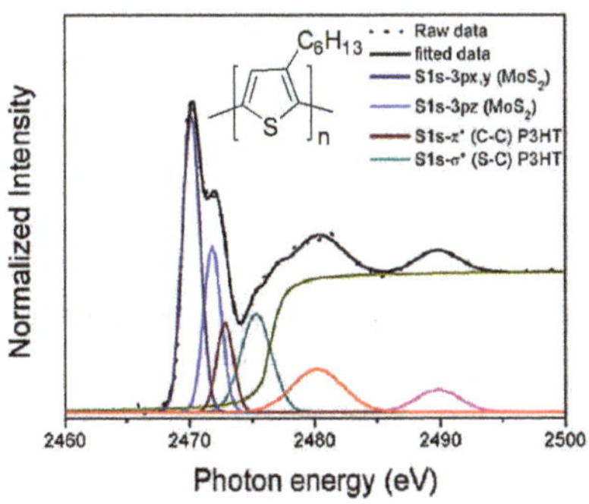

(a)

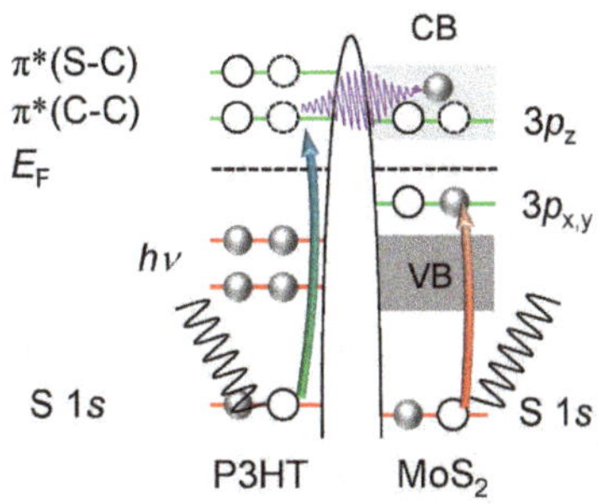

(b)

Fig. 9. (Color online) (a) Sulfur K-edge NEXAFS spectrum and corresponding fitting curves for P3HT on ML MoS_2/SiO_2. (b) Schematic of the relative alignment between unoccupied orbitals of core-excited P3HT molecules and CB edge of MoS_2. S-$KL_{2,3}L_{2,3}$ RAS spectra and corresponding deconvolution results for MoS_2/SiO_2, P3HT/SiO_2, and P3HT/MoS_2/SiO_2 collected at (c) 2470.5 eV (S $3p_{x,y}$), (d) 2472.1 eV (S $3p_z$), (e) 2472.9 eV (π^*(C–C)), and (f) 2474.5 eV (σ^*(S–C)). Reproduced from Ref. 88 with permission.

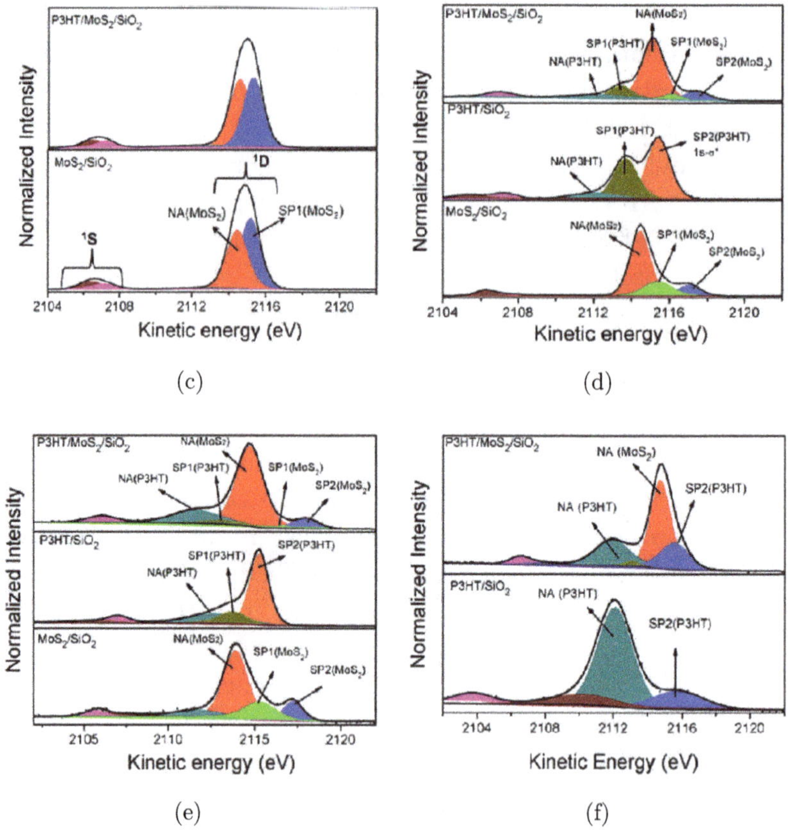

Fig. 9. (*Continued*)

P3HT is thus energetically forbidden, whereas the electron transfer from higher lying electronic states between P3HT and MoS_2 is energetically favorable.

The RAES spectra conducted at four resonant energies and corresponding fitting curves are shown in Figs. 9(c)– 9(f). The SP and NA components are extracted, whereas the participator component is ignored because SP decay dominates over participator decay with a small cross-section. The red and dark green filled lines represent the NA features for MoS_2 and P3HT, respectively. The corresponding SP component(s) (guided by colored filled lines) are lying at high kinetic energy side of NA feature. The charge transfer time scale is

Table 1. Charge transfer timescales for MoS_2/SiO_2, P3HT/SiO_2, and P3HT/MoS_2/SiO_2. Standard deviation values are shown in parentheses. The values are adopted from Ref. 88.

			P3HT/MoS_2/SiO_2	
Electronic transitions	MoS_2/SiO_2(fs)	P3HT/SiO_2(fs)	MoS_2 (fs)	P3HT (fs)
S $1s \rightarrow$S $3p_{x,y}$	1.32 (5)	—	1.25 (4)	—
S $1s \rightarrow$S $3p_z$	0.62 (4)	11.3 (6)	0.34 (3)	2.41 (4)
S $1s \rightarrow \pi^*$ (C–C)	0.50 (5)	4.13 (5)	0.20 (3)	0.45 (6)
S $1s \rightarrow \sigma^*$(S–C)	—	0.36 (4)	—	0.32 (3)

determined from the integral intensities of SP and NA components by using Eq. (3) with $\tau_{CH} = 1.27$ fs for S $1s$ and summarized in Table 1. It is worth noting that the charge transfer dynamics exhibit site (MoS_2 versus P3HT) dependence. The charge transfer time associated with S $3p_{x,y}$ states for isolated MoS_2/SiO_2 (1.32 fs) is comparable with P3HT/MoS_2/SiO_2 (1.25 fs), reflecting the electron delocalization nature within the MoS_2 due to the electronic coupling between S $3p_{x,y}$ and unoccupied Mo $4d$ states.[90] This delocalization nature also explains the sub-femtosecond (0.62 fs) charge transfer time for the S $3p_z$ resonance for MoS_2/SiO_2 isolated film. The formation of P3HT/MoS_2 interface leads to a decrease of charge transfer time to 0.34 fs, indicating the opening of a new sub-femtosecond electron delocalization pathway from MoS_2 to P3HT. The charge transfer time at S $1s \rightarrow \pi^*$(C–C) resonance for P3HT/MoS_2/SiO_2 heterojunction (0.45 fs) is one order of magnitude smaller than that for isolated P3HT/SiO_2 thin film (4.13 fs) indicating that MoS_2 facilitates sub-femtosecond electron transfer from P3HT to the substrates. In addition, the charge transfer time from MoS_2 species to P3HT at S $1s \rightarrow \pi^*$(C–C) resonance also improves from 0.50 fs in isolated MoS_2/SiO_2 film to 0.20 fs in P3HT/MoS_2/SiO_2 heterojunction, indicating the bidirectional interfacial electron transfer between P3HT to MoS_2. In contrast, comparable timescales are determined at S $1s \rightarrow \sigma^*$(S–C) resonance for both P3HT/MoS_2/SiO_2 heterojunction and isolated P3HT/SiO_2 thin film, suggesting that interfacial charge transfer do not occur owing to a low spatial overlap between σ^*(S–C) states and

MoS_2 unoccupied states. This orbital dependence implies that the charge transfer from π^*(C–C) orbital is much more efficient than the σ^*(S–C) orbital. The study of charge transfer dynamics with both site and orbital specificity at organic–2D material hybrid heterojunction provides important insights into understanding the relevant optoelectronic device properties, such as photovoltaic devices.

6. X-ray Magnetic Circular Dichroism

The electronic state hybridization or mixing between organic molecules and underlying substrates could result in the magnetic behaviors distinct from the individual materials.[92] On one hand, at organic/ferromagnetic interface with strong electronic states hybridization, the spinterface effect, in particular spin polarized interfacial states, leads to a plethora of interfacial magnetic phenomena, which is essential to device functionality.[93] For example, the spin filter effect resulting from a spin-dependent trapping of electrons at the tris-(8-hydroxyquinoline)aluminum (Alq_3)/Co interface can enhance magnetoresistance of corresponding organic spin valve devices.[94,95] On the other hand, at magnetic molecules/magnetic substrates interfaces, the electronic state hybridization influence or even quench the magnetic moment of magnetic molecules and/or magnetic substrates.[96,97] The insertion of 2D materials, in particular graphene, between organic layer and magnetic substrate as a buffer layer will decouple or mediate the magnetic coupling,[98–101] causing various magnetic properties, such as ferromagnetic,[102] antiferromagnetic[103] and magnetic anisotropy.[103] Magnetic characterization at organic–2D interfaces is essential for addressing the fundamental phenomena behind the wide variation of magnetic properties. Synchrotron based XMCD, a variant of XAS employing circularly polarized X-rays, is an element-specific and surface sensitive probe, which is useful for unraveling the magnetic properties at organic–2D interfaces.

Figure 10(a) illustrates the resonant excitation processes. The left circularly polarized photon (−1 helicity) excites spin polarized electron with $m_s = -1/2$ from $2p_{3/2}$ (L_3 edge) and $2p_{1/2}$(L_2 edge) core levels to unoccupied states determined by dipole selection rules

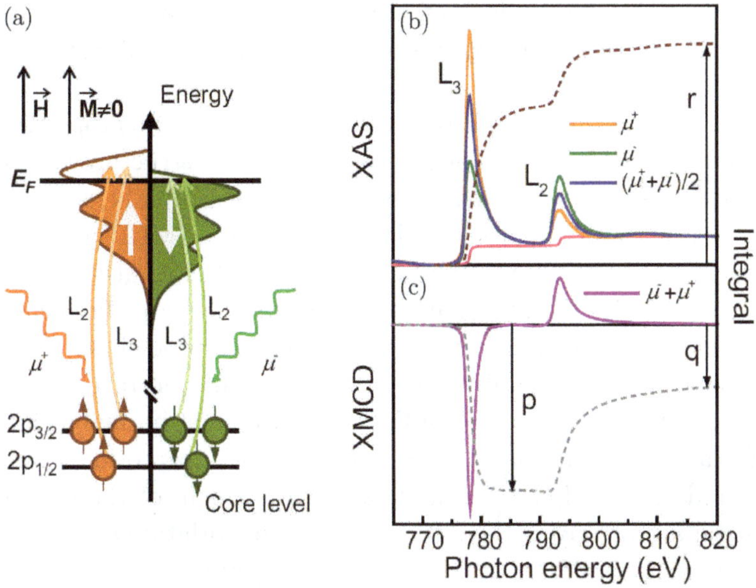

Fig. 10. (a) Schematic diagram describing the resonant excitation processes for magnetic materials. (b) Co $L_{2,3}$-edge XAS spectra recorded with left and right circular polarized photon and (c) corresponding XMCD spectra. Reproduced with permission from Refs. 106 and 107.

$\Delta m_j = -1$, whereas right circularly polarized photon (+1 helicity) excites transition of spin polarized electron with $m_s = 1/2$ governed by $\Delta m_j = 1$. The transition probability is proportional to the unoccupied DOS. In a magnetic material, the exchange splitting leads to the difference in unoccupied DOS for different spin orientation. Consequently, under a magnetic field inverted spin polarization induced by change in helicity leads to distinct absorption spectra as shown in Fig. 10(b). XMCD spectrum (Fig. 10(c)) can be obtained by subtracting the two absorption spectra recorded by left and right circularly polarized light. The spin and orbital magnetic moment of the absorbing atom can be separately quantified from the integrated intensities of XAS and XMCD signals by using the sum rules. The derivation of the sum rules is detailed in Refs. 104 and 105. The spin moment μ_S and orbital moment μ_L can be separately estimated by using $\mu_S = -(6p - 4q)/rn_h$ and $\mu_L = -4q/3rn_h$, in which n_h is the number of d holes. The p, q and r values are extracted from integrated intensities illustrated in Figs. 10(b) and 10(c).

XMCD has been utilized to investigate the magnetic properties for iron (II) phthalocyanine (FePc, with a molecular structure shown in Fig. 11(a) inset) molecules absorbed on graphene/Ir(111) surface.[108] Figures 11(a) and 11(b) show the Fe $L_{2,3}$-edge XMCD spectra for single layer (SL) and thick film (TF) of FePc collected at normal (0°) and grazing (70°) X-ray incidence, respectively. At grazing incidence with which the in-plane magnetic moments are probed, the XMCD magnitudes are comparable to SL and TF. In contrast, a strong reduction of dichroism for SL of FePc is observed at normal incidence suggesting a reduction in the out-of-plane magnetic moments compared to bulk FePc. To understand the origin of the enhanced magnetic anisotropy in SL FePc on graphene, orbital (μ_L) and spin (μ_S) moments for SL and TF FePc and its dependence as a function of incidence angles are estimated from XMCD by using the sum rules and summarized in Figs. 11(d) and 11(e), respectively. The angular dependence of μ_S displays a very similar trend for SL and TF of FePc

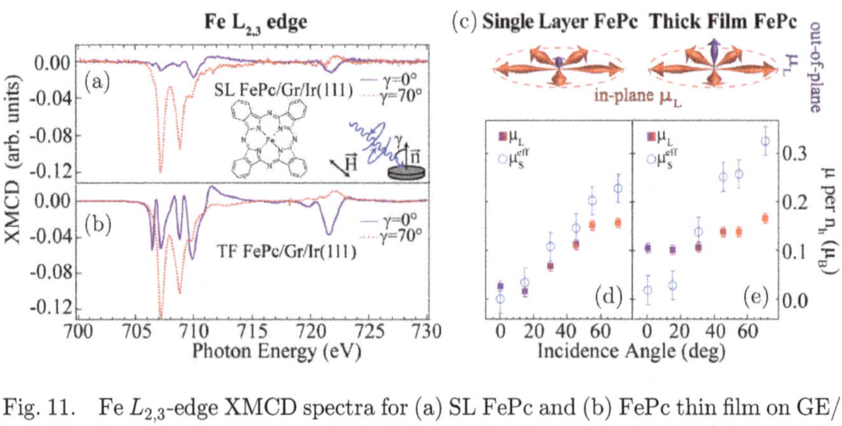

Fig. 11. Fe $L_{2,3}$-edge XMCD spectra for (a) SL FePc and (b) FePc thin film on GE/Ir(111) collected with magnetic field of 5 T at 8 K. The inset of panel (a) shows the measurement geometry. (c) Schematic illustration of the orbital-associated magnetic anisotropy. The evolution of orbital (μ_L) and spin (μ_S) moments determined by sum rules as a function of incidence angles (γ) for (d) SL and (e) TF FePc. The magnetic moments are normalized to Fe atoms and $3d$ hole number. Reproduced from Ref. 108 with permission.

on graphene with a predominant in-plane orientation. In contrast, enhanced angular dependence of μ_L is observed for SL FePc molecules in which the in-plane orbital magnetic moment is strongly quenched, whereas μ_L for TF FePc tends to be isotropic. The significant increase of orbital-associated magnetic anisotropy for the SL FePC, as illustrated in Fig. 11(c), can be attributed to the graphene buffer layer which decouples the magnetic coupling between FePc and Ir (110) and localizes the electron due to the break of D_{4h} symmetry for 2D SL FePc. Designing heterojunctions using magnetic molecule and 2D materials in a hybrid vdW heterostructure therefore opens a new route for tailoring the magnetic order for device application.

7. Summary and Outlook

The organic–2D materials hybrid heterostructures provide a vibrant playground for the exploration of both fundamental properties and exciting new applications through a coherent understanding of the electronic structures underpinning the heterointerfaces. SR-SXS, by harnessing the superb excitation source from a soft X-ray beamline at a synchrotron facility that combines high resolution, high brilliance, tunable energy, and polarization, serves as an invaluable and versatile tool to unravel the rich electronic structures and interfacial properties buried at the heterointerface. In this review, we have given a brief account of the application of several SR-SXS techniques to probe a variety of interfacial properties, including charge transfer, energy level alignment, chemical composition, molecular orientation, and ultrafast charge transfer dynamics, and spin configuration at a few exemplary organic–2D materials heterointerfaces at which the electronic, chemical, charge transport/transfer dynamics and magnetic properties of 2D materials can be modulated through the interaction with organic molecules.[109–111]

With the development of fourth generation SR sources, the energy resolution, spatial resolution, temporal resolution, brightness and coherence of synchrotron will be significantly improved, which will spur a rapid development of synchrotron based SXS in surface science. On one hand, this will help to reveal richer information on spatial or temporal dimensions. For instance, when the spatial resolution is comparable with typical channel lengths of electronic devices, *in situ* characterization of interfacial electronic structures in operating device could be realized. High temporal resolution on the sub-picosecond or even femtosecond time scale enables to explore the ultrarapid phenomena and ultrafast dynamics on the catalytic reactions and magnetization. On the other hand, this will allow the examination of subtle effects that go beyond current research practice. For instance, high energy resolution and brightness X-rays with left- and right-hand circular polarization enables the measurement of spin-dependent ultrafast charge transfer dynamics at organic–magnetic 2D interface with spin hybridization and spintronic application. However, one should be mindful that the high brightness is a double-edged sword, and may cause material damage during synchrotron exposure, which in particular is the case for organic materials. Avenues to mitigate such radiation damage while retaining the benefit of high resolution and versatility of SR-SXS need to be pursued.

Acknowledgments

L. C. acknowledges support from the National Key Research and Development Program of China (No. 2017YFA0402900), the National Natural Science Foundation of China (No. 12074385), and the High Magnetic Field Laboratory of Anhui Province. D. Q. acknowledges the support of the Australian Research Council (No. FT160100207). D. Q. also acknowledges continued support from the Queensland University of Technology (QUT) through the Centre for Materials Science. The authors thank beamline BL11U at NSRL for providing the beam time.

References

1. K. S. Novoselov *et al.*, *Science* **306** (2004) 666.
2. K. S. Novoselov *et al.*, *Science* **353** (2016) aac9439.
3. Y. Liu, Y. Huang and X. Duan, *Nature* **567** (2019) 323.
4. Y. Liu *et al.*, *Nat. Rev. Mater.* **1** (2016) 16042.
5. A. K. Geim and I. V. Grigorieva, *Nature* **499** (2013) 419.
6. X. Duan *et al.*, *Chem. Soc. Rev.* **44** (2015) 8859.

7. H. Schmidt, F. Giustiniano and G. Eda, *Chem. Soc. Rev.* **44** (2015) 7715.
8. A. Avsar *et al.*, *Rev. Mod. Phys.* **92** (2020) 021003.
9. N. Rohaizad *et al.*, *Chem. Soc. Rev.* **50** (2021) 619.
10. A. Bolotsky *et al.*, *ACS Nano* **13** (2019) 9781.
11. J. Azadmanjiri *et al.*, *ACS Appl. Nano Mater.* **3** (2020) 3116.
12. D. B. Sulas-Kern, E. M. Miller and J. L. Blackburn, *Energy Environ. Sci.* **13** (2020) 2684.
13. L. Wang *et al.*, *Nat. Mater.* **19** (2020) 861.
14. A. Nimbalkar and H. Kim, *Nano-Micro Lett.* **12** (2020) 126.
15. Y. Cao *et al.*, *Nature* **556** (2018) 43.
16. Y. Cao *et al.*, *Nature* **556** (2018) 80.
17. K. Pei and T. Zhai, *Cell Rep. Phys. Sci.* **1** (2020) 100166.
18. P. C. Y. Chow and T. Someya, *Adv. Mater.* **32** (2020) 1902045.
19. Y. Lin *et al.*, *Adv. Mater.* **24** (2012) 3087.
20. K. Cho *et al.*, *ACS Nano* **13** (2019) 9713.
21. J. Sun *et al.*, *Adv. Mater.* **31** (2019) 1803831.
22. S. Bertolazzi *et al.*, *Chem. Soc. Rev.* **47** (2018) 6845.
23. Y. L. Huang *et al.*, *Chem. Soc. Rev.* **47** (2018) 3241.
24. R. Tai *et al.*, in *Synchrotron Radiation in Materials Science: Light Sources, Techniques, and Applications*, eds. C. Fan and Z. Zhao (Wiley-VCH Verlag GmbH & Co. KGaA, Germany, 2018), pp. 61–121.
25. T.-K. Sham, Photon-in photon-out spectroscopic techniques for materials analysis: Some recent developments, in *Synchrotron Radiation in Materials Science: Light Sources, Techniques, and Applications*, eds. C. Fan and Z. Zhao (Wiley-VCH Verlag GmbH & Co. KGaA, Germany, 2018), pp. 123–135.
26. W. Yang and Z. Liu, Techniques and demonstrations of synchrotron-based *In situ* soft X-ray spectroscopy for studying energy materials, in *Synchrotron Radiation in Materials Science: Light Sources, Techniques, and Applications*, eds. C. Fan and Z. Zhao (Wiley-VCH Verlag GmbH & Co. KGaA, Germany, 2018), pp. 511–562.
27. Y. Zhu *et al.*, Synchrotron-based bioimaging in cells and *In vivo*, in *Synchrotron Radiation in Materials Science: Light Sources, Techniques, and Applications*, eds. C. Fan and Z. Zhao (Wiley-VCH Verlag GmbH & Co. KGaA, Germany, 2018), pp. 563–596.
28. L. Cao *et al.*, *Environ. Sci. Technol.* **51** (2017) 2643.
29. A. Sharma and D. Hesterberg, Synchrotron radiation-based spatial methods in environmental biogeochemistry, in *Multidimensional Analytical Techniques in Environmental Research*, eds. R. M. B. O. Duarte and A. C. Duarte (Elsevier, B. V., 2020), pp. 231–265.
30. W.-Q. Shi *et al.*, Exploring actinide materials through synchrotron radiation techniques, in *Synchrotron Radiation in Materials Science: Light Sources, Techniques, and Applications*, eds. C. Fan and Z. Zhao (Wiley-VCH Verlag GmbH & Co. KGaA, Germany, 2018), pp. 389–509.
31. S. Quartieri, Synchrotron radiation in the earth sciences, in *Synchrotron Radiation*, eds. S. Mobilio, F. Boscherini and C. Meneghini (Springer, Berlin, 2015), pp. 677–695.

32. M. P. Seah and W. A. Dench, *Surf. Interface Anal.* **1** (1979) 2.
33. G. Hähner, *Chem. Soc. Rev.* **35** (2006) 1244.
34. J. Stöhr, *J. Magn. Magn. Mater.* **200** (1999) 470.
35. T. Funk *et al.*, *Coord. Chem. Rev.* **249** (2005) 3.
36. D.-C. Qi, W. Chen and A. T. S. Wee, NEXAFS studies of molecular orientations at molecule-substrate interfaces, in *The Molecule–Metal Interface*, eds. N. Koch, N. Ueno and A. T. S. Wee (Wiley-VCH Verlag GmbH & Co. KGaA, Germany, 2013), pp. 119–151.
37. F. de Groot and A. Kotani, *Core Level Spectroscopy of Solids* (CRC Press, Boca Raton, 2008).
38. J. J. Yeh and I. Lindau, *At. Data Nucl. Data Tables* **32** (1985) 1.
39. M. Oehzelt, N. Koch and G. Heimel, *Nat. Commun.* **5** (2014) 4174.
40. M. Oehzelt *et al.*, *Sci. Adv.* **1** (2015) e1501127.
41. Y. Wang *et al.*, *Nano Res.* **14** (2021) 1682.
42. X. Zhang *et al.*, *Adv. Mater.* **28** (2016) 10409.
43. M. T. Edmonds *et al.*, *J. Chem. Phys.* **136** (2012) 124701.
44. W. Chen *et al.*, *J. Am. Chem. Soc.* **129** (2007) 10418.
45. W. Gao and A. Kahn, *Appl. Phys. Lett.* **79** (2001) 4040.
46. W. Gao and A. Kahn, *Org. Electron.* **3** (2002) 53.
47. C. Coletti *et al.*, *Phys. Rev. B* **81** (2010) 235401.
48. A. Tuxen *et al.*, *ACS Nano* **4** (2010) 4677.
49. V. Kumar *et al.*, *J. Phys. Chem. C* **122** (2018) 21076.
50. C. R. Ryder *et al.*, *Nat. Chem.* **8** (2016) 597.
51. H. Qiu *et al.*, *Nat. Commun.* **4** (2013) 2642.
52. J. Shim *et al.*, *Adv. Mater.* **28** (2016) 6985.
53. E. Liu *et al.*, *Adv. Funct. Mater.* **26** (2016) 1938.
54. S. Wang, A. Robertson and J. H. Warner, *Chem. Soc. Rev.* **47** (2018) 6764.
55. J. Jiang *et al.*, *Research* **2019** (2019) 4641739.
56. D. Rhodes *et al.*, *Nat. Mater.* **18** (2019) 541.
57. Z. Yu *et al.*, *Nat. Commun.* **5** (2014) 5290.
58. D. Voiry *et al.*, *Nat. Chem.* **7** (2015) 45.
59. K. C. Knirsch *et al.*, *ACS Nano* **9** (2015) 6018.
60. X. Zhou *et al.*, *J. Mater. Chem. C* **8** (2020) 4432.
61. X. Zhu *et al.*, *Nat. Commun.* **9** (2018) 4177.
62. D. M. Sim *et al.*, *ACS Nano* **9** (2015) 12115.
63. S. Niyogi *et al.*, *Nano Lett.* **10** (2010) 4061.
64. G. Ambrosio *et al.*, *Nanoscale* **12** (2020) 9032.
65. R. E. Brandt *et al.*, *Chem. Mater.* **29** (2017) 4667.
66. K.-J. Kim, H. S. Kim and H. Lee, *Sci. Rep.* **10** (2020) 2119.
67. K. V. Emtsev *et al.*, *Phys. Rev. B* **77** (2008) 155303.
68. Z. Wang *et al.*, *ACS Nano* **12** (2018) 12619.
69. A. Nefedov and C. Wöll, Advanced applications of NEXAFS spectroscopy for functionalized surfaces, in *Surface Science Techniques*, eds. G. Bracco and B. Holst (Springer, Berlin, 2013), pp. 277–303.
70. J. L. Solomon, R. J. Madix and J. Stöhr, *Surf. Sci.* **255** (1991) 12.
71. J. Stöhr, *NEXAFS Spectroscopy* (Springer, Berlin, 1992).

72. E. Ludwig *et al.*, *Phys. Chem. Chem. Phys.* **15** (2013) 20272.
73. W. Zhang *et al.*, *Phys. Chem. Chem. Phys.* **14** (2012) 10125.
74. A. Vilan, D. Aswal and D. Cahen, *Chem. Rev.* **117** (2017) 4248.
75. S. Duhm *et al.*, *Nat. Mater.* **7** (2008) 326.
76. L. Cao *et al.*, *J. Phys. Chem. C* **115** (2011) 24880.
77. L. Cao *et al.*, *Nanoscale Adv.* **1** (2019) 1991.
78. B. Kempgens *et al.*, *J. Phys. B, At. Mol. Opt. Phys.* **29** (1996) 5389.
79. M. O. Krause and J. H. Oliver, *J. Phys. Chem. Ref. Data* **8** (1979) 329.
80. M. Coville and T. D. Thomas, *Phys. Rev. A* **43** (1991) 6053.
81. P. A. Brühwiler, O. Karis and N. Mårtensson, *Rev. Mod. Phys.* **74** (2002) 703.
82. L. Wang, W. Chen and A. T. S. Wee, *Surf. Sci. Rep.* **63** (2008) 465.
83. R. Friedlein *et al.*, *J. Electron Spectrosc. Relat. Phenom.* **183** (2011) 101.
84. L. Cao *et al.*, *Adv. Mater.* **26** (2014) 7880.
85. M. Zharnikov, *J. Electron Spectrosc. Relat. Phenom.* **200** (2015) 160.
86. A. Föhlisch *et al.*, *Nature* **436** (2005) 373.
87. F. Blobner *et al.*, *Phys. Rev. Lett.* **112** (2014) 086801.
88. Y. Garcia-Basabe *et al.*, *Phys. Chem. Chem. Phys.* **21** (2019) 23521.
89. D. Guay *et al.*, *Chem. Mater.* **6** (1994) 614.
90. Y. Garcia-Basabe *et al.*, *Phys. Chem. Chem. Phys.* **19** (2017) 29954.
91. U. Aygül *et al.*, *J. Phys. Chem. C* **116** (2012) 4870.
92. K. V. Raman *et al.*, *Nature* **493** (2013) 509.
93. M. Galbiati *et al.*, *MRS Bull.* **39** (2014) 602.
94. S. Steil *et al.*, *Nat. Phys.* **9** (2013) 242.
95. A. Droghetti *et al.*, *Nat. Commun.* **7** (2016) 12668.
96. P. Gargiani *et al.*, *Phys. Rev. B* **87** (2013) 165407.
97. M. S. Jagadeesh *et al.*, *Appl. Phys. Lett.* **115** (2019) 082404.
98. M. Scardamaglia *et al.*, *J. Phys. Chem. C* **117** (2013) 3019.
99. G. Avvisati *et al.*, *J. Chem. Phys.* **150** (2019) 054704.
100. G. Avvisati *et al.*, *Phys. Rev. B* **98** (2018) 115412.
101. H. Adler *et al.*, *J. Phys. Chem. C* **119** (2015) 8755.
102. A. Candini *et al.*, *J. Phys. Chem. C* **118** (2014) 17670.
103. V. Corradini *et al.*, *J. Appl. Phys.* **125** (2019) 142904.
104. B. T. Thole *et al.*, *Phys. Rev. Lett.* **68** (1992) 1943.
105. P. Carra *et al.*, *Phys. Rev. Lett.* **70** (1993) 694.
106. G. van der Laan and A. I. Figueroa, *Coord. Chem. Rev.* **277–278** (2014) 95.
107. A. I. Figueroa, X-ray absorption spectroscopy: Element-selective tools to characterize magnetic nanoparticles, in *Complex Magnetic Nanostructures: Synthesis, Assembly and Applications*, ed. S. K. Sharma (Springer, Cham, 2017), pp. 173–193.
108. S. Lisi *et al.*, *J. Phys. Chem. Lett.* **6** (2015) 1690.
109. X. Arramel *et al.*, *ACS Appl. Mater. Interfaces* **9** (2017) 5566.
110. K. Greulich *et al.*, *J. Phys. Chem. C* **124** (2020) 16990.
111. M. L. Ng *et al.*, *Phys. Rev. B* **81** (2010) 115449.

https://doi.org/10.1142/9789811247859_0008

Chapter 8

Recent Advances in Bond-resolved Scanning Tunneling Microscopy

Shaotang Song*[,‡], Jie Su, Xinnan Peng,
Xinbang Wu and Mykola Telychko[†,‡]

Department of Chemistry,
National University of Singapore,
3 Science Drive 3, Singapore 117543, Singapore
**chmv234@nus.edu.sg*
†chmtm@nus.edu.sg

Recent advances in bond-resolved scanning tunneling microscopy (BRSTM) have demonstrated the tremendous potential of this characterization technique to attain an ultra-high spatial resolution at the level of a single chemical bond. Due to such a unique ability to visualize chemical bonds, BRSTM has been recognized as a valuable characterization tool in the rapidly developing field of on-surface chemistry. In this paper, we discuss the recent experimental advances in BRSTM imaging techniques and their applications in the characterization of a wide scope of functional nanostructures, including individual molecules, elusive nanographenes fabricated by means of surface-assisted synthetic strategies and extended supramolecular self-assemblies.

Keywords: Bond-resolved STM; On-surface chemistry; Nanographene.

1. Introduction

Rapidly the invention of,[1] scanning probe imaging techniques, including noncontact atomic force microscopy (ncAFM) and scanning tunneling microscopy (STM), have become highly valuable

‡Corresponding authors.

characterization tools in the fields of materials science and nanotechnology, allowing for the real-space imaging of surfaces of solid-state materials and molecular systems with unprecedented resolution.[2,3] A broad range of systems, suitable for STM and ncAFM imaging, span from the nanostructures comprised of individual atoms[4] to complex supramolecular assemblies[5] and novel two-dimensional materials.[6–8]

It has long been known that functionalization of the tip apex with an atom or a molecule may lead to a remarkable enhancement of the spatial resolution.[9] The first experimental attempt to achieve enhanced spatial resolution on Cu(211) surface relied on tip functionalization with Xe atom.[10] Later, it has been demonstrated that tip functionalization with a single Cl atom results in a higher spatial resolution of the NaCl(100) layer due to the presence of highly localized *p*-wave states on a Cl atom.[11] Nonetheless, the feasibility of spatial resolution at the level of "atomic bond" had long been a challenge until the pioneering work by Temirov *et al.*, who demonstrated that trapping a molecular hydrogen in the tunneling junction allows for the visualization of intramolecular features, resembling the chemical structure of the molecule.[12]

In 2009, Gross *et al.*[13] achieved an submolecular resolution on pentacene molecule using ncAFM technique via the controllable functionalization of tip apex with a single carbon monoxide (CO) molecule. Subsequent experimental advancements demonstrate that submolecular resolution can be also achieved using controllable decoration of the tip apex with various atoms and molecules, including Xe,[14] H,[15] N_2O,[16] O[17] and Si.[18] Experimental breakthroughs in high-resolution imaging of various nanostructures using tip-functionalized ncAFM measurements have been summarized in recent review articles.[19–22]

Later, it has been realized that such a tip functionalization strategy can be also utilized to obtain submolecular spatial resolution solely with an STM setup. Owing to the ease of technical realization and the wealth of offered structural information comparable to ncAFM, bond-resolved STM (BRSTM) has been rapidly adopted as a frequently used characterization tool by the scientific community working in the fields of surface science and nanotechnology.

2. Scanning Hydrogen Tunneling Microscopy

Historically, scanning tunneling hydrogen microscopy (STHM) represents a seminal imaging technique allowing for the real-space visualization of a chemical bond. This technique was pioneered by Temirov and co-workers, who employed it to resolve the internal chemical structure of 3,4,9,10-perylenetetracarboxylic dianhydride (PTCDA) molecules (Fig. 1(a)).[12,23] STHM is realized by introducing

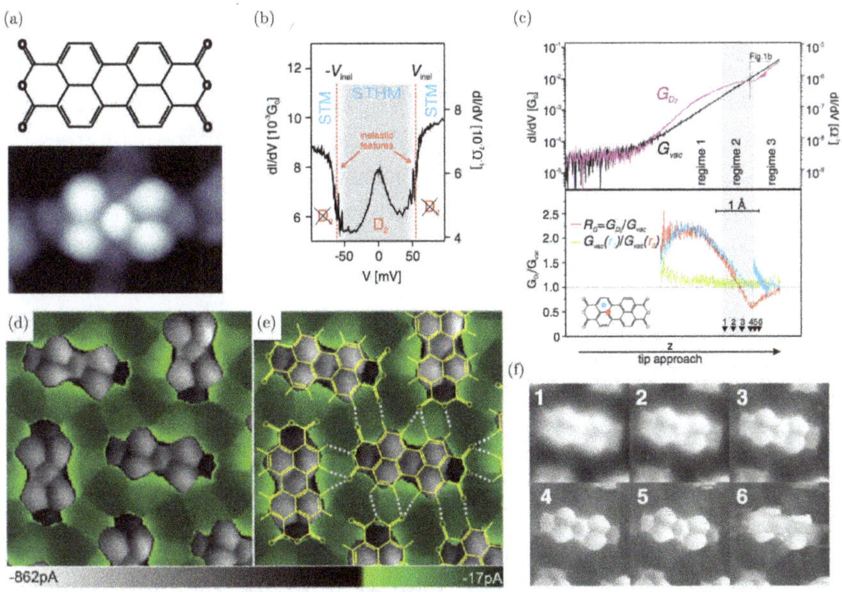

Fig. 1. (Color online) (a) The chemical structure of PTCDA. STHM image of PTCDA on Au(111), $V = -5\,\text{mV}$. The chemical structure of PTCDA. (b) dI/dV spectrum measured in the center of PTCDA–Au(111) and (c) $dI/dV(z)$ spectra recorded at fixed bias, measured with D_2 on PTCDA–Au(111) (z-axis scale given by the scale bar). $G_0 = 2e^2/h$ is the quantum of conductance. G_{D2} (magenta): $V = -5\,\text{mV}$. G_{vac} (black): $V = -130\,\text{mV}$. (d) STHM image of herringbone phase of PTCDA molecules on Au(111). (e) Superposition of the STHM image from panel (d) with the structure of the herringbone phase. (f) STHM images measured at different z indicated by black triangles in panel (c).

Source: Adapted from Refs. 12, 23 and 24.

hydrogen (H_2) or deuterium (D_2) molecules into the STM chamber at low temperature (~10 K), leading to spontaneous trapping of the molecules within the tunneling junction. Constant-height STM imaging acquired in a bias voltage window of ±50 mV (Fig. 1(b)) reveals the chemical structure of the molecule consisting of five bright rings associated with the perylene cores and slightly darker C_5O heterorings (Fig. 1(a)). Interestingly, STHM has also revealed its ability to provide information about local noncovalent intermolecular interactions between molecular layers. As shown in the STHM image constant-height dI/dV image of the PTCDA–Ag(111) interface, clear sharp lines are observed in the space between the molecules. These lines coincidently connect oxygen atoms and hydrogen atoms between neighboring molecules (Fig. 1(d) and 1(e)).

The origin of STHM contrast has been attributed to the variations of differential conductance ($G = dI/dV$) at bias voltage close to zero. Specifically, the trapped H_2 molecule experiences different degrees of repulsive Pauli interaction over distinct sites of the PTCDA molecule, leading to the spatial-dependent modulation of differential conductance (Figs. 1(c) and 1(f)). The H_2 molecule thus can be viewed as a transducer,[24] which effectively probes the magnitude of the short-range repulsive Pauli interaction between H_2 and PTCDA, thereby transforming the magnitude of a force signal into the variations of differential conductance. The largest magnitude of Pauli repulsion occurs over areas of increased electron density (i.e. at the atoms or chemical bonds of the molecule), which is essential for visualization of the intramolecular structure.

STHM can be viably implemented into existing low-temperature STM systems by the dosage of H_2 or D_2 into the vacuum chamber. However, the working mechanism of STHM is not yet fully understood due to complexity of the imaging process. The unclear relationship between H_2–sample interaction and the obtained dI/dV signal makes the image interpretation challenging. To date, STHM has only been used to study molecular self-assemblies and experiments on the individual isolated molecules have not yet been performed.

3. BRSTM Imaging with a CO-Functionalized Tip

3.1. *Mechanism of BRSTM imaging with a CO-functionalized tip*

Shortly after the pioneering ncAFM experiments performed by Gross *et al.*, functionalization of the tip apex with a single CO molecule has emerged as a viable route to reach submolecular resolution.[13] Later experimental advancements demonstrate that the wealth of structural information offered by BRSTM (acquired using CO-functionalized tip) is practically identical to the one offered by ncAFM. Furthermore, the relatively simpler feedback loop of the STM setup allows for a remarkably reduced time of BRSTM image acquisition (i.e. 2–3 min per $3 \times 3\text{-nm}^2$ image), compared to ncAFM setup, wherein the operation of phase-locked loop requires an acquisition time of 25–30 min. Therefore, BRSTM with CO-functionalized tip has gained increased popularity in the scientific community working in the field of on-surface chemistry.

Conventional tip functionalization procedure relies on the dosage of CO gas onto the sample held at low temperature (less than 10 K), followed by the controllable transfer of an individual CO molecule onto the tip apex. The BRSTM imaging is subsequently conducted in the Pauli repulsion regime. In this regime, the CO molecule at the tip apex undergoes lateral bending with the largest deflection magnitude over the electron-rich areas presented by a chemical bond (Fig. 2). The relaxation mechanism of CO molecule has been theoretically quantified by the numerical probe particle model developed by Jelínek and co-workers,[25–27] who demonstrated that the CO molecule tends to tilt over the ridges of the potential energy landscape associated with the chemical structure of the molecule (Figs. 2(a) and 2(b)). Furthermore, the same authors also demonstrated that a lateral deflection of the CO molecule effectively modulates the overall differential conductance of the tip–sample junction (Figs. 2(c)–2(f))[28] (analogously to the aforementioned working principle of STHM), leading to the appearance of sharp line features over the positions of chemical bonds in the tunneling current image (constant-height mode), topography image (constant-current mode) or differential conductance (dI/dV channel).

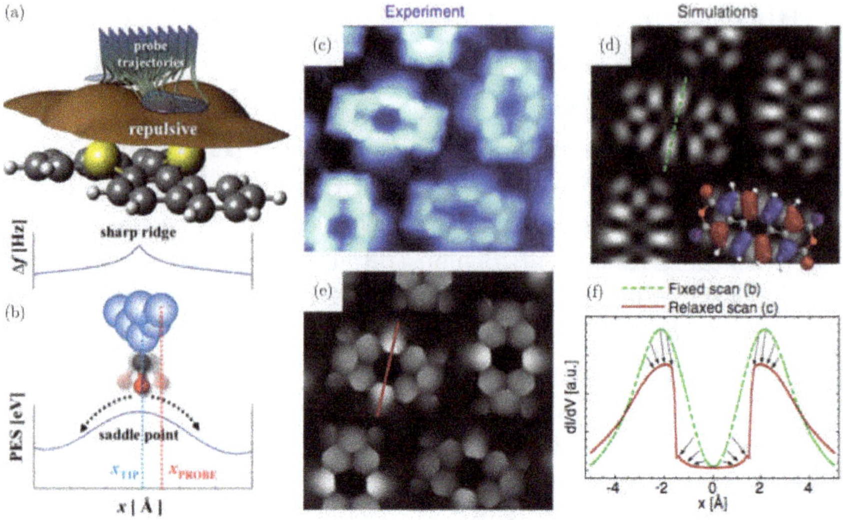

Fig. 2. (Color online) (a) Schematic view of the probe particle trajectory during tip approach represented by individual threads dominated by the Pauli repulsion energy. (b) Appearance of sharp edge in the frequency-shift Δf signal due to convex shape of the potential energy surface (PES) landscape causing probe particle relaxation at close distances. (c) Experimental constant-height BRSTM dI/dV image of PTCDA–Au(111) obtained with CO tip. Calculated constant-height dI/dV images of PTCDA–Au(111) obtained with the probe particle STM code for (d) fixed CO and (e) relaxed one. (f) Profiles taken along the green and red lines in panels (d) and (e), respectively.

Source: Adapted from Refs. 25, 26 and 28.

The notable difference between ncAFM and BRSTM images manifests in contrast variation over the chemical bonds (or atoms). In ncAFM image, the chemical bond (hosting larger electron density) appears brighter compared to empty regions, while BRSTM features darker contrast over chemical bonds. This can be attributed to the site-specific variations of the differential conductance of BRSTM and peculiar p-like orbital nature of the CO-decorated tip.[29] Specifically, the dI/dV signal between the tip and the sample is lower for the bent CO molecule (over chemical bonds) and higher for a unrelaxed CO molecule (over empty regions).

3.2. *BRSTM for on-surface synthesis and characterization*

Over the last decades, surface-assisted synthesis has emerged as a powerful strategy for the bottom–up fabrication of low-dimensional nanographenes (NGs) with tailored structural, electronic and magnetic properties unattainable by conventional wet-chemistry synthesis. The core concept of the surface-assisted synthetic strategies relies on the preliminary rational design of a precursor molecule, which can subsequently undergo chemical transformation towards desired products[30,31] (Figs. 3(a), 3(c) and 3(e)). The versatility of this synthetic strategy has been exemplified by the recent seminal fabrications of a library of open-shell and closed-shell NGs.[32] In addition, the electronic, magnetic and optical properties of NGs can be precisely tuned via engineering their chemical structure and/or concomitant π-electron topology. Such a rapid progress of the on-surface synthetic paradigm has largely benefitted from the development of the BRSTM imaging technique with a CO-decorated tip, which allows for the precise and efficient identification of the surface-assisted reaction intermediates and products. In this subsection, we highlight the recent experimental breakthroughs in the field of on-surface synthesis, wherein BRSTM imaging with a CO-functionalized tip has been extensively employed to unveil the structure of on-surface synthetized graphene-like nanostructures, including zero-dimensional (0D) nanographenes, one-dimensional (1D) graphene nanoribbons (GNRs), two-dimensional (2D) porous graphene and supramolecular self-assemblies.

BRSTM imaging of 0D NGs and their assemblies. A few exemplar on-surface synthetized NGs are highlighted in Fig. 3. BRSTM imaging has been employed to unveil the structure of large aromatic NGs (containing as many as 64 rings), as shown in Fig. 3(b).[33] The molecule was fabricated by the trimerization of a precursor with 12 benzene rings (Fig. 3(a)). Thermal annealing of the precursor triggers Ullmann coupling and cyclodehydrogenation processes, wherein two conjoined cove regions fuse into an azulene motif. The seven-membered rings are easily discernible in the BRSTM image due to their darker contrast compared to those of the hexagonal and pentagonal rings.

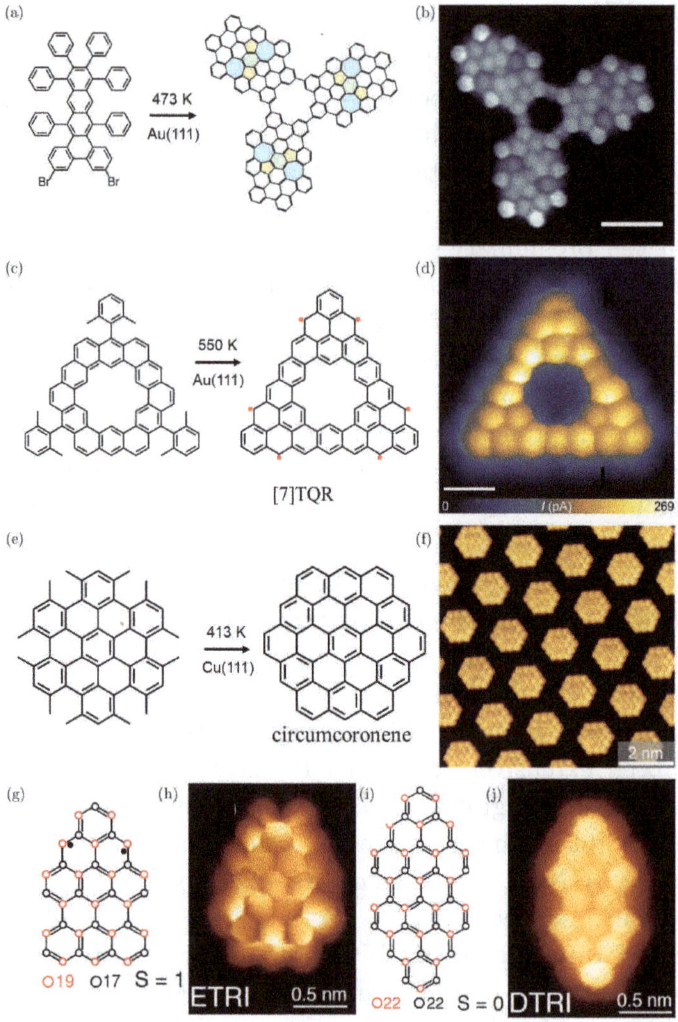

Fig. 3. (Color online) BRSTM imaging of NGs. (a) Reaction scheme of the propeller-shaped C_3-symmetric trimer. (b) BRSTM image of the trimer obtained from dI/dV channel scanning at constant-height mode with CO-functionalized tip. (c) The reaction scheme of [7]TQR. (d) BRSTM image of [7]TQR, $V = 5\,\text{mV}$. (e) Reaction scheme of circumcoronene. (f) BRSTM image of self-assembled cirumcoronene on Cu(111). (g) Molecular structure of ETRI. Red and black empty circles represent two different sublattices. Sublattice imbalance induces a triplet open-shell ground state. (h) BRSTM image of ETRI, $V = 2\,\text{mV}$. (i) Molecular structure of DTRI. (j) BRSTM image of DTRI, $V = 2\,\text{mV}$.

Source: Adapted from Refs. 33–36.

Recently, Su *et al.* reported the surface-assisted synthesis and BRSTM characterization of the [7]triangulene quantum ring ([7]TQR) molecule with a high-spin ground state (Figs. 3(c) and 3(d)).[34] The precursor molecule contains a kekulene core attached by three 2,6-dimethylphenyl substituents at the *meso*-positions of three alternating edges. This design allows for the efficient production of [7]TQR under thermal annealing on the precursor-decorated Au(111) surface. The BRSTM image (Fig. 3(d)) unambiguously resolves the molecular skeleton consisting of 12 fused benzene rings with a hexagonal antidot in the center. Furthermore, Telychko *et al.* recently reported a surface-assisted route towards ultra-high-yield synthesis of circumcoronene molecules (Figs. 3(e) and 3(f)) on Cu(111) surface.[35] This work not only highlights a novel synthetic approach, which is the methyl radical coupling within a rationally designed precursor molecule, but also underpins the ability of BRSTM imaging technique to probe the large-scale assemblies of NGs with submolecular resolution.

In addition to the investigation of closed-shell molecules, BRSTM has been extensively utilized to probe the structure of open-shell NGs with a magnetic ground state. In contrast to conventional magnetism that is usually associated with the *d*- or *f*-orbitals of heavier elements, the magnetism in open-shell NGs originates from the presence of unpaired π-electrons. Li *et al.* performed the BRSTM imaging of open-shell extended triangulenes (ETRI) NGs as shown in Fig. 3(g).[36] ETRI hosts a triplet ground state (net spin $S = 1$) arising from the sublattice imbalance in the molecular backbone. The presence of radicals also leads to Kondo effect, manifested by a pronounced resonance at Fermi level in the dI/dV spectrum. As a result, BRSTM image of ETRI (Fig. 3(h)), collected at a bias voltage of 2 mV, shows additional bright features surrounding the backbone structure, which indicates the localization of the π-radicals in one sublattice [red circles in Fig. 3(g)]. These features are absent in the BRSTM image of closed-shell DTRI molecule, with a sublattice-balanced backbone (Figs. 3(i) and 3(j)). However, it is important to note that BRSTM imaging alone is not sufficient enough to determine whether the nanographene possesses open- or closed-shell configuration as well as reveal the spin density distribution within open-shell molecule. Complementary scanning tunneling spectroscopy measurements and density functional theory (DFT) calculations are needed to answer these questions.

Post-processing of the BRSTM images using Laplace filtering is often employed to resolve ultimately fine features of the molecular backbone. Recently, Fasel and co-workers used Laplace-filtered BRSTM to characterize the open-shell graphene nanostructures exemplified in Fig. 4. It is evident that Laplace-filtered images unambiguously discern pentagonal, hexagonal and heptagonal rings. For example, Fig. 4(a) shows the azulene-embedded NGs which adopt an aromatically driven open-shell diradical configuration.[37] Another notable example of magnetic NG is [5]rhombene (Figs. 4(g) and 4(h)), wherein the magnetic ground state emerges as a result of the competition between electron hybridization and electrostatic Coulomb repulsion.[38] This mechanism is size-dependent; so [5]rhombene hosts two unpaired spins, whereas it is absent in [4]rhombene. Another different mechanism was reported by Mishra *et al.*, who demonstrated the seminal synthesis of open-shell Clar's goblet molecule, envisioned in 1972 by Erich Clar (Figs. 4(e) and 4(f)).[39] The magnetism of this bowtie-shaped NG originates from the topological frustration in its π-electron network, which makes it impossible to draw the classical Kekulé structure without leaving unpaired electrons. Aromatic stabilization, namely the formation of six Clar's sextets in the closed-shell Kekulé form, versus nine Clar's sextets in the open-shell non-Kekulé form, is considered to be the driving force for its open-shell diradical ground state.[40] In contrast to the all-benzenoid topology of NG in Fig. 4(i), the presence of a pentagonal ring (formed by a methyl group detachment during intramolecular dehydrogenation) in the NG in Fig. 4(j) disrupts the bipartite symmetry of the underlying honeycomb lattice. This non-Kekulé structure possesses one unpaired electron localized at the triangulene-like termini. Therefore, the Laplace-filtered BRSTM images in Figs. 4(k), 4(l) and 4(n) show "tail" features in the vicinity of the zigzag terminus, attributed to the presence of a radical, whereas these features are absent in Fig. 4(m).

BRSTM imaging of 1D GNRs and 2D nanoporous graphene. Fig. 5 highlights the BRSTM images of few notable

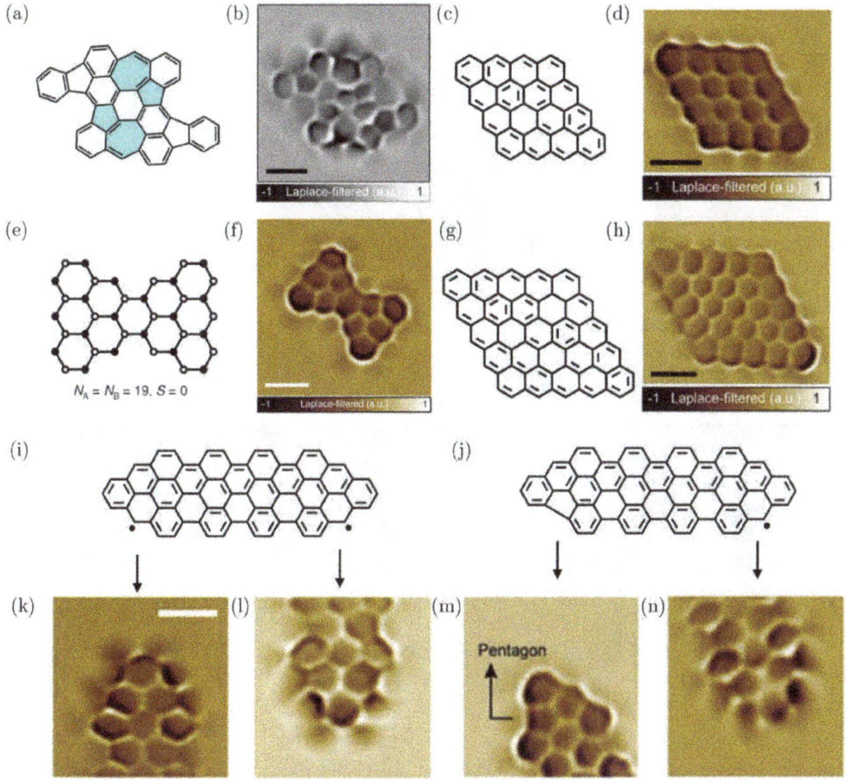

Fig. 4. Laplace-filtered BRSTM imaging of open-shell NGs. (a) Molecular structure of the azulene-embedded nanographene. (b) Laplace-filtered BRSTM image of the nanographene as shown in panel (a). (c), (g) Molecular structures of [4]rhombene and [5]rhombene. (d), (h) Laplace-filtered BRSTM images of [4]rhombene and [5] rhombene. (e) Molecular structure of Clar's goblet. Empty and filled circles represent two different sublattices. (f) Laplace-filtered BRSTM image. (i), (j) Molecular structures of the defect-free and pentagon-defect NGs. (k), (l) Laplace-filtered BRSTM images of termini of the defect-free nanographene. (m), (n) Laplace-filtered BRSTM images of termini of the nanographene with pentagon defect at one of the termini.

Source: Adapted from Refs. 37–39.

examples of the recently reported GNRs,[41] GNR heterojunctions and nanoporous graphene formed by the lateral fusion of GNRs. Moreno *et al.* fabricated ultra-long GNRs, arranged into superlattice arrays templated by the herringbone reconstruction of Au(111).[42] Fig. 5(a)

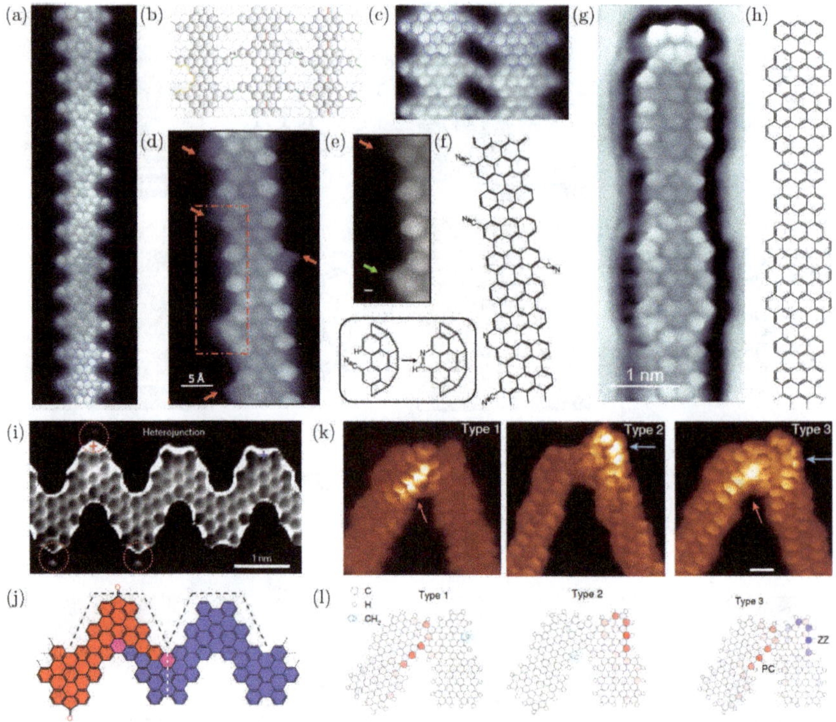

Fig. 5. (Color online) BRSTM imaging of GNRs. (a) BRSTM image of an ultra-long GNR. (b) Molecular structure of the nanoporous graphene formed by H^3 bonds interconnecting via dehydrogenative cross-coupling of the ultra-long GNR. (c) BRSTM image of nanoporous graphene. (d) BRSTM image of CN group-functionalized GNR obtained from the dI/dV channel at constant-height mode, $V = 5\,\text{mV}$. (e) Zoomed-in image of the dot–dashed region in panel (d), highlighting the pyridine ring. Red arrow indicates the CN group, green arrow points the linear bright contrast of the edge nitrogen. Scale bar is 0.1 nm. Inset is the reaction scheme from nitrile cycloisomerization into pyridine. (f) Molecular structure of the GNR in panel (d). (g) BRSTM image of 7/9-AGNR obtained with a tip that was functionalized via an unknown molecule from the surface ($V = 20\,\text{mV}$, $I = 80\,\text{pA}$ and $V_{ac} = 12\,\text{mV}$). (h) Molecular structure of the 7/9-AGNR. (i) BRSTM image of a fluorenone-unfunctionalized chevron heterojunction obtained from the dI/dV channel ($V = 40\,\text{mV}$, $I = 10\,\text{pA}$ and $V_{ac} = 20\,\text{mV}$). The red dashed boxes indicate the fluorenone carbonyl groups. (j) Molecular structure of the heterojunction. (k) BRSTM images of three different types of GNR junction structures showing the open-shell ground states. (l) Spin-density distribution of the corresponding three types of GNR junction structures as shown in panel (k).

Source: Adapted from Refs. 42–47.

shows the BRSTM image of the representative individual GNR. As the coverage of GNRs increases, the periodicity of the parallel GNR array decreases, subsequently leading to the lateral fusion of densely packed GNRs towards the formation of nanoporous graphene (Fig. 5(b)), as revealed by the BRSTM image (Fig. 5(c)).[43] This unique approach enables the construction of atomically precise nanopores with highly anisotropic electronic structures. The emergence of unique orthogonal one-dimensional electronic bands with a bandgap of $\sim 1\,\mathrm{eV}$ makes the nonporous graphene a highly versatile semiconductor candidate for nanodevice applications. Some also fabricated nitrogen-doped GNRs by replacing a benzene ring in the precursor with a nitrile group.[44] As shown in Figs. 5(d) and 5(e), the nitrile groups can react with their neighboring carbon atoms [as depicted in the inset of Fig. 5(e)] to form new pyridine rings as shown by the green arrows in the BRSTM images.

Engineering the heterojunction in GNRs represents a feasible route to tune the electronic and optical properties of GNRs.[45] The majority of GNR heterojunctions reported so far are fabricated by introducing two different molecular precursors for on-surface synthesis. However, Nguyen *et al.* recently demonstrated a new synthetic approach that allows for the fabrication of the GNR heterojunctions using a single molecular precursor.[46] This new type of GNR precursor incorporates a sacrificial carbonyl group which can be converted into carbonyl-decorated fluorenone GNRs by thermal annealing or STM tip-assisted manipulation. Interestingly, BRSTM allows for unambiguous discrimination between the fluorenone oxygen atoms and chevron GNR segments. As shown in Fig. 5(i), fluorenone oxygen appears as bright protrusions labeled with red-dashed-line circles. Such a BRSTM resolution allows to derive the molecular structure model of the GNR heterojunction depicted in Fig. 5(j).

In addition to the regular GNR heterojunctions derived from rationally designed molecular precursors, random junction nanostructures formed by uncontrollable inter-ribbon fusion have also gained an increased interest. Li *et al.* recently reported a study of three different types of open-shell GNR junctions.[47] BRSTM images in Fig. 5(k) show the termini of the three GNR junctions fused by the reorganization of carbon atoms in the vicinity of the joint positions.

Interestingly, the junctions show brighter contrast in the BRSTM image, reflecting an enhancement of the dI/dV signal close to Fermi level, due to the aforementioned Kondo effect. This indicates that the GNR junctions host localized magnetic moments at its specific sites. The features of BRSTM images resemble the pattern of the calculated spin-density distribution (Fig. 5(l)).

Although heteroatoms, such as oxygen and nitrogen, may show different contrasts from carbon atoms in the BRSTM images due to their distinct electronic properties, it is challenging to discriminate the species and position of heteroatoms solely from the BRSTM image. For instance, the contrast of nitrogen dopant at the edge of GNR in Fig. 5(e) is almost identical to the rest of the carbons. Therefore, correct interpretation of the BRSTM features, particularly determination of the heteroatom sites, must be corroborated by careful analysis of the on-surface reaction pathways and bolstered by DFT calculations.

Structural identification of molecular species using BRSTM. Recently, our group demonstrated the ability of BRSTM to investigate the on-surface transformations of a monoradical at the single-molecule level.[48] Organic radicals have long fascinated chemists because they are not only fundamentally interesting but also possess numerous potential applications in diverse fields, including organic magnetism and spintronics. The ability to understand the complex chemical transformations of organic radicals at the single-molecule level not only sheds light into fundamental radical chemistry but also provides new insights into the design of stable organic radicals for future applications.

The formation of a monoradical can be triggered by thermal annealing of precursor molecule 1 (Figs. 6(a) and 6(b)) on the Au (111) substrate. The chemical structure of monoradical molecule was characterized by BRSTM (Fig. 6(c)). The dI/dV spectra acquired at different positions over the generated monoradical reveal a Kondo resonance centered at zero sample bias (Fig. 6(d)), which indicates the existence of the 1/2 spin delocalized over the conjugated system. A subsequent sample annealing induced a cascade structural rearrangement of monoradical into various products, which were

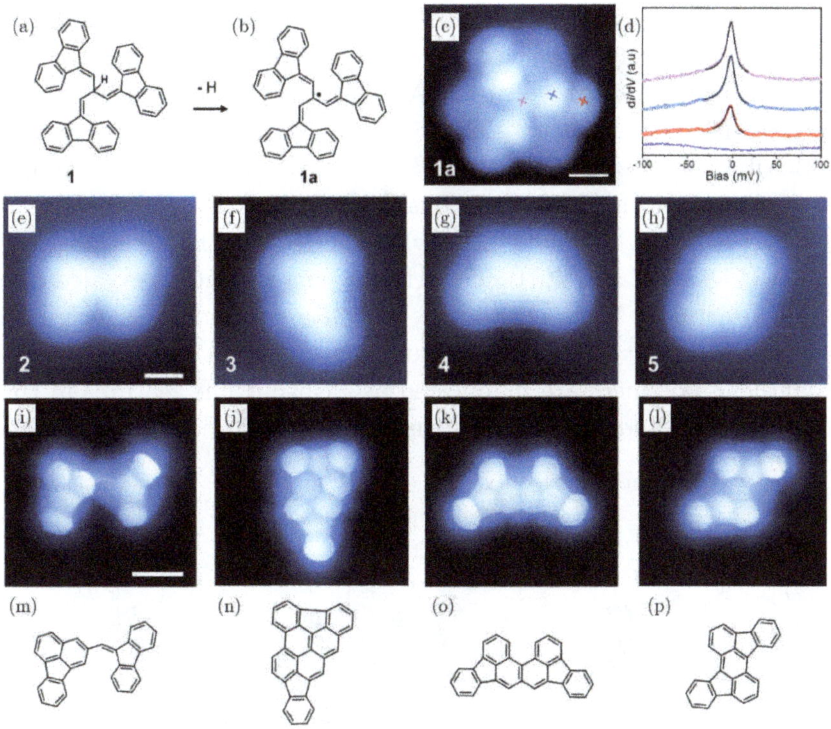

Fig. 6. Real-space imaging of monoradical reaction on Au(111) using BRSTM. (a) Dehydrogenation of precursor 1 during thermal evaporation to generate a monoradical 1a. (c) Constant-current STM image of the monoradical 1a on Au (111) ($I = 100\,\text{pA}$, $V = 50\,\text{mV}$). (d) The dI/dV spectra taken at the positions labeled in panel (c) showing Kondo resonance. (e), (f) Constant-current STM images of the four products 2, 3, 4 and 5 via annealing the monoradical 1a on Au(111) ($I = 100\,\text{pA}$, $V = 1.5\,\text{V}$). (i)–(l) BRSTM images of the four products corresponding to panels (e) and (f), respectively (CO tip, $V = 50\,\text{mV}$). (m)–(p) The molecular structures of the four products corresponding to (i)–(l), respectively.

Source: Adapted from Ref. 48.

directly captured by constant-current STM (Figs. 6(e)–6(h)) and BRSTM (Figs. 6(i)–6(l)) imaging. Therefore, the molecular backbones are resolved as shown in Fig. 6(m)–6(p). These results highlight the tremendous potential of BRSTM for the study of complex radical transformations on surface, which in turn guides the design and synthesis of stable radical species towards material processing and quantum device fabrication.

Discrimination of the chemical bond order using BRSTM. It has been recently demonstrated that discrimination of the chemical bond order can be realized by ncAFM imaging.[49,50] Analogously to ncAFM, BRSTM is also a useful tool for analyzing a bond order via noticeable variations of the BRSTM contrast. Specifically, Écija and co-workers performed simultaneous BRSTM and ncAFM imaging of the on-surface synthetized ethynylene-bridged anthracene polymer chains (Figs. 7(a) and 7(d)). The triple chemical bond is revealed as a bright protrusion in ncAFM image and as a slight depression in the BRSTM image.[51,52] In contrast, double bonds within cumulene-linked bisanthene polymers are manifested as clear-cut line features in the ncAFM image and as bright protrusions in BRSTM image (Figs. 7(d) and 7(e)).[53]

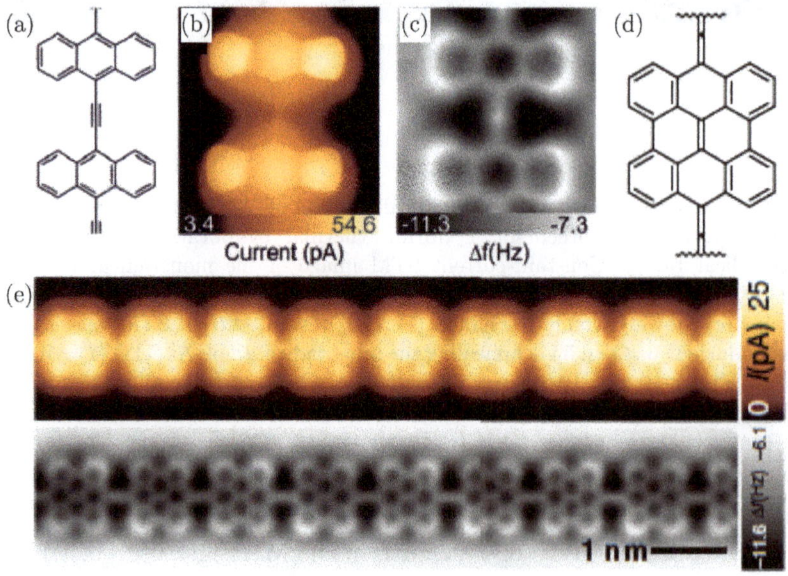

Fig. 7. Bond-order analysis using BRSTM. (a) Chemical structure of the ethynylene-bridged anthracene polymer chain, and the corresponding (b) BRSTM and (c) ncAFM images of the triple-bond linked chain. (d) Chemical structure of the cumulene-linked bisanthene polymers linked by double bonds. (e) The corresponding ncAFM and BRSTM images of panel (d).

Source: Adapted from Refs. 51 and 53.

BRSTM imaging of 2D supramolecular self-assemblies. In addition to resolving the products and intermediates of the on-surface chemical reactions, BRSTM has also been used to identify the intermolecular interactions within two-dimensional supramolecular self-assemblies. Intermolecular interactions, such as halogen bonding and hydrogen bonding, usually steer the molecular self-organization on surface. Lawrence recently reported the identification of self-assembled 3,9-dibromo-peri-xanthenoxanthene (3,9-Br_2PXX) molecules (Fig. 8(a)) on Au(111).[54] The DFT-calculated electrostatic potential (ESP) clearly reveals a positive potential sign on the top of Br atoms, a negative electrostatic potential belt around the C–Br bonds and a negative potential of O atoms (Fig. 8(b)). Therefore,

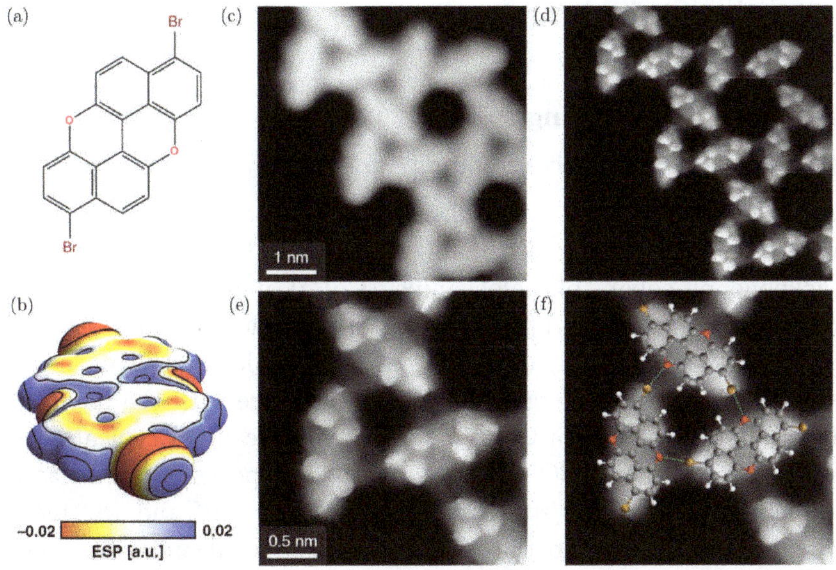

Fig. 8. (Color online) BRSTM imaging of halogen bonding interaction. (a) Molecular structure of 3,9-Br_2PXX. (b) Electrostatic potential mapping of 3,9-Br_2PXX showing the sigma hole and electron-rich belt on the Br atoms. (c) Constant-current STM image of the Kagome assembly of 3,9-Br_2PXX (CO tip, $I = 160\,\text{pA}$ and $V = 510\,\text{mV}$). (d) Constant-height BRSTM image of the same region as panel (c), $V = 30\,\text{mV}$. (e) Zoomed-in BRSTM image of the same region in panel (d), $V = 30\,\text{mV}$. (f) Molecular model overlay of panel (e). The halogen bonds are indicated by green dot lines.

Source: Adapted from Ref. 54.

both halogen and hydrogen bond may occur in the 3,9-Br2PXX-based self-assembly (Fig. 8(c)).

The BRSTM image (Fig. 8(d)) unambiguously visualizes the backbone of individual molecules. Furthermore, the zoomed-in BRSTM image, judiciously overlaid with a DFT-relaxed self-assembly model (Figs. 8(e) and 8(f)), reveals that the Br atoms of one molecule directly face the O atoms of a neighboring molecule, which validates the formation of a halogen bond.

This work demonstrated the ability of BRSTM to provide valuable insight into the nature of intermolecular interactions within molecular self-assemblies. Lastly, we note that sharp intermolecular line features do not indicate the presence of a halogen bond,[55] but appear as a result of an artificial bending of the CO molecule over a ridge of potential energy landscape [see Fig. 2(a)].

4. Inelastic Tunneling Probe

4.1. *ItProbe for intramolecular resolution*

The inelastic tunneling probe, or itProbe, has been pioneered by Ho and co-workers,[56] who used it for the visualization of the chemical backbone of cobalt phthalocyanine (CoPc) molecule on Ag(110) surface (Figs. 9(a) and 9(c)). The underlying mechanism of itProbe relies on the utilization of the vibrational modes of CO molecule, which undergoes an energetic shift in accordance with the potential energy landscape over a CoPc molecule. Specifically, a typical STM tip, decorated with a CO molecule, reveals two peaks in the inelastic tunneling spectroscopy spectrum (IETS) at 2.1 meV and 18.2 meV, which are attributed to the hindered translational and vibrational modes of the CO molecule, respectively (Fig. 9(d)). Owing to the higher probe sensitivity, the hindered translational mode of CO molecule was further exploited for the imaging process. The IETS spectra collected at different positions over a CoPc molecule (Fig. 9(e)) show that the hindered translational mode of CO probe undergoes an energetic shift. Such a site-specific energetic shift of translational mode is further reflected in the spatial imaging of the molecule at a small bias voltage (<2.5 meV). As shown in

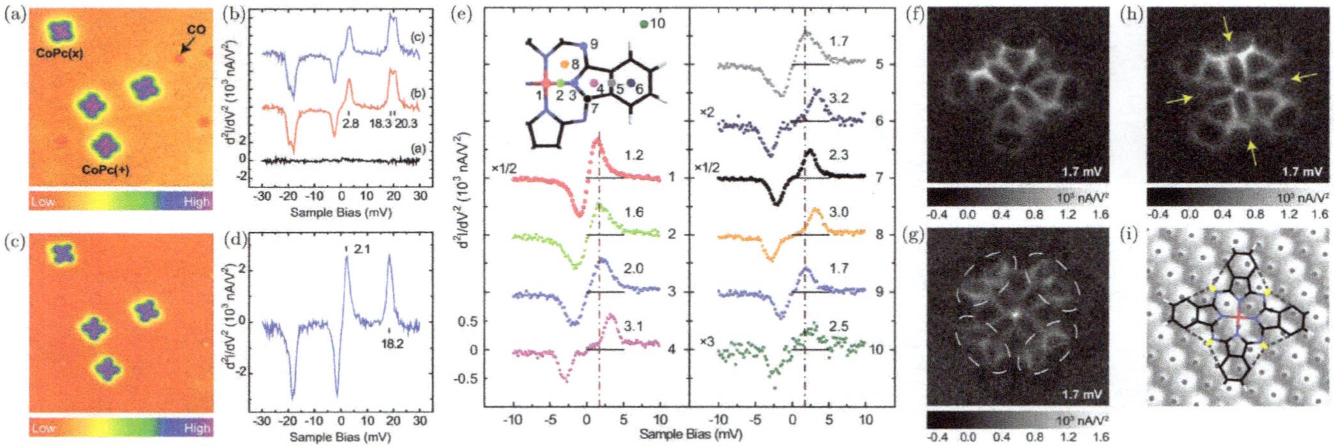

Fig. 9. Imaging the chemical structure of a single CoPc molecule using itProbe. (a) Constant-current topography of CO coadsorbed with CoPc(Å∼) and CoPc(+) on Ag(110) at 600 mK. (b) Vibrational STM-IETS d^2I/dV^2 spectra taken with a bare tip over the (a) Ag surface, (b) CO molecule and (c) background-subtracted spectrum. (c) Constant-current topography of the same area as in panel (a) after transferring a CO [the one indicated by the arrow in panel (a)] to the tip. CO is imaged as a protrusion instead of a depression. (d) Vibrational STM-IETS d^2I/dV^2 spectrum of CO-terminated tip taken over the Ag surface. (e) Spatially resolved d^2I/dV^2 vibrational spectra taken at locations indicated in the schematic of the right-hand side of CoPc. (f) Constant-height image of CoPc(+) on Ag(110). (g) Constant-current image of CoPc(+) on Ag(110). (h) Constant-height image of CoPc(×) on Ag(110). (i) Schematic diagram showing the skeletal structure of CoPc(+) and the intramolecular hydrogen bonds highlighted in dashed ellipses in panel (b).

Source: Adapted from Ref. 56.

(Figs. 9(f)–9(h)), the images of CoPc by constant-height and constant-current modes reveal intramolecular sharp line features, which coincide with the chemical backbone of a CoPc molecule (Fig. 9(i)).

4.2. *Imaging the halogen bond by itProbe*

The noncovalent halogen bond is highly relevant in different fields, including crystal engineering, supramolecular self-assembly and drug design. Formation of a halogen bond can be captured by X-ray crystallography, while the feasibility of real-space imaging of the halogen bond by SPM has been a long-lasting debate. In 2017, Han *et al.* reported the real-space imaging of the halogen bonding of hexafluorobenzene (C_6F_6), hexabromobenzene (C_6Br_6) and 1,3,5-trifluorobenzene ($C_6H_3F_3$) molecules on Au(110) surface at 600 mK using itProbe.[57] A layered structure of C_6F_6 can be formed on the Ag (110) surface after deposition. Fig. 9(a) illustrates the itProbe image of the C_6F_6 molecule, showing a characteristic quasi-hexagonal ring feature with the C–C and C–F bonds resolved, consistent with the previous report. Surprisingly, additional line features were observed in between the C_6F_6 molecules, forming a "network" among the C_6F_6 molecules. Intuitively, such observations indicate the existence of intermolecular halogen bonds. Upon detailed examination, it can be revealed that three neighboring C_6F_6 molecules are organized into a windmill structure, which is identical to a typical halogen-3 synthon of the halogen bonding. Notably, the angle of C–F $\cdots$ F is not 180°C, which can be attributed to the interaction between the C_6F_6 molecules and the underlying Ag(110) substrate (supported by DFT calculations). In addition to the F-3 synthon feature, a trans-type-I F $\cdots$ F interaction is also revealed by the itProbe image [indicated by orange dashed line in the right of Fig. 10(a)]. In contrast to C_6F_6, the Br atoms are recognized to have a larger polarizability, leading to the standard pattern of halogen bonding among the C_6Br_6 molecules (Fig. 10(b)). $C_6H_3F_3$ molecules were also studied on Ag(110) surface (Fig. 10(c)). From the magnified itProbe images in Figs. 10(d)–10(f), the V-shaped bifurcation structures with different origins (one F atom interacting with two H atoms, or one H atom interacting with two

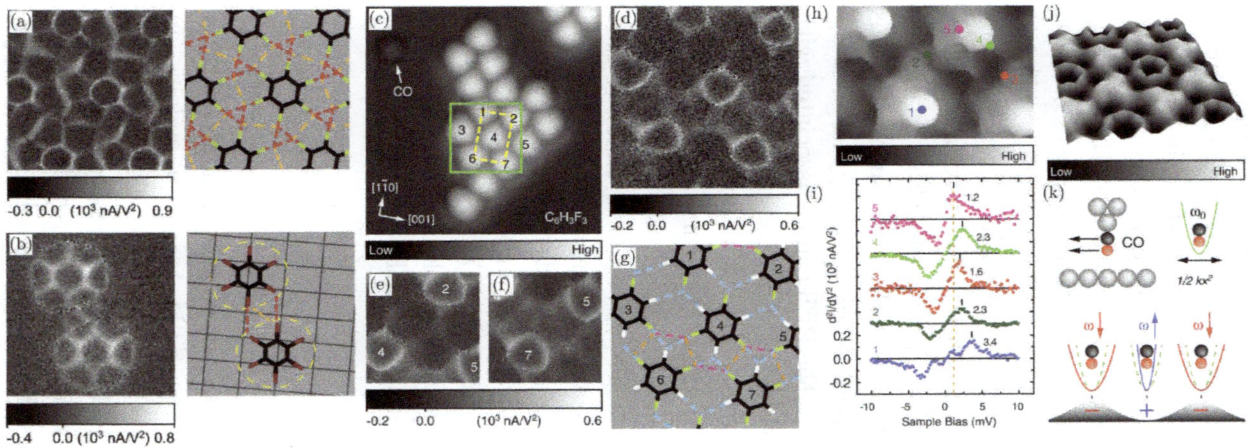

Fig. 10. (Color online) Imaging halogen bonds by using itProbe. (a) Constant-current itProbe imaging at 1.2 mV of C_6F_6 molecules (left) and its corresponding schematic diagram. (b) Constant-height itProbe imaging at 1.5 mV of the C_6F_6 dimers and its corresponding schematic diagram. (c) Constant-current topography of a $C_6H_3F_3$ island scanned with a CO tip. (d) Constant-current itProbe imaging at 1.5 mV over the green square area marked in panel (c). (e), (f) Magnified constant-current itProbe imaging at 1.75 mV over the intermolecular region in panel (c). (g) The intermolecular bonding network is highlighted by blue, orange and magenta dashed lines, representing the C–H ··· F interaction and the cis- and trans-configurations of type-I F ··· F interactions, respectively (F in green and H in white). (h) Constant-current topography of C_6F_6 island scanned with a CO tip. (i) d^2I/dV^2 spectra of the CO tip measured over five different locations of the C_6F_6 island specified in panel (h). (j) Calculations of the 2D PES of the Ag–CO tip and the underlying C_6F_6 lattice without the substrate. (k) Schematic diagram of the mechanism for itProbe, showing the downshift of CO tip-hindered translational vibration over two ridges [negative (–) curvature] and the upshift of CO tip-hindered translational vibration over a valley [positive(+)curvature] of the PES. Here, k is the local curvature of the vibrational well and w is the vibrational frequency.

Source: Adapted from Ref. 57.

F atoms) can be revealed. Furthermore, typical trans- and cis-type-I halogen interactions can be observed from the itProbe image of the molecular island (Fig. 10(g)), which hints the halogen bonding nature of the interaction of self-assembled C_6F_6, C_6Br_6 and $C_6H_3F_3$ molecules.

Combined IETS measurements and DFT calculations were made to gain an in-depth understanding of the origin of the imaging contrast. As presented in Fig. 10(i), the CO probe interacts with the C_6F_6 molecule underneath, leading to vibrational energy shifts at different positions of the molecule (Fig. 10(h)). Such differences can be understood from landscape of the PES, wherein the ridges and valleys correspond to different regions of the C_6F_6 molecules (Fig. 10(j)). The CO probe can sense the spatial variations of PES, and such variations can modify the hindered translational vibration energy of CO. As illustrated in Fig. 10(k), the ridges in the PES can be associated with the regions of negative curvature, while the valley can be attributed to the positive curvature of the vibrational potential well of CO molecule. Specifically, over the ridges (valleys) of PES the curvature of the CO vibrational potential well decreases (increases), leading to downshift (upshift) of the hindered translational energy of CO.

The itProbe imaging significantly expands the capability of STM, while it is noted that such an imaging technique is conducted at ultra-low temperature (600 mK), which may not be applicable in standard low-temperature instruments. However, in 2017, Torre *et al.* have proved that itProbe imaging can also be applied at liquid helium temperature (4.5 K) in combination with BRSTM and ncAFM imaging.[58]

5. Conclusion and Outlook

Recent experimental and theoretical advancements demonstrate that controllable functionalization of the STM tip apex endows an extraordinary imaging capability to the BRSTM technique to resolve chemical bonds. To date, the BRSTM has been extensively utilized in the field of on-surface chemistry, to resolve on-surface reaction products and intermediates with submolecular resolution. Furthermore, recent experimental breakthroughs pinpoint the ability of

BRSTM to provide information beyond merely the determination of molecular structure. For instance, BRSTM images of the open-shell nanographene reveal not only the molecular backbone, but also the spatial localization of π-radicals. The ability of BRSTM to quantify the intermolecular interactions in molecular self-assemblies has also been demonstrated. It can be envisioned that increased popularity of BRSTM will push the boundaries of imagination in surface chemistry and nanotechnology.

Acknowledgments

Telychko acknowledges the support from A*STAR under its AME YIRG Grant (Project No. A20E6c0098).

References

1. G. Binnig, H. Rohrer, C. Gerber and E. Weibel, *Appl. Phys. Lett.* **40** (1982) 178.
2. R. Wiesendanger, *Scanning Probe Microscopy and Spectroscopy: Methods and Applications* (Cambridge University Press, Cambridge, 1994).
3. P. Sutter, Scanning tunneling microscopy in surface science, in *Science of Microscopy*, eds. P. W. Hawkes and J. C. H. Spence (Springer, New York, 2007), pp. 969–1024.
4. A. A. Khajetoorians, D. Wegner, A. F. Otte and I. Swart, *Nat. Rev. Phys.* **1** (2019) 703.
5. K. S. Mali *et al.*, *Chem. Soc. Rev.* **46** (2017) 2520.
6. J. L. Zhang, *Nano Lett.* **19**(8) (2019) 5340.
7. R. Pawlak, C. Drechsel, P. D'Astolfo, M. Kisiel, E. Meyer and J. I. Cerda, *Proc. Natl. Acad. Sci. USA* **117**(1) (2020) 228.
8. M. Telychko and J. Lu, *Nano Mater. Sci.* **1**(4) (2019) 260.
9. B. Neu, G. Meyer and K.-H. Rieder, *Mod. Phys. Lett. B* **9**(15) (1995) 963.
10. D. Eigler, P. Weiss, E. Schweizer and N. Lang, *Phys. Rev. Lett.* **66**(9) (1991) 1189.
11. Z. Li, K. Schouteden, V. Iancu, E. Janssens, P. Lievens, C. Van Haesendonck and J. I. Cerdá, *Nano Res.* **8**(7) (2015) 2223.
12. R. Temirov, S. Soubatch, O. Neucheva, A. Lassise and F. Tautz, *New J. Phys.* **10**(5) (2008) 053012.
13. L. Gross, F. Mohn, N. Moll, P. Liljeroth and G. Meyer, *Science* **325**(5944) (2009) 1110.
14. F. Mohn, B. Schuler, L. Gross and G. Meyer, *Appl. Phys. Lett.* **102**(7) (2013) 073109.
15. H. Labidi, M. Koleini, T. Huff, M. Salomons, M. Cloutier, J. Pitters and R. A. Wolkow, *Nat. Commun.* **8** (2016) 14222.

16. T. Chutora, B. de la Torre, P. Mutombo, J. Hellerstedt, J. Kopeček, P. Jelínek and M. Švec, *Beilstein J. Nanotechnol.* **10** (2019) 315.
17. H. Mönig *et al.*, *Nat. Nanotechnol.* **13**(5) (2018) 371.
18. E. J. Spadafora, J. Berger, P. Mutombo, M. Telychko, M. Švec, Z. Majzik, A. B. McLean and P. Jelínek, *J. Phys. Chem. C* **118**(29) (2014) 15744.
19. N. Pavliček and L. Gross, *Nat. Rev. Chem.* **1**(1) (2017) 0005.
20. F. J. Giessibl, *Rev. Sci. Instrum.* **90**(1) (2019) 011101.
21. L. Gross, B. Schuler, N. Pavliček, S. Fatayer, Z. Majzik, N. Moll, D. Peña and G. Meyer, *Angew. Chem., Int. Ed.* **57**(15) (2018) 3888.
22. L. Gross, *Nat. Chem.* **3**(4) (2011) 273.
23. C. Weiss, C. Wagner, R. Temirov and F. S. Tautz, *J. Am. Chem. Soc.* **132**(34) (2010) 11864.
24. C. Weiss, C. Wagner, C. Kleimann, M. Rohlfing, F. Tautz and R. Temirov, *Phys. Rev. Lett.* **105**(8) (2010) 086103.
25. P. Jelínek, *J. Phys., Condens. Matter* **29** (2017) 343002.
26. P. Hapala *et al.*, *Phys. Rev. B* **90** (2014) 085421.
27. P. Hapala, R. Temirov, F. S. Tautz and P. Jelínek, *Phys. Rev. Lett.* **113** (2014) 226101.
28. O. Krejčí, P. Hapala, M. Ondráček and P. Jelínek, *Phys. Rev. B* **95** (2017) 045407.
29. L. Gross, N. Moll, F. Mohn, A. Curioni, G. Meyer, F. Hanke and M. Persson, *Phys. Rev. Lett.* **107** (2011) 086101.
30. S. Clair and D. G. Oteyza, *Chem. Rev.* **119** (2019) 4717.
31. Q. Shen, H.-Y. Gao and H. Fuchs, *Nano Today* **13** (2017) 77.
32. L. Talirz, P. Ruffieux and R. Fasel, *Adv. Mater.* **8** (2016) 6222.
33. J. Hieulle *et al.*, *Nano Lett.* **18**(1) (2018) 418.
34. J. Su, W. Fan, P. Mutombo, X. Peng, S. Song, M. Telychko, P. Jelínek, J. Wu and J. Lu, *Nano Lett.* **21** (2020) 861.
35. M. Telychko, G. Li, P. Mutombo, D. Soler-Polo, X. Peng, J. Su, S. Song, M. J. Koh, M. Edmonds, P. Jelínek, J. Wu and J. Lu, *Sci. Adv.* **7** (2021) eabf0269.
36. J. Li *et al.*, *Phys. Rev. Lett.* **124**(17) (2020) 177201.
37. J. Liu *et al.*, *J. Am. Chem. Soc.* **141**(30) (2019) 12011.
38. S. Mishra, *Nat. Chem.* **13** (2021) 581.
39. S. Mishra *et al.*, *Nat. Nanotechnol.* **15**(1) (2020) 22.
40. S. Mishra *et al.*, *J. Am. Chem. Soc.* **142**(3) (2020) 1147.
41. J. Cai *et al.*, *Nature* **466**(7305) (2010) 470.
42. C. Moreno *et al.*, *Chem. Commun.* **54**(68) (2018) 9402.
43. C. Moreno *et al.*, *Science* **360**(6385) (2018) 199.
44. E. Carbonell-Sanromà *et al.*, *ACS Nano* **11**(7) (2017) 7355.
45. G. D. Nguyen *et al.*, *Nat. Nanotechnol.* **12**(11) (2017) 1077.
46. D. J. Rizzo *et al.*, *Nature* **560**(7717) (2018) 204.
47. J. Li *et al.*, *Nat. Commun.* **10**(1) (2019) 200.
48. S. Song *et al.*, *J. Am. Chem. Soc.* **142**(31) (2020) 13550.
49. L. M. Scriven *et al.*, *J. Am. Chem. Soc.* **142** (2020) 12921.
50. K. Kaiser *et al.*, *Science* **365** (2019) 1299.
51. A. Sánchez-Grande *et al.*, *Angew. Chem., Int. Ed.* **58** (2019) 6559.

52. K. Biswas *et al.*, *Chem. Commun.* **56** (2020) 15309.
53. B. de la Torre *et al.*, *Nat. Commun.* **11** (2020) 4567.
54. J. Lawrence, *Nat. Commun.* **11**(1) (2020) 2103.
55. A. Gallardo, J. Fanfrlík, P. Hobza and P. Jelínek, *J. Phys. Chem. C* **123** (2019) 8379.
56. C.-L. Chiang, C. Xu, Z. Han and W. Ho, *Science* **344** (2014) 885.
57. Z. Han *et al.*, *Science* **358** (2017) 206.
58. B. de la Torre *et al.*, *Phys. Rev. Lett.* **119** (2017) 166001.

https://doi.org/10.1142/9789811247859_0009

Chapter 9

Holographic Convergent Electron Beam Diffraction (CBED) Imaging of Two-dimensional Crystals

T. Latychevskaia[*,**], S. J. Haigh[†,‡] and K. S. Novoselov[†,§,¶,||]

[*]*Institute of Physics, Laboratory for Ultrafast Microscopy and Electron Scattering (LUMES), École Polytechnique Fédérale de Lausanne (EPFL), CH-1015 Lausanne, Switzerland*
Paul Scherrer Institute, Forschungsstrasse 111 5232 Villigen, Switzerland
[†]*National Graphene Institute, University of Manchester, Oxford Road, Manchester M13 9PL, UK*
[‡]*Department of Materials, University of Manchester, Oxford Road, Manchester M13 9PL, UK*
[§]*Department of Materials Science and Engineering, National University of Singapore, Singapore 117575, Singapore*
[¶]*Centre for Advanced 2D Materials, National University of Singapore, 117546 Singapore, Singapore*
[||]*Chongqing 2D Materials Institute, Liangjiang New Area, Chongqing 400714, P. R. China*
[**]*tatiana@physik.uzh.ch*

Convergent beam electron diffraction (CBED) performed on two-dimensional (2D) materials recently emerged as a powerful tool to study structural and stacking defects, adsorbates, atomic 3D displacements in the layers, and the interlayer distances. The formation of the interference patterns in individual CBED spots of 2D crystals can be considered as a

[**]Corresponding author.

hologram, thus the CBED patterns can be directly reconstructed by conventional reconstruction methods adapted from holography. In this study, we review recent results applying CBED to 2D crystals and their heterostructures: holographic CBED on bilayers with the reconstruction of defects and the determination of interlayer distance, CBED on 2D crystal monolayers to reveal adsorbates, and CBED on multilayered van der Waals systems with moiré patterns for local structural determination.

Keywords: Graphene; two-dimensional materials; van der Waals structures; electron holography; convergent beam electron diffraction.

1. Introduction

Convergent beam electron diffraction (CBED) was first demonstrated by Kossel and Möllenstedt in 1939.[1] Conventionally, CBED is performed on three-dimensional (3D) crystals by focussing a convergent electron beam on a small area (about 10 nm in diameter) of a 3D crystal and acquiring the diffraction CBED pattern. Unlike conventional selected area electron diffraction, in which the planar illumination produces a diffraction pattern consisting of sharp peaks, in CBED mode the pattern is formed of finite-size disks, whose diameter is determined by the convergence semi-angle of the beam (Fig. 1). The CBED disks exhibit intensity variations related to the atomic structure and local atomic displacements in the crystal. CBED has been applied for studying crystallographic structure and lattice parameters,[2–6] specimen thickness,[7] measurements of strain,[4,8] and crystallographic deformations.[9,10] In large-angle convergent-beam electron diffraction (LACBED) regime, the CBED spots strongly overlap, which allow for an easier and more precise extraction of information on the structure and defects.[11,12] A good overview of CBED techniques applied for 3D crystal structure determination can be found in Refs. 13–16. In general, direct interpretation of CBED patterns is not possible and simulations are required to deduce the crystal structure, which often limits wider application of the technique.

Recently, CBED imaging was applied to study van der Waals heterostructures.[17–22] For 2D crystals, the interpretation of their CBED patterns is significantly simpler than in the case of 3D crystals because the intensity distributions in individual CBED spots directly map to the atomic arrangement, including local displacements in defects. In CBED, the intensity distribution of the scattered wave in the far field can be interpreted as being acquired from a sample that is being illuminated at different angles. This allows capturing of 3D information about atomic positions in the sample. The formation of the interference patterns within the individual CBED spots of 2D crystals can be described by holographic principles, and therefore CBED patterns can be used to directly reconstruct the real space coordinates *via* conventional reconstruction methods adapted from holography. To emphasize the difference in the data analysis, we therefore call the technique holographic CBED (HCBED). The reconstructed distributions provide information about the 3D arrangement of atoms in individual layers and in the stack, including the local strain, lattice orientations, local vertical separation between the layers, etc., which is not accessible by conventional TEM imaging.[23,24] In this study, we review the recent results in CBED on 2D crystals: holographic CBED on bilayer (BL) with the reconstruction of defects and the measurement of interlayer distance,[19,22] CBED on 2D crystal monolayers (MLs) with imaging of adsorbates,[20] and CBED on multilayer van der Waals systems.[21]

2. Principle of CBED on 2D Crystals

2.1. *CBED spots parameters*

In a CBED experiment, the sample z position, or defocus Δf, can be relatively easily changed to move the sample along the optical axis, which allows imaging of the sample with a convergent or divergent electron beam. The CBED arrangement in a convergent wavefront mode ($\Delta f < 0$) is shown in Fig. 1. A CBED pattern from an ML of graphene or hexagonal boron nitride (hBN) consists of finite-sized spots arranged into a six-fold symmetrical pattern. The centers of the spots have the same positions as the corresponding diffraction peaks, defined by the crystal periodicity

$$\sin\vartheta = \lambda/a, \tag{1}$$

where λ is the wavelength, ϑ is the diffraction angle, and a is the period of the crystallographic planes. The diameter of the probing beam on the sample can be evaluated as scaling with the diameter of the limiting aperture as:

$$D = 2|\Delta f| \tan \alpha, \tag{2}$$

where α is the semi-convergence angle, as shown in Fig. 1, and where chromatic instabilities and geometric lens aberrations are negligible. For small defocus ($|\Delta f| \approx 0$), the diameter of the probing beam is then given by diffraction on the aperture:

$$D = 1.22 \frac{\lambda}{\tan \alpha}, \tag{3}$$

where

$$\tan \alpha = \frac{D_{\mathrm{A}}}{2\Delta f_{\mathrm{A}}}, \tag{4}$$

where D_A is the limiting aperture diameter and Δf_A is the distance between the limiting aperture and the virtual source plane, as shown in Fig. 1. By changing the z position of the sample in the

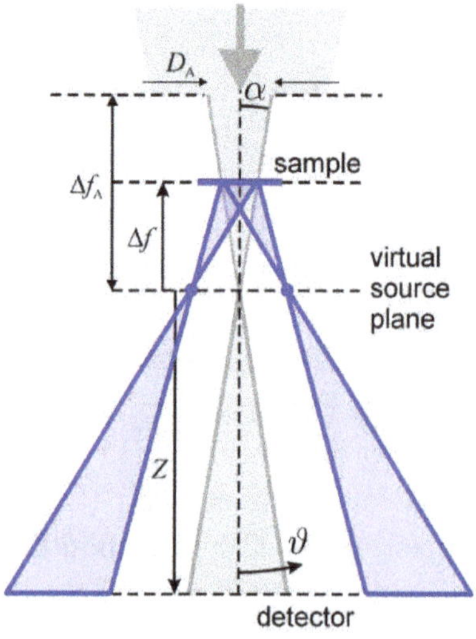

Fig. 1. (Color online) CBED arrangement.

convergent electron beam, the defocus Δf is changed, which in turn changes the diameter of the probing electron beam according to Eqs. (2) and (3). The diameters of the CBED spots however do not change, because they are defined by the limiting aperture angular size or $\tan \alpha$, as given by Eq. (4). For example, the diameter of the zero-order CBED spot size can be derived from geometrical considerations as shown in Fig. 1. Thus, the semi-convergence angle α, or the diameter of the limiting aperture, defines the diameter of the CBED spots and for HCBED it should be selected such that the CBED spots from a single crystal structure do not

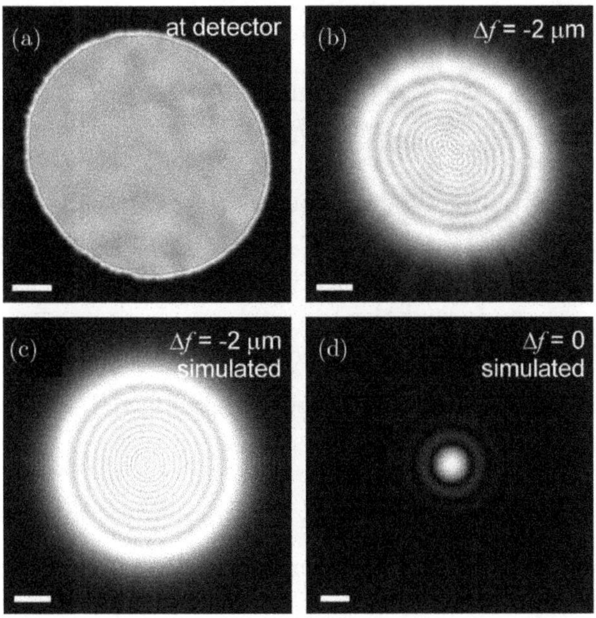

Fig. 2. Experimental and simulated images of the probing wave at different focal planes in the transmission electron microscope. (a) Experimentally measured intensity distribution in the zero-order CBED spot at $\Delta f = -2\,\mu\text{m}$. (b) Amplitude of the probing wavefront at $\Delta f = -2\,\mu\text{m}$, obtained by backward propagation of the wavefront from the detector plane (shown in (a)). (c) Amplitude of the simulated wavefront at $\Delta f = -2\,\mu\text{m}$. (d) Amplitude of the simulated wavefront at $\Delta f = 0$. The scale bars correspond to (a) 100 nm, (b) and (c) to 5 nm, and (d) to 0.5 nm. The experimental images were obtained with a Titan ChemiSTEM microscope operated at 80 kV, with an ultra-stable high-brightness Schottky FEG source. Note that the experimental data show evidence of a slight astigmatism. Adapted from Ref. 19.

overlap. Then, the size of the probing beam, or the probed area, can be regulated by selecting Δf. Note that this is equivalent to adjusting the sample z position where lens aberrations are neglected.

2.2. *Probing wavefront distribution*

Where the probing wavefront is formed by a diffraction on the limiting aperture, the intensity distribution is described by Fresnel diffraction on a round aperture:

$$\psi_0(\boldsymbol{r}) \propto \int\int a(\boldsymbol{r}_0)\frac{\exp(-ikr_0)}{r_0}\frac{\exp(ik|\boldsymbol{r}_0-\boldsymbol{r}|)}{|\boldsymbol{r}_0-\boldsymbol{r}|}\mathrm{d}\boldsymbol{r}_0, \tag{5}$$

where $a(\boldsymbol{r}_0)$ is the aperture function, $\boldsymbol{r}_0 = (x_0, y_0, z_0)$ is the coordinate in the aperture plane and $\boldsymbol{r} = (x, y, z)$ is the coordinate in the sample plane. For defocus values $\Delta f \approx 0$, the probing wavefront distribution is described by Fraunhofer diffraction on a round aperture, that is, by Fourier transform (FT) of the aperture distribution. Examples of probing wavefront distributions are shown in Fig. 2.

3. CBED on 2D Crystal Monolayers

CBED on 2D crystal MLs is studied in detail in Ref. 20, and here we present the main results from the study.

3.1. *In-plane and out-of-plane ripples*

3.1.1. *Phase shifts caused by atomic displacements, geometrical approach*

A 2D ML crystal is a perfect test sample to explain the basics of the intensity distributions observed in individual CBED spots. Atomic misalignments in the form of out-of-plane and in-plane shifts are illustrated in Fig. 3.

From the geometrical considerations, shown in Fig. 3, the phase differences due to atomic shifts are:

$$\Delta\varphi_z = \frac{2\pi}{\lambda}\Delta z(1-\cos\vartheta) \tag{6}$$

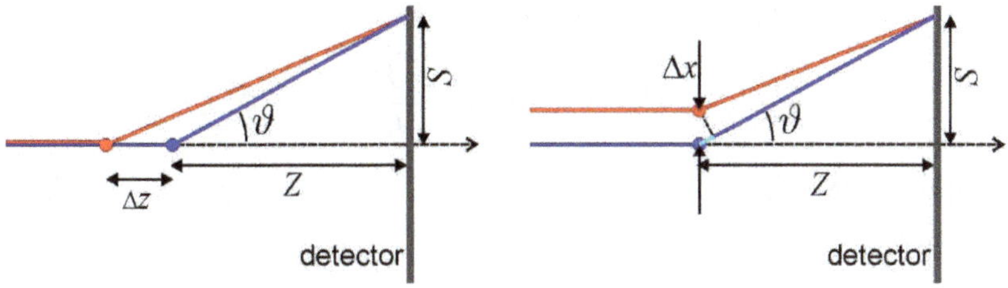

(a) (b)

Fig. 3. (Color online) Geometrical arrangement of scattering from two atoms separated by (a) Δz and (b) Δx distances. (a) Where two atoms are separated by Δz, the path difference is given by $\Delta l = l_{\text{blue}} - l_{\text{red}} \approx \Delta z(1 - \cos\vartheta)$. (b) Where two atoms are separated by Δx, the optical path difference (shown in cyan) is given by $\Delta l = \Delta x \sin\vartheta$.

and

$$\Delta\varphi_x = \frac{2\pi}{\lambda}\Delta x \sin\vartheta \tag{7}$$

for out-of-plane Δz and in-plane Δx atomic shifts, respectively. From Eqs. (6) and (7), we see that atomic Δx shifts cause stronger phase shifts because $\Delta\varphi_x \propto \vartheta$ while $\Delta\varphi_z \propto \vartheta^2$. Thus, in-plane shifts introduce stronger intensity variations into HCBED spots interference patterns, and therefore they are easier to detect and reconstruct.

3.1.2. *Phase shifts caused by atomic displacements, wave theory approach*

Alternatively, using optical wave theory, the phase shifts of the waves scattered from misaligned atoms can be derived as follows. The distribution of the scattered wave in the far-field $U(\boldsymbol{R})$ is given by:

$$\begin{aligned} U(\boldsymbol{R}) &\approx \int \frac{\exp(-ikr)}{r} t(\boldsymbol{r}) \frac{\exp(ik|\boldsymbol{r}-\boldsymbol{R}|)}{|\boldsymbol{r}-\boldsymbol{R}|} \mathrm{d}\boldsymbol{r} \\ &\propto \frac{\exp(ikR)}{R} \int \frac{\exp(-ikr)}{r} t(\boldsymbol{r}) \exp\left(-ik\frac{\boldsymbol{r}\boldsymbol{R}}{R}\right) \mathrm{d}\boldsymbol{r}, \end{aligned} \tag{8}$$

where we consider a convergent probe wavefront $\frac{\exp(-ikr)}{r}$, $t(\boldsymbol{r})$ is the transmission function of the sample, $\boldsymbol{R} = (X, Y, Z)$ is the coordinate in the detector plane, and using $R \gg r$ the approximation

$$|\boldsymbol{r} - \boldsymbol{R}| \approx R - \frac{\boldsymbol{rR}}{R} \tag{9}$$

is applied. We introduce K-coordinates to simplify the algebraic expression:

$$\begin{aligned} \boldsymbol{K} &= (K_x, K_y, K_z) = k\frac{\boldsymbol{R}}{R} = \frac{2\pi}{\lambda R}(X, Y, Z), \\ |\boldsymbol{K}| &= k = \frac{2\pi}{\lambda}, \quad K_z = \sqrt{K^2 - K_x^2 - K_y^2} \end{aligned} \tag{10}$$

and re-write

$$\begin{aligned} U(K_x, K_y) \approx\ & \exp(ikR) \int \exp(-ikr) t(x, y, z) \\ & \times \exp[-i(K_x x + K_y y)] \exp(-izK_z) \mathrm{d}x\mathrm{d}y\mathrm{d}z. \end{aligned} \tag{11}$$

The wavefront scattered by an atom at $\boldsymbol{r} = (x, y, z)$ is given by:

$$\begin{aligned} U(K_x, K_y) \propto\ & \exp(ikR) \exp(-ikr) \\ & \times \exp[-i(xK_x + yK_y)] \exp(-izK_z). \end{aligned} \tag{12}$$

3.1.2.1. Phase shift caused by an out-of-plane displacement

Wavefronts scattered by atoms positioned at $\boldsymbol{r}_1 = (0, 0, -|\Delta f|)$, $r_1 = |\Delta f|$ and $\boldsymbol{r}_2 = (0, 0, -|\Delta f| + \Delta z)$, $r_2 = |-|\Delta f| + \Delta z|$, are given by:

$$\begin{aligned} U_1(K_x, K_y) &\propto \exp(ikR) \exp(-ik|\Delta f|) \exp[iK_z|\Delta f|] \\ U_2(K_x, K_y) &\propto \exp(ikR) \exp[ik(-|\Delta f| + \Delta z)] \\ & \quad \times \exp[-iK_z(-|\Delta f| + \Delta z)]. \end{aligned} \tag{13}$$

The corresponding phases of the wavefronts are:

$$\begin{aligned} \varphi_1 &= kR - k|\Delta f| + \Delta f K_z \\ \varphi_2 &= kR + k(-|\Delta f| + \Delta z) - (-|\Delta f| + \Delta z)K_z, \end{aligned} \tag{14}$$

and the phase difference is given by:

$$\Delta\varphi_z = \varphi_2 - \varphi_1 = \Delta z(k - K_z) = \Delta z \frac{2\pi}{\lambda}(1 - \cos\vartheta), \tag{15}$$

where we applied $K_z = \frac{2\pi}{\lambda} \cos \vartheta$. The obtained phase shift agrees with the phase shift derived from geometrical consideration, Eq. (6) and Fig. 3(a). The CBED pattern and phase shifts produced by an ML with out-of-plane atomic shifts forming a "bulge" are shown in Figs. 4(a)–4(c).

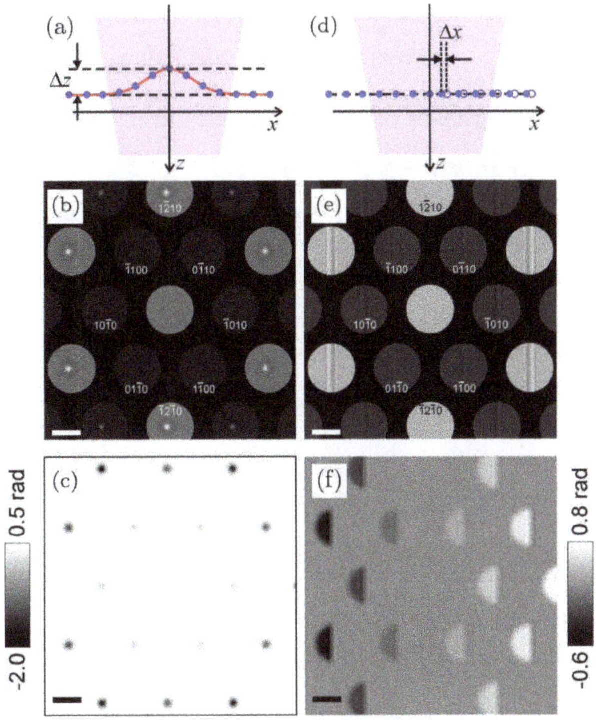

Fig. 4. CBED patterns of a graphene monolayer with (a)–(c) an out-of-plane "bulge" and (d)–(f) an in-plane lateral displacement both simulated for $\Delta f = -2\,\mu\text{m}$ (underfocus). (a) Sketch of the side view of a graphene layer with atoms displaced out of plane in form of a "bulge." (b) Corresponding simulated CBED pattern, the "bulge" height is $|\Delta z| = 2\,\text{nm}$. (c) Phase shift introduced by the lattice deformation into the probing electron wave, calculated at the detector plane. (d) Sketch of the side view of a graphene layer with atoms displaced in plane in form of a lateral shift. (e) The corresponding simulated CBED pattern, where the atoms positioned at $x > 0$ are displaced by $\Delta x = -10\,\text{pm}$. (f) Phase shift introduced by the lattice deformation into the probing electron wave, calculated at the detector plane. For these simulations the probed area is about 30 nm in diameter. The scale bars in (b), (c), (e) and (f) correspond to $2\,\text{nm}^{-1}$. Adapted from Ref. 20.

3.1.2.2. Phase shift caused by an in-plane displacement

Wavefronts scattered by two adjacent atoms positioned at $\boldsymbol{r}_1 = (0, 0, -|\Delta f|)$ and $\boldsymbol{r}_2 = (a + \Delta x, 0, -|\Delta f|)$, where $r_1 = |\Delta f|$ and $r_2 = \sqrt{(\Delta f)^2 + (a + \Delta x)^2} \approx |\Delta f|$, and a is the lattice period, are given by:

$$\begin{aligned} U_1(K_x, K_y) &\propto \exp(ikR)\exp(-ik|\Delta f|)\exp[iK_z|\Delta f|] \\ U_2(K_x, K_y) &\propto \exp(ikR)\exp(-ik|\Delta f|) \\ &\quad \times \exp[-iK_x(a + \Delta x)]\exp(iK_z|\Delta f|). \end{aligned} \tag{16}$$

The corresponding phases of the wavefronts are:

$$\begin{aligned} \varphi_1(K_x, K_y) &= kR - k|\Delta f| + K_z|\Delta f| \\ \varphi_2(K_x, K_y) &= kR - k|\Delta f| - K_x(a + \Delta x) + K_z|\Delta f|, \end{aligned} \tag{17}$$

and the phase difference is given by:

$$\begin{aligned} \Delta\varphi_x(K_x, K_y) &= \varphi_2(K_x, K_y) - \varphi_1(K_x, K_y) \\ &= -K_x(a + \Delta x). \end{aligned} \tag{18}$$

When $\Delta x = 0$, the phase shift $\Delta\varphi = 2\pi$, and we obtain:

$$\Delta\varphi_x(K_x, K_y) = -K_x^{(1)}a = 2\pi, \tag{19}$$

which corresponds to the position of the nth-order diffraction peak:

$$K_x^{(n)} = n\frac{2\pi}{a}. \tag{20}$$

The phase shift due to a lateral shift Δx is given by:

$$\Delta\varphi_x(K_x, K_y) = -K_x\Delta x = -\frac{2\pi}{\lambda}\Delta x \sin\vartheta, \tag{21}$$

which is an odd function of K_x. Thus, for $\Delta x \neq 0$ there will be an additional phase shift in opposite CBED spots, as, for example, in the spots $(\bar{1}010)$ and $(10\bar{1}0)$, and these phase shifts will be of opposite sign. The obtained phase shift agrees with the phase shift derived from geometrical considerations, Eq. (7) and Fig. 3(b). The CBED pattern and phase shifts produced by an ML with in-plane atomic shifts forming a linear lateral shift is illustrated in Figs. 4(d)–4(f).

The results shown in Figs. 4(b) and 4(e) demonstrate that it is easy to distinguish between out-of-plane and in-plane atomic displacements even without performing a reconstruction simply by comparing the intensity contrast in opposite CBED spots: An out-of-plane defect will always result in symmetric variations in intensity distribution between mirror-symmetric CBED spots, and an in-plane defect will result in antisymmetric variations.

3.2. *Imaging adsorbates on MLs*

Lattice deformations such as strain or rippling do not cause noticeable intensity variations in the zero-order CBED spot and cause increasing intensity variations as the order of the CBED spot increases (spots are further from the zero-order spot and correspond to higher spatial frequencies). In contrast, adsorbates exhibit strong intensity variations in all CBED spots. The zero-order CBED spot is in fact an in-line hologram of the sample, and in-line holography is known to exhibit high intensity contrast that is very valuable for imaging of weak phase objects that are not detectable for conventional TEM at Gaussian focus. The zero-order CBED spot of an ML with phase adsorbates on one surface displays an inversion of contrast when acquired at $\Delta f > 0$ and $\Delta f < 0$ defocus values, respectively.[20] In the case of an adsorbate which is not a phase object and also absorbs electrons, the zero-order CBED spot displays a dark feature.

Experimental CBED patterns of a hBN sample acquired at $\Delta f = 5\,\mu$m and $\Delta f = -5\,\mu$m are shown in Fig. 5. The corresponding high angle annular dark field (HAADF) scanning transmission electron microscope (STEM) images acquired before and after CBED imaging demonstrate no visible radiation damage to the sample (Figs. 5(a) and 5(b)). CBED patterns acquired at $\Delta f > 0$ and $\Delta f < 0$ (Figs. 5(c) and 5(e), respectively) show inversion of the contrast, as expected for weak phase objects.[20] In addition, the intensity distribution within a selected CBED spot demonstrates centro-symmetrical flipping upon transition from $\Delta f > 0$ and $\Delta f < 0$ (see the magnified zero-order CBED spots shown in Figs. 5(d) and 5(f)). The zero-order CBED spot is an in-line hologram of the probed sample region formed by

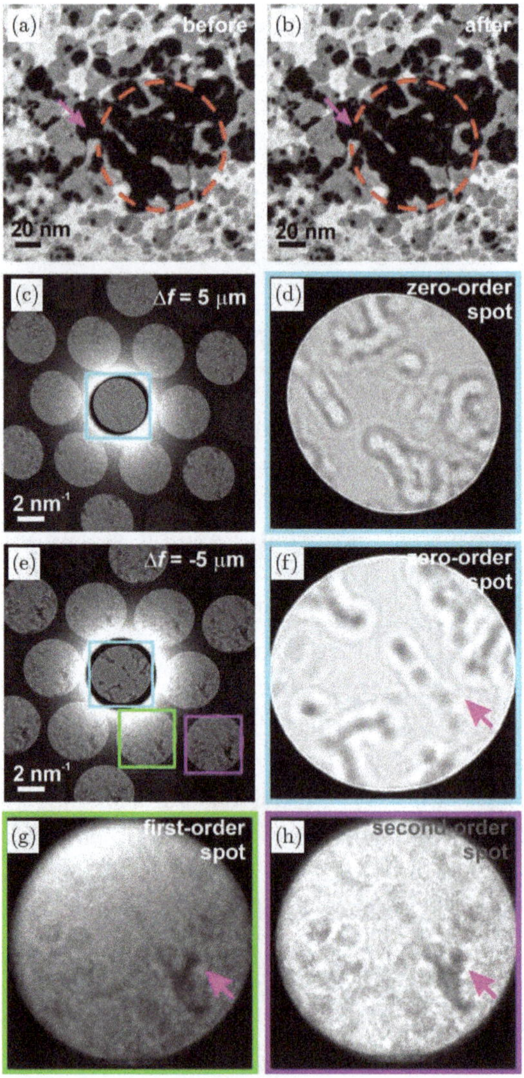

Fig. 5. (Color online) CBED patterns of ML hBN acquired at $\Delta f > 0$ and $\Delta f < 0$. (a)–(b) High angle annular dark field (HAADF) scanning transmission electron microscope (STEM) images of the sample before and after CBED imaging; the imaged area is marked by the red dashed circle. (c) CBED pattern acquired at $\Delta f = 5\,\mu\text{m}$ and (d) the intensity distribution in the zero-order CBED spot. (e) CBED pattern acquired at $\Delta f = -5\,\mu\text{m}$ and magnified images of the intensity distributions of (f) the zero-order, (g) the first-order, and (h) the second-order CBED spots. Adapted from Ref. 20.

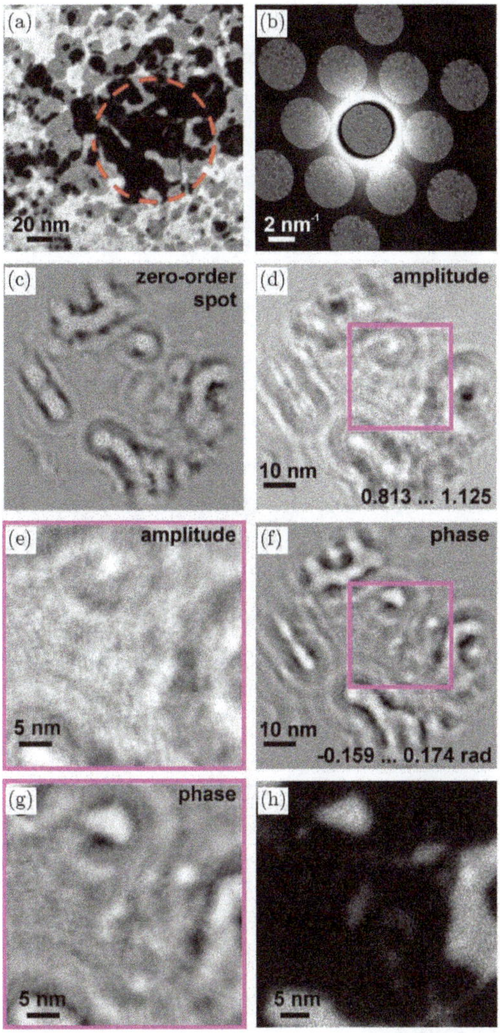

Fig. 6. (Color online) Reconstruction of zero-order CBED spot as in-line hologram. (a) High angle annular dark field (HAADF) images of the sample before CBED imaging; the imaged area is marked by the red dashed circle. (b) CBED pattern. (c) Zero-order CBED spot. (d) and (e) The amplitude distribution reconstructed from the zero-order CBED spot at $\Delta f = 5\,\mu\text{m}$. (f) and (g) The phase distribution reconstructed from the zero-order CBED spot at $\Delta f = 5\,\mu\text{m}$. (h) Zoomed-in region in the HAADF image for comparison with the reconstructed amplitude and phase distributions of the same region (marked by the magenta square). The values given at the bottom of (d) and (f) are the range of the reconstructed amplitude and phase values, respectively. Adapted from Ref. 20.

interference of the scattered and non-scattered waves. The higher-order CBED spots are formed only by the scattered wave and therefore show different intensity distributions than the zero-order CBED spot. The higher-order CBED spots are highly sensitive to the atomic misalignments as discussed earlier, and their intensity distribution reflects the ripples in the 2D crystal. For example, in the first- and second-order CBED spots, there is a darker feature as indicated by magenta arrow in Figs. 5(f)–5(h). Such a dark feature is not observed in the zero-order CBED spot and thus cannot be attributed to an adsorbate. The HAADF STEM images of the sample also confirm that there are no adsorbates at this location, as indicated by the magenta arrows in Figs. 5(a) and 5(b). Thus, the dark feature can be explained by a surface deformation in that location, probably because of the stress between the two adjacent adsorbates.

The zero-order CBED spot can be treated as an in-line hologram, and the amplitude and phase distributions of the transmission function of the sample can be reconstructed by available algorithms.[25] An example of such reconstruction is shown in Fig. 6. Adsorbate patches of sub-nanometer sizes are observed in the reconstructed phase distributions and can be cross-validated with the corresponding HAADF STEM images. Also here, the reconstructed amplitude distributions exhibit blurred structure when compared with the reconstructed phase distributions, which suggests that the imaged objects are phase objects without any absorption contrast.

4. Holographic CBED of BLs

4.1. *Formation of interference pattern in CBED spots*

For CBED of a BL sample, the two sets of electron beams diffracted on each layer interfere in the detector plane, creating interference patterns at the positions of the overlapping CBED spots, as illustrated in Fig. 7. Such interference patterns contain rich information about the local inter-atomic spacing (local strain), the vertical distance between the layers, the relative orientation between the layers, etc. The period and tilt of the fringes in the interference pattern can be explained by considering the position of the sources in

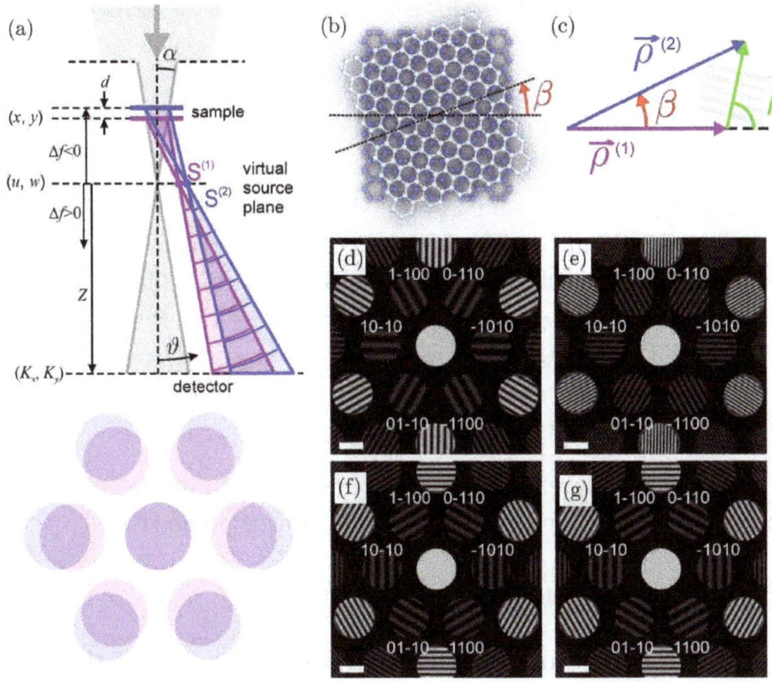

Fig. 7. (Color online) CBED on a bilayer system. (a) Top: Schematics of the HCBED experimental arrangement. Δf is the sample z-position counted from the focus of the electron beam (in this particular case underfocus, $\Delta f < 0$, CBED conditions are shown), Z is the distance from the virtual source plane to the detector, $S^{(1)}$ and $S^{(2)}$ are the virtual sources for the first-order CBED spots of bottom (1) and top (2) crystals in the heterostructure stack, respectively. ϑ is the angular coordinate on the detector. Bottom: Distribution of CBED spots on a detector in the case of two aligned crystals (no relative twist $\beta = 0$), but with slightly different lattice constants. (b) Real-space distribution of a BL sample with a difference in lattice parameter and a relative in-plane twist angle β. (c) Arrangement of the vectors in the virtual source plane for the situation shown in (b). (d)–(g) Simulated CBED patterns for: (d) and (e) BL graphene in AA stacking with twist angles 1° and 2°, respectively; (f) and (g) graphene-hBN BL stacking without twist, in AA and AB stacking configurations, respectively. The scale bars in (d)–(g) correspond to $2\,\text{nm}^{-1}$.

the virtual source plane, as sketched in Fig. 7(a). A real-space illustration of a BL sample with an in-plane twist angle β is shown in Fig. 7(b). In the virtual source plane, the Bragg diffraction peaks create virtual sources, which are rotated by the twist angle β relative to each other, as depicted in Fig. 7(c).

The interference pattern in a CBED spot is described by the formula[22]:

$$I(\boldsymbol{K}) \propto 2 + 2\cos\left\{\boldsymbol{K}\Delta\boldsymbol{\rho} + \frac{\pi}{\lambda}\left[\frac{(\boldsymbol{\rho}^{(2)})^2}{|\Delta f + d|} - \frac{(\boldsymbol{\rho}^{(1)})^2}{|\Delta f|}\right] - \Delta\gamma\right\} \quad (22)$$

where $\Delta\boldsymbol{\rho} = \boldsymbol{\rho}^{(1)} - \boldsymbol{\rho}^{(2)}$ and $\Delta\gamma = \gamma^{(1)} - \gamma^{(2)}$, $\boldsymbol{\rho}^{(1)}$ and $\boldsymbol{\rho}^{(2)}$ are the positions of the virtual sources in the virtual source plane, whereas $\gamma_m^{(1)}$ and $\gamma_m^{(2)}$ are the constant phases of the virtual sources. The first term in the argument of the cosine term describes the interference fringes distribution. The period of the interference fringes is given by:

$$T = \frac{2\pi}{\Delta\rho}. \quad (23)$$

The tilt of the interference fringes μ can be found from the geometrical arrangement of the vectors in the virtual source plane (Fig. 7(c)):

$$\tan\mu = \frac{\rho^{(2)}\sin\beta}{\rho^{(2)}\cos\beta - \rho^{(1)}}. \quad (24)$$

The higher the twist angle, the larger $\Delta\rho$ and the smaller is the period of the fringes, as can be seen in the simulations of BL graphene at twist angles of 1° and 2° in Figs. 7(d) and 7(e), respectively. The second term in the argument of cosine in Eq. (22) is a constant offset. The third term $\Delta\gamma$ depends on the local stacking of the layers (for example, AA or AB crystal stacking) and defines the position of the center of the interference pattern. If the local stacking under the center of the electron beam is AA, then $\Delta\gamma = 0$ and the interference pattern within an overlapping CBED spot has a maximum at the center of the CBED spot. For AB stacking, $\Delta\gamma \neq 0$ and the interference pattern within an overlapping CBED spot does not display a maximum at the center of the CBED spot, as can be seen in the simulations of a graphene-hBN system for AA and AB stacking in Figs. 7(f) and 7(g), respectively.

4.2. *Holographic reconstruction*

4.2.1. *Protocol of the reconstruction procedure*

The atomic displacements relative to their position in perfect lattices for a BL system can be reconstructed from the corresponding CBED

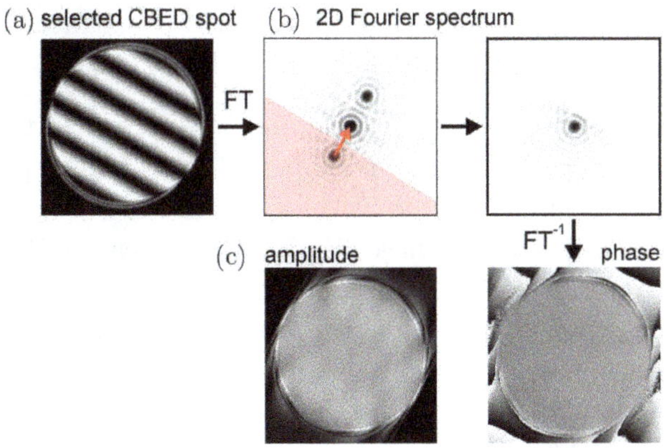

Fig. 8. (Color online) Reconstruction of a CBED spot as off-axis holograms. (a) Selected CBED spot. (b) Amplitude of its Fourier spectrum. The area shaded red is selected and shifted so that the sideband peak is in the center. (c) Inverse 2D FT gives the complex-valued distribution, where the amplitude and phase are extracted.

pattern when treating the CBED spots as off-axis holograms. The reconstruction procedure is described in Refs. 19 and 22. Here we provide the main reconstruction steps, also illustrated in Fig. 8:

(1) For holographic reconstruction, a CBED spot is selected at the center at an arithmetic average of the centers of the individual CBED spots.
(2) The 2D Fourier spectrum of the selected region is calculated. In the obtained complex-valued spectrum, one zero-order and two sidebands are observed (Figs. 8(a) and 8(b)).
(3) One of the sidebands is selected while the zero-order and remaining sideband are set to zero (Fig. 8(b)). Note, it is crucially important that the sidebands in different CBED spots are chosen consistently. The chosen sidebands in different CBED spots have to be related by the same symmetry transformations as the CBED spots themselves (rotation by an integer number of $\pi/3$).
(4) The whole spectrum is shifted so that the maximum of the sideband is located at the origin of the reciprocal plane (Fig. 8(b)).
(5) The inverse 2D Fourier transform of the resulting distribution is calculated.

(6) The amplitude and phase distributions are extracted from the obtained complex-valued distribution (Fig. 8(c)).

In steps (3) and (4), the right sideband is selected at the position defined by the fringes tilt angle μ and period T, as defined by Eqs. (23) and (24) and explained in Ref. 22.

After the amplitude and phase distributions for each CBED spot are reconstructed, only the phase distributions are considered since only these carry the information about the atomic positions. The individual reconstructed phase distributions are averaged, that is, all six distributions are added together and divided by six, and the out-of-plane atomic shifts are calculated by Eq. (15). The in-plane atomic shifts are then calculated from the reconstructed phases of the opposite CBED spots by applying Eq. (21).

4.2.2. *Simulated example of reconstruction in-plane and out-of-plane displacement*

An example of the simulated and reconstructed CBED pattern for a graphene-hBN system with both out-of-plane and in-plane shifts is shown in Fig. 9.

4.2.3. *Experimental example of reconstruction in-plane and out-of-plane displacement*

An experimental CBED pattern from a twisted BL system (graphene-hBN) is shown in Fig. 10. CBED spots originating from the graphene layer are found at a slightly larger diffraction angle than CBED spots from the hBN layer (due to differences in the in-plane lattice parameter), which allows an easy assignment of the CBED spots corresponding to different layers. A stacking fault between the layers is evident by the presence of a distinctive ridge in the interference patterns in the first- and higher-order CBED spots (Figs. 10(a) and 10(b)). The defect causes no significant contrast in the zero-order CBED spot meaning that the defect consists of only atomic position misalignments, which introduce a significant additional phase shift between the electron waves scattered from the two layers. Note that in diffraction imaging mode, where only diffraction peaks are

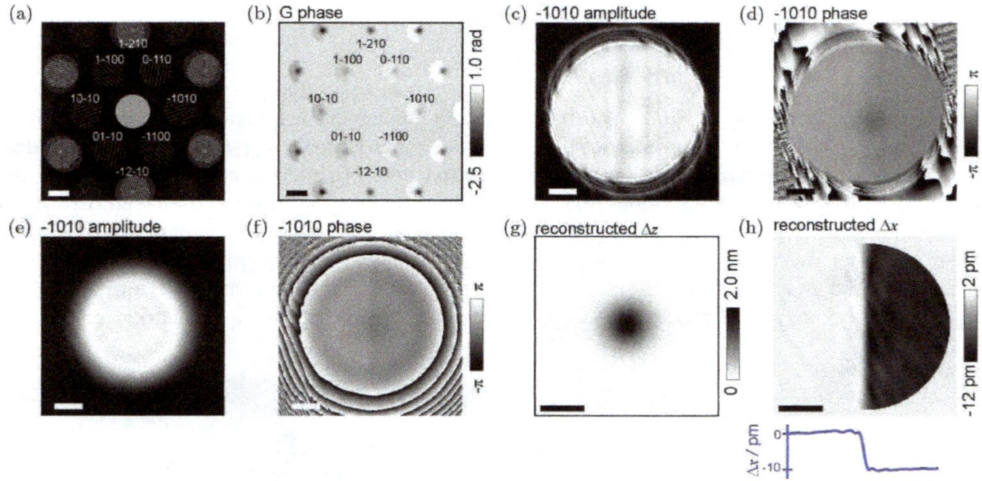

Fig. 9. Simulated CBED pattern of a graphene-hBN bilayer where the graphene lattice is deformed, the twist angle 4°, and the reconstructed in-plane and out-of-plane graphene lattice deformations. (a) Simulated CBED pattern where the atoms in the graphene layer are mis-positioned as follows: the atoms positioned at $x > 0$ are displaced by $\Delta x = -10\,\text{pm}$, and the atomic z-positions are shifted by $\Delta z = -A_{\text{B}} \exp(-\frac{x^2+y^2}{2\sigma_{\text{B}}^2})$, $A_B = 2\,\text{nm}$, $\sigma_B = 2\,\text{nm}$. For these simulations $\Delta f = -2\,\mu\text{m}$, the distance between the layers is 3.35 Å, the probed area is about 28 nm in diameter. (b) The difference of the phases of the wavefronts scattered by the graphene layer with and without atomic displacement is calculated as $\Delta\varphi = \varphi_{\text{with}} - \varphi_{\text{without}}$; only the phase difference is shown, the phase distribution φ_{without} (not shown) exhibits constant values. (c) Amplitude of the reconstructed wavefront at the (−1010) CBED spot. (d) Phase of the reconstructed wavefront at the (−1010) CBED spot. (e) Amplitude of the reconstructed complex-valued distribution from the (−1010) CBED spot after correction for defocus. (f) Phase of the reconstructed complex-valued distribution from the (−1010) CBED spot after correction for defocus. (g) Reconstructed distribution of the atomic out-of-plane displacement, Δz, in the graphene layer. (h) Reconstructed distribution of the atomic in-plane displacement, Δx, in the graphene layer. The scale bars in (a) and (b) correspond to $2\,\text{nm}^{-1}$. The scale bars in (c)–(h) correspond to 5 nm. Adapted from Ref. 19.

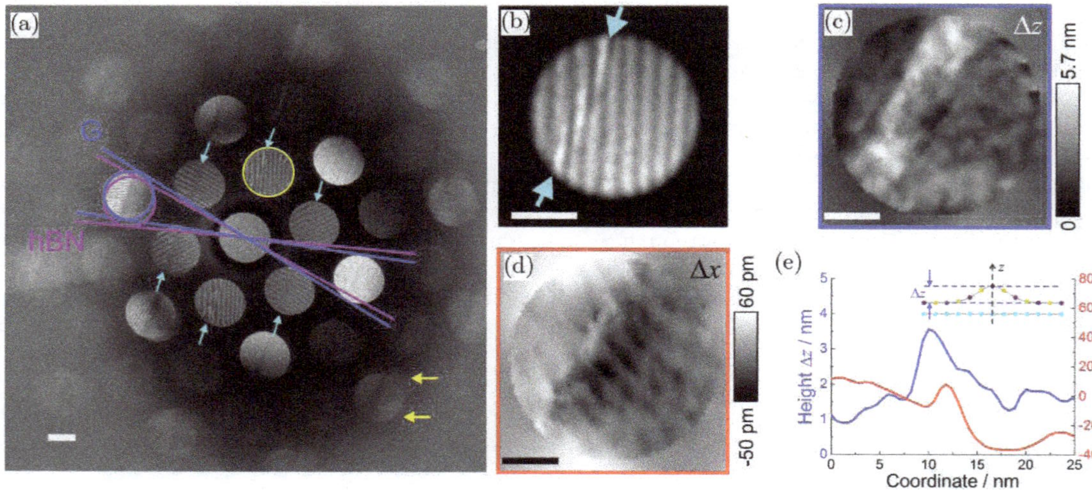

Fig. 10. (Color online) Extracting the shape of an out-of-plane ripple from a CBED pattern. (a) Experimental CBED pattern acquired at defocus $\Delta f = -3\,\mu$m, with irregularity in the interference patterns marked by the arrows. The blue and purple lines indicate the relative rotation between graphene and hBN layers, which amounts to 3°. The cyan arrows indicate an out-of-plane ripple observed in the first-order CBED spots. The yellow arrows indicate the separation of CBED spots originating from graphene and hBN layers, where it becomes clear that the ripple is in the hBN layer. The intensity of the central spot is reduced by a factor of 0.1 for clarity. The scale bar corresponds to $2\,\text{nm}^{-1}$. (b) Magnified selected CBED spot (circled yellow in (a)) where irregularities of the fringe pattern can be seen. The scale bar corresponds to $1\,\text{nm}^{-1}$. (c) The reconstructed distribution of the ripple height Δz. (d) The reconstructed distribution of the lateral shift Δx. (e) Profiles for the magnitude of Δz and Δx profiles perpendicular to the ripple in (c) and (d). The scale bars in (c) and (d) correspond to 10 nm in real space. Adapted from Ref. 19.

detected, imaging of such a defect would not be possible. The area of overlap between CBED spots from the graphene and hBN layers is less in the higher order diffraction spots. The intensity contrast caused by the corrugation is more pronounced in the higher order CBED spots, in accordance with Eqs. (6) and (7). The CBED pattern was holographically reconstructed by the procedure described earlier. Figures 10(c) and 10(d) show the recovered out-of-plane Δz and in-plane Δx atomic displacements. Figure 10(e) compares the out-of-plane and in-plane atomic shifts along the ripple. The retrieved height of the out-of-plane ripple in hBN layer is about 2 nm, which agrees well with the observed out-of-plane ripples in graphene/hBN stacks due to self-cleansing effects.[26]

4.3. *Reconstruction of interlayer distance*

HCBED can be applied for reconstruction of interlayer distances from a single CBED pattern as was demonstrated in Ref. 19 and explained in more detail in Ref. 22. The reconstruction procedure consists of the steps provided above, and the reconstructed phase distribution allows recovery of the interlayer distance according to the formula[22]:

$$\Delta\varphi = \frac{\pi d}{\lambda} \tan\vartheta^{(2)} \tan\vartheta^{(1)} \cos\beta, \tag{25}$$

where $\vartheta^{(1)}$ and $\vartheta^{(1)}$ are the diffraction angles of the two layers. For BLs where both layers are identical $\vartheta^{(1)} = \vartheta^{(2)} = \vartheta$ and for small twist angles, we obtain:

$$\Delta\varphi \approx \frac{\pi d\lambda}{a^2}. \tag{26}$$

Figure 11 shows the simulated 2D distributions and one-dimensional (1D) profiles of the reconstructed interlayer distances for BL graphene (BLG) with the interlayer distances of 0 Å and 10 Å at three different twist angles of 0.5°, 2°, and 4°. From Fig. 11, we see that the reconstructed interlayer distances obtained from CBED patterns with smaller twist angles exhibit a smoother appearance, whereas at larger twist angles, artifact due to moiré structure becomes more pronounced in the reconstructions. The precision of the reconstructed interlayer distance is about ±0.5 Å.[22]

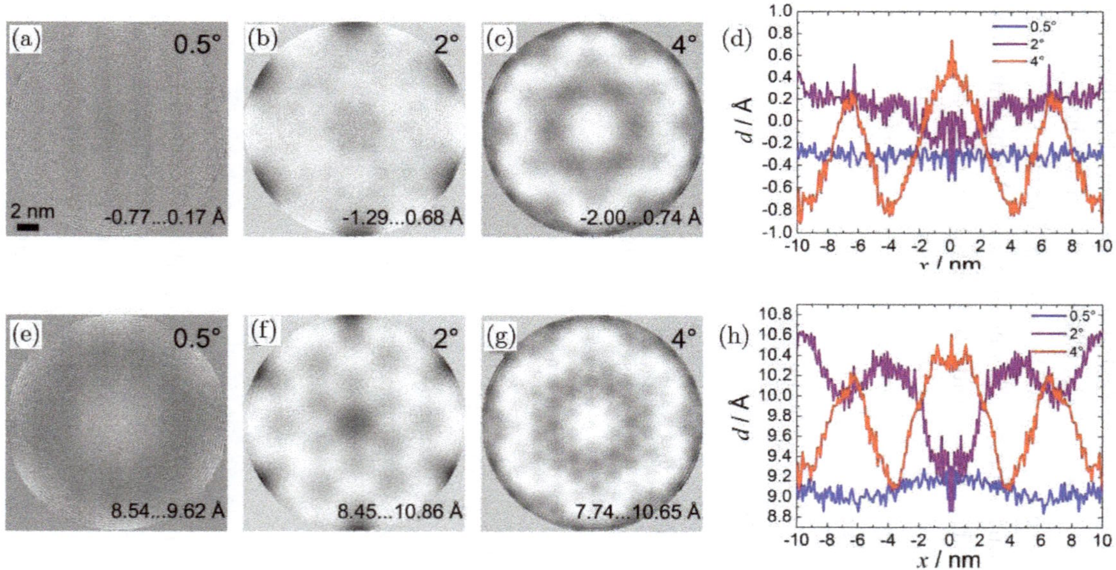

Fig. 11. (Color online) 2D distributions ((a)–(c) and (e)–(g)) and 1D profiles through the middle of the 2D distributions ((d) and (h)) of the reconstructed interlayer distances from CBED patterns of BLG with the interlayer distances of 0 Å (a)–(d) and 10 Å (e)–(h), for three different twist angles of 0.5°, 2°, and 4°. The values given in bottom right in (a)–(c) and (e) and (f) refer to the interlayer distance. Defocus distance is $\Delta f = -2\,\mu m$. Adapted from Ref. 22.

5. HCBED on Multilayer Systems, CBED Moiré

Twisted multilayer systems exhibit extra modulations of the interference fringes in CBED patterns, i.e. a CBED moiré, as shown in Fig. 12. These extra modulations are coming from the interference between the moiré CBED spots. Such moiré CBED spots are the results of a second order process, which involves electrons being scattered by both layers sequentially.[27] Due to its second order nature, such moiré peaks have much weaker intensity than the main diffraction peaks,[27] and therefore the moiré CBED spots are also much weaker than the main CBED spots. This is the reason why they are practically invisible in twisted BL samples. Moreover, moiré CBED spots strongly overlap with the main CBED spots. For these two reasons, moiré CBED spots are not directly visible in the CBED patterns but they manifest themselves in the additional modulations of the CBED interference pattern — the CBED moiré, which is

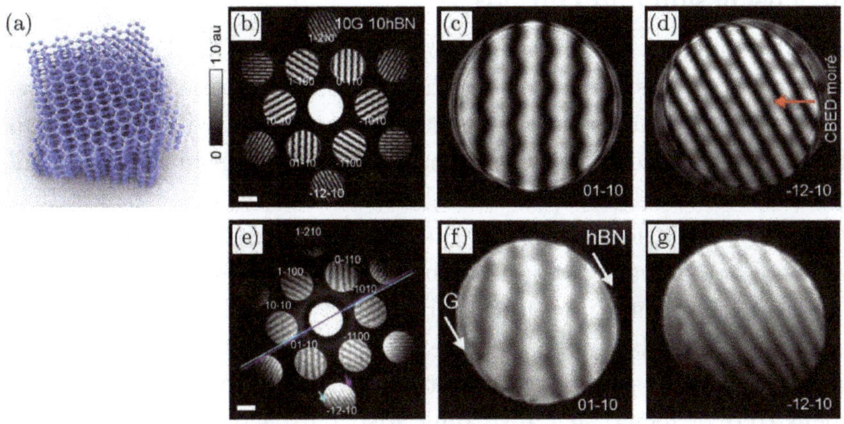

Fig. 12. (Color online) CBED on twisted multilayer systems. (a) Illustration of twisted multilayer graphene. (b) Simulated CBED patterns for a multilayer van der Waals structure consisting of 10 graphene and 10 hBN layers, the twist angle between graphene and hBN layers is 2°, at a defocus, $|\Delta f| = 2.0\,\mu\text{m}$. (c) and (d) magnified CBED spots where a CBED moiré is observed in CBED interference fringes. (e) Experimental CBED patterns acquired at a defocus, $|\Delta f| = 2.0\,\mu\text{m}$, (f) and (g) magnified CBED spots. The scalebars in (b) and (e) are $2\,\text{nm}^{-1}$. Adapted from Ref. 21.

created by the interference between the moiré CBED spots and the major CBED spots.

A simple and robust method for evaluation of the composition and the number of layers from a single-shot CBED pattern was demonstrated in Ref. 21. The composition and the relative number of layers can be evaluated from the intensity distribution in the non-overlapping regions of a CBED spot. Number of layers can be evaluated by inspecting a 2D Fourier spectrum of a CBED spot, where the presence of peaks due to the CBED moiré indicates that there are five or more layers in the sample. Although the precision of such sample characterisation is very modest when compared to cross-sectional TEM imaging,[28] the presented approach has the advantage that it is non-destructive, requires only a single CBED pattern, and CBED is relatively easy to realize in a conventional TEM.

6. Discussion and Outlook

HCBED offers a lot of information in one single CBED image such as atomic 3D displacements in the layers, the interlayer distances, and imaging of adsorbates. Adsorbates can be clearly distinguished from 3D displacements of atoms: an adsorbate is identified in all orders of CBED spots, while atomic displacements produce intensity variations only in the first and higher order CBED spots. The contrast of the adsorbate image allows us to determine whether the adsorbate is an absorbing or phase object. Phase adsorbates display opposite intensity contrast when the probing wave is changed from convergent to a divergent wavefront (under focus to over focus). An absorbing adsorbate is displayed as a dark feature in both imaging regimes. The adsorbate distribution can be reconstructed from the zero-order CBED spot by applying an in-line hologram reconstruction routine. 3D displacements of atoms, on the other hand, display minimal intensity variations in the zero-order CBED spot, but they cause significant contrast in the higher-order CBED spots. Moreover, the type of displacement can be clearly identified just by comparing opposite CBED spots: an in-plane displacement leads to an opposite intensity contrast in opposite CBED spots (antisymmetric pattern), whereas an out-of-plane displacement leads to the same intensity contrast in the opposite CBED spots (pattern is symmetric).

For BL samples, the interference patterns formed in overlapping CBED spots can be treated as off-axis holograms and the phase of the interfering waves, and with this the 3D positions of the scattering atoms can be retrieved. By using this approach, in-plane and out-of-plane ripples and the interlayer distances can be quantitatively reconstructed. The resolution at which the atomic shifts are recovered exceeds the intrinsic resolution provided by the classical resolution criteria. The lateral and axial (along the z axis) resolutions evaluated from a CBED pattern k-value range are given by $d_{x,y} = \frac{\lambda}{\sin \vartheta_{\max}}$ and $d_z = \frac{\lambda}{1-\cos \vartheta_{\max}}$, respectively, where $\vartheta_{\max}$ is the maximal detected diffraction angle in the CBED pattern. According to these formulas, for a BL graphene CBED pattern acquired only up to the first-order CBED spots, the lateral resolution is $d_{x,y} = 2.13\,\text{Å}$ and the axial resolution is $d_z = 217.2\,\text{Å}$. It is therefore a remarkable result that the holographic CBED approach allows reconstruction of interlayer distances of a few Angstroms at 0.5 Å accuracy, which is more than 400 times the diffraction-defined z-resolution.[19,22]

From the experimental point of view, the best probing beam for acquiring CBED patterns would be a perfect convergent electron beam. However, holographic techniques in general show very high tolerance to the probing beam imperfections, because the resulting interference pattern (hologram) is formed due the difference of the phases in the scattered and reference waves. In holographic CBED, the interference contrast is formed due to the relative phase shift between the waves scattered of atoms in different 2D layers. This relative phase shift is given by the local arrangement of the atoms in the structure, and the atoms are probed with the same local distribution of the probing beam. Thus, the probing beam imperfections, even when present, should have a minimal effect on the CBED pattern formation, and with this, on the resulting reconstruction. However, a study can be performed to quantitatively evaluate the effect of beam aberrations, in particular, imperfections in the phase distribution, on the resulting reconstructions.

To conclude, HCBED has already demonstrated capability of providing high-resolution information about atomic arrangement from a single CBED pattern. We expect that HCBED can be further developed to become a practical tool for studying 2D materials at atomic resolution in 3D for complex heterostructures and arbitrary numbers of layers.

Acknowledgment

The authors acknowledge support from EU Graphene Flagship Program (contract CNECTICT-604391), European Research Council Synergy Grant Hetero2D (contract 319277), European Research Council Starting Grant EvoluTEM (contract 715502), the Royal Society, EPSRC grants EP/S019367/1, EP/P026850/1 and EP/N010345/1, EP/P009050/1, EP/S021531/1, FLAG-ERA project TRANS2DTMD. This project was supported by the Ministry of Education (Singapore) through the Research Centre of Excellence program (grant EDUN C-33-18-279-V12, I-FIM).

Appendix A

Simulation of CBED Patterns

Monolayer samples. Simulation of CBED patterns was performed as follows. For calculation of the CBED pattern of an ML with displaced atoms, the input data are an array of coordinates of all atoms (x_n, y_n, z_n). The far-field wavefront distribution of the scattered wave is simulated by summing up the waves scattered by each individual atom:

$$U(K_x, K_y) = \sum_n \psi_0(\boldsymbol{r}_n) \exp[-i(K_x x_n + K_y y_n)] \times \exp\left(-iz_n\sqrt{K^2 - K_x^2 - K_y^2}\right),$$

where $\psi_0(\boldsymbol{r}_n)$ is the complex-valued value of the probing wave at the nth atom location, $\psi_0(\boldsymbol{r})$ is calculated by Eq. (5). No fast Fourier transforms are applied in the simulations to avoid sampling artifacts.

Bilayer and multilayer samples. For simulations of CBED patterns on multilayer samples, it is assumed that all atoms in one layer have the same z position and the following simulation procedure is applied.[22] The transmission functions of MLs are calculated as:

$$t(x, y) = \exp[i\sigma v_z(x, y) \otimes l(x, y)], \tag{A.1}$$

where $v_z(x, y)$ is the projected potential of an individual atom, $l(x, y)$ is the function describing positions of the atoms in the lattice, and $\otimes$ denotes convolution. The projected potential of a single carbon atom is simulated in the form:

$$v_z(r) = 4\pi^2 a_0 e \sum_{i=1}^{3} a_i K_0(2\pi r \sqrt{b_i}) + 2\pi a_0 e \sum_{n=1}^{3} \frac{c_i}{d_i} \exp(-\pi^2 r^2/d_i),$$

where $r = \sqrt{x^2 + y^2}$, a_0 is the Bohr radius, e is the elementary charge, $K_0(\ldots)$ is the modified Bessel function, and a_i, b_i, c_i, d_i are parameters that depend on the chemical origin of the atoms and are tabulated elsewhere.[29] In $v_z(r)$, the singularity at $r = 0$ is replaced by the value of $v_z(r)$ at $r = 0.1$ Å. The convolution $v_z(x, y) \otimes l(x, y)$ in Eq. (A.1) is calculated as $\mathrm{FT}^{-1}\{\mathrm{FT}[v_z(x, y)]\mathrm{FT}[l(x, y)]\}$, where FT denotes Fourier transform. $\mathrm{FT}[l(x, y)]$ is simulated without applying Fast Fourier transform (FFT) to avoid artifacts associated with FFT; it is calculated as $\mathrm{FT}[l(x, y)] = \sum_n \exp[-i(k_x x_n + k_y y_n)]$, where (x_n, y_n) are the exact atomic coordinates and not pixels. The inverse FT is calculated by applying inverse FFT to the product of $\mathrm{FT}[v_z(x, y)]$ and $\mathrm{FT}[l(x, y)]$.

Each ML is assigned a transmission function calculated as described above. The incident convergent wave distribution $\psi_0(\boldsymbol{r})$ is calculated by simulating the diffraction of a spherical wavefront on an aperture (the second condenser aperture). No weak phase object approximation is applied in the simulations. The exit wave after passing through the first layer is given by the product of the incident wave and the transmission function of the first layer $u_1(x_1, y_1) = \psi_0(x_1, y_1)t_1(x_1, y_1)$. Next, this wave is propagated to the second layer. The propagation is calculated by the angular spectrum method.[25] The distribution of the propagated wave in the plane (x_2, y_2) is

$u_{2,0}(x_2, y_2)$. The exit wave after passing the second ML is calculated as $u_2(x_2, y_2) = u_{2,0}(x_2, y_2)t_2(x_2, y_2)$ and so forth. The CBED pattern is then simulated as the square of the amplitude of the FT of the exit wave after passing through the last layer.

References

1. W. Kossel and G. Möllenstedt, Elektroneninterferenzen im konvergenten Bündel, *Annal. der Physik* **36** (1939) 113.
2. P. Goodman, *Acta Crystallogr. A* **31** (1975) 804.
3. B. F. Buxton, J. A. Eades, J. W. Steeds and G. M. Rackham, *Philos. Trans. Royal Soc. A* **281** (1976) 171.
4. P. M. Jones, G. M. Rackham and J. W. Steeds, *Proc. R. Soc. Lond. A* **354** (1977) 197.
5. M. Tanaka, H. Sekii and T. Nagasawa, *Acta Crystallogr. A* **39** (1983) 825.
6. M. Tanaka, R. Saito and H. Sekii, *Acta Crystallogr. A* **39** (1983) 357.
7. P. M. Kelly, A. Jostsons, R. G. Blake and J. G. Napier, *Phys. Status Solidi A* **31** (1975) 771.
8. R. W. Carpenter and J. C. H. Spence, Three-dimensional strain-field information in convergent beam electron diffraction patterns, *Acta Crystallogr. A* **38** (1982) 55.
9. L. J. Wu, Y. M. Zhu and J. Tafto, *Phys. Rev. Lett.* **85** (2000) 5126.
10. L. J. Wu, Y. M. Zhu, J. Tafto, D. O. Welch and M. Suenaga, *Phys. Rev. B* **66** (2002) 104517.
11. M. Tanaka, R. Saito, K. Ueno and Y. Harada, *J. Electron Microsc.* **29** (1980) 408.
12. J. P. Morniroli, CBED and LACBED characterization of crystal defects, *J. Microsc.-Oxf.* **223** (2006) 240.
13. M. Tanaka, M. Terauchi, K. Tsuda and K. Saitoh, *Convergent-Beam Electron Diffraction*, Vol. I–IV, JEOL (1985).
14. P. E. Champness, *Miner. Mag.* **51** (1987) 33.
15. J. C. H. Spence and J. M. Zuo, *Electron Microdiffraction* (Plenum Press, 1992).
16. J. M. Zuo and J. C. H. Spence, *Advanced Transmission Electron Microscopy* (Spinger, 2017).
17. J. C. Meyer, A. K. Geim, M. I. Katsnelson, K. S. Novoselov, D. Obergfell, S. Roth, C. Girit and A. Zettl, *Solid State Commun.* **143** (2007) 101.
18. T. Latychevskaia, W.-H. Hsu, W.-T. Chang, C.-Y. Lin and I.-S. Hwang, *Nat. Commun.* **8** (2017) 14440.
19. T. Latychevskaia, C. R. Woods, Y. B. Wang, M. Holwill, E. Prestat, S. J. Haigh and K. S. Novoselov, *Proc. Natl Acad. Sci.* **115** (2018) 7473.
20. T. Latychevskaia, C. R. Woods, Y. B. Wang, M. Holwill, E. Prestat, S. J. Haigh and K. S. Novoselov, *Front. Phys.* **14** (2019) 13606.
21. T. Latychevskaia, C. R. Woods, Y. B. Wang, M. Holwill, E. Prestat, S. J. Haigh and K. S. Novoselov, *Ultramicroscopy* **212** (2020) 112976.

22. T. Latychevskaia, Y. Zou, C. R. Woods, Y. B. Wang, M. Holwill, E. Prestat, S. J. Haigh and K. S. Novoselov, *Ultramicroscopy* **219** (2020) 113020.
23. D. Gabor, *Nature* **161** (1948) 777.
24. D. Gabor, *Proc. R. Soc. Lond. A* **197** (1949) 454.
25. T. Latychevskaia and H.-W. Fink, *Appl. Opt.* **54** (2015) 2424.
26. A. V. Kretinin, Y. Cao, J. S. Tu, G. L. Yu, R. Jalil, K. S. Novoselov, S. J. Haigh, A. Gholinia, A. Mishchenko, M. Lozada, T. Georgiou, C. R. Woods, F. Withers, P. Blake, G. Eda, A. Wirsig, C. Hucho, K. Watanabe, T. Taniguchi, A. K. Geim and R. V. Gorbachev, *Nano Lett.* **14** (2014) 3270.
27. T. Latychevskaia, C. Escher and H.-W. Fink, *Ultramicroscopy* **197** (2019) 46.
28. S. J. Haigh, A. Gholinia, R. Jalil, S. Romani, L. Britnell, D. C. Elias, K. S. Novoselov, L. A. Ponomarenko, A. K. Geim and R. Gorbachev, *Nat. Mater.* **11** (2012) 764.
29. E. J. Kirkland, *Advanced Computing in Electron Microscopy* (Springer, 2010).

https://doi.org/10.1142/9789811247859_0010

Chapter 10

Electronic and Optical Modification of Organic-hybrid Perovskites

Silmi Kaffah

Department of Computer Systems Engineering,
Universitas Prasetiya Mulya,
Kavling Edutown I.1, Jl. BSD Raya Utama, BSD City,
Tangerang 15339, Indonesia
silmi.kaffah@prasetiyamulya.ac.id

Lina Jaya Diguna*

Department of Renewable Energy Engineering,
Universitas Prasetiya Mulya,
Kavling Edutown I.1, Jl. BSD Raya Utama, BSD City,
Tangerang 15339, Indonesia
lina.diguna@prasetiyamulya.ac.id

Suriani Abu Bakar

Nanotechnology Research Centre,
Faculty of Science and Mathematics,
Universiti Pendidikan Sultan Idris,
35900 Tanjung Malim, Perak, Malaysia
suriani@fsmt.upsi.edu.my

Muhammad Danang Birowosuto

Department of Renewable Energy Engineering,
Universitas Prasetiya Mulya,
Kavling Edutown I.1, Jl. BSD Raya Utama, BSD City,
Tangerang 15339, Indonesia

*Corresponding authors.

CNRS-International-NTU-THALES
Research Alliances (CINTRA)/UMI 3288,
Research Techno Plaza, 50 Nanyang Drive,
Border X Block, Level 6, Singapore 637553, Singapore
muhammad.birowosuto@prasetiyamulya.ac.id

Arramel *

Department of Physics,
National University of Singapore,
2 Science Drive 3, Singapore 117551, Singapore
phyarr@nus.edu.sg

Renewed interest has brought significant attention to tune coherently the electronic and optical properties of hybrid organic–inorganic perovskites (HOIPs) in recent years. Tailoring the intimate structure–property relationship is a primary target toward the advancement of light-harvesting technologies. These constructive progresses are expected to promote staggering endeavors within the solar cells community that needs to be revisited. Several considerations and strategies are introduced mainly to illustrate the importance of structural stability, interfacial alignment, and photo-generated carriers extraction across the perovskite heterostructures. Here, we review recent strides of such vast compelling diversity in order to shed some light on the interplay of the interfacial chemistry, photophysics, and light-emitting properties of HOIPs via molecular engineering or doping approach. In addition, we outline several fundamental knowledge processes across the role of charge transfer, charge carrier extraction, passivation agent, bandgap, and emission tunability at two-dimensional (2D) level of HOIPs/molecule heterointerfaces. An extensive range of the relevant work is illustrated to embrace new research directions for employing organic molecules as targeted active layer in perovskite-based devices. Ultimately, we address important insights related to the physical phenomena at the active molecules/perovskites interfaces that deserve careful considerations. This review specifically outlines a comprehensive overview of surface-based interactions that fundamentally challenges the delicate balance between organic materials and perovskites, which promotes bright future of desired practical applications.

Keywords: 2D-HOIPs; X-ray and ultraviolet photoelectron spectroscopy; optical properties; photovoltaics; photoluminescence.

1. Introduction

The emergence of hybrid organic-inorganic perovskites (HOIPs) plays a pivotal role in energy applications such as three-dimensional (3D)

HOIPs solar cells absorbers, e.g. $MAPbI_3$ (MA: methylammonium), and two-dimensional (2D) HOIPs light emitters.[1–8] The current reports in this fields indicated that the optoelectronic devices using 3D hybrid-based perovskite have a tradeoff concerning their structural stability. For example, moisture instabilities, halide vacancies, ion migrations, and interstitials have hampered its progress as an alternative of silicon.[9,10] In comparison to the 3D analogs, 2D-HOIPs have an improved processability and stability toward water and light.[11–13] These could originate from the fact that the substitution of a volatile and hydrophilic methylammonium in 3D perovskite by other organic cations in 2D-HOIPs could serve as a protective layer from the harsh environment.

In terms of structural diversity in hybrid perovskites and its advancements, 2D-HOIPs have evolved their progress into vast collection of examples with compelling chemical composition and multiple platforms. In recent years, diverse classes of hybrid perovskites have been investigated extensively. Several comprehensive reviews illustrated the structural guidelines of the reduced dimensionality perovskites, focusing on 2D perovskite derivatives which hold promise in their tunability and excellent photophysical properties. For example, these hybrid materials can be considered as quantum wells system that composed of semiconducting inorganic sheets separated by the insulating organic barriers.[7,14–16] In this review, we mention several members of the groups that essentially can be classified into three categories based on the arrangement of the inorganic layers and its connectivity with respect to the insulating organic moiety barrier. These structural models can be rationalized based on the cleavage direction along the $\langle 100 \rangle$, $\langle 110 \rangle$, and $\langle 111 \rangle$ crystallographic orientations of the parent 3D hybrid perovskite crystal structure. The richest database in the literature focuses on the $\langle 100 \rangle$-oriented perovskites.[4,17,18] Chemical diversity is highly promising in this hybrid material since there are numerous structural variations driven by the choice of capping cation and their relative position between the inorganic sheets. Thus, 2D-HOIPs can be classified further into three categories: Aurivillius phase,[19] Dion–Jacobson (DJ) phase,[20,21] and Ruddlesden–Popper (RP) phase.[3,4,22]

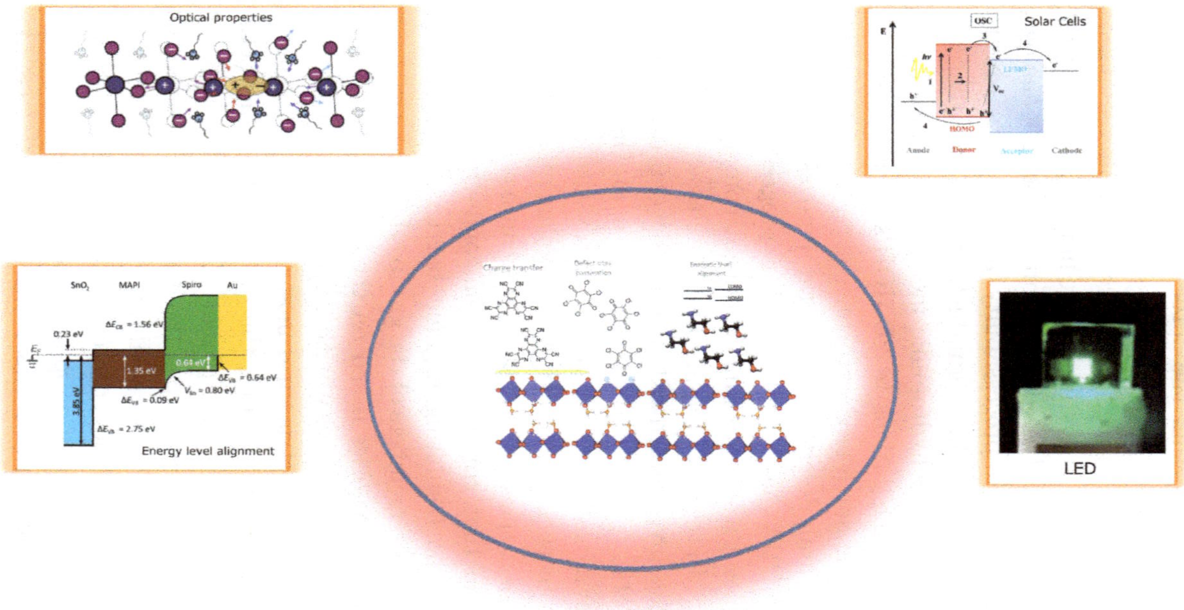

Fig. 1. (Color online) Schematic illustration of the organic molecule/2D-HOIPs heterointerfaces and their practical applications. The energy level alignment is reproduced with permission from Ref. 28. Copyright 2020, Wiley-VCH. The exciton representation is reproduced with permission from Ref. 29. Copyright 2019, Elsevier Inc. The charge carriers collection process is reproduced with permission from Ref. 30. Copyright 2017, Elsevier Inc. The green EL device is reproduced with permission from Ref. 31. Copyright 2019, The Royal Society of Chemistry.

The latter one can be ascribed as a slab that is interleaved with cations generally found in the halide perovskites (HPs) family.[4,23] Structurally, the chemical compound of the perovskite can be presented as $L_2A_{n-1}B_nX_{3n+1}$, where L corresponds to a large cation, A is a small cation (for example, Cs^+ and MA^+), B is a divalent metal cation (typically Pb^{2+} and Sn^{2+}) and X is a halide. The large variety of such hybrid perovskites promotes a fine tuning of their photophysical and electronic properties across solar cells, light-emitting diodes (LEDs), and field effect transistor applications. Further descriptions on the structure–property relationship of the emergent 2D materials have been comprehensively reported elsewhere.[1–8,24–27]

This review summarizes recent literature on the role of organic semiconductors as the electronic dopants and trap passivation agents of hybrid perovskite absorbers, and their corresponding hetero-interfaces are illustrated in Fig. 1. In particular, Sec. 2 presents a short discussion on the fundamental principles of the interfacial nature of perovskite absorbers and photoemission spectroscopy (PES) as surface-sensitive technique to analyze the chemical and electronic properties of 2D-HOIPs surfaces and interfaces. We carry on by reviewing recent progresses on the application of organic molecular structures in engineering the interfacial electronic structure of 2D-HOIPs. In Sec. 3, we review several ideas to tailor the optical properties of 2D-HOIPs upon the addition of molecular materials. Section 4 summarizes the defect passivation strategy to enhance the performance of photovoltaics and LEDs. In the end, Sec. 5 includes a summary and perspectives.

2. Turning the Knob of Electrical Control in Organic-Perovskite Heterointerfaces

In a similar fashion as the conventional semiconductor crystals, several types of defects in ABX_3 perovskite system are commonly identified from the intrinsic to extrinsic surface defects. The surrounding environment or unsaturated surface bonds can be the deciding factor in determining hybrid perovskite as active material in solar-based devices. Theoretically, Yin *et al.* postulated three types of possible intrinsic point defects across $MAPbI_3$, namely the atomic constituent vacancies (MA or Pb or I), interstitials [I interstitials (I_i) or MA interstitials (MA_i)], and substitutions.[32] The third case covers anti-site substitutions [MA-on-Pb substitutions (MA_{Pb}), Pb-on-MA substitutions (Pb_{MA}), and MA-on-I substitutions (MA_I)]. These defect sites exhibit very shallow transition energy levels near the band edges.[32,33] On the other hand, the implication of chemical composition within the $MAPbI_3$ led to deep-level defects, such as I-on-MA substitutions (I_{MA}), Pb-on-I substitutions (Pb_I), I-on-Pb substitutions (I_{Pb}), and Pb interstitials (Pb_i), which have high formation energies. These in turn are responsible for the existence of the non-radiative recombination centers in perovskite under Pb-poor (I-rich) or Pb-rich (I-poor) conditions.[34–36]

In regard to 2D-HOIPs, the origin of such inefficient radiative recombination is not yet fully understood. The presence of fast exciton quenching at room temperature led to high trap density that promote such undesired event.[15,37,38] To shed some light in this subject, Liu *et al.* discussed the defect properties and carrier recombination dynamics of the $BA_2MA_3Pb_4I_{13}$ perovskite film and device. It turns out that the defects in 2D perovskite are mainly governed by the deep-level defects with an activation energy (E_a) of 0.398 eV. This value is much higher than that of 3D perovskite (0.075 eV).[39] Ghosh *et al.* described that the non-radiative carrier recombination events affected by the electron–phonon interactions mediated from the structural fluctuations and instantaneous charge localization in the RP and DJ phases two-dimensional HP materials.[40] Several interesting reviews summarized the current knowledge of the morphology and electronic structure of defects across hybrid perovskite materials at atomic-level resolution using versatile technique such as scanning probe microscopy.[41–43] Based on these aforementioned examples, the current advancement of the 2D-HOIPs research is mostly oriented to tailor their diverse chemical composition toward the realization of hybrid-based devices with better moisture and illumination stability.

A closer topological impact of the solution-processed hybrid perovskites thin films is mainly dominated by the grainy and polycrystalline networks of the surface. Thus, it creates surface imperfections such as intrinsic (or native) point defects that promote the charge carrier entrapment, thereby reducing the device performance. For example, Fig. 2(a) shows the implication of halogen-terminating surface establishing undesired traps caused by the insufficient local stoichiometric composition and improper surface bonding (left). This undesired condition can be improved using the so-called passivator (i.e. surface chemical treatment). In order to acquire an effective passivation treatment (indicated by the blue circle), chemical bonds are interfaced at the Pb-X derived surface orbitals using molecular adlayer that shifts the defect level progressing into or close to the band edges, minimizing the negative impact of such undesired sites. Further comprehensive description on the nature of trap discussion can be found in the recent excellent reviews.[44,46–49]

In Fig. 2(b) (top panel), a representative energy scheme illustrates the relative energetic position of shallow traps that typically show a small energy difference with respect to their conduction/valence band edges ($\Delta E \leq k_B T$). On the other hand, a deep trap energetic position is commonly found inside the semiconductor bandgap ($\Delta E \geq k_B T$). Based on current knowledge, shallow traps mostly bound the free charges displacement through trapping and detrapping processes, while deep traps avoid detrapping and consequently promote the non-radiative recombination pathways. As a result, the imperfections lead to confined electronic states at energy levels which are different than the carrier transport bands, promoting for photogenerated free carriers to energetically fall into this undesired site (Fig. 2(b), bottom panel).

The current challenges of 2D-HOIPs solar cell investigation mainly involve the interplay between the hysteresis effect and long-term stability. It turns out that they are closely related to the quality of the corresponding surfaces by taking into account the atomically smooth interfaces. A convenient and powerful technique to probe the chemical state and electronic properties of surfaces and interfaces of 2D-HOIPs is by using PES. Since the hybrid compounds comprise of different elements, then it is of importance to assess their chemical state to characterize their energetic band offsets upon interfacing with

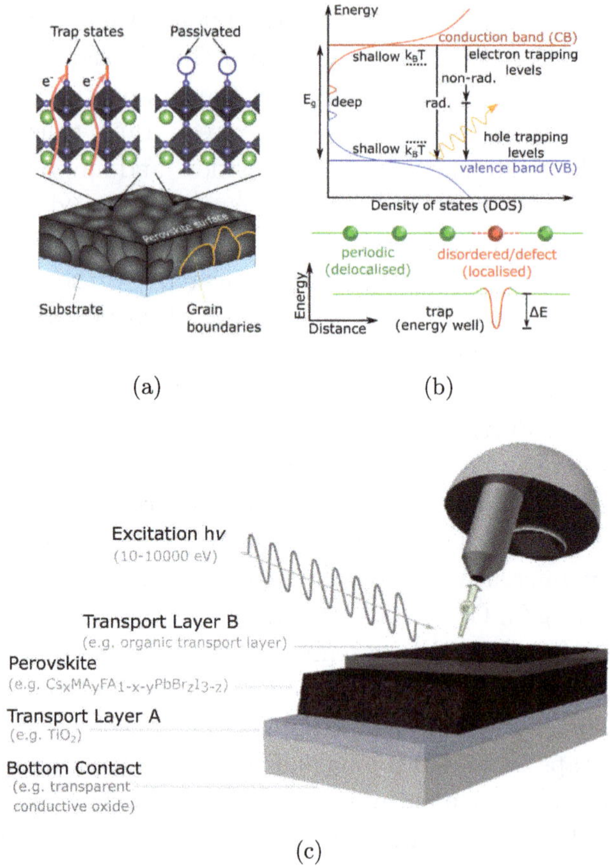

Fig. 2. (Color online) (a) An illustration of parasitic recombination at surface traps (and subsurface traps) that are passivated via additional atoms/molecules (blue circle). This attempt intended to improperly bound orbitals manifesting in the presence of trap states. (b) Top panel: An overview of state density region in a disordered semiconductor. In this scenario, the coexistence of band-to-band radiative (rad.) and non-band-to-band non-radiative (non-rad.) recombination (vertical arrows) is proposed. Bottom panel: Non-periodic disorder and defect in a semiconductor lattice mediates new energetic levels due to charge localization. Reproduced with permission from Ref. 44. Copyright 2020, The Royal Society of Chemistry. (c) PES of HP thin films on top of a transport layer. Reproduced with permission from Ref. 45. Copyright 2020, Wiley-VCH.

the other solar cell components such as hole/electron transporting layers. In a nutshell, PES can be approximated as a non-destructive analytical surface characterization on diverse materials based on the well-known photoelectric effect.[50–52]

In general, the experimental procedure is mainly performed under ultrahigh vacuum condition with typical base pressure in the order of 10^{-10} mbar, to 10^{-11} mbar, as schematically shown in Fig. 2(c). This surface-sensitive technique is versatile to determine the chemical state of specific element according to their specific binding energy region. According to the classical approach, the surface sample is usually impinged with high-energy photons (X-ray non-monochromatic source, a gas discharge lamp, laser excitation, or a synchrotron radiation source) such that the electron bounded to the nucleus are emitted toward the surface. Some of the photo-emitted electrons that successfully escaped with certain kinetic energies are detected by the energy analyzer, and those photo-emitted electrons can be evaluated upon correction by the work function (WF) of spectrometer. Further description on the instrumentation and knowledge of general PES technique can be found elsewhere.[52,53]

Given the intrinsic nature of 2D-HOIPs, multiple components and the intrinsic softness of organic spacer could be quite complex and result in poor electrical conductivity (charging) or photochemical instabilities.[53–55] In addition, de-excitations and charge compensations are considerably quite challenging for the PES experiment. In a recent review, Béchu *et al.* illustrated a comprehensive summary that cover several factors to acquire an accurate PES measurement on these metal HPs system.[45] These photoelectron-based studies comprise of X-ray (XPS),[56–58] ultraviolet (UPS),[57,59–61] inverse (IPES),[61,62] and angle-resolved (ARPES) photoemission spectroscopies.[63–65]

2.1. *The chemistry of energetic alignment processes at the interfaces between the hybrid perovskites and molecular layers*

The emergence of hybrid organic–inorganic perovskite as an active layer for solar cells applications has greatly overcome several issues targeted a reliable device with high power conversion efficiency (PCE) performance. The underlying question in these fields is how to obtain a good interfacial nature when the hybrid perovskite materials are intimately coupled onto charge carrier transporting layers or electrodes. Furthermore, the improvement on the charge separation and reduction of the undesired charge recombination would be the main interest toward the ideal energy level alignment (ELA) at the hybrid interfaces.[46,55,57,66–72]

A simplified electron transfer that facilitates the charge carriers is feasible when the conduction band minimum (CBM) of perovskite materials is higher than that of the electron transfer layers (e.g. TiO_2 or ZnO); and conversely, to facilitate hole transfer, the valence band minimum (VBM) of hybrid perovskite materials should be lower than that of hole transfer materials.[67] One should consider that a simplified energetic model with a perfect vacuum-level alignment across the energy levels of perovskite and electron/hole transfer layers is not entirely true. This hypothesis on the interfacial states, interfacial dipoles and band bending could be influenced by the interplay between charge transfer and ionic migration at the perovskite interface. Thus, it might promote the Fermi level (E_F) pinning across the two layers, and this event promotes the generation of band onsets (or gap states) positions with respect to the vacuum level.

Earlier reviews suggested the full description of the energetic details to minimize the energy loss at the perovskite/charge transport layer heterointerfaces. Several initial cases in 3D-HOIPs hetero-interfaces are formed with four possible profiles, namely staggered gap (type I), straddling gap (type II), reverse staggered gap (type III) with the deep lying of hole transport layer (HTL), and reverse straddling-gap heterointerface in contact with a HTM of large energy gap (type IV).[66] For example, the energy differences between the perovskite VBM and the HTL's highest occupied molecular orbital (HOMO) determine the injection barrier or extraction barrier for holes. In the other energy level, the energy differences between the perovskite CBM and the lowest unoccupied molecular orbital (LUMO) determine the injection barrier or extraction barrier for electrons. The former will limit the open-circuit voltage (V_{OC}) due to thermionic loss for holes, while the latter will strongly affect the fill factor (FF) with a negligible effect on V_{OC}.[73]

2.2. *Implication of the energy mismatch and role of the substrate toward an efficient interfacial perovskite/organic heterointerfaces*

Schulz *et al.*, in 2014, provided a fundamental knowledge on how to minimize the energy loss in the solar cell device, namely, a small energy mismatch of the electronic energy levels in the perovskite layer with respect to the adjacent transport layers is essentially needed.[74] Figure 3(a) outlines the approximate energy diagram across a stack of hybrid perovskite incorporated between the electron collecting cathode (TiO_2) and organic-based HTL on top part of the device. A successful and efficient operational device is achieved with certain criteria in which electrons were extracted from the perovskite into the conduction band of TiO_2, whereas holes are obtained from the HOMO of spiro-OMeTAD. In this work, a combination UPS and IPES is intended to extract the occupied states at the interface and resulted the WF determination as shown in the left panel of Figs. 3(b) ($MAPbI_3$) and 3(c) ($MAPbI_{3-x}Cl_x$), respectively. These findings suggested that the organic layer deposition does not change substantially the secondary electron cutoff, whereas for the $MAPbBr_3$ a downward vacuum-level shift about 0.15 eV was observed at the interface.

In terms of VBM region, all the reported perovskite/organic heterointerface did not affect significantly. Interestingly, the IPES spectra indicate that the CBM positions are about 0.3 eV above E_F in $MAPbI_3$ and 0.4 eV in $MAPbI_{3-x}Cl_x$ (right panel, Figs. 3(b) and 3(c)). These values suggest that the perovskite CBM established a significant energy barrier for electrons and therefore prevented any undesired electron being captured in that particular active layer. In a similar fashion, Chen *et al.* reported that the HOMO position of CuPc is found to be closer to the E_F than the VBM of $CH_3NH_3PbI_3$. Thereby, it promotes hole transfer from $CH_3NH_3PbI_3$ to CuPc.[75] These findings were corroborated from the evolution of C1s core-level spectra (Fig. 3(d)) and the incremental change of barrier energy of the

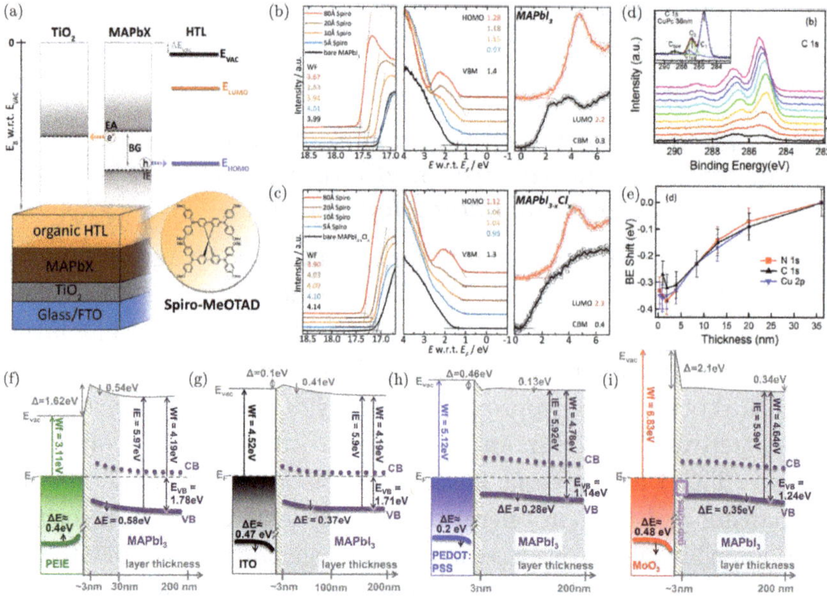

Fig. 3. (Color online) (a) Schematic of device structure and its corresponding energy diagram of a perovskite absorber (MAPbX)-based solar cell with an organic HTL. Left panels: photoemission cut-off for the bare perovskite and as a function of HTL thickness, from which the vacuum level of the film is extracted. The occupied states evolution of $MAPbI_3$ (b) and $MAPbI_{3-x}Cl_x$ (c) upon the incremental change spiro-OMeTAD layers. The WFs of the bare perovskite films are in the range of 4.0–4.1 eV. Middle panels: UPS spectra of the top of occupied states, i.e. VBM for the bare perovskites and HOMO for spiro-OMeTAD/perovskites as a function of HTL thickness. Right panels: IPES spectra of the bottom of unoccupied states of bare perovskites and 80 Å spiro-OMeTAD/perovskites. Reproduced with permission from Ref. 74. Copyright 2018, Royal Society of Chemistry. (d) The evolution of C1s core-level spectra as the function of CuPc molecular thickness. (Inset) Detailed peak fitting of 36 nm CuPc film. (e) Binding energy shift of the three elements in CuPc film. Reproduced with permission from Ref. 75. Copyright 2014, American Institute of Physics. Energetic alignment at the interfaces $MAPbI_3$ thin films on the four different substrates, namely (f) PEIE, (g) ITO, (h) PEDOT:PSS, and (i) MoO_3, respectively. Reproduced with permission from Ref. 76. Copyright 2017, the Nature Publishing Group.

three atomic constituents of CuPc films as a function of molecular thickness (Fig. 3(e)).

To illustrate the role of substrate, Olthof and Meerholz found that tuning the energetic alignment of the hybrid perovskite $MAPbI_3$ on four different substrates PEIE, ITO, PEDOT:PSS, and MoO_3 (Figs. 3(f)–(i)) was accompanied by the interface dipoles and band bending within the perovskite films.[76] In addition, the WF of the perovskite is influenced by the nature of the substrate that can be manipulated from 4.2 eV for the two low WF substrates, such as

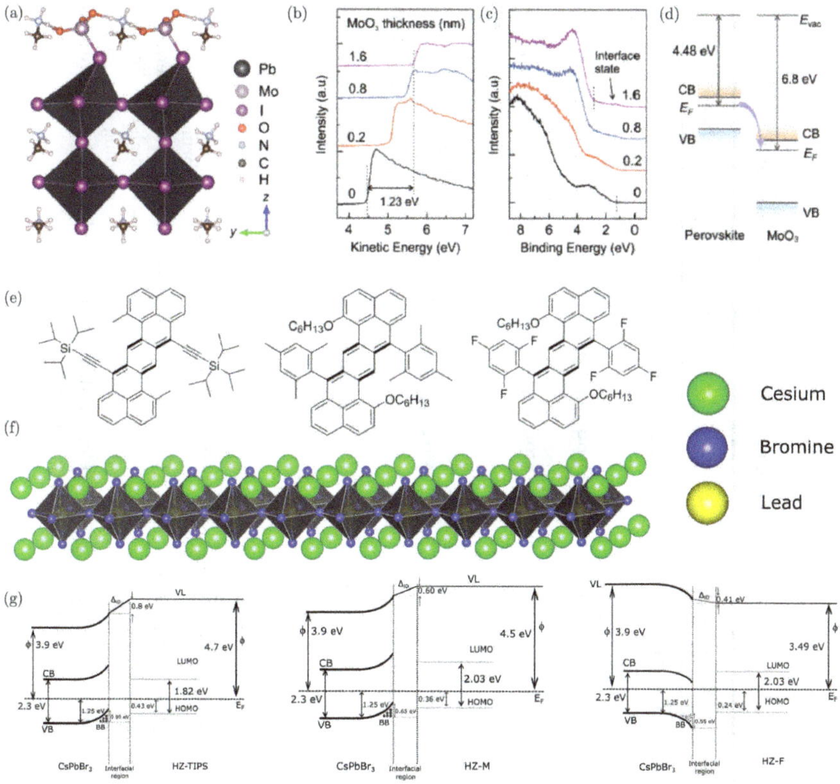

Fig. 4. (Color online) (a) Relaxed geometric structure of the MoO_3-doped $MAPbI_3$ perovskite surface. (b) The secondary electron cut-off regime as a function of MoO_3 coverage. (c) Low binding energy region close to E_F. (d) Schematic energy-level diagram depicted the electron transfer from perovskite to MoO_3. Reproduced with permission from Ref. 77. Copyright 2018, Wiley-VCH. (e) Molecular structure of heptazethrene derivatives HZ-TIPS (left), HZ-M (middle), and HZ-F (right). (f) Representation of cubic structure $CsPbBr_3$. Reproduced with permission from Ref. 78. Copyright 2019, Tsinghua University Press and Springer-Verlag GmbH Germany. (g) The corresponding energy band diagram across the heterointerface of the HZ-TIPS (left), HZ-M (middle), and HZ-F (right). Reproduced with permission from Ref. 79. Copyright 2020, Royal Society of Chemistry.

Table 1. Summary of organic molecules/hybrid perovskites study.

Perovskite	Organic molecules	Techniques	Finding	Refs.
$CH_3NH_3PbI_3$, $CH_3NH_3PbI_{3-x}Cl_x$, and $CH_3NH_3PbBr_3$	Spiro-OMeTAD	UPS and IPES	Band bending of 0.1–0.3 eV in the Spiro-OMeTAD layer	Schulz *et al.*[74]
$CH_3NH_3PbI_3$	CuPc	XPS and UPS	0.3 eV band bending was found for the CuPc layer	Chen *et al.*[75]
$CH_3NH_3PbI_3$	C_{60}	UPS and IPES	A high electron-blocking barrier and only a small loss in hole potential energy of 0.2 eV	Hellmann *et al.*[28] and Schulz *et al.*[80]
$CH_3NH_3PbI_3$	PTCDI-C5	XPS and UPS	The electron extraction with a moderate VBM-HOMO offset of $\sim$ 0.7 eV	Zhang *et al.*[81]
$(FA_{0.83}MA_{0.17})_{0.95}Cs_{0.05}Pb(I_{0.6}Br_{0.4})3$	Indene-C_{60} bisadduct (ICBA)	Current density–voltage $(J-V)$	Passivation by Lewis acid	Lin *et al.*[82]
$FASnI_3$	C_{60}	UPS, IPES, and XPS	Surface treatment with carboxylic acids and bulky ammonium improve the PV performance	Boehm *et al.*[83]
$CH_3NH_3PbI_{3-x}Cl_x$	Thiophene and pyridine	Current density–voltage $(J-V)$	Passivation by Lewis bases through coordinate bonding	Noel *et al.*[84]
$CH_3NH_3PbI_3$	L-α-phosphatidylcholine, choline chloride	Current density–voltage $(J-V)$	Passivation by zwitterions	Zheng *et al.*[85]

$CH_3NH_3PbI_{3-x}Cl_x$ and $CH_3NH_3PbI_3$	HATCN and F6TCNNQ	XPS and UPS	Initial perovskite surface state density promote different interface ELA (variations over 0.5 eV)	Zu *et al.*[86]
$CH_3NH_3PbI_3$	$Co(C_5H_5)_2$	XPS and photoconductivity	*n*-type charge transfer doping by directly depositing the organic small molecule cobaltocene	Perry *et al.*[87]
Ethylammonium (EA), butylammonium (BA), and phenethylammonium (PEA) of PbI_4	HATCN	Current density–voltage $(J-V)$, EL and PL measurements	Organic cation in the 2D perovskite layer plays a critical role in the CT process	Zhao *et al.*[88]
EA_2PbI_4 and PEA_2PbI_4	Ethanolamine (EA, $HOCH_2CH_2NH_3^+$)	Photocurrent measurement	The smooth 2D perovskite is less sensitive to ambient moisture and exhibits a considerably low dark current	Ho *et al.*[89]
A_2PbI_4	Pyrene-$C_4H_8NH_3^+$ and TCNQ or TCNB	Absorption and PL measurements	The incorporation of charge-transfer complexes has a significant influence on the absorption in the visible wavelength region	Van Gompel *et al.*[90]
NAP_2PbI_4	TCBQ	Temperature-dependent ultraviolet–visible absorption spectra	Lowering the exciton binding energy relative to the undoped perovskite	Passarelli *et al.*[91]

(*Continued*)

Table 1. (*Continued*)

Perovskite	Organic molecules	Techniques	Finding	Refs.
$(C_4H_9NH_3)_2$ $(CH_3NH_3)_3Pb_4I_{13}$ ($(BA)_2(MA)_3Pb_4I_{13}$, $n = 4$)	Graphene	Current density–voltage ($J - V$), STM, femtosecond pump–probe spectroscopy	Plane-contacted perovskite and graphene interface presents a lower barrier than gold for charge injection	Leng *et al.*[92]
(HEA)-based quasi-2D perovskite materials ($n = 9$)	Halogenated 4-fluorobenzyl-ammonium (FBA) and 4-chloro-benzylammonium (CBA)	UV–Vis absorption, XPS, current density–voltage ($J - V$)	0.1 CBA-based quasi-2D PSCs achieve a superior PCE of 18.75%	Liu *et al.*[93]

PEIE and ITO, to 4.7 eV for the high WF substrates, like PEDOT: PSS and MoO_3. In addition, the work by Miller *et al.* suggested that the insulating and n-type substrates increase the VBM of $MAPbI_3$ toward E_F, indicating the extent of pinning the conduction band to the Fermi level. In the contrary, the p-type substrates decrease the valence band to Fermi energy difference.[56]

Ou and coworkers have employed transition metallic oxide, MoO_3 as a surface dopant to locally tailor the charge dynamics of $CH_3NH_3PbI_3$ perovskite nanosheets, as schematically shown in Fig. 4(a). A very broad depletion region up to a few micrometers in the perovskite $p-n$ junction was established based on the complementary detail of SECO (Fig. 4(b)) and low binding energy region (Fig. 4(c)). The interface state upon MoO_3 was incrementally introduced to the perovskite in combination with the high WF of MoO_3, which then led to electron transfer from the perovskite to MoO_3 at the MoO_3/perovskite interface, as schematically depicted in Fig. 4(d). For comparison, Arramel *et al.* used several zethrenes (Fig. 4(e)) as molecular dopants in which n-type or p-type doping were realized by coating them onto all-inorganic perovskite absorber such as $CsPbBr_3$ (crystal structure is shown in Fig. 4(f)).[78] The WF modulation at the heterointerface is realized such that the efficient electron/hole transfer can be regulated via chemical design as illustrated in Fig. 4(g).[79] In summary, a collection of several papers that put an emphasis to tune the interfacial electronic structure on 3D- and 2D-HOIPs is presented in Table 1.

3. Functionalization of Molecules in Tuning the Optical Properties of Perovskites

The optical properties in 2D-HOIP materials are important features in determining the characteristics desirable in high performance photovoltaic and optoelectronic devices.[23] For photovoltaics, it is important to highlight the absorption of the materials while for the optoelectronics, we also need to consider the emission wavelengths and lifetimes. In principle, for 2D-HOIP materials, there are already many reports in photovoltaic applications with the respect of their stability and efficiency,[94–97] while for perovskite LEDs (PeLEDs), the attractiveness is the higher quantum efficiencies than those of 3D-HOIPs.[98] Although some optical properties of those 2D-HOIP materials are already beyond their siblings of 3D-HOIP materials, there is still room for improvements toward better devices. In this section, we discuss how the functionalization of molecules can be attractive to enhance the optical characteristics of 2D materials.

3.1. *Tuning the reflection, the transmission, and the absorption*

The important goal for tuning of all these optical properties is to optimize how photovoltaic devices can absorb more lights in such perovskite thin materials. The common methods are by introducing anti-reflection coating[99] or nanostructures[100] on the top of the perovskite photovoltaic films. However, there is currently no report on the anti-reflection coating made by functionalization of atomic-thick molecules. Most of the cases, tuning of the absorption is performed through controlling the spacer organic cations through adding molecules in RP phase,[4,96] see Fig. 5. In this method, we can add more overlaps with the solar spectrum creating more infrared photons to be absorbed by the solar cell.

3.2. *Tuning the emission and the energy transfer*

Unlike 3D ones, 2D-HOIP materials have various profiles in the emission. Figures 5(a) and 5(b) exhibit the photoluminescence (PL) spectra and time-resolved PL (TRPL) decay curves from various perovskite bromides and iodides, respectively, while they are related to their structures in Fig. 5(c). Perovskite bromides, PL spectra shown in Fig. 5(a), have narrow and broad emission features. The broad emission spectra shown in (API)$PbBr_4$, (AEIU)$PbBr_4$, and (EDBE)$PbBr_4$ originate from their (110)-corrugated crystal structures, which is commonly found in the white-emitted perovskites.[101–103] We note that those structures also have very large Stokes shift as large as 1 eV.[104] Such broad emission is related to self-trapped exciton. Another structure with (100)-flat $(CHA)_2PbBr_4$ has

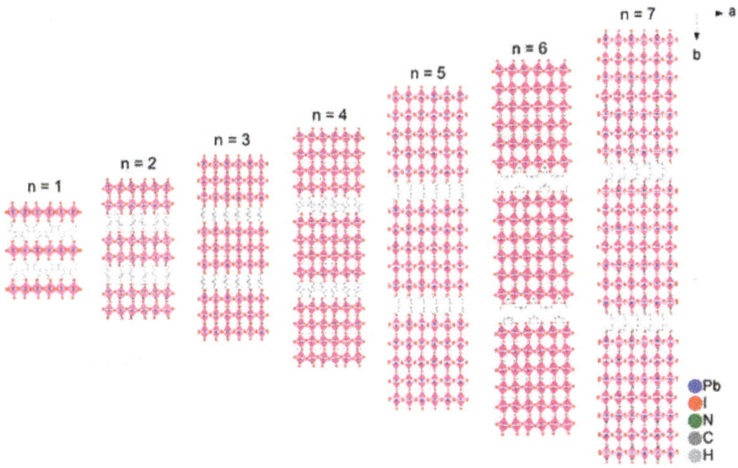

(a)

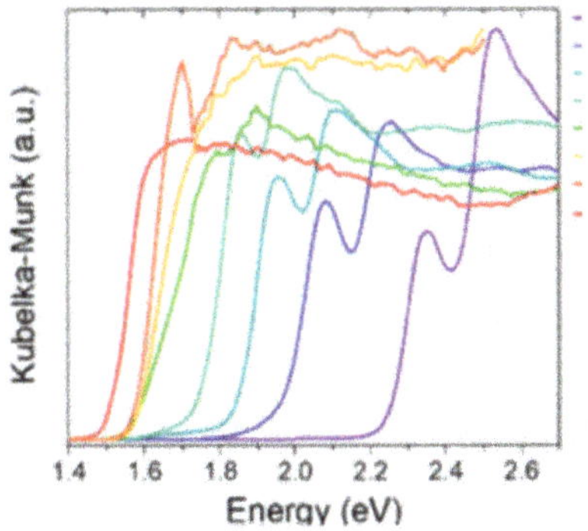

(b)

Fig. 5. (Color online) (a) Crystal structures and (b) optical absorption spectra of the RP phase $(BA)_2(MA)_{n-1}Pb_nI_{3n+1}$ ($n = 1-7$). Figures reproduced from Ref. 4. Copyright 2018, American Chemical Society.

two emission bands. Besides that at 520 nm related with self-trapped exciton,[11] it has a narrow band at 425 nm and is assigned to the free excitonic emission. $(PEA)_2PbBr_4$ has similar relatively bulky-ring structures as $(CHA)_2PbBr_4$ but it has much more contributions from the narrow band emission. Structures with a linear alkyl chain like $(DAB)_2PbBr_4$, $(BA)_2PbBr_4$, and $(OA)_2PbBr_4$ have the same narrow and broad emission patterns, especially the narrower self-trapped exciton emission in comparison to those of (110)-corrugated and (100)-flat crystal structures.

For perovskite iodides, the emission seems similar to their bromide siblings. In addition, we observe evidence of dual bandgap structures, which is characterized by double peaks, in $(BA)_2PbBr_4$ and $(BA)_2PbI_4$ as it was also previously reported in the literature.[105] Decay curves in Fig. 6(b) were fitted with three-exponential functions and the average decay times and components are shown in Table 2.

Most perovskite materials show a fast decay, as fast as 0.6 ns, except $(EDBE)PbBr_4$, whose decay is up to 12.4 ns. The slower decay for $(EDBE)PbBr_4$ is related to the substantially long lifetime of the exciton self-trapping process which also explains its broad PL spectrum. In addition, the lifetimes of perovskite iodides are much faster in comparison with those of perovskite bromides. Under similar excitation power, it turns out that the bromide-based compounds have shown a brighter emission compared to the iodide counterparts. The latter has a large contribution from the fast non-radiative decay processes, consequently led to a PL quenching and fast decay time.

Adding functionalized molecules in those perovskite materials is useful in terms of tuning the emission spectra. Such method has been shown previously for tuning the absorption via controlling the spacer organic cation.[4,96] Besides tuning the emission, we also expect that through the charge and energy transfers, we may also optimize the emission into free excitonic path rather than the self-trapped path due to its fast nanosecond emission. However, beside 3D-HOIP materials,[78,79] there is no report for observation of charge and energy transfers from molecule functionalization of 2D-HOIP materials.

Figure 4 shows the effect of molecular functionalization of all-inorganic perovskite materials to the emission.[78] Previously, we used kinetically blocked heptazethrene triisopropylsilyl ethynylene (HZ-TIPS) derivatives and its homologue octazethrene (OZ-TIPS) illustrated in Fig. 4(e) as the doping molecules for $CsPbBr_3$ crystals with its structure depicted in Fig. 4(f).

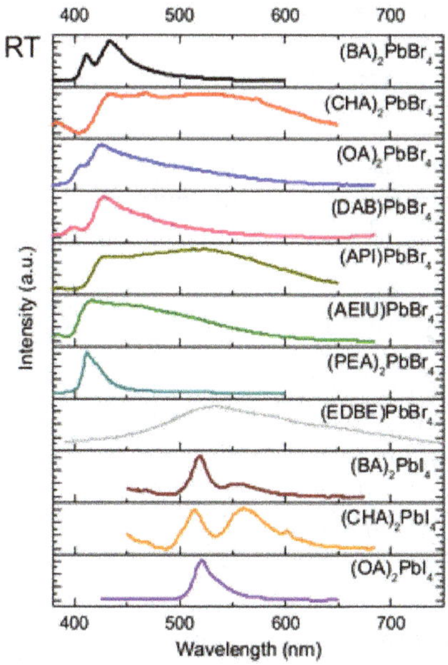

(a)

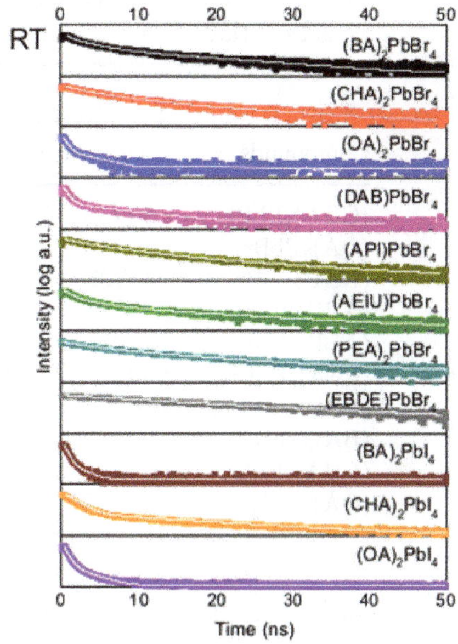

(b)

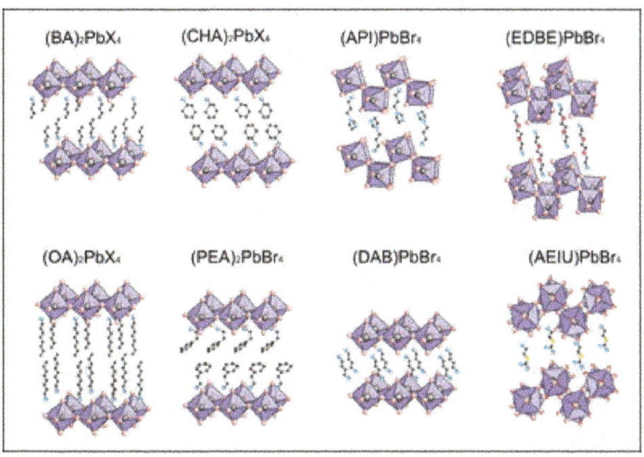

(c)

Fig. 6. (Color online) (a) PL spectra, (b) TRPL decay curves, and (c) schematic representation structures of the 2D perovskite bromides and iodides. Figures reproduced from Ref. 101. Copyright 2020, American Chemical Society.

Table 2. Decay components, their averages and their contributions from three-exponential fits of TRPL.

Compounds	Results of fitting: Decay components (ns (%))	Results of fitting: Average (ns)
$(CHA)_2PbBr_4$	1.0 (13%), 2.9 (68%), 9.2 (19%)	3.9
$(OA)_2PbBr_4$	0.3 (43%), 0.4 (43%), 2.5 (14%)	0.7
$(DAB)PbBr_4$	0.4 (43%), 0.5 (44%), 4.9 (13%)	1.0
$(API)PbBr_4$	1.5 (26%), 4.0 (59%), 9.6 (15%)	4.2
$(AEIU)PbBr_4$	0.7 (46%), 2.5 (37%), 10.3 (17%)	3.0
$(PEA)_2PbBr_4$	2.6 (37%), 6.9 (54%), 19.2 (9%)	6.4
$(EDBE)PbBr_4$	0.3 (2%), 0.4 (2%), 12.9 (96%)	12.4
$(BA)_2PbI_4$	0.3 (3%), 0.5 (89%), 1.0 (8%)	0.5
$(CHA)_2PbI_4$	0.3 (9%), 1.0 (85%), 9.7 (6%)	1.4
$(OA)_2PbI_4$	0.4 (36%), 0.6 (59%), 1.6 (5%)	0.6

PL spectra of $CsPbBr_3$ crystals before and after doping with HZ-TIPS are presented in Fig. 7(a). The HZ-TIPS molecular doping on $CsPbBr_3$ shows three emission peak (424, 446, and 473 nm) that can be associated with the signature of HZ-TIPS. These additional peaks were not observed when we measured the bare perovskite crystal. However, we found that the perovskite PL intensity decreases by 10 folds and the excitonic emission of the perovskite emission band is shifted to shorter emission wavelength of about 520 nm with the appearance of two peaks. We attribute the blue-shifted emission peak in thermally deposited HZ-TIPS sample is due to highly efficient charge-transfer excited state that provides efficient intramolecular π–π stacking interactions.[78] To explain the PL intensity decrease, the TRPL decay curves for both bare and doped crystals are shown in Figs. 7(b) and 7(c), respectively. At a glance, the decay curve from doped crystals yields faster decay times in comparison to that from bare crystals. From the two-exponential fits,[78] the average lifetimes for both bare and doped crystals are reported to be 8.3 and 3.6 ns, respectively. The emission in doped crystal is much faster by 2.3 times than that of bare crystal showing a consistent PL quenching upon HZ-TIPS molecular deposition. In the same paper,[78] we also noticed that the OZ-TIPS molecule functionalization is responsible for the

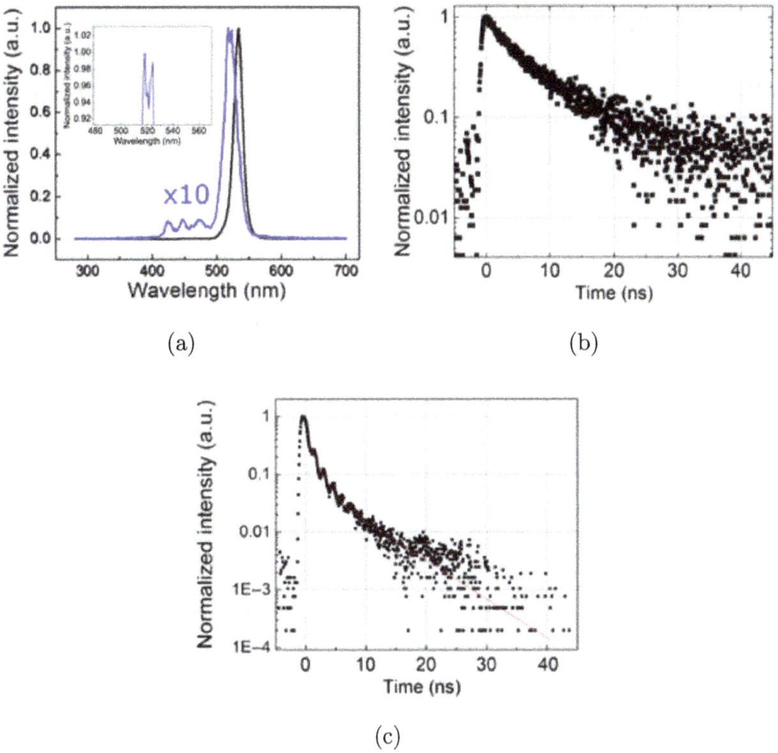

Fig. 7. (Color online) (a) PL emission spectra of bare (black solid line) versus HZ-TIPS doped $CsPbBr_3$ (blue solid line). The latter curvature is multiplied by 10 for the sake of guidance for the eye. The inset indicates the enlargement of PL peak that corresponds to the splitting of emission peak. TRPL decay of (b) bare $CsPbBr_3$ and (c) its corresponding HZ-TIPS doped sample. A fitting curve is indicated by the solid red line. Reproduced with permission from Ref. 78. Copyright 2018, Tsinghua University Press and Springer-Verlag GmbH Germany.

decrease of the emission intensity and the lifetime by factors of 12 and 2.9 times, respectively. Using data from XPS, UPS, and NEXAFS measurements, we later conclude that this quenching is related to the charge-transfer event at the interface and this may affect the excitonic emission at the interface.[78,79] However, those studies are still in the infancy level and therefore further investigations and improvements are needed so that we can suppress the non-radiative recombination in the molecular functionalization while at the same time, we can optimize the charge transfer for the fast radiation recombination route. Such faster emission properties will be useful for the fast scintillator in high-resolution positron emission tomography,[101] particularly for slow response materials,[106] and the fast-modulated LED of light fidelity in telecommunication application.[107]

4. Applications

Perovskite materials have excellent optoelectronic properties for the applications of solar cells and LEDs.[108,109] In this section, the applications of HOIPs, particularly 2D perovskites, are discussed in solar cells for enhancing light absorption and improving photo-generated carrier extraction for solar cells, and in LEDs for improving extraction efficiencies and tuning the color.

4.1. *Enhancing light absorption in photovoltaic cells*

In the past decade, perovskite materials particularly for organic–inorganic halides have attracted much interest in photovoltaic applications due to their excellent optical and electrical properties.[110] As reducing the perovskite dimension from 3D to 2D, unique properties appear due to quantum confinement effect and the enhancement of stability offers new approaches to stabilizing perovskite-based photovoltaic devices.[12,96,111,112] Figure 8 illustrates the crystalline structure of 2D perovskites, mixed dimensionality, and 3D perovskites.[113] In 2D perovskites, the optoelectronic properties are greatly impacted by its tunable structure of organic layers based on the organic spacer specifications such as alkyl chain length, insertion of π-conjugated segment to facilitate charge transport, and ammonium dications, while its superior air stability is improved by the hydrophobic nature of organic spacer and the van der Waals force interactions between the adjacent layers.[114] The pioneering work of hybrid solar cells based on $PEA_2MA_2Pb_3I_{10}$ has enabled the fabrication under ambient humidity levels and enhanced moisture stability of the perovskite film.[115] Moreover, 2D $BA_2MA_{n-1}Pb_nI_{3n+1}$ ($n = 1, 2, 3$, and 4) perovskites were reported to have the bandgaps of the series decreasing with increasing n values, from 2.24 eV for $n = 1$ to 1.52 eV

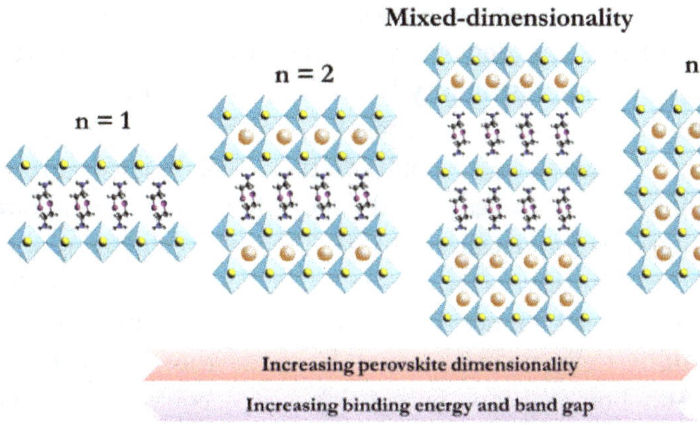

Fig. 8. (Color online) Illustration of crystalline structure of 2D perovskites ($n = 1$ and 2, where n represents the layered number of inorganic sheets), mixed dimensionality perovskites and 3D perovskites ($n = \infty$). Reproduced with permission from Ref. 113. Copyright 2016, Wiley-VCH.

for $n = \infty$ ($MAPbI_3$), while the semi-2D perovskite films displayed an ultrahigh surface coverage due to the orientation of $[Pb_3I_{n+1}]^-$ layers perpendicular to the substrates with the maximum PCE of 4.02% featuring V_{OC} of 929 mV and short-circuit current density (J_{SC}) of 9.42 mA/cm^2 for $BA_2MA_2Pb_3I_{10}$($n = 3$) remained its performance upon extensive high-humidity exposure.[97] To overcome the insulating properties of organic cations that hindered the out-of-plane charge transport, near-single-crystalline quality $BA_2MA_3Pb_4I_{13}$ layered perovskite was grown through hot-casting technique promoting a preferred out-of-plane alignment toward the contact of the planar solar cells, and resulted in an efficient charge transport with PCE up to 12.52% without hysteresis.[12] In comparison to films casted at room temperature, the hot casted 2D perovskite films showed the improved crystal quality of substantially larger grains (400 nm) and much lower pinholes, and additionally the absence of Urbach tails in their optical absorption spectrum accompanied by a small Stokes shift. In addition, the prominent feature of the absorption and PL spectral signature dictates the perovskite films with an intrinsic direct-bandgap semiconductor, excellent crystallinity, low trap concentration, and little disorder-induced density of states in the bandgap.[12] Using the short branched chain spacer cations (iso-BA^+) compared to conventional linear n-BA^+ has demonstrated significant impact in their optical absorption and crystallinity.[116] In fact, the performance of 2D perovskite still lags behind 3D counterpart, particularly with the low V_{OC} indicating the presence of non-radiative recombination process caused by inhomogeneous energy landscape, which leads to undesired energy loss, imposes band tail states and consequently reduces the evolution of minority carrier transport.[117] By using (aminomethyl) piperidinium quasi-two-dimensional (quasi-2D) perovskites and controlling carefully their crystallization kinetics, the lowest energy distribution and homogeneous energy landscape could be achieved with the significant reduction of the Shockley–Read–Hall recombination and suppression of the quasi-Fermi-level splitting, resulting in excellent PCE of 16.53% and V_{OC} of 1.21 V.[117] For further enhancement of optical absorption, tandem solar cell architecture offers the opportunity to match the spectral distribution of solar light, in which the wider bandgap of semiconductor is suitable as higher bandgap absorber and vice versa. A mixed Sn–Pb narrow-bandgap perovskite anchored with ultrathin 2D layer for defect passivation at the interface and grain boundaries, and the combination of this passivated narrow bandgap with a wide-bandgap perovskite in two terminal tandem solar cells, could achieve the stabilized PCE of 23.5%.[118] Moreover, the employment of wide-bandgap perovskite composed of 2D/3D perovskite heterostructure in four terminal tandem configurations, i.e. perovskite/c-Si and perovskite/CIGS solar cells showed the stabilized PCEs of up to 25.7% and 25.0%, respectively.[119] Further utilization of 2D/3D perovskite heterostructure for better solar cells is also described in the following section.

4.2. *Improving photo-generated carrier extraction in photovoltaic cells*

Interfacing 2D perovskites with an organic semiconductor has been found to form donor/acceptor charge-transfer states at this heterojunction, which dissociates excitons in 2D perovskites and thus increases the photocurrent production since changing the organic cation size could govern the charge-transfer event.[88] For efficient solar cells, the presence of long-lived charge carriers is needed so that the charges can be extracted before significant recombination occurs, in which the diffusion length should be longer than the required thickness for complete light absorption.[120] To enhance the charge transport, using small propane-1,3-diammonium (PDA) cation in 2D perovskite could reduce the interlayer distance in 2D perovskite and thus yield the PCE of 13.0%.[121] Recently, in-plane carrier diffusion length in 2D $(BA)_2PbI_4$ perovskite has been determined to be 1.82 μm.[122] Furthermore, doping 2D $BA_2MA_3Pb_4I_{13}$ perovskite with Cs by replacing MA^+ with Cs^+ has been proposed to improve the optoelectronic properties in terms of trap density, charge-carrier mobility, and charge-transfer kinetics, in which the PCE could increase from 12.3% (without Cs^+) to 13.7% (with 5% Cs^+).[94] Controlling the crystallization kinetics by appropriate partial substitution of MA cation with formamidinium (FA) cation in quasi-2D $BA_2MA_3Pb_4I_{13}$ perovskite could enlarge the dimensionality of the grain such that reduce the non-orientated phase, leading to less non-radiative recombination centers.[123] Furthermore, interfaces also dominate the charge–charge recombination and initiate the degradation process, thus affecting the photovoltaic performance. Therefore, modification on perovskite layer surface is promising to minimize the implication of interfacial recombination and improve its stability. The donor-π-acceptor in (D-π-A) organic molecule can be used to modulate the electron density distribution interfacing with a perovskite solar cells toward an efficient defect-passivation treatment.[124] Phenethylammonium iodide (PEAI) treatment of $MAPbI_3$ perovskite reduces the surface defects and suppressed the non-radiative recombination resulting in higher photovoltaic performances with 23.56% efficiency achieved.[125] On the contrary, PEAI treatment at the $MAPbI_3/C_{60}$ interface showed the decrease in electron affinity and ionization energy, thus hindering electron transfer from $MAPbI_3$ to C_{60} electron transport layer (ETL) and decreasing the performance.[126] This surface energetic change upon ligand application indicates that an appropriate selection of surface ligand is necessary for favorable energy alignment based on where the treatment will be carried out either at perovskite/electron transfer layer or perovskite/hole transfer layer interface.[126] At the interface between $FASnI_3$ and C_{60}, favorable energy landscape for electron transfer from $FASnI_3$ to C_{60} has been reported with and without the surface ligand, likely due to iodide diffusion out of $FASnI_3$ into C_{60} resulting in n-doping of C_{60}, while surface ligand would greatly improve the stability of $FASnI_3$ solar cells.[83] In the interface engineering, doped charge-transporting layers in planar perovskite solar cells have been regarded as an efficient pathway to enhance photovoltaic performances.[127] In addition to surface ligand, 2D perovskites have also been considered as passivating materials in enhancing the performance and stability of lead HP solar cells. With 2D/3D $(HOOC(CH_2)_4NH_3)_2PbI_4/MAPbI_3$ perovskite junction interface engineering and aminovaleric acid iodide ($HOOC(CH_2)_4NH_3I$, AVAI) as organic precursor, one-year stable perovskite device could be achieved and thus opens the way for the realization of highly efficient and stable perovskite solar cells for extensive practical deployment, shown in Figs. 9(a) and 9(b).[128] By spin-coating a phenylformamidinium iodide (PFAI) solution on the FA-based perovskite film surface of $Cs_{0.05}Rb_{0.05}(FA_{0.83}MA_{0.17})Pb(I_{0.95}Br_{0.05})_3$, the unwanted charge recombination and surface defect can be effectively suppressed by the mixed-cation 2D perovskites that serve as "patches" on a solar cell surface, thus resulting in the improved photovoltaic performance and stability.[130] Mixed 2D/3D hybrid perovskite system offers a promising pathway to obtain chemical/photochemical stability compared to 3D counterpart while gathering the significant properties of 3D in terms of bandgap tuning through number of layers.[131] As shown in Figs. 9(c)–(e), further introduction of molecule additive (i.e. 2,3,5,6-tetrafluoro-7,7,8,8-tetracyanoquinodimethane or F4TCNQ) into the 2D layer in 3D/2D perovskite solar cells, facilitated charge extraction and reduced charge recombination with the maximum PCE of 20.1% compared with 18.2% for unmodified 2D capping layers.[129] Here, the strong electron-acceptor molecule such as F4TCNQ enables it to mediate partial charge transfer with the perovskite layer to provide a passivation effect, promoting an enlargement in the domain size. 2D/3D perovskite heterostructure on top of the 3D $MA_{1-y}FA_yPbI_{3-x}Cl_x$

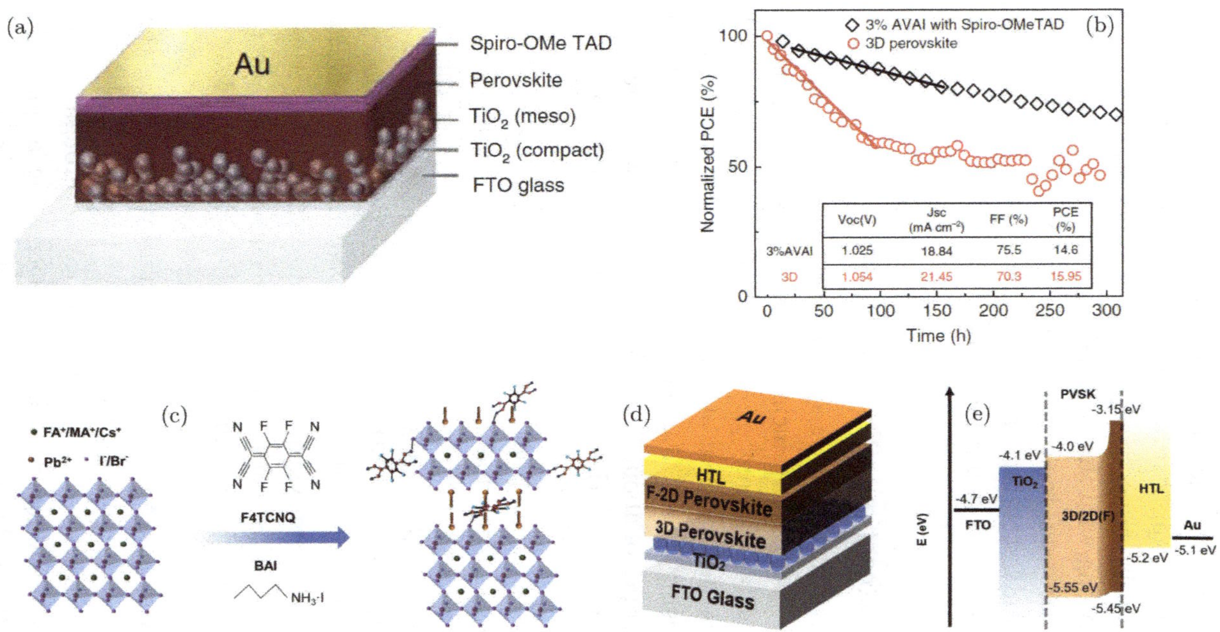

Fig. 9. (Color online) (a) A layout of 2D/3D mesoporous solar cell device and (b) the module stability test of 2D/3D perovskite with 3% AVAI compared to the standard 3D $CH_3NH_3PbI_3$ under AM 1.5 G illumination and stabilized temperature of 45°C with the champion device parameters listed in the inset. Reproduced with permission from Ref. 128. Copyright 2017, the Nature Publishing Group. (c) Schematics of 3D/F4TCNQ-2D perovskite deposition, (d) solar cell device architecture, and (e) corresponding energy-level landscape. Reproduced with permission from Ref. 129. Copyright 2020, American Chemical Society.

perovskite with modulated diffusion passivation by introducing PEAI and *N*, *N*-dimethylformamide (DMF) as an additive could achieve a high PCE of up to 22.18% due to a significant increase of V_{OC} and FF.[132] Some reported applications of organic-hybrid perovskite devices including their respective photovoltaic performances are shown in Table 3. Along with the material and device developments, advanced characterization tools have also become necessary to provide further understanding and to overcome the existing challenges on the road to commercialization.[133]

4.3. *Improving extraction efficiencies in LED*

Driven by an intrinsically diverse structure, chemical, optical properties of hybrid perovskites including their low-cost fabrication and solution processability, many groups have shown a great deal of attention toward the development of PeLEDs and research activities are rapidly increasing.[134–138] Particularly, 2D perovskites have exciton binding energies of about 300 meV. Not only they are greater than those of 3D perovskites, the amount is also an advantage in the field of optoelectronics.[5,139,140]

HP platelets possess good optical and electrical properties suitable for LED applications. Yet, due to the challenge in producing a uniform film, only a small number of reports recorded their role as LED materials. Ling *et al.* used $MAPbBr_3$ nanoplatelets to create green LEDs.[141] The 2D $FAPbBr_3$ dielectric quantum well was investigated by Kumar *et al.*, who managed to improve the photoluminescence quantum yield (PLQY) up to 92%.[142] The same material also yielded LEDs with a maximum current efficiency of 13.02 cd A^{-1} and external quantum efficiency (EQE) of 3.04%.[142] A success in creating blue LEDs has also been reported. Yang *et al.* made blue LEDs with sharp emission at 480 nm by controlling the thickness of $CsPbBr_3$ with high precision at the atomic scale.[143]

One way to improve the LED's performance is through multi-dimensional HPs. By adding more layers and creating a quasi-2D perovskites, the EQE can be improved. The multiple quantum wells in the quasi-2D perovskites facilitate exciton formation and lower the

Table 3. List of some reported applications in photovoltaic cells.

Perovskite	Device structure	Finding	Performances				Refs.
			J_{SC} (mA/cm^2)	V_{OC} (V)	FF (%)	PCE (%)	
2D BA_2PbI_4	ITO/poly-TPD/ Perovskite/HATCN/ BCP/Al	Donor/acceptor charge-transfer states at perovskite/organic heterojunctions	0.11	0.37	31.3	0.012	Zhao *et al.*[88]
2D $PEA_2MA_2Pb_3I_{10}$	FTO/TiO_2/Perovskite/ Spiro-OMeTAD/Au	Enhanced moisture stabiliy compared with 3D counterpart	11.5	1.21	50	7.02	Smith *et al.*[115]
2D (iso-BA)$_2$(MA)$_3$Pb$_4$I$_{13}$ and (n-BA)$_2$(MA)$_3$Pb$_4$I$_{13}$	FTO/C_{60}/2D Perovskite/ Spiro-OMeTAD/Au	Short branched-chain spacer cations enhancing optical absorption and crystallinity	14.87	1.14	52.04	8.82	Chen *et al.*[116]
2D PDAMA$_{n-1}$Pb$_n$I$_{3n+1}$ ($n = 2$, 3, and 4) and $BA_2MA_3Pb_4I_{13}$	ITO/PEDOT:PSS/ Perovskite/C_{60}/BCP/ Ag	Short interlayer distance enhancing charge transport	18.0	0.97	74.3	13.0	Ma *et al.*[121]
Quasi-2D (BA)$_2$(MA,FA)$_4$Pb$_4$I$_{13}$	ITO/PEDOT:PSS/ Perovskite/PCBM/ BCP/Ag	Controlled crystallization kinetics by appropriate partial substitution of MA with FA cation	18.12	0.999	70.79	12.81	Zhou *et al.*[123]
$MAPbI_3$ passivated with D-π-A molecule having N,N-dibutylaminophenyl unit on the donor part	FTO/NiO$_x$/Perovskite : D-π-A molecule/ PCBM/BCP/Ag	Importance of electronic configuration of passivation molecules	22.69	1.146	78.55	20.43	Wu *et al.*[124]

(Continued)

Table 3. (*Continued*)

Perovskite	Device structure	Finding	Performances				Refs.
			J_{SC} (mA/cm^2)	V_{OC} (V)	FF (%)	PCE (%)	
$MAPbI_3$	ITO/SnO_2/Perovskite/PEAI/Spiro-OMeTAD/Au	Surface passivation by PEAI treatment on $MAPbI_3$	24.9	1.16	81.4	23.56	Jiang *et al.*[125]
$MAPbI_3$ treated by various surface ligands (PEAI, PTS, PPA, BrBA, *p*-TA, TP, AnCl, OPA, OCA, OT, OAI)	ITO/PTAA/$MAPbI_3$/Surface ligand/C_{60}/BCP/Al	Shift of ionization energy and electron affinity by surface ligands	21.29	1.047	76.3	17.03	Park *et al.*[126]
2D/3D $(HOOC(CH_2)_4NH_3)_2PbI_4/MAPbI_3$	FTO/c-TiO_2/m-TiO_2/Perovskite/Spiro-OMeTAD/Au	Interface engineering at perovskite junction towards long-term device stability	18.84	1.025	75.5	14.6	Grancini *et al.*[128]
2D $Cs_{0.05}Rb_{0.05}(FA_{0.83}MA_{0.17})Pb(I_{0.95}Br_{0.05})_3$ with PFAI treatment	FTO/SnO_2/FA-based perovskite/PFAI-induced layer/Spiro-OMeTAD/Ag	Surface passivation of mixed PFA and FA cations in 2D FA-based perovskite	24.11	1.098	80.8	21.38	Yangi *et al.*[130]
3D $Cs_{0.05}FA_{0.79}MA_{0.16}PbI_{2.49}Br_{0.51}$/ 2D $BA_2(FA/MA/Cs)_{n-1}Pb_n(I/Br)_{3n+1}$ with F4TCNQ additive	FTO/TiO_2/m-TiO_2/3D Perovskite/F-2D Perovskite/Spiro-OMeTAD/Au	Capping 2D perovskite layer with F4TCNQ molecule	22.8	1.16	76.0	20.1	Sun *et al.*[129]
2D/3D $MA_{1-y}FA_yPbI_{3-x}Cl_x$ with PEAI and DMF additives	ITO/NiO_x/Perovskite/PCBM/BCP/Ag	Modulating diffusion passivation	23.6	1.12	83.6	22.18	Zhou *et al.*[132]

possibility of exciton dissociation. The EQE of quasi-2D PeLEDs have achieved around 10% and highest EQE of 11.7% in near-infrared region was obtained.[144,145]

HPs have a promising functionality to emit a photon that can be extensively improved via surface passivation. The effect of trioctylphosphine oxide (TOPO) deposition, spin-coated onto the perovskite layer, has been proven to be able to significantly increase the PLQY of an $n = 3$ composition of $PEA_2(FAPbBr_2)_{n-1}PbBr_4$ perovskite from 57.3% to 73.8%. The average fluorescence lifetime of the $n = 3$ composition film is also increased from 0.17 μs to 0.36 μs after TOPO treatment.[146] Figure 10 shows the band alignment and the cross-section of a PeLED structure where the perovskite layer is TOPO coated as well as the electroluminescence (EL) characteristics of the resulting device.

ETLs and HTLs are utilized in photovoltaic cells to optimize the charge carrier injection and blocking functions. Similarly in PeLEDs, the improvements in EQE can be made through better choice and quality of the ETL and HTL.[147,148]

4.4. *Tuning color in LED*

PeLEDs have attracted intense research interest due to their remarkable properties, including a tunable bandgap, high PLQY, and high color purity. Although green- and near-infrared-emitting PeLEDs have recently arrived at an impressive level of EQEs,[145,149–151] fabricating efficient and color-stable blue PeLEDs useful for lighting and displays still requires further extensive study.

The electronic bandgap plays a big role in determining the emission wavelength of a LED. In a conventional three-dimensional perovskite structure, it can easily be tuned monotonically by adjusting the halogen component. The emission wavelength between 450 and 490 nm can be achieved through incorporating Cl:Br-mixed HPs as the emissive material. Unfortunately, illumination and voltage bias both lead to a side effect, namely undesirable shifts in EL wavelength from blue to green during device operation.[147,152]

Quasi-2D-HOIPs follow a different procedure to deliver blue PeLEDs. They are smaller in terms of structure dimension and

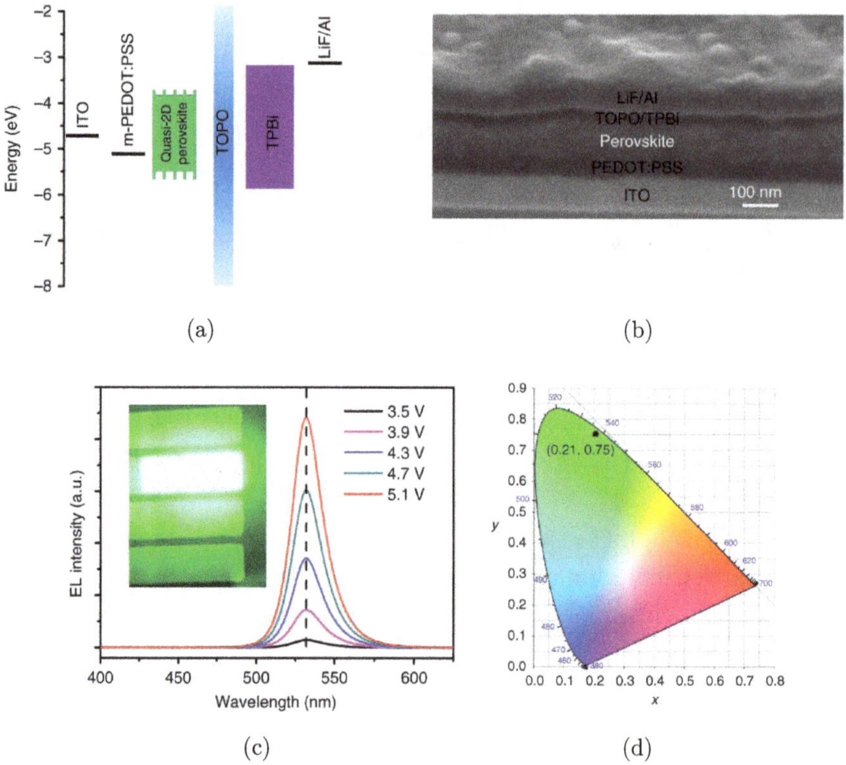

Fig. 10. (Color online) Perovskite LEDs structure and EL. (a) Band alignment of each function layer in the devices. (b) Cross-section scanning electron microscopy (SEM) image of the device, the scale bar is 100 nm. (c) Typical EL spectra of $PEA_2(FAPbBr_3)_{n-1}PbBr_4$ ($n = 3$ composition)-based PeLEDs under different voltage bias. Inset shows the EL image of PeLEDs. (d) The corresponding Commission Internationale de l'Eclairage (CIE) coordinate of typical PeLEDs based on $PEA_2(FAPbBr_3)_{n-1}PbBr_4$ ($n = 3$ composition). Reproduced with permission from 146. Copyright 2018, the Nature Publishing Group.

possess a long-chain ligand such as BA or PEA as B in the standard formula of $B_2A_{n-1}M_nX_{3n+1}$. Since quasi-2D perovskites consist of a finite amount of inorganic layers (n), we can tune their bandgap simply through reducing the number of total layers. Quan *et al.* recorded the tuning of bandgap in quasi-2D perovskites from approximately 2.6 eV ($n = 4$), to 2.7 eV ($n = 3$), 2.9 eV ($n = 2$), and 3.1 eV ($n = 1$).[153] Still, reported quasi-2D blue PeLEDs exhibit poor color stability under LEDs operation.[154–158] The undesired low stability arises due to contribution from the multiple emissions owing to mixed phases in a LED device, leading to inefficient charge transfer and various induced-emissions by a bias voltage. To balance out the different phases, hoping to achieve even crystallization, long ligands PEA are substituted by shorter ones (isopropylammonium).[147]

The maximum luminescence and the maximum EQE were measured from varying devices. Luminance (L) is proportional to the product of current density (J) and efficiency (EQE), signaling that luminance is affected by both current density and efficiency. Although thinner films lead to higher J, their EQEs are not always higher. The perovskite thickness reduction would result in a higher chance of charge carrier leakage and hence produce lower EQE.

Apart from experimenting with the emissive layer, the selection of HTL material can also contribute to the peak wavelength of the PeLED. Recent study suggests that NiO_x, PVK, and PEDOT:PSS help provide better emission color stability during device operation.[147,148]

Apart from the intriguing excitonic profile of 2D-HOIPs, a broadband emission is typically manifested from their rich chemical tunability in this hybrid system. This type of emission is triggered by the rich chemical tunability in 2D-HOIPs. If the emission covers the entire visible spectrum, the emitters can act as single-source white-light emitters superior to current mixed phosphor-based technology in terms of energy efficiency. The 2D perovskite-based single-source white-light emitters and their development belong to the emerging research field. Since the work from Karunadasa *et al.* in 2014,[102,159] which demonstrated bright broadband visible emission in 2D perovskite (EDBE) PbX_4 and (N-MEDA) $PbBr_{4-x}Cl_x$, (N-MEDA: N^1-methylethane-1,2-diammonium; EDBE: 2,2-(ethylenedioxy)bis (ethylammonium); X: Cl, Br), several other compositions have been investigated for broadband emission. Some research groups are convinced that the broadband emission is due to crystal defects, but in general, it is believed that the exciton self-trapping phenomenon is behind the broadband emission mechanism. Yu *et al.* recently proved that introducing impurity through Sn doping in PEA_2PbI_4 can provoke extrinsic self-trapping, yielding broadband red-to-near-infrared emission at room temperature.[160] This as well as a similar observation by Li *et al.* on modification of the optical and electrical properties using metal impurities, led to a novel strategy in the development of broadband-emitting — that is, dopant-induced extrinsic exciton self-trapping approach.[161]

5. Summary and Future Perspectives

Recent efforts to negate the effect of unintentional defects that reside within the hybrid perovskites structure have led the community to implement diverse candidates of organic molecules as passivating agents. While 2D-HOIPs intrinsically have a promising solution to overcome the perturbation from the light and moisture instabilities, these emergent materials mostly suffer ionic migration that pinpointed to the internal defects at the vicinity of topmost perovskite surface or within the inorganic sheets.

In this review, we have highlighted several examples to avoid large energy offset, which is feasible by selecting suitable hole/electron transport materials in solar cell devices. In addition, the current knowledge to combat a structural deformation within 2D-HOIPs that alter the perovskite surface is underway by depositing organic adlayers with different molecular functionalization. Here, we also conceptually revisit several important knowledge such as interfacial dipole, band bending and charge transfer that requires careful interpretation for understanding the 2D-HOIPs/organic molecules heterointerfaces.

In terms of tuning the electronic and optical properties at the perovskite/organic heterointerfaces, one should consider a tradeoff, which must be taken into account when ultrathin organic adlayers are adsorbed onto the desired hybrid perovskite surface. A favorable interfacial band alignment is pursued to achieve efficient charge transfer at the interface without creating a considerable parasitic effects. Furthermore, our forefront outlook systematically relates to two intriguing optical aspects, namely, how to suppress non-radiative recombination and promote an effective fast recombination, which trigger the PL lifetime enhancement of active layer of perovskite absorber. To this end, finding a coherent active organic layer that promotes superior charge carrier emission properties is highly desirable toward the advancement of light fidelity in the emerging communication technology.

The realization of tunable compositional engineering of the cations and anions of the perovskite constituents has embarked upsurge in the development of hybrid-based perovskite solar cells and LEDs. The incorporation of large conjugated organic ligands that impose spatially confined inorganic sheets has invoked great challenges in the hybrid-based solar cells. Nevertheless, 2D-HOIPs have been proven as versatile material due to their fascinating environmental stability with respect to their inorganic analogs. This in turn increases the PCEs with a tradeoff, outlining the remaining significant hysteresis as an unresolved task caused by the poor extraction efficiency at the interface of organic molecules/2D-HOIPs. Therefore, future research directions must have a clear pathway to tackle these obstacles such that an efficient and yet stable organic-hybrid-based solar cells can be attainable. This way, the approaches discussed in this review are very applicative and stimulate not only in the development of solar cell devices but also showed great potential in the future of PeLED technology.

Acknowledgments

The authors acknowledge the financial support from Universitas Prasetiya Mulya, Indonesia Toray Science Foundation, and Ministry of Education, Singapore (award no. MOE2018-T2-1-088).

ORCID

Silmi Kaffah https://orcid.org/0000-0002-7468-7170

Suriani Abu Bakar https://orcid.org/0000-0002-6541-7291

Muhammad Danang Birowosuto https://orcid.org/0000-0002-9997-6841

Arramel: https://orcid.org/0000-0003-4125-6099

References

1. D. B. Mitzi, *J. Chem. Soc. Dalton Trans.* (2001) 1.
2. D. B. Mitzi, *Hybrid Organic–Inorganic Electronics* (John Wiley & Sons, Ltd, 2003), Chap. 10, pp. 347–386.
3. C. C. Stoumpos, D. H. Cao, D. J. Clark, J. Young, J. M. Rondinelli, J. I. Jang, J. T. Hupp and M. G. Kanatzidis, *Chem. Mater.* **28** (2016) 2852.
4. L. Mao, C. C. Stoumpos and M. G. Kanatzidis, *J. Am. Chem. Soc.* **141** (2019) 1171.
5. D. B. Straus and C. R. Kagan, *J. Phys. Chem. Lett.* **9** (2018) 1434.
6. C. Ortiz-Cervantes, P. Carmona-Monroy and D. Solis-Ibarra, *Chem. Sus. Chem.* **12** (2019) 1560.
7. C. Katan, N. Mercier and J. Even, *Chem. Rev.* **119** (2019) 3140.
8. Y. Gao, Z. Wei, S.-N. Hsu, B. W. Boudouris and L. Dou, *Mater. Chem. Front.* **4** (2020) 3400.
9. T. A. Berhe, W.-N. Su, C.-H. Chen, C.-J. Pan, J.-H. Cheng, H.-M. Chen, M.-C. Tsai, L.-Y. Chen, A. A. Dubale and B.-J. Hwang, *Energy Environ. Sci.* **9** (2016) 323.
10. A. Oranskaia, J. Yin, O. M. Bakr, J.-L. Brédas and O. F. Mohammed, *J. Phys. Chem. Lett.* **9** (2018) 5474.
11. M. D. Smith, B. A. Connor and H. I. Karunadasa, *Chem. Rev.* **119** (2019) 3104.
12. H. Tsai, W. Nie, J.-C. Blancon, C. C. Stoumpos, R. Asadpour, B. Harutyunyan, A. J. Neukirch, R. Verduzco, J. J. Crochet, S. Tretiak, L. Pedesseau, J. Even, M. A. Alam, G. Gupta, J. Lou, P. M. Ajayan, M. J. Bedzyk, M. G. Kanatzidis and A. D. Mohite, *Nature* **536** (2016) 312.
13. C. C. Stoumpos, C. M. M. Soe, H. Tsai, W. Nie, J.-C. Blancon, D. H. Cao, F. Liu, B. Traoré, C. Katan, J. Even, A. D. Mohite and M. G. Kanatzidis, *Chem* **2** (2017) 427.
14. T. Ishihara, J. Takahashi and T. Goto, *Solid State Commun.* **69** (1989) 933.
15. X. Hong, T. Ishihara and A. V. Nurmikko, *Phys. Rev. B* **45** (1992) 6961.
16. J.-C. Blancon, J. Even, C. C. Stoumpos, M. G. Kanatzidis and A. D. Mohite, *Nat. Nanotechnol.* **15** (2020) 969.
17. L. Dou, A. B. Wong, Y. Yu, M. Lai, N. Kornienko, S. W. Eaton, A. Fu, C. G. Bischak, J. Ma, T. Ding, N. S. Ginsberg, L.-W. Wang, A. P. Alivisatos and P. Yang, *Science* **349** (2015) 1518.
18. B. Saparov and D. B. Mitzi, *Chem. Rev.* **116** (2016) 4558.
19. S. Zhao, C. Lan, H. Li, C. Zhang and T. Ma, *J. Phys. Chem. C* **124** (2020) 1788.
20. L. Mao, W. Ke, L. Pedesseau, Y. Wu, C. Katan, J. Even, M. R. Wasielewski, C. C. Stoumpos and M. G. Kanatzidis, *J. Am. Chem. Soc.* **140** (2018) 3775.
21. X. Li, W. Ke, B. Traoré, P. Guo, I. Hadar, M. Kepenekian, J. Even, C. Katan, C. C. Stoumpos, R. D. Schaller and M. G. Kanatzidis, *J. Am. Chem. Soc.* **141** (2019) 12880.
22. X. Tian, Y. Zhang, R. Zheng, D. Wei and J. Liu, *Sustain. Energy Fuels* **4** (2020) 2087.
23. F. Zhang, H. Lu, J. Tong, J. J. Berry, M. C. Beard and K. Zhu, *Energy Environ. Sci.* **13** (2020) 1154.

24. X. Zhang, J. Jiang, A. A. Suleiman, B. Jin, X. Hu, X. Zhou and T. Zhai, *Adv. Funct. Mater.* **29** (2019) 1906585.
25. B. Li, Y. Wu, N. Li, X. Chen, X. Zeng, Arramel, X. Zhao and J. Jiang, *ACS Appl. Mater. Interfaces* **12** (2020) 9261.
26. J. Jiang, N. Li, J. Zou, X. Zhou, G. Eda, Q. Zhang, H. Zhang, L.-J. Li, T. Zhai and A. T. S. Wee, *Chem. Soc. Rev.* **48** (2019) 4639.
27. J. Zou, S. Wu, Y. Liu, Y. Sun, Y. Cao, J.-P. Hsu, A. T. Shen Wee and J. Jiang, *Carbon* **130** (2018) 652.
28. T. Hellmann, C. Das, T. Abzieher, J. A. Schwenzer, M. Wussler, R. Dachauer, U. W. Paetzold, W. Jaegermann and T. Mayer, *Adv. Energy Mater.* **10** (2020) 2002129.
29. C. M. Mauck and W. A. Tisdale, *Trends Chem.* **1** (2019) 380.
30. N. Marinova, S. Valero and J. L. Delgado, *J. Colloid Interface Sci.* **488** (2017) 373.
31. W. Cai, Z. Chen, D. Chen, S. Su, Q. Xu, H.-L. Yip and Y. Cao, *RSC Adv.* **9** (2019) 27684.
32. W.-J. Yin, T. Shi and Y. Yan, *Adv. Mater.* **26** (2014) 4653.
33. K. X. Steirer, P. Schulz, G. Teeter, V. Stevanovic, M. Yang, K. Zhu and J. J. Berry, *ACS Energy Lett.* **1** (2016) 360.
34. A. Walsh, D. O. Scanlon, S. Chen, X. G. Gong and S.-H. Wei, *Angew. Chem. Int. Ed.* **54** (2015) 1791.
35. C. Eames, J. M. Frost, P. R. F. Barnes, B. C. O'Regan, A. Walsh and M. S. Islam, *Nat. Commun.* **6** (2015) 7497.
36. Z. Ni, C. Bao, Y. Liu, Q. Jiang, W.-Q. Wu, S. Chen, X. Dai, B. Chen, B. Hartweg, Z. Yu, Z. Holman and J. Huang, *Science* **367** (2020) 1352.
37. K. Gauthron, J.-S. Lauret, L. Doyennette, G. Lanty, A. A. Choueiry, S. Zhang, A. Brehier, L. Largeau, O. Mauguin, J. Bloch and E. Deleporte, *Opt. Express* **18** (2010) 5912.
38. W. K. Chong, K. Thirumal, D. Giovanni, T. W. Goh, X. Liu, N. Mathews, S. Mhaisalkar and T. C. Sum, *Phys. Chem. Chem. Phys.* **18** (2016) 14701.
39. N. Liu, P. Liu, H. Zhou, Y. Bai and Q. Chen, *J. Phys. Chem. Lett.* **11** (2020) 3521.
40. D. Ghosh, D. Acharya, L. Pedesseau, C. Katan, J. Even, S. Tretiak and A. J. Neukirch, *J. Mater. Chem. A* **8** (2020) 22009.
41. J. Hieulle, C. Stecker, R. Ohmann, L. K. Ono and Y. Qi, *Small Methods* **2** (2018) 1700295.
42. S. Sahare, P. Ghoderao, S. B. Khan, Y. Chan and S.-L. Lee, *Nanoscale* **12** (2020) 15970.
43. M. Telychko and J. Lu, *Nano Mater. Sci.* **1** (2019) 260.
44. H. Jin, E. Debroye, M. Keshavarz, I. G. Scheblykin, M. B. J. Roeffaers, J. Hofkens and J. A. Steele, *Mater. Horiz.* **7** (2020) 397.
45. S. Béchu, M. Ralaiarisoa, A. Etcheberry and P. Schulz, *Adv. Energy Mater.* **10** (2020) 1904007.
46. B. Chen, P. N. Rudd, S. Yang, Y. Yuan and J. Huang, *Chem. Soc. Rev.* **48** (2019) 3842.
47. J. Xue, R. Wang and Y. Yang, *Nat. Rev. Mater.* **5** (2020) 809.

48. L. K. Ono, S. F. Liu and Y. Qi, *Angew. Chem. Int. Ed.* **59** (2020) 6676.
49. J. Chen and N.-G. Park, *ACS Energy Lett.* **5** (2020) 2742.
50. H. Hertz, *Ann. Phys.* **267** (1887) 983.
51. A. Einstein, *Ann. Phys.* **322** (1905) 891.
52. P. van der Heide, *X-Ray Photoelectron Spectroscopy* (John Wiley & Sons, Ltd, 2011).
53. J. F. W. J. Wolstenholme, *Electron Spectroscopy* (John Wiley & Sons, Ltd, 2019), Chap. 1, pp. 1–18.
54. R. L. Z. Hoye, P. Schulz, L. T. Schelhas, A. M. Holder, K. H. Stone, J. D. Perkins, D. Vigil-Fowler, S. Siol, D. O. Scanlon, A. Zakutayev, A. Walsh, I. C. Smith, B. C. Melot, R. C. Kurchin, Y. Wang, J. Shi, F. C. Marques, J. J. Berry, W. Tumas, S. Lany, V. Stevanović, M. F. Toney and T. Buonassisi, *Chem. Mater.* **29** (2017) 1964.
55. B. Philippe, B.-W. Park, R. Lindblad, J. Oscarsson, S. Ahmadi, E. M. J. Johansson and H. Rensmo, *Chem. Mater.* **27** (2015) 1720.
56. E. M. Miller, Y. Zhao, C. C. Mercado, S. K. Saha, J. M. Luther, K. Zhu, V. Stevanović, C. L. Perkins and J. van de Lagemaat, *Phys. Chem. Chem. Phys.* **16** (2014) 22122.
57. P. Schulz, D. Cahen and A. Kahn, *Chem. Rev.* **119** (2019) 3349.
58. J. Emara, T. Schnier, N. Pourdavoud, T. Riedl, K. Meerholz and S. Olthof, *Adv. Mater.* **28** (2016) 553.
59. X. Wu, M. T. Trinh, D. Niesner, H. Zhu, Z. Norman, J. S. Owen, O. Yaffe, B. J. Kudisch and X. Y. Zhu, *J. Am. Chem. Soc.* **137** (2015) 2089.
60. M. Kollár, L. Ćirić, J. H. Dil, A. Weber, S. Muff, H. M. Ronnow, B. Náfrádi, B. P. Monnier, J. S. Luterbacher, L. Forró and E. Horváth, *Sci. Rep.* **7** (2017) 695.
61. F. Zhang, S. H. Silver, N. K. Noel, F. Ullrich, B. P. Rand and A. Kahn, *Adv. Energy Mater.* **10** (2020) 1903252.
62. S. Silver, Q. Dai, H. Li, J.-L. Brédas and A. Kahn, *Adv. Energy Mater.* **9** (2019) 1901005.
63. D. Niesner, *APL Mater.* **8** (2020) 090704.
64. J.-P. Yang, M. Meissner, T. Yamaguchi, X.-Y. Zhang, T. Ueba, L.-W. Cheng, S. Ideta, K. Tanaka, X.-H. Zeng, N. Ueno and S. Kera, *Sol. RRL* **2** (2018) 1800132.
65. T. Komesu, X. Huang, T. R. Paudel, Y. B. Losovyj, X. Zhang, E. F. Schwier, Y. Kojima, M. Zheng, H. Iwasawa, K. Shimada, M. I. Saidaminov, D. Shi, A. L. Abdelhady, O. M. Bakr, S. Dong, E. Y. Tsymbal and P. A. Dowben, *J. Phys. Chem. C* **120** (2016) 21710.
66. Q.-K. Wang, R.-B. Wang, P.-F. Shen, C. Li, Y.-Q. Li, L.-J. Liu, S. Duhm and J.-X. Tang, *Adv. Mater. Interfaces* **2** (2015) 1400528.
67. T. C. Sum, S. Chen, G. Xing, X. Liu and B. Wu, *Nanotechnology* **26** (2015) 342001.
68. J.-M. Yang, Y. Luo, Q. Bao, Y.-Q. Li and J.-X. Tang, *Adv. Mater. Interfaces* **6** (2019) 1801351.
69. L. Qiu, S. He, L. K. Ono and Y. Qi, *Adv. Energy Mater.* **10** (2020) 1902726.
70. S. Shao and M. A. Loi, *Adv. Mater. Interfaces* **7** (2020) 1901469.

71. L. K. Ono and Y. Qi, *J. Phys. Chem. Lett.* **7** (2016) 4764.
72. S. Wang, T. Sakurai, W. Wen and Y. Qi, *Adv. Mater. Interfaces* **5** (2018) 1800260.
73. W. Tress, K. Leo and M. Riede, *Adv. Funct. Mater.* **21** (2011) 2140.
74. P. Schulz, E. Edri, S. Kirmayer, G. Hodes, D. Cahen and A. Kahn, *Energy Environ. Sci.* **7** (2014) 1377.
75. S. Chen, T. W. Goh, D. Sabba, J. Chua, N. Mathews, C. H. A. Huan and T. C. Sum, *APL Mater.* **2** (2014) 081512.
76. S. Olthof and K. Meerholz, *Sci. Rep.* **7** (2017) 40267.
77. Q. Ou, Y. Zhang, Z. Wang, J. A. Yuwono, R. Wang, Z. Dai, W. Li, C. Zheng, Z.-Q. Xu, X. Qi, S. Duhm, N. V. Medhekar, H. Zhang and Q. Bao, *Adv. Mater.* **30** (2018) 1705792.
78. Arramel, H. Pan, A. Xie, S. Hou, X. Yin, C. S. Tang, N. T. Hoa, M. D. Birowosuto, H. Wang, C. Dang, A. Rusydi, A. T. S. Wee and J. Wu, *Nano Res.* **12** (2019) 77.
79. Arramel, P. Hu, A. Xie, X. Yin, C. S. Tang, K. Ikeda, M. H. Mahyuddin, M. F. Sahdan, D. Wang, K. Yoshizawa, H. Wang, M. D. Birowosuto, C. Dang, A. Rusydi, A. T. S. Wee and J. Wu, *J. Mater. Chem. C* **8** (2020) 12587.
80. P. Schulz, L. L. Whittaker-Brooks, B. A. MacLeod, D. C. Olson, Y.-L. Loo and A. Kahn, *Adv. Mater. Interfaces* **2** (2015) 1400532.
81. X. Zhang, Z. Su, B. Zhao, Y. Yang, Y. Xiong, X. Gao, D.-C. Qi and L. Cao, *Appl. Phys. Lett.* **113** (2018) 113901.
82. Y. Lin, B. Chen, F. Zhao, X. Zheng, Y. Deng, Y. Shao, Y. Fang, Y. Bai, C. Wang and J. Huang, *Adv. Mater.* **29** (2017) 1700607.
83. A. M. Boehm, T. Liu, S. M. Park, A. Abtahi and K. R. Graham, *ACS Appl. Mater. Interfaces* **12** (2020) 5209.
84. N. K. Noel, A. Abate, S. D. Stranks, E. S. Parrott, V. M. Burlakov, A. Goriely and H. J. Snaith, *ACS Nano* **8** (2014) 9815.
85. X. Zheng, B. Chen, J. Dai, Y. Fang, Y. Bai, Y. Lin, H. Wei, X. C. Zeng and J. Huang, *Nat. Energy* **2** (2017) 17102.
86. F. Zu, P. Amsalem, M. Ralaiarisoa, T. Schultz, R. Schlesinger and N. Koch, *ACS Appl. Mater. Interfaces* **9** (2017) 41546.
87. E. E. Perry, J. G. Labram, N. R. Venkatesan, H. Nakayama and M. L. Chabinyc, *Adv. Electron. Mater.* **4** (2018) 1800087.
88. L. Zhao, Y. L. Lin, H. Kim, N. C. Giebink and B. P. Rand, *ACS Energy Lett.* **3** (2018) 2708.
89. K.-T. Ho, S.-F. Leung, T.-Y. Li, P. Maity, B. Cheng, H.-C. Fu, O. F. Mohammed and J.-H. He, *Adv. Mater.* **30** (2018) 1804372.
90. W. T. M. Van Gompel, R. Herckens, K. Van Hecke, B. Ruttens, J. D'Haen, L. Lutsen and D. Vanderzande, *Chem. Commun.* **55** (2019) 2481.
91. J. V. Passarelli, C. M. Mauck, S. W. Winslow, C. F. Perkinson, J. C. Bard, H. Sai, K. W. Williams, A. Narayanan, D. J. Fairfield, M. P. Hendricks, W. A. Tisdale and S. I. Stupp, *Nat. Chem.* **12** (2020) 672.
92. K. Leng, L. Wang, Y. Shao, I. Abdelwahab, G. Grinblat, I. Verzhbitskiy, R. Li, Y. Cai, X. Chi, W. Fu, P. Song, A. Rusydi, G. Eda, S. A. Maier and K. P. Loh, *Nat. Commun.* **11** (2020) 5483.

93. G. Liu, X.-X. Xu, S. Xu, L. Zhang, H. Xu, L. Zhu, X. Zhang, H. Zheng and X. Pan, *J. Mater. Chem. A* **8** (2020) 5900.
94. X. Zhang, X. Ren, B. Liu, R. Munir, X. Zhu, D. Yang, J. Li, Y. Liu, D.-M. Smilgies, R. Li, Z. Yang, T. Niu, X. Wang, A. Amassian, K. Zhao and S. F. Liu, *Energy Environ. Sci.* **10** (2017) 2095.
95. B.-E. Cohen, Y. Li, Q. Meng and L. Etgar, *Nano Lett.* **19** (2019) 2588.
96. J. Hu, I. W. H. Oswald, S. J. Stuard, M. M. Nahid, N. Zhou, O. F. Williams, Z. Guo, L. Yan, H. Hu, Z. Chen, X. Xiao, Y. Lin, Z. Yang, J. Huang, A. M. Moran, H. Ade, J. R. Neilson and W. You, *Nat. Commun.* **10** (2019) 1276.
97. D. H. Cao, C. C. Stoumpos, O. K. Farha, J. T. Hupp and M. G. Kanatzidis, *J. Am. Chem. Soc.* **137** (2015) 7843.
98. B. Sun, Y. Xu, Y. Chen and W. Huang, *APL Mater.* **8** (2020) 040901.
99. J. Wang, H. Zhang, L. Wang, K. Yang, L. Cang, X. Liu and W. Huang, *ACS Appl. Energy Mater.* **3** (2020) 4484.
100. M. M. Tavakoli, K.-H. Tsui, Q. Zhang, J. He, Y. Yao, D. Li and Z. Fan, *ACS Nano* **9** (2015) 10287.
101. A. Xie, F. Maddalena, M. E. Witkowski, M. Makowski, B. Mahler, W. Drozdowski, S. V. Springham, P. Coquet, C. Dujardin, M. D. Birowosuto and C. Dang, *Chem. Mater.* **32** (2020) 8530.
102. E. R. Dohner, A. Jaffe, L. R. Bradshaw and H. I. Karunadasa, *J. Am. Chem. Soc.* **136** (2014) 13154.
103. D. Cortecchia, S. Neutzner, A. R. Srimath Kandada, E. Mosconi, D. Meggiolaro, F. De Angelis, C. Soci and A. Petrozza, *J. Am. Chem. Soc.* **139** (2017) 39.
104. D. Cortecchia, J. Yin, A. Bruno, S.-Z. A. Lo, G. G. Gurzadyan, S. Mhaisalkar, J.-L. Brédas and C. Soci, *J. Mater. Chem. C* **5** (2017) 2771.
105. T. Sheikh, A. Shinde, S. Mahamuni and A. Nag, *ACS Energy Lett.* **3** (2018) 2940.
106. H. Hardhienata, F. Ahmad, Arramel, M. Aminah, D. Onggo, L. J. Diguna, M. D. Birowosuto, M. E. Witkowski, M. Makowski and W. Drozdowski, *J. Phys. D, Appl. Phys.* **53** (2020) 455303.
107. Q. Shan, C. Wei, Y. Jiang, J. Song, Y. Zou, L. Xu, T. Fang, T. Wang, Y. Dong, J. Liu, B. Han, F. Zhang, J. Chen, Y. Wang and H. Zeng, *Light Sci. Appl.* **9** (2020) 163.
108. J. Y. Kim, J.-W. Lee, H. S. Jung, H. Shin and N.-G. Park, *Chem. Rev.* **120** (2020) 7867.
109. X.-K. Liu, W. Xu, S. Bai, Y. Jin, J. Wang, R. H. Friend and F. Gao, *Nat. Mater.* **20** (2021) 10.
110. P. Gao, M. Grätzel and M. K. Nazeeruddin, *Energy Environ. Sci.* **7** (2014) 2448.
111. S. Chen and G. Shi, *Adv. Mater.* **29** (2017) 1605448.
112. J. Jagielski, S. Kumar, W.-Y. Yu and C.-J. Shih, *J. Mater. Chem. C* **5** (2017) 5610.
113. T. M. Koh, V. Shanmugam, J. Schlipf, L. Oe-singhaus, P. Müller-Buschbaum, N. Ramakrishnan, V. Swamy, N. Mathews, P. P. Boix and S. G. Mhaisalkar, *Adv. Mater.* **28** (2016) 3653.

114. Y. Chen, Y. Sun, J. Peng, J. Tang, K. Zheng and Z. Liang, *Adv. Mater.* **30** (2018) 1703487.
115. I. C. Smith, E. T. Hoke, D. Solis-Ibarra, M. D. McGehee and H. I. Karunadasa, *Angew. Chem. Int. Ed.* **53** (2014) 11232.
116. Y. Chen, Y. Sun, J. Peng, W. Zhang, X. Su, K. Zheng, T. Pullerits and Z. Liang, *Adv. Energy Mater.* **7** (2017) 1700162.
117. T. He, S. Li, Y. Jiang, C. Qin, M. Cui, L. Qiao, H. Xu, J. Yang, R. Long, H. Wang and M. Yuan, *Nat. Commun.* **11** (2020) 1672.
118. M. Wei, K. Xiao, G. Walters, R. Lin, Y. Zhao, M. I. Saidaminov, P. Todorović, A. Johnston, Z. Huang, H. Chen, A. Li, J. Zhu, Z. Yang, Y.-K. Wang, A. H. Proppe, S. O. Kelley, Y. Hou, O. Voznyy, H. Tan and E. H. Sargent, *Adv. Mater.* **32** (2020) 1907058.
119. S. Gharibzadeh, I. M. Hossain, P. Fassl, B. A. Nejand, T. Abzieher, M. Schultes, E. Ahlswede, P. Jackson, M. Powalla, S. Schäfer, M. Rienäcker, T. Wietler, R. Peibst, U. Lemmer, B. S. Richards and U. W. Paetzold, *Adv. Funct. Mater.* **30** (2020) 1909919.
120. N. K. Noel, S. D. Stranks, A. Abate, C. Wehrenfennig, S. Guarnera, A.-A. Haghighirad, A. Sadhanala, G. E. Eperon, S. K. Pathak, M. B. Johnston, A. Petrozza, L. M. Herz and H. J. Snaith, *Energy Environ. Sci.* **7** (2014) 3061.
121. C. Ma, D. Shen, T.-W. Ng, M.-F. Lo and C.-S. Lee, *Adv. Mater.* **30** (2018) 1800710.
122. C. Zhou, W. Chen, S. Yang, Q. Ou, Z. Gan, Q. Bao, B. Jia and X. Wen, *ACS Appl. Mater. Interfaces* **12** (2020) 26384.
123. N. Zhou, Y. Shen, L. Li, S. Tan, N. Liu, G. Zheng, Q. Chen and H. Zhou, *J. Am. Chem. Soc.* **140** (2018) 459.
124. T. Wu, Y. Wang, X. Li, Y. Wu, X. Meng, D. Cui, X. Yang and L. Han, *Adv. Energy Mater.* **9** (2019) 1803766.
125. Q. Jiang, Y. Zhao, X. Zhang, X. Yang, Y. Chen, Z. Chu, Q. Ye, X. Li, Z. Yin and J. You, *Nat. Photonics* **13** (2019) 460.
126. S. M. Park, A. Abtahi, A. M. Boehm and K. R. Graham, *ACS Energy Lett.* **5** (2020) 799.
127. Z.-K. Wang and L.-S. Liao, *Adv. Opt. Mater.* **6** (2018) 1800276.
128. G. Grancini, C. Roldán-Carmona, I. Zimmermann, E. Mosconi, X. Lee, D. Martineau, S. Narbey, F. Oswald, F. De Angelis, M. Graetzel and M. K. Nazeeruddin, *Nat. Commun.* **8** (2017) 15684.
129. J. Sun, N. Chandrasekaran, C. Liu, A. D. Scully, W. Yin, C. K. Ng and J. J. Jasieniak, *ACS Appl. Energy Mater.* **3** (2020) 8205.
130. L. Yangi, Y. Li, Y. Pei, J. Wang, H. Lin and X. Li, *J. Mater. Chem. A* **8** (2020) 7808.
131. A. Krishna, S. Gottis, M. K. Nazeeruddin and F. Sauvage, *Adv. Funct. Mater.* **29** (2019) 1806482.
132. L. Zhou, Z. Lin, Z. Ning, T. Li, X. Guo, J. Ma, J. Su, C. Zhang, J. Zhang, S. Liu, J. Chang and Y. Hao, *Sol. RRL* **3** (2019) 1900293.
133. M.-C. Kim, S.-Y. Ham, D. Cheng, T. A. Wynn, H. S. Jung and Y. S. Meng, *Adv. Energy Mater.* (2020) 2001753.
134. Y. Zhao and K. Zhu, *Chem. Soc. Rev.* **45** (2016) 655.

135. K. Wang, S. Wang, S. Xiao and Q. Song, *Adv. Opt. Mater.* **6** (2018) 1800278.
136. M. P. Arciniegas, A. Castelli, S. Piazza, S. Dogan, L. Ceseracciu, R. Krahne, M. Duocastella and L. Manna, *Adv. Funct. Mater.* **27** (2017) 1701613.
137. H. Huang, Q. Xue, B. Chen, Y. Xiong, J. Schneider, C. Zhi, H. Zhong and A. L. Rogach, *Angew. Chem. Int. Ed.* **56** (2017) 9571.
138. Q. A. Akkerman, G. Rainò, M. V. Kovalenko and L. Manna, *Nat. Mater.* **17** (2018) 394.
139. L. Etgar, *Energy Environ. Sci.* **11** (2018) 234.
140. N. Kitazawa, M. Aono and Y. Watanabe, *Thin Solid Films* **518** (2010) 3199.
141. Y. Ling, Z. Yuan, Y. Tian, X. Wang, J. C. Wang, Y. Xin, K. Hanson, B. Ma and H. Gao, *Adv. Mater.* **28** (2016) 305.
142. S. Kumar, J. Jagielski, N. Kallikounis, Y.-H. Kim, C. Wolf, F. Jenny, T. Tian, C. J. Hofer, Y.-C. Chiu, W. J. Stark, T.-W. Lee and C.-J. Shih, *Nano Lett.* **17** (2017) 5277.
143. D. Yang, Y. Zou, P. Li, Q. Liu, L. Wu, H. Hu, Y. Xu, B. Sun, Q. Zhang and S.-T. Lee, *Nano Energy* **47** (2018) 235.
144. M. Yuan, L. N. Quan, R. Comin, G. Walters, R. Sabatini, O. Voznyy, S. Hoogland, Y. Zhao, E. M. Beauregard, P. Kanjanaboos, Z. Lu, D. H. Kim and E. H. Sargent, *Nat. Nanotechnol.* **11** (2016) 872.
145. N. Wang, L. Cheng, R. Ge, S. Zhang, Y. Miao, W. Zou, C. Yi, Y. Sun, Y. Cao, R. Yang, Y. Wei, Q. Guo, Y. Ke, M. Yu, Y. Jin, Y. Liu, Q. Ding, D. Di, L. Yang, G. Xing, H. Tian, C. Jin, F. Gao, R. H. Friend, J. Wang and W. Huang, *Nat. Photonics* **10** (2016) 699.
146. X. Yang, X. Zhang, J. Deng, Z. Chu, Q. Jiang, J. Meng, P. Wang, L. Zhang, Z. Yin and J. You, *Nat. Commun.* **9** (2018) 570.
147. J. Xing, Y. Zhao, M. Askerka, L. N. Quan, X. Gong, W. Zhao, J. Zhao, H. Tan, G. Long, L. Gao, Z. Yang, O. Voznyy, J. Tang, Z.-H. Lu, Q. Xiong and E. H. Sargent, *Nat. Commun.* **9** (2018) 3541.
148. E. Yoon, K. Y. Jang, J. Park and T.-W. Lee, *Adv. Mater. Interfaces* **8** (2021) 2001712.
149. Z.-K. Tan, R. S. Moghaddam, M. L. Lai, P. Docampo, R. Higler, F. Deschler, M. Price, A. Sadhanala, L. M. Pazos, D. Credgington, F. Hanusch, T. Bein, H. J. Snaith and R. H. Friend, *Nat. Nanotechnol.* **9** (2014) 687.
150. L. Zhang, X. Yang, Q. Jiang, P. Wang, Z. Yin, X. Zhang, H. Tan, Y. M. Yang, M. Wei, B. R. Sutherland, E. H. Sargent and J. You, *Nat. Commun.* **8** (2017) 15640.
151. F. Yan, J. Xing, G. Xing, L. Quan, S. T. Tan, J. Zhao, R. Su, L. Zhang, S. Chen, Y. Zhao, A. Huan, E. H. Sargent, Q. Xiong and H. V. Demir, *Nano Lett.* **18** (2018) 3157.
152. G. Li, F. W. R. Rivarola, N. J. L. K. Davis, S. Bai, T. C. Jellicoe, F. de la Peña, S. Hou, C. Ducati, F. Gao, R. H. Friend, N. C. Greenham and Z.-K. Tan, *Adv. Mater.* **28** (2016) 3528.
153. L. N. Quan, Y. Zhao, F. P. García de Arquer, R. Sabatini, G. Walters, O. Voznyy, R. Comin, Y. Li, J. Z. Fan, H. Tan, J. Pan, M. Yuan, O. M. Bakr, Z. Lu, D. H. Kim and E. H. Sargent, *Nano Lett.* **17** (2017) 3701.

154. S. Kumar, J. Jagielski, S. Yakunin, P. Rice, Y.-C. Chiu, M. Wang, G. Nedelcu, Y. Kim, S. Lin, E. J. G. Santos, M. V. Kovalenko and C.-J. Shih, *ACS Nano* **10** (2016) 9720.
155. Q. Wang, J. Ren, X.-F. Peng, X.-X. Ji and X.-H. Yang, *ACS Appl. Mater. Interfaces* **9** (2017) 29901.
156. Z. Chen, C. Zhang, X.-F. Jiang, M. Liu, R. Xia, T. Shi, D. Chen, Q. Xue, Y.-J. Zhao, S. Su, H.-L. Yip and Y. Cao, *Adv. Mater.* **29** (2017) 1603157.
157. D. N. Congreve, M. C. Weidman, M. Seitz, W. Paritmongkol, N. S. Dahod and W. A. Tisdale, *ACS Photonics* **4** (2017) 476.
158. L. Cheng, Y. Cao, R. Ge, Y.-Q. Wei, N.-N. Wang, J.-P. Wang and W. Huang, *Chin. Chem. Lett.* **28** (2017) 29.
159. E. R. Dohner, E. T. Hoke and H. I. Karunadasa, *J. Am. Chem. Soc.* **136** (2014) 1718.
160. J. Yu, J. Kong, W. Hao, X. Guo, H. He, W. R. Leow, Z. Liu, P. Cai, G. Qian, S. Li, X. Chen and X. Chen, *Adv. Mater.* **31** (2019) 1806385.
161. T. Li, X. Chen, X. Wang, H. Lu, Y. Yan, M. C. Beard and D. B. Mitzi, *ACS Energy Lett.* **5** (2020) 347.

https://doi.org/10.1142/9789811247859_0011

Chapter 11

Technology and Applications of Graphene Oxide Membranes

Xuanye Leng[*,§], Siyu Chen[*,§], Kou Yang[*,§], Musen Chen[*,§], Majid Shaker[*,†], Evgenii E. Vdovin[‡], Qi Ge[†], Kostya S. Novoselov[*,†,§] and Daria V. Andreeva[*,§,¶]

[*]Department of Materials Science and Engineering, National University of Singapore, Singapore 117575, Singapore

[†]Chongqing 2D Materials Institute, Liangjiang New Area, Chongqing, 400714, P. R. China

[‡]Institute of Microelectronics Technology RAS, Chernogolovka, 142432, Russia

[§]Institute for Functional Intelligent Materials, National University of Singapore, 117544, Singapore

[¶]daria@nus.edu.sg

Graphene oxide (GO) is an amphiphilic, water dispersible, chemical derivative of graphene. Widely used as a pathway to obtain graphene, it also has a number of interesting applications by itself due to its ability to form covalently and non-covalently bonded organic–inorganic hybrids and polymer composites. Thus, GO-based composites are used in numerous applications in membrane and coating technologies. It is important that due to the presence of functional acidic groups, GO possesses tunable physicochemical properties like a negatively charged polyelectrolyte and can be used as stimuli responsive membranes, membranes that can interact with environment and switch their properties on demand. Thus, ionic/molecular separation, water purification, selective sensing, and stimuli responsive properties have already been demonstrated in the laboratory. Good mechanical strength and conductivity (in its partially reduced form) make it attractive for the construction of the membranes for energy devices and sensors. However, concentration and distribution of the functional

[¶]Corresponding author.

groups on GO molecules is difficult to control. It makes GO materials difficult to standardize, produce, and apply in industry. To this end, it is important to highlight recent achievement in the synthesis of GO as well as in design of GO-based energy devices, corrosion inhibiting coatings, and biomedical devices with improved working performances to evoke interest on mass production of GO with improved formulation.

Keywords: 2D materials; GO membranes; ion/molecular separation; water treatment; transistors; corrosion protection.

1. Introduction

Graphene oxide (GO) is a chemical derivative of graphene obtained by oxidation with successive exfoliation of graphite. Figure 1 shows the schematic structures of GO. While maintaining the main backbone of graphene, GO possesses multiple oxygen-contained groups such as hydroxyl (–OH), carboxyl (–COOH), and epoxyl (–O–), which provide tunable physicochemical properties and stimuli responsive behavior. Such functional groups make GO hydrophilic and allow it to form stable aqueous suspensions.

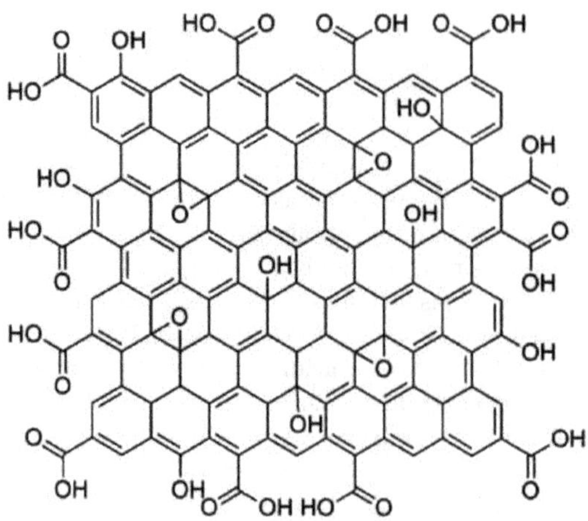

Graphene oxide

Fig. 1. Schematic structure of GO, adapted with permission.[6] Copyright © 2010, American Chemical Society.

Along with the quite facile fabrication process, GO has been widely studied for multipurpose applications such as membranes for ionic/molecular separation, water treatment, biomedical usage, energy conversion and storage, and functional coatings.[1] As membranes, 2D structures of GO are very attractive due to their good mechanical properties. The Young's Modulus of monolayer GO and GO membranes are measured to be 207.6 and 32 GPa, respectively,[2,3] while a small bandgap of 0.72 eV was found in multilayer GO and the thermal conductivity of GO was tested to be $3.77\,W\,m^{-1}\,K^{-1}$.[4,5]

The preparations of GO usually include chemical assisted exfoliations. In bulk layered materials, the layers could be expanded via intercalation of strong oxidative agents, thus leading to the production of 2D counterparts.[7] Hummers and Offeman's method to get the graphite oxide is a classical chemical assisted exfoliation method.[8] In this method, powdered graphite flakes are treated with $NaNO_3$, H_2SO_4, $KMnO_4$, and H_2O_2 to obtain graphite oxide suspension. GO can be obtained from such graphite oxide suspension via exfoliation in water (typically requires mild sonication) and removing of the unexfoliated graphite flakes with the sedimentation steps.[9] Furthermore, reduced graphene oxide (rGO) can be produced by reduction of GO by reducing agents (such as N_2H_4 and $NaBH_4$).[10] Till now, the conventional Hummers' method is still the most common way to produce GO.

Amphiphilic nature of GO makes it easy to form laminated structures, known as GO paper or GO membranes. A facile method to fabricate GO membranes is vacuum-assisted filtration of GO suspensions. Dikin *et al.*[3] used Anodisc membrane filter to filtrate 3 mg/mL GO water dispersion and obtained GO membranes, which exhibited excellent mechanical strength. Evaporation method to fabricate GO membranes has also been reported. By heating condensed GO hydrosol at 353 K under a thermostated water bath, Chen *et al.*[11] obtained freestanding GO membranes with good mechanical and optical properties. Vacuum assisted suction pressed method was developed by Huang *et al.*[12] They utilized a ceramic hollow tube as the substrate and immersed it into GO aqueous solution. When the ceramic hollow tube was vacuumed by an external vacuum pump, GO in the solution could be pressed onto the surface of the ceramic hollow tube and thus the GO membrane was fabricated.

Within abundant carboxyl groups anchored on graphene skeleton, structures and properties of GO are thought pH dependent with the protonation/deprotonation of the carboxyl groups.[13,14] GO tends to be more hydrophilic in high pH condition where carboxy groups deprotonate and GO is more negatively charged while more hydrophobic in low pH where carboxy groups protonate and GO is less negatively charged.[13,15] Moreover, pH also determines the size of GO sheets in aqueous solution where smaller size GO sheets have better solubility than the larger ones under the same pH for their high charge density by large edge-to-area ratio.[14] The pH-dependent characteristic of GO contributes to customizing GO sheets with demanded structure and property and is of great significance to tune the corresponding membrane performance on the applications such as drug delivery,[16] separation,[17] and biomimetics.[18]

The enticing features of GO sheets, such as large specific surface area, presence of oxygen-contained radicals (carboxyl, hydroxyl, and epoxy groups), and good mechanical property, enable GO membranes to be utilized in ionic and molecular separation applications.[19–23] GO-based membranes have shown great potential in solving water pollution and scarcity issues and have become promising candidates for applications such as seawater desalination and wastewater purification.[24–26]

GO laminates also facilitate gas selectivity and transport rate, and thus play an important role in gas separation technology.[27] The edges or defects on GO surface allow gas molecules to enter into the membrane, and nanochannels formed by adjacent GO sheets provide pathway for them to permeate. Consequently, the membrane performance is greatly affected by stacked structure of GO such as pore sizes, local defects, tortuosity (which is determined by GO sheet size and thickness), interlayer distances, and degree of functionalization.[27,28]

Compared with water treatment and gas separation, the research on GO-based membranes for solvent separation is still being developed.[29]

Pervaporation dehydration is a promising membrane-based technology for organic solvent separation application.[30] However, traditional pervaporation membranes confront the tradeoff between permeability and selectivity.[31] GO membranes provide the advantages of good mechanical properties, fast water permeation, and high selectivity and are regarded as the highly promising materials for pervaporation application. The interlayer spacing as well as the operation conditions are the dominant factors which influence the pervaporation dehydration.

GO-based membranes have been widely used in sensors and biomedical applications owing to their low cost and facile fabrication process. Taking advantage of its abundant functional oxygen-contained groups,[32] GO shows excellent performance in gas absorption and gas detection. Therefore, gas sensors containing GO have been extensively studied. Meanwhile, due to the good biocompatibility and chemical stability, GO membranes have also been widely explored in biomedical devices.[33–35] In addition, GO/biopolymers complex membranes have also been reported to demonstrate excellent performance in biomedical applications. To form GO-based composite membranes for biomedical application, various bio-based polymers including chitosan are intensely utilized.[36] In addition, GO also exhibits good *in vivo* biological performance while maintains exceptional cytocompatibility and biological suitability.[37]

In energy conversion and storage, GO membranes have been explored as flexible electrodes, ion exchange membrane (IEM), and separators. Thanks to their reasonable conductivity, good mechanical strength, flexibility, abundant chemical reactive sites and high specific area, GO membranes are also promising candidates for intelligent and wearable electronic devices.[38,39]

GO also demonstrates great performance in anticorrosion coatings on biomaterials due to their barrier effects. As a result of the enhanced corrosion resistance of the biomaterials coated with 2D materials, their lifetime is prolonged in body and their biocompatibility is improved. Nevertheless, there is still a long way to pave for developing anticorrosion coatings employing GO, since more research on biocompatibility is required.

2. Membranes for Ion/Molecular Separation

Nair *et al.*[40] firstly reported that GO membranes allow ultrafast permeation for water molecules while remaining impermeable for other liquids and gases. The unimpeded transport of water could be explained by low-friction flow of water monolayer through unoxidized graphene nanochannels. In the presence of water, other liquids and gases are blocked by limited size of capillaries in low humidity and intercalating water in high humidity. Hong *et al.*[41] demonstrated ultrahigh charge selectivity and size selectivity of GO membrane, which can be ascribed to electrostatic repulsion between negatively charged GO and Cl^- as well as compression of hydration shells of ions, as shown in Fig. 2.

Water and ion permeation mechanism through GO membranes has been studied intensively, focusing on the simulation methods, investigation of the effect of regulated interlayer distance (d) of GO sheets, varying of such external parameters such as applied pressure, salt concentration, and pH.

Based on first-principles pseudopotential calculations, Boukhvalov *et al.*[42] demonstrated that hexagonal ice monolayers can be formed

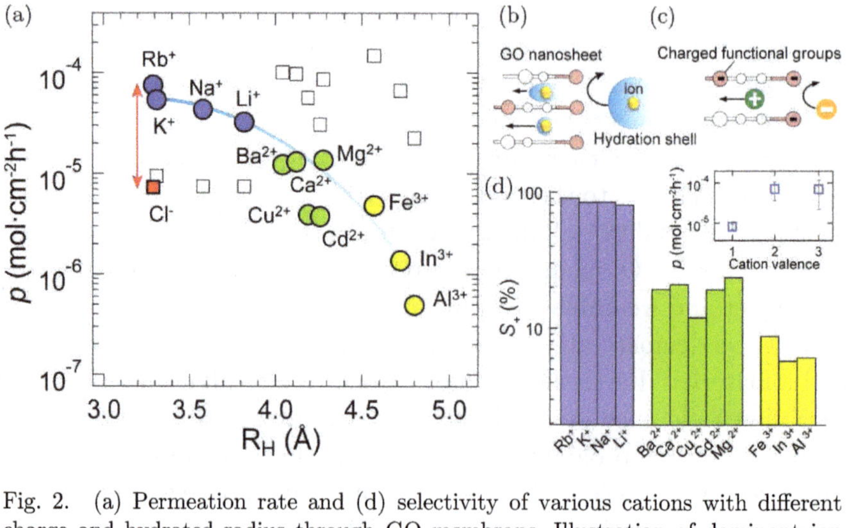

Fig. 2. (a) Permeation rate and (d) selectivity of various cations with different charge and hydrated radius through GO membrane. Illustration of dominant ion rejection mechanisms: (b) size exclusion and (c) electrostatic repulsion. Adapted with permission.[41] Copyright © 2017 American Chemical Society.

between graphene layers with interlayer distance $d = 6$ Å. For larger d of 7–10 Å, the formation of ice bilayer and its destruction at GO edges give rise to anomalous water penetration through GO paper. In addition, the structure of water for different channel widths and the interaction of group I and II cations with GO membrane were investigated by Dix.[43] The d of 7, 9, and 11 Å correspond to monolayer, bilayer, and trilayer of water molecules, respectively. The molecular dynamic simulation parameters are modified to include the cation–π interaction between cations and delocalized π electrons of graphene. It suggests that Na^+ and K^+ prefer to reside near graphene surface while Li^+, Mg^{2+}, and Ca^{2+} distribute at the center or just slightly off-center of graphene channel. Such studies open new perspectives in regulated pumping of water and the formation of biomimetic ionic channels using GO.

Meanwhile, various methods have been employed to control the interlayer spacing of GO membranes, resulting in control over the water flux and salt rejection rate. Based on strong cation–π interaction between hydrated cations and graphene channel and the interaction between hydrated cations and oxygen-containing groups on GO surface, Chen *et al.*[44] demonstrated a possibility to regulate d-spacing by immersing GO membranes in solution of KCl, NaCl, $CaCl_2$, LiCl, or $MgCl_2$. KCl-treated membrane could reject Na^+, Mg^{2+}, Ca^{2+}, and even K^+ itself at a high efficiency. Moreover, density functional theory computation illustrates that the most stable adsorption position of K^+, Na^+, and Li^+ is the region where oxygen-containing groups and aromatic rings coexist.

Abraham *et al.*[45] stored GO laminates at different relative humidities, leading to varying of d-spacing from 6.4 Å to 9.8 Å. Salts permeation results show that the permeation rate and cutoff diameter for ions decrease as d-spacing decreases and that no ions permeate through GO membrane with $d = 6.4$ Å. The permeation rates of K^+ and Na^+ have exponential dependence on d, whereas that for water permeance only vary slightly.

Another method to control water and ionic transport is to incorporate a controllable amount of graphene, which is proposed by Hung *et al.*[46] Since graphene provides frictionless capillary action in GO–graphene composite membrane, the permeability of water molecules is improved and salt rejection is also enhanced due to decreased *d*-spacing. External pressure regulation is employed by Lai *et al.*[46] to control *d*-spacing. They demonstrated that interlayer distance decreases with increasing external pressure from 2 to 6 MPa, which results in decreasing water permeance and increasing salts rejection.

Intriguingly, a rare lattice expansion of GO under pressure was revealed by Talyzin *et al.*[47] As the applied pressure gradually increases to 1.3–1.5 GPa, the interlayer distance also exhibits a remarkable increase because of the insertion of liquid water molecules. The sign of compressibility then becomes positive when pressure is higher than 1.4–1.5 GPa, which is attributed to phase transition of liquid water to ice VI structure. Other external stimuli such as pressure, salt concentration, and pH were studied by Huang *et al.*[48] by studying the performance of GO nanochannel. At the pressure range of 0.3–1.3 MPa, nanochannels begin to shrink and simultaneously the electrostatic repulsion between negatively charged GO sheets becomes stronger, resulting in slower growth of water flux and increasing rejection of some organic compounds, such as Evans blue.

Because of protonation of carboxylic groups and charge screening effect, low pH and high salt concentration leads to the passivation of electrostatic repulsions between GO sheets as well as decreasing interlayer distance. Thus, lower water flux and higher rejection rate have been achieved.

The pH-responsive phenomenon of GO/poly(allylamine hydrochloride) (GO/PAH) membrane was also investigated by Oh *et al.*[17] The water transport gets impeded as pH increases because increasing pH slightly enhanced channel size and significantly increased occupied pore volume by hydration of –COO–, which is different from aforementioned Huang *et al.*'s[48] work. At pH 4, due to electrostatic interaction between cations or anions and positively charged PAH molecules, the rejection of $MgCl_2$ ranks the highest and that of Na_2SO_4 ranks the lowest. With pH increasing from 7 to 10, PAH becomes unprotonated and, thus, less charged. Thus, the selectivity towards $MgCl_2$ decreases sharply.

Electrostatic interaction plays an important role for the removal of charged organic dyes. The rejection of positively charged methylene blue decreases from 97% at pH 7 to 62% at pH 10, and that of negatively charged rhodamine-WT decreases from 100% at pH 7 to 70% at pH 4, due to the fact that GO in the GO/PAH composites is negatively charged at pH 10 while lose most of the negative charges at pH 4. It is also noted that π–π interaction plays an important role for the removal of neutral organic molecules. The mobility of such dyes such as, for example, triclocarban is hindered by π–π interaction between its two benzene rings and the unoxidized graphene regions.

Apart from diffusion driven ionic transport, ion transport in GO nanochannels is also photo-responsive. Yang *et al.*[49] reported that upon light illumination, the electric potential difference is generated due to the different diffusivity and mobility of holes and electrons, which induces ions flow from non-illuminated region to illuminated ones. With the increasing light density, photo-induced current can exceed the opposite diffusion current, exhibiting the anti-gradient cation transport phenomenon. Based on this concept in combination with the fact that light intensity decays with increasing depth, Feng *et al.*[50] further designed geometry-asymmetric GO membranes with thickness gradient. A cationic flow is generated from thin side to thick side under full-area light illumination, revealing the photo-responsiveness of GO membranes. Unlike pore size and pressure controlled ion transport in common GO membranes, recently, our group also reported an ion–membrane interaction determined adaptive biomimetic GO–polyamine composite membrane, where competing interactions between H^+/ions and functional radicals of GO–PA enable it with programmable permselectivity for various ions.[18]

Solvents also have been indicated to impact on the separation capability for GO membranes. Zheng *et al.*[51] reported a solvent-influenced interlayer spacing of GO membrane, based on the Hansen solubility distance discrepancy between GO and different solvents. Hansen solubility distance is a solubility parameter used to judge the swelling effect of GO in varying solvent. Zheng *et al.* found that the swelling of GO membranes decrease as the solubility distance of the solvents increase, and thus resulting in decrease of interlayer spacing and better separation of GO membranes in solvents with large solubility distance. Consequently, GO membranes in ethanol and acetone (having large solubility distance) were found to perform minor swelling and have good separation performance of 90% for rhodamine B and methylene blue, which is superior to GO membranes in dimethylformamide, N-methyl-2-pyrrolidone (having small solubility distance and severe swelling).

The trade-off between the permeability and selectivity of GO membrane is one of the challenges for separation applications. Recently, Zhang *et al.*[52] illustrated a novel approach to break this limit by rotating nanoporous graphene membrane, based on the molecular dynamic simulation. Within this manner, the surface slip occurred at the graphene/liquid interface leads to the time discrepancy between ion's penetration and ion's crossing, which enables this particular membrane with good selectivity of nearly 100% for salt rejection while maintaining high permeation even when its pore size is larger than the aimed hydrated ions.

One notorious drawback of pure GO membrane is the swelling behavior in aqueous solution due to GO's hydrophilicity endowed by oxygen-containing groups. Swelling leads to an increase in the size of nanochannels and impair membrane selectivity. In order to mitigate this problem, GO can be mixed with other functional materials to enhance the stability and selective properties of the composite membranes. Because of the introduction of GO nanosheets, the synthesized membrane also exhibits strengthened mechanical property and chemical resistance.[53,54]

3. Membranes for Water Treatment

GO membranes are actively investigated for applications in forward osmosis (FO), reverse osmosis (RO), nanofiltration (NF), and electrodialysis (ED).[55] Advantages and disadvantages of these processes for applications are summarized in Table 1. The main challenges are the trade-off between high water flux, high solute rejection, and membrane fouling.

Table 1. GO-based membrane processes for water desalination and wastewater purification.

Membrane processes	Driven force	Advantages	Disadvantages
Forward osmosis (FO)	Osmotic pressure difference	Low energy consumption Less membrane fouling Enabling treatment of high salinity solution	Concentration polarization
Reverse osmosis (RO)	External hydraulic pressure	High solute rejection efficiency	Membrane fouling High pressure demand
Nanofiltration (NF)	External hydraulic pressure	Lower operation pressure than RO	Membrane fouling Unsatisfactory solute rejection efficiency
Electrodialysis (ED)	Applied electrical potential difference	Less membrane fouling Long membrane lifetime	Invalid for non-charged species

As for water desalination, novel polyamide (PA)-crosslinked GO membranes for FO were synthesized by using m-xylylenediamine (MXDA) as intra-crosslinker and trimethyl chloride (TMC) as inter-crosslinker.[56] The water flux and solute flux for 0.25 M trisodium citrate (TSC) reach $18\,L\,m^{-2}\,h^{-1}$ and $0.1\,mol\,m^{-2}\,h^{-1}$, and those for 0.33 M Na_2SO_4 are $11\,L\,m^{-2}\,h^{-1}$ and $0.3\,mol\,m^{-2}\,h^{-1}$, respectively. GO/PA membrane exhibits better performance than a commercial cellulose triacetate (CTA) FO membrane. Chae *et al.*[57] also fabricated a GO/PA membrane for RO mode. After gradually incorporating GO up to 38 ppm, the water flux of the membranes increases to $16.6\,L\,m^{-2}\,h^{-1}$ and NaCl rejection constantly maintains at 99.5% under 225 psi. Another work demonstrated that incorporating GO into polysulfone (PSf) matrix enhances surface hydrophilicity and porosity, thus leading to higher water flux as GO content increases. The rejection for Na_2SO_4 through 2000 ppm GO/PSf membrane is 72% under 4 bar pressure. Nan *et al.*[58] reported a positively charged GO/polyethyleneimine (PEI) NF membrane for refusing multivalent ions, which possesses a water flux of $4.2\,L\,m^{-2}\,h^{-1}$, 93.9% $MgCl_2$ rejection, 80% $MgSO_4$ rejection, 38.1% NaCl rejection and 28% Na_2SO_4 rejection.

Because of charged functional groups, GOs have also been incorporated into polymer matrices to serve as IEMs.[59] A normal schematic configuration of ED process is illustrated in Fig. 3(a). During ED process, cations move through negatively charged cation exchange membrane (CEM) from the dilute chamber to the concentrated one and then are blocked by positively charged anion exchange membrane (AEM). Hence, ion conductivity and permselectivity are significant factors that affect the performance of IEMs.[60] Gahlot *et al.*[61] fabricated GO/sulfonated polyethersulfone (GO/SPES) nanocomposite IEMs for ED application. The ionic conductivity for pure SPES membrane increases from $2.34 \times 10^{-2}\,S\,cm^{-1}$ to $6.4 \times 10^{-2}\,S{\cdot}cm^{-1}$ for 10 wt.% GO/SPES membrane due to the presence of carboxyl and hydroxyl groups on GO surfaces. ED experiments indicate that the concentration of NaCl in dilute compartment decreases at a higher rate as the incorporated amount of GO increases, which can contribute to enhanced ionic conductivity. The salt flux, current efficiency, and power consumption at the

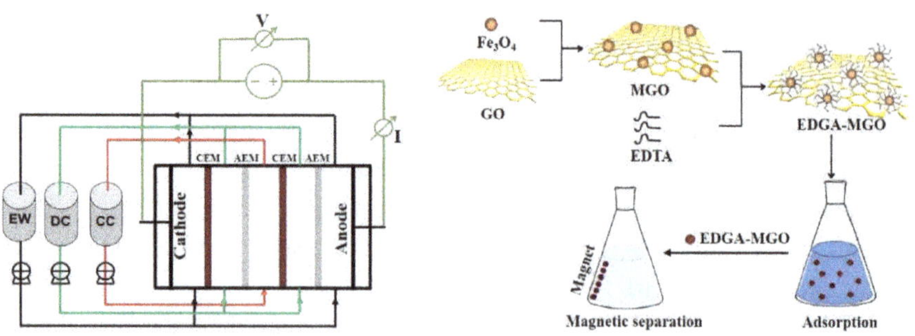

(a) (b)

Fig. 3. (a) Schematic illustration of ED system, adapted with permission.[61] Copyright © The Royal Society of Chemistry 2014. (b) Synthesis and work flow of ethylenediaminetetraacetic acid functionalized magnetic GO (EDTA–mGO) for removal of Pb(II), Hg(II), and Cu(II), adapted with permission.[64] Copyright © 2015 Elsevier B.V.

constant potential are calculated to be 3.71 mol m^{-2} h^{-1}, 97.4%, and 4.3 kW h kg^{-1}, respectively, exhibiting great potential for water desalination.

In 2017, Hosseini *et al.*[62] designed a series of heterogeneous CEMs based on polyvinylchloride (PVC)/GO nanoplates (GONs)/TiO_2 nanoparticles. Benefiting from the synergistic effect between high specific surface area of GO and antifouling properties of TiO_2 nanoparticles, the modified membranes generally have better performance than pristine PVC membrane. Among them, the one consisting of 3 wt.% GONs and 1 wt.% TiO_2 nanoparticle has the highest permselectivity over 99%, highest Na^+ flux of 9.3×10^5 mol m^{-2} s^{-1}, and lowest electrical resistance of 4–5 Ω cm^2.

However, as pressure-driven GO membranes for desalination based on salt rejection mechanism have been intensively researched, concerns over the actual desalination capability of these membranes for practical applications have also been proposed since most of the experiments were conducted in a dead-end system without water flow. Hu *et al.*[63] found that after prolonging the measuring time, the salt rejection capability for Na_2SO_4, $MgCl_2$, NaCl, $MgSO_4$, and KCl of GO membrane surprisingly drops down, which can be accounted by the shield effects of positive ions accumulating on the negative surface of GO laminates when extending test time. This finding urges the attention needed to be paid on the criterion and feasibility of salt rejection-based GO membrane applications for desalination.

Meanwhile, industrial wastewater treatment is of great concern since heavy metal ions and organic dyes are detrimental both to natural environment and humans. In many recent studies, GO has been investigated and proved to be excellent adsorbents for metal ions and dyes removal.[65] The mechanisms of heavy metal contaminants adsorption are based on electrostatic attraction between negatively charged GO and positively charged metal ions, ion exchange between the proton in –COOH, –OH groups, and heavy metal ions, and surface complexation between heavy metal ions and oxygen-containing groups.[66,67] As for organic dyes, the adsorption is dominated by electrostatic interaction, Van der Waals force, or π–π stacking interaction between GO sheets and organic dyes.[68] Furthermore, as abundant aromatic domains GO still possesses, it is worth nothing that cation–π interaction also plays a significant role in GO-based separation and adsorption, where cation–π interplay induced cross-linking effects promote the stability and permselectivity of GO membrane and corresponding adsorption capability.[69]

To enhance adsorption capacity, GO has also been functionalized and coalesced with other materials. In 2011, Zhao *et al.*[26] used GO nanosheets as adsorbents for the removal of Cd(II) and Co(II) ions. It indicates that the solution pH can affect the adsorption of these two ions since pH affects GO surface charge and the consequent electrostatic interaction. At pH 6.0 and 303 K, the maximum adsorption capacities of Cd(II) and Co(II) ions could reach 106.3 and $68.2\,mg\,g^{-1}$, respectively. As illustrated in Fig. 3(b), Cui *et al.*[64] functionalized GO with Fe_3O_4 nanoparticles and ethylene diamine tetraacetic acid (EDTA) and studied the effect of adsorbent dosage, pH, ionic strength, and contact time on the removal efficiency. The maximum adsorption capacities are found to be $508.4\,mg\,g^{-1}$ for Pb(II), $268.4\,mg\,g^{-1}$ for Hg(II), and $301.2\,mg\,g^{-1}$ for Cu(II), respectively.

GO/TiO_2 films were fabricated by Xu *et al.*[70] via a simple vacuum filtration method and showed 100% removal efficiency for 8 mL Methyl Orange (MO) and 14 mL Rhodamine B, respectively. Natural polysaccharides (PS) including inulin, xylan, and κ-carrageenan were also employed to form plywood-like structure with GO for four organic dyes removal.[68] It elucidates that GO/PS composites have better adsorption performance for cationic dyes (methylene blue and rhodamine 6G) than anionic dyes (orange II and acid fuchsin). Notably, the adsorption capacity of methylene blue could even reach $800\,mg\,g^{-1}$, which is much higher than pure GO.

4. Gas Separation Membranes

Kim *et al.*[28] studied gas separation behavior through GO membranes fabricated by different methods. By controlling adaptive nanopores and flow channels within GO membrane, gas permeation is affected. The applied transmembrane pressure, water content, and temperature were also confirmed to influence gas permeation. These membranes possess high CO_2 selectivity because of the interaction between CO_2 and carboxylic acid groups, which can be used for practical CO_2 separation process. A series of GO membranes with different thickness ranging from 1.8 to 180 nm were fabricated by Li *et al.*[71] using vacuum filtration method. The highest selectivity for H_2/CO_2 and H_2/N_2 mixtures reaches 3400 and 900, respectively, which are much higher than other advanced microporous separation membranes.

The gas separation performance of polymer membranes could be improved with the incorporation of nanofillers. Among them, GO has received extensive attention because of their abundant functional groups and stable dispersion property. Zahri *et al.*[72] combined glassy polysulfone (PSf) polymer with GO to fabricate hollow fiber mixed matrix membranes (MMM). Since thinner, denser, and defect free surface layer is formed by additive GO, the permeance for relatively larger gas molecules, CH_4 and N_2, decreases from 3.88 GPU ($1\text{GPU} = 1 \times 10^{-6}\,\text{cm}^3/\text{cm}^2$ sec cmHg at standard temperature and pressure) through pure PSf membrane to 1.96 GPU through GO/PSf membrane and from 3.90 GPU through pure PSf membrane to 2.63 GPU through GO/PSf membrane, respectively. However, the permeation of CO_2 increases from 65.24 GPU to 74.47 GPU due to its affinity to GO π–π stacking bonds.

Recently, in order to regulate the size of permeation channels, Cheng *et al.*[73] incorporated a small molecule, cysteamine, with GO (CGO membrane) via the interaction between amino and thiol in cysteamine and oxygen-containing groups on GO sheets. After incorporating 7.6 wt.% cysteamine, the interlayer distance reduces from 0.98 nm to 0.94 nm. The CGO membrane with 7.6 wt.% cysteamine possesses the H_2 permeance of 51.5 GPU and a high H_2/CO_2 selectivity of 21.3 which is more than twice that for unmodified GO membrane.

Metal-incorporation in rGO membranes for gas separation has also been investigated by Jin *et al.*[74] recently. Transition metals in the incorporated membranes are thought to preferably interact with specific gas molecule and thus contributes to the selective permeation for gas mixtures. Particularly, Fe-intercalated rGO membrane shows a reproducible high selectivity for N_2/CO_2 of 97% with pressure at 110 mbar, based on the preferred interaction between Fe active sites and N_2. Additionally, coexistence of Maxwell–Stefan diffusion mode in low transmembrane pressure and Knudsen diffusion mode in high transmembrane pressure were revealed in the process of N_2 permeation across Fe-incorporated membrane.

In general, the gas transport through nanopores on 2D materials surface gains great interest, and many factors such as pore size, pore geometry, and the interaction between pore and gas have been proved to influence gas separation.[75] Both theoretical and experimental work have been extensively conducted to understand the mechanism and optimize the performance of nanoporous graphene and GO membrane.[75,76]

Jiang *et al.*[77] used first-principle density functional theory calculations to investigate the permeability and selectivity of graphene with passivated holes, where pore width plays an important role. They found that nitrogen functionalized pores and hydrogen passivation pores exhibit extremely high H_2/CH_4 selectivity of 10^8 and 10^{23}, respectively, demonstrating the superiority of nanoporous graphene membrane compared with traditional polymer and silica membranes.

The separation of Kr-85, a radioactive gas, from air was predicted by Fatemi *et al.*[78] using the molecular dynamic stimulation. A critical pore size with seven graphene ring units removed on GO sheets allows the permeation of Kr-85 with relatively high gas flow of $3500\,\mathrm{mol\,m^{-2}\,s^{-1}}$ but is too small for N_2 and O_2. Liu *et al.*[79] investigated pure gases (H_2, CH_4, N_2, and CO_2) and H_2/CO_2 mixture through double layer GO membrane and elucidated that gas permeation increases as the interlayer distance and pore offset increase. As their affinity to GO membranes decreases in the order of $CO_2 > N_2 \approx CH_4 > H_2$, the permeance changes in a reverse way, exhibiting $11 \times 10^{-5}\,\mathrm{mol\,m^{-2}\,s^{-1}\,Pa^{-1}}$ for CO_2, $14 \times 10^{-5}\,\mathrm{mol\,m^{-2}\,s^{-1}\,Pa^{-1}}$ for N_2, $18 \times 10^{-5}\,\mathrm{mol\,m^{-2}\,s^{-1}\,Pa^{-1}}$ for CH_4 and $60\times 10^{-5}\,\mathrm{mol\,m^{-2}\,s^{-1}\,Pa^{-1}}$ for H_2. For H_2/CO_2 mixture, CO_2 is selected over H_2 at lower interlayer distance, which is dominated by CO_2 adsorption on GO surface. However, at larger interlayer distance, H_2 is selected over CO_2 because sieving effect becomes dominant.

5. Membranes for Solvent Dehydration

Ever since the pioneering work of Nair *et al.*[80] on water permeation through GO membranes, it became clear that GO laminates can be used for solvent dehydration. Tang *et al.*[29] employed a pressurized ultrafiltration system to fabricate freestanding GO membranes for ethanol dehydration. Pervaporation experiments show that as the water content in feed solution increases, the permeability of water and ethanol both increase due to membrane swelling. But the permeability of ethanol is much lower than that of water, which resulted from greater partitioning and intercalation (i.e. channel occupation) of water in the transport channels compared with ethanol. For 85 wt.% ethanol solution, the water permeability and water/ethanol selectivity reach 13,800 Barrer (1 Barrer $= 3.348 \times 10^{-19}\,\mathrm{kmol\,m\,m^{-2}\,s^{-1}\,Pa^{-1}}$) and 227, respectively. Similarly, Li *et al.*[81] firstly modified GO and ceramic support surface to enhance GO adhesion to porous Al_2O_3 and then deposited GO layer under pressure.

The GO membrane exhibits extraordinary separation performance towards the mixtures of water and organic solvents including ethanol, n-propanol (NPA), isopropyl alcohol (IPA), and ethyl acetate (EA), with the corresponding separation factors of 250, 1293, 2942, and 4400, respectively. This is in accordance with the increasing size of organic molecules. The incompatibility between hydrophilic nature of GO and hydrophobic nature of organic solvents is responsible for the separation effect of smaller solvents like ethanol.

The polyvinyl alcohol (PVA)–GO–TiO_2 MMMs were used for IPA dehydration.[82] The authors observed that increasing amount of water in feed solution is due to more hydrophilic behavior of PVA–GO–TiO_2 membrane and more flexibility of polymer chains, contributing to increased flux and decreased selectivity. Moreover, as GO–TiO_2 concentration gradually increases to 0.75%, permeate flux decreases from 271.4 $g\,m^{-2}\,h^{-1}$ to 190.92 $g\,m^{-2}\,h^{-1}$ and selectivity dramatically increases from 95.5 to 631.6 due to denser membrane structure.

Halakoo and Feng[30] designed a crosslinked membrane composed of PEI and GO with glutaraldehyde as the crosslinking agent to restrict membrane swelling tendency. GO/PEI membrane with seven bilayers demonstrates a total flux of 1770 $g\,m^{-2}\,h^{-1}$ and a separation factor of 77 for 98 wt.% ethanol/water mixture, and those of 1494 $g\,m^{-2}\,h^{-1}$ and 197 for 98 wt.% isopropanol/water mixture.

As a trans-membrane protein, aquaporin (AQP) possesses both excellent permeability and high selectivity of water molecules due to the formation of a hydrophobic and positively charged narrow pore and its adjacent hydrophilic region. Inspired by the hourglass structure of AQP, Mao *et al.*[83] employed imidazole-ureido (IU) with water binding extremities and hydrophobic alkane chains and designed a novel pervaporation membrane based on IU and GO (IU/GO). The total flux and separation factor of IU/GO membrane reach 3506 $g\,m^{-2}\,h^{-1}$ and 4454 for 80 wt.% butanol/water mixture.

In summary, the ionic and molecular separation performances of GO-based membranes have been discussed from the above three aspects of water treatment, gas separation, and solvent dehydration. The selectivity of ions and molecules is principally governed by the size exclusion and the various interaction between functional groups of GO and target ions or molecules. The unique characteristics of GO membranes such as good mechanical strength, good chemical stability, controllable size of nanochannels, and the ability of diverse functionalization make them the next-generation materials for various separation processes.

6. Membranes in Sensor and Biomedical Field

GO was used to construct sensors for a number of different applications including biomedical. Lee and Wang.[84] reported a NH_3 gas sensor by incorporating rGO into a polyaniline (PANI) matrix. In these PANI–rGO nanocomposite films, rGO nanosheets act as active nucleation sites to the homogeneously distributed PANI particles.[85] As the amount of NH_3 gas absorbed on the surface of rGO increases, the conductivity of the composite films will decrease due to carrier transference, thus it is possible to use such films for sensor applications. Also, utilizing PANI as the polymer matrix, Peng *et al.*[86] developed an optical fiber dual Fabry–Pérot interferometric carbon monoxide (CO) gas sensor based on PANI/Co_3O_4/GO (PCG) sensing film coated on the end face of the optical fiber. The functional groups on the surface of GO improve the interaction between the composite membrane and CO molecules, especially promote the formation of hydrogen bonds, resulting in variation of the composite film thickness. In this way, the sensor can respond extremely fast to CO.

Another example of GO membrane used as a gas sensor was reported by Miyamoto *et al.*[87] in 2017. The authors found that the GO membrane shows good proton-conducting properties when exposed to humid air. Thus, they proposed a solid gas sensor based on proton-conducting GO membrane, which can detect combustible gases at low concentrations continuously.

Besides gas sensors, GO-based membranes have shown great promise in temperature sensors, humidity sensors, and chemical sensors. Thus, the fact that adsorbed water will significantly change the dielectric properties of GO has been used to construct wireless humidity sensors by coating an RFID antenna with GO layer.[88] To detect temperature, Khan and Kang.[89] mixed GO with polyvinylpyrrolidone polymer and constructed an interdigitated capacitor (IDC)-based temperature sensor. When the temperature increases, the electron–hole concentration of GO will change due to thermal excitation. As a result, the dynamic conductivity and the dielectric constant of the IDC sensing film is changed. For humidity detection, a novel quartz crystal microbalance (QCM) humidity sensor using a GO membrane as a filter layer was demonstrated by Liu *et al.*[90] Because of the unique nature of GO membranes to selectively permeate water molecules, sensors incorporating GO filter membranes greatly reduce the cross-sensitivity of the humidity sensor to other gases.

Furthermore, GO membrane also plays an important role in chemical sensing. Taking the dopamine sensor as an example,

Jia *et al.*[91] reported a novel Mo-doped rGO/polyimide (Mo–rGO/PI) composite membrane, which exhibits reliable detection of dopamine signals in biological samples. rGO not only improves the electrical conductivity, but also enhances the thermal stability and mechanical properties of the composite membrane.

On the other hand, GO- or rGO-based membranes have attracted considerable interests in biomedical field, which is attributed to the good chemical stability and biocompatibility of GO.[33–35] Many studies have validated that the composite membranes composed of GO- and bio-based polymers are promising candidates for biomedical applications. With the integration of chitosan, Jeon *et al.*[92] reported a nano-biopolymer membrane produced by pendent sulfonated chitosan and functionalized GO, which can be used for energy-efficient artificial muscles, biomimetic robots and disposable biomedical devices in the future. GO is introduced to improve strong ionic interactions and increase in-plane mechanical rigidity and electro-chemical-mechanical properties. Additionally, the profound effect of GO nanosheets on the healing process of open wounds were noticed,[93] and chitosan/GO composite membranes were found to show great potential in tissue engineering, wound healing, and drug delivery.[94]

Ioniţă *et al.*[37] used a non-solvent induced phase separation technique and generated a cellulose acetate/ammonia functionalized GO membrane. By conjugating GO, *in vitro* biological performance is greatly improved, and the resulting composite material shows excellent cytocompatibility and biological suitability. Recently, Yahia *et al.*[95] further introduced hydroxyapatite to the cellulose/GO system and reported an antibacterial agent. An antibacterial hybrid GO/hydroxyapatite/cellulose composite membrane containing silver nanoparticles (AgNPs) was successfully fabricated. In Yahia's work, GO promotes AgNPs to deposit on cellulose fibers and prevent leaking of Ag ions, resulting in effective antimicrobial activity. In addition to cellulose/GO composite membrane, Jin *et al.*[96] modified rGO films with fibronectin and proposed the potential of rGO films for application in cell culture and tissue engineering. Moreover, collagen/GO composite membrane and glucomannan/GO composite membrane were found to show potential in skin or wound dressings as well as in other engineering and biomedical applications.[97,98]

GO shows excellent compatibility with different hydrophilic and hydrophobic polymers. Very interesting for biomedical application are the composites of GO with biodegradable/bioresorbable polymers. In poly(l-lactide) (PLL)/GO films,[99] for example, GO enhanced mechanical flexibility and increased antibacterial activity of biocompatible PLL films. Reported by Mao *et al.*,[100] the addition of GO to PLL/GO nanofiber membranes significantly improves thermal stability and promotes drug release rate. In the work of Thampi *et al.*,[101] GO was dispersed in electrospun poly(carbonate urethane) matrix, contributing to an increase in the mechanical performance of the composite biocompatible membranes. By confining GO to a sulfonated polyanion co-doped hydrogel thin film, He *et al.*[102] performed *in vitro* evaluation and found that the composite showed good anti-coagulant activity, suppressed platelet adhesion and activation, low inflammation potential, and good red blood cell compatibility. Such properties are highly desirable for the development of new bioink formulations for scaffold-based 3D bioprinting and tissue engineering.

To conclude this section, the application of GO-based membranes for sensors and biomedical applications has been reported extensively. With the addition of GO, thermal and mechanical properties of membranes are greatly improved. To date, GO-based membranes act as gas, humidity and temperature sensors have been explored.

On the other hand, integration of GO in bio-based and biocompatible polymers shows outstanding performance in cell culture, drug delivery, and tissue engineering. However, up to now only a few efforts have been done towards bioengineering, bioimaging, and biosensing applications of GO. Furthermore, a main challenge of GO-based membranes is *in vivo* evaluation, as membrane behavior in living matter, *in vivo* stability and performance, interactions with cells and membrane proteins are rarely reported. Understanding of *in vivo* behavior of GO and GO-based composites can be proposed for further investigations.

7. Membranes for Energy Conversion and Storage

In the electrochemical systems, like secondary batteries, fuel cells (FCs), and supercapacitors, GO membranes can act as separators and/or solid-state electrolytes (ion conductors). Unlike graphene and graphite, GO is an electrical insulator, which is the primary requirement for separators.[103] The functional groups in GO make it behave as a weak polyacid with ion conducting property and good polymer electrolytes compatibility. Such natures make GO a promising material for the designs of inexpensive GO nanocomposite membranes with synergistically enhanced properties for high-performance electrochemical applications and biocompatible ionic transistors. It is worth illustrating that the boundary between separators and ion conductors mentioned here is not strict. GO membranes can simultaneously serve as ion conductors and separators based on their high ionic/molecular selectivity.

GO membranes have been applied to different types of secondary batteries, such as lithium-ion,[104–107] lithium–oxygen (Li–O_2),[108,109] lithium–sulfur (Li–S),[110,111] vanadium redox flow,[112–114] zinc-ion,[115] and Zn/Ce redox flow,[116] etc. The GO membranes are exploited as separators, IEMs, and flexible electrodes.

(i) As flexible electrodes. The flexible electrodes can be used to fabricate flexible battries.[15,21] For example, Du *et al.*[115] designed a flexible aqueous zinc-ion battery (see Fig. 4(a)) with freestanding PANI–GO/carbon nanotubes film as a cathode. The excellent film-forming property of GO makes the preparation process quite easy. They attributed the high reversible discharge capacity, rate performance, and long cycle stability of the prepared battery to the proton doping impact of GO molecules.

(ii) As separators. High quality separators are urgently required for Li–S batteries, which are one of the most promising candidates for next generation secondary batteries, because they might have much higher energy capacity and density than commercial Li-ion batteries.[110,111] Before the commercialization can be achieved, the shuttle effect (dissolution and migration of polysulfide toward anode side during cycling) in Li–S batteries must be overcome to avoid severe capacity fading.[117] Introduction of GO membranes as separators is an

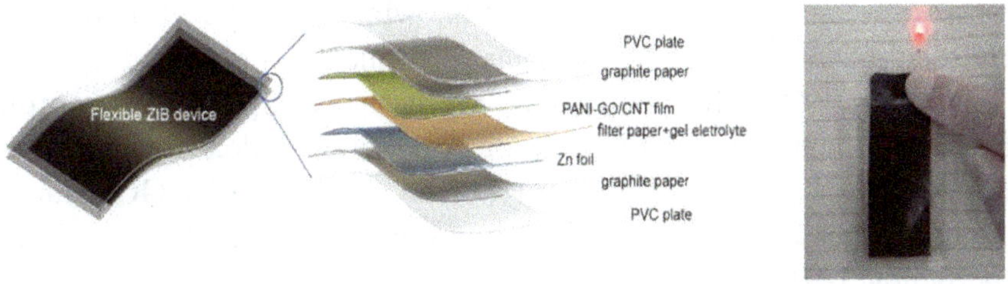

(a)

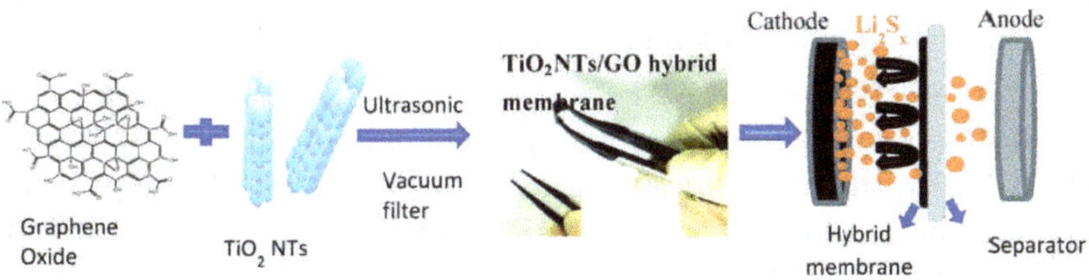

(b)

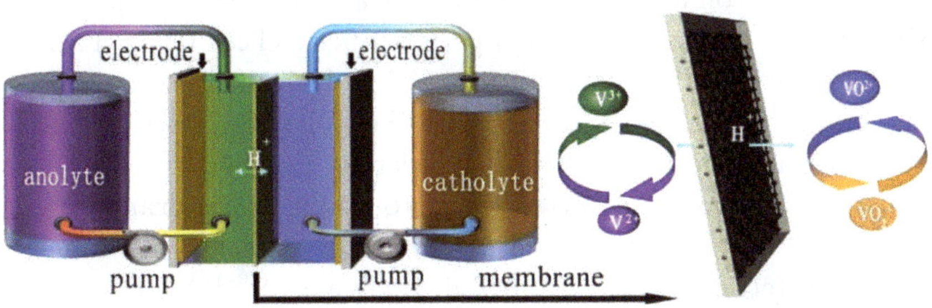

(c)

Fig. 4. Schematic of (a) a flexible zinc-ion battery configuration and corresponding image of lighting up a LED, adapted with permission.[115] Copyright © 2020 Elsevier B.V. (b) Preparation and working of an ultrathin TiO_2 nanotubes/GO hybrid membrane as an interlayer separator for Li–S battery, adapted with permission.[110] Published by The Royal Society of Chemistry. (c) A vanadium redox flow battery single cell, adapted with permission.[113] Copyright © 2020 Wiley-VCH Verlag GmbH & Co. KGaA, Weinheim.

effective way to inhibit shuttle effect by blocking the migration pathways of polysulfide. For instance, Song *et al.*[110] prepared an ultrathin TiO_2 nanotubes/GO hybrid membrane as an interlayer separator for Li–S batteries, as illustrated in Fig. 4(b). The hybrid membrane blocks polysulfides while allowing the permeation of Li^+, thus, significantly improved cycling performance is obtained.

(iii) As IEMs. GO membranes are actively exploited as proton exchange membranes (PEMs) for vanadium redox flow batteries.[112–114] Lou and co-authors.[113] reported a cost-effective GO/Nafion composite membrane as effective vanadium-ion barrier for vanadium redox flow batteries, as illustrated in Fig. 4(c). In addition, Nicotera *et al.*[104] incorporated sulfonated GO (sGO–Li^+) into lithiated Nafion to successfully fabricate a novel nanocomposite membrane as single Li^+ conducting polyelectrolyte for lithium-ion batteries.

More applications of GO as IEMs can be found in FCs. The schematic assembly of a proton exchange membrane fuel cell (PEMFC) with GO membrane is given in Fig. 5. In fact, the pristine GO membranes have been used as PEMs to design the so-called GO membrane FCs.[118,119] The obtained maximum power density is $\sim 34\,\mathrm{mW\,cm^{-2}}$ (at 30°), approaching that of a commercial Nafion under similar conditions.[118] This promisingly indicates the potential of GO for the preparation of high-performance fuel cell IEMs. Therein the fabrication of functionalized GO/polyelectrolyte composite membranes is the best strategy. For the polyelectrolytes, a future trend is using the low-cost and eco-friendly or even bio-based and biodegradable polyelectrolytes (refer to Table 2 for examples) to replace the costly, non-biodegradable plastics like Nafion (a sulfonated tetrafluoroethylene-based copolymer).[120–123]

For better presentation, some selected cutting-edge examples of functionalized GO/polyelectrolyte IEMs are summarized in Table 2. The focus is on functionalization strategy of GO, polyelectrolytes, IEM types, and the corresponding FC categories. Actually, the IEMs should not be limited to the tested FC types, as these IEMs can be tuned to meet the needs of different fuels and conditions. In FCs, IEMs simultaneously serve as ion conductor and separator, owing to their high ion/molecule selectivity, which is exactly the property of GO, and can be further improved by chemical modification of GO. The blocking of fuel crossover is indeed as important as gaining high targeted-ion conductivity, as the fuel crossover can cut down the cell performance and even lead to explosion.

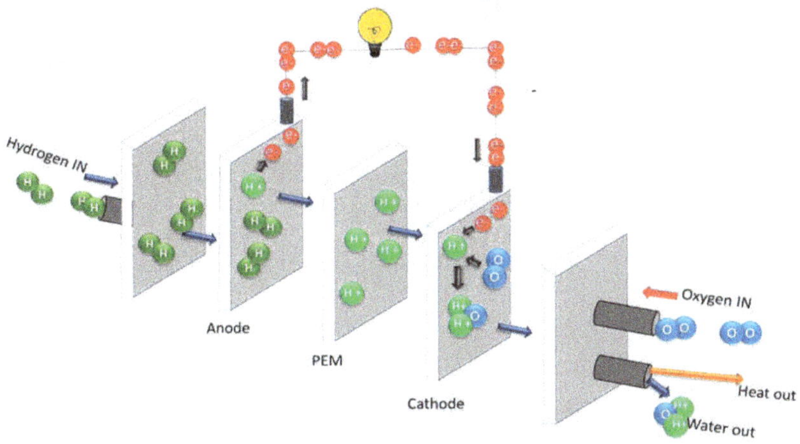

(a)

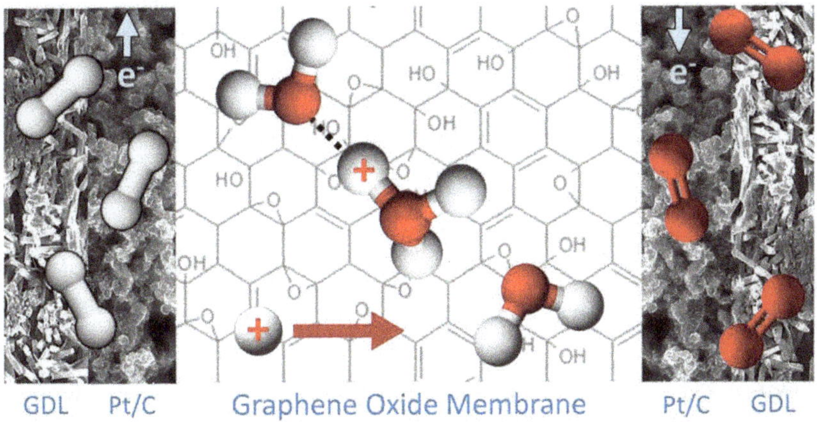

(b)

Fig. 5. (a) A sketch of PEMFC assembly, adapted with permission.[123] Copyright © 2018, American Chemical Society. (b) A schematic illustration of GO membrane with proton transportation (GDL: gas diffusion layer, Pt/C: catalyst layer), adapted with permission.[118] Copyright © 2014 Elsevier B.V.

Supercapacitors store energy by electrostatically charging the double layer and thus require conducting materials with high surface area.[38] In recent years, researches paid intensive attention to the mechanically flexible solid-state supercapacitors.[39,131] An example is

Table 2. The selected cutting-edge examples of functionalized GO/polyelectrolyte IEMs for FC applications.

Polyelectrolyte	Functionalized GO	IEM type	FC	Reference
Nafion	Fe_3O_4 nanoparticles anchored sulfonated GO	PEM	HT-LH-FC	124
Polybenzimidazole (PBI)	Phosphonated GO	PEM	HT-FC	125
Sulfonated poly(ether ether ketone) (SPEEK)	Ag@GO	PEM	MFC	126
Sulfonated poly (arylene ether sulfone) (SPAES)	Sulfonated poly(arylene thioether sulfone)–grafted GO	PEM	PEMFC	127
Sulfonated polyimide (SPI)	Sulfonated imidized GO	PEM	DMFC	128
Chitosan (CS)	Quaternary phosphonium polymer brush functionalized GO	AEM	DEFC	129
Polyvinyl alcohol (PVA)	Sulfonated GO	CEM	DBFC	130

Notes: (1) HT: high temperature, (2) LH: low humidity, (3) MFC: microbial fuel cell, (4) DMFC: direct methanol fuel cell, (5) DEFC: direct ethanol fuel cell, (6) DBFC: direct borohydride fuel cell, (7) AEM: anion exchange membrane (for OH^- in this case), (8) CEM: cation exchange membrane (for Na^+ in this case).

shown in Fig. 6(a). Such supercapacitors consist of flexible electrodes and polymer gel electrolyte.

In general, the fabrication of flexible electrodes using graphene and rGO has become one of the most exciting tasks for flexible electronics. Due to their high specific surface area, flexibility, high conductivity, and excellent mechanical properties, as well as rich chemical modification possibilities 2D materials are becoming more and more popular components for the design of flexible electrodes.[38,39] Especially, such electrodes are important for new self-powered bionic device. For instance, Feng *et al.*[132] reported a novel strategy to fabricate high-performance flexible solid-state supercapacitors via ionic liquid gel polymer electrolyte and hierarchically graphene

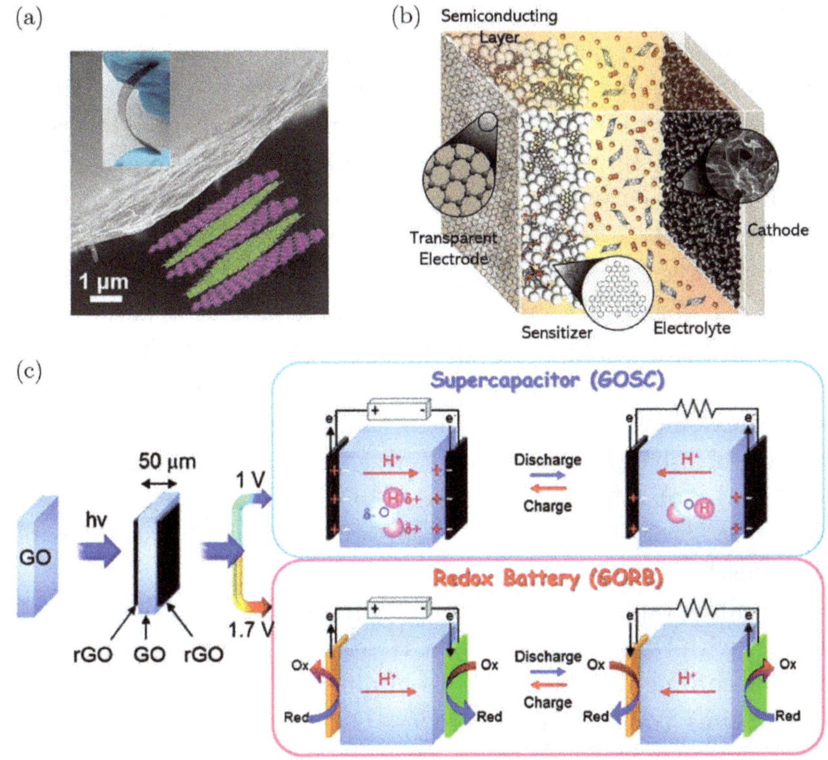

Fig. 6. (a) A schematic representation and lateral SEM image of exfoliated 2D graphene-thiophene heterostructure for high performance all-solid-state pseudocapacitor, adapted with permission.[137] Copyright © 2016 WILEY-VCH Verlag GmbH & Co. KGaA, Weinheim. (b) A sketch of preparation of an all-GO device and its applications to supercapacitors and redox battery under different working voltage, adapted with permission.[136] Copyright © 2016 The Royal Society of Chemistry. (c) A dye-sensitized solar cell incorporating graphene-based 2D materials in each part of the device, adapted with permission.[138] Copyright © 2014 American Chemical Society.

nanocomposite (by carbonization of GO@PANI nanocomposite) electrode material. The polymer gel electrolyte usually contains polymer and ion conductor. GO can also be used for such purposes due to its functional groups. Thereof high-performance GO-doped gel electrolytes have been prepared to be used in supercapacitors.[103,133]

The versatility of graphene-based 2D materials in this context indeed realizes the fabrication of flexible solid-state all-GO-supercapacitors.[134–136] An example is illustrated in Fig. 6(b). The all-GO-supercapacitors have a unique membrane structure, i.e. rGO/GO/rGO, and the preparation process is quite simple. Hereof GO layer serves as both electrolyte and separator, and rGO layers act as an electrode. Interestingly, the rGO/GO/rGO device can exhibit both supercapacitor and battery (via redox reactions) behaviors depending on the working voltage. This provides a solution to bridge the gap between supercapacitors and batteries.

Graphene-based 2D materials also exhibit promising performance in the construction of solar cells (SCs), benefiting from their superior optoelectronic properties, transparency, conductivity, chemical stability, flexibility, cost-effectiveness, and compatibility.[38,139] Such research mainly focuses on the new generation SCs such as dye-sensitized solar cells (DSSCs),[138,140] perovskite solar cells (PSCs),[141–143] organic (polymer) solar cells (OSCs),[144,145] and flexible solar cells.[38,139]

Thin film graphene, GO, and rGO are used in SCs as sketched in Fig. 6(c) with DSSC as an example. Here, we summarize different uses of GO according to their functions in different parts of SCs.

(i) (Semi) transparent electrodes. Owing to high transparency, flexible and conductive nature, graphene-based 2D materials show great potential to replace the traditional indium tin oxide (ITO) or fluorine tin oxide (FTO). Sensitivity to pH, high cost, patterning difficulty, and brittleness of ITO and FTO greatly limit their applications to flexible SCs, particularly for OSCs.[38,146]

(ii) Carrier (hole/electron) transport layers. Graphene-based 2D materials have been successfully utilized to enhance the hole/electron transport properties of SCs due to their high charge mobility and tunable energy bandgap/work function. The p- and n-type graphene can be created for hole and electron transfer respectively by doping.[139]

(iii) Interface/buffer layer. Graphene-based materials can also be exploited as interface layer between different parts of SCs to reduce current leakage via their capability to be either p- or n-doped.

(iv) Stabilizer/protection layers. Their molecular separation properties are utilized to improve the lifetime of SCs by protecting air-sensitive layers. For example, rGO films act as gas barrier for

OSCs.[147] Graphene can be used for encapsulation of perovskites to protect them from moisture and photo-oxidation, and to prevent the diffusion of halide.[38]

One more interesting example is that Phan *et al.*[148] managed to use GO-coated substrate as template to prepare reconfigurable loliginid-inspired biomimetic infrared camouflage coated thin films. They obtained dynamically tunable reflectance over more than 600 nm. The fabrication strategy is inexpensive and robust, and the coatings can be employed on arbitrary surfaces.

Based on the above discussion, we can conclude that, promisingly, graphene-based 2D materials would continue playing an important role in energy conversion and storage applications and in the design of metamaterials. More strategies should be exploited to utilize the versatility of graphene-based 2D materials, for on the one hand, further improvement of performances of the prepared devices, on the other hand, fabrication of low-cost, flexible solid-state electronic devices without sacrificing the electronic performances and switchable properties.

8. GO and Other 2D Materials for Corrosion Protection

Transportation industry (automotive, aerospace, and railroad) is one of the major contributors to corrosion expenditure.[149] The coatings containing pure 2D anticorrosion materials or a combination of them with other materials are relatively new techniques for corrosion protection of steel, aluminum, and magnesium alloys.

Pure graphene is used for surface coating and preventing its contact with a corrosive environment. According to Böhm,[150] graphene protects the substrate by retarding the diffusion of corrosive species. Additionally, high electrical conductivity of graphene impedes electrons reaching cathodic sites to complete the corrosion reaction. Ye *et al.*[151,152] benefited anticorrosion properties of graphene coating on the carbon steel surface and obtained a lower annual corrosion rate of 0.05 mm y^{-1} in comparison with 0.09 mm y^{-1} of the uncoated carbon steel.

Although pure graphene coatings have been reported to be successful in preventing and slowing down the corrosion process, they are not flawless. As graphene possesses high electrical conductivity and could be actually more cathodic in practical applications compared to most common metals and alloys such as Cu, Zn, Ni, Brass, 316 stainless steel, Al alloys, and Mg alloys, there would be a risk that graphene coatings can produce galvanic cell effects and thus cause severe corrosion instead.[153] For instance, Hsieh *et al.*[154] founded that the defects in the grain boundaries of graphene can lead to a less efficient passivation. Consequently, the created microgalvanic cells accelerated the metal substrate's corrosion. Regarding this concern, perfect continuous and pinhole-free graphene coating is demanded for good anticorrosion, while this prerequisite on the one hand is high cost and on the other hand not realistic in many real-life applications. Besides, the local oxidation of graphene can also deteriorates the mechanical properties of coatings.[155]

One of the strategies to improve the quality of graphene-containing coatings is to combine graphene with organic and inorganic compounds in order to combine in one coating materials with different mechanisms of corrosion protection. Graphene-containing composite coatings can relieve corrosion than pure graphene. On this basis, GO and rGO appear to be good alternatives with their abundant oxygen containing groups, which enable them to form target composite coatings with various molecules and polymers. Furthermore, jointed with preserving the majority of graphene backbone, GO and rGO can provide similar barrier properties, but with improved compatibility.

Indeed, both graphene and GO have a high surface area and surface energy, which enhances their tendency to agglomerate; however, the surface moieties of GO can facilely undergo covalent modifications to decline the surface energy and to obtain a uniform and stable dispersion in polymeric coatings.[156–159] For example, Mo *et al.*[160] benefited from the surface functional groups of GO such as hydroxyl and carboxyl, and Si–OH active bonds of a hydrolyzed silane coupling agent to modify GO and improve its dispersibility and compatibility with polyurethane matrix. As a result, the corrosion current on the white cast iron substrate was decreased by 10 times with a high protection efficiency of 91.84%.

GO can also be modified by means of loading inorganic materials including metal and metal oxide particles for anticorrosion purposes.[161–165] Ramezanzadeh *et al.*[166] synthesized GO nanosheets modified with two corrosion inhibitors, PANI nanofibers, and CeO_2 (GO–PANI–CeO_2). GO–PANI–CeO_2 particles improved the coating barrier properties. PANI–CeO_2 particles served as corrosion inhibitors for mild steel.

GO/rGO can improve the efficiency of cathodic protection with rGO being generally more efficient than GO.[167] rGO can extend the cathodic protection time and anticorrosion capability of epoxy with metallic zinc coating (Fig. 7(a)). rGO nanosheets improve the electrical conductivity between zinc particles and the substrate causing the maximal use of zinc sacrificial anode.[156] Additionally, rGO/GO sheets hinder the permeation of corrosive species through the coating to a high extent.[167,168] These factors resulted in excellent anticorrosion performance and high impedance modulus as shown in Fig. 7(b).[167]

Graphene and GO are also effectively used in anticorrosion coatings for preserving aluminum and magnesium alloys. To enhance

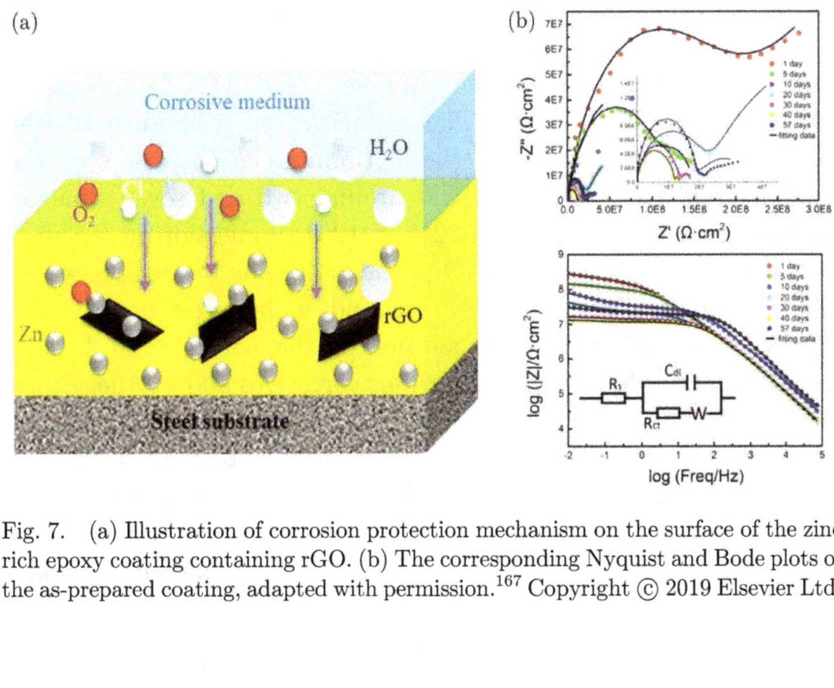

Fig. 7. (a) Illustration of corrosion protection mechanism on the surface of the zinc rich epoxy coating containing rGO. (b) The corresponding Nyquist and Bode plots of the as-prepared coating, adapted with permission.[167] Copyright © 2019 Elsevier Ltd.

the corrosion resistance of 6063 aluminum alloy, Chen *et al.*[169] added graphene particles to γ-Al_2O_3 and coated this material by micro-arc oxidation technique on the substrate. The presence of graphene in the coating resulted in a higher corrosion potential (E_{corr}) and a lower corrosion current density (i_{corr}) in a 3.5 wt.% NaCl solution, which indicates the advantageous capability of graphene to isolate the aluminum substrate. In the optimum concentration, the coatings were crack free and exhibited -0.561 V E_{corr} and 5.094×10^{-8} A cm^{-2} i_{corr}. In a different approach, Hikku *et al.*[170] mixed graphene with PVA to fabricate an anticorrosion nanocomposite coating on Aluminum-2219 alloy. Post-corrosion EDX mapping results at the voids affirmed that the graphene sheets in the coating did not permit any corrosive Cl^-.

GO has also demonstrated anticorrosion properties in protection of other aluminum alloys. To this end, GO was modified with 3-glycidoxypropyl-trimethoxysilane (GPTMS) to form a film that can protect AA2024-T3 aluminum alloy.[171] The silane–GO coating provided a suitable anticorrosion layer on the substrate with protective efficiency of 95%. This coating demonstrated excellent anticorrosion performance due to sealing of the defects by perfectly adhered GO particles. Moreover, Nezamdoust and Seifzadeh.[172] modified GO nanoplates through silanization with 3-aminopropyl-triethoxysilane (APTES) and dicyclohexylcarbodiimide (DCC) via epoxide and carboxylic groups to produce rGO@APTES nanoplates for 2024 aluminum alloy. The polarization resistance of the nanocomposite coating with the optimum concentration of the functionalized GO nanoplates was around two and seven times of the pure sol–gel coating after 72 and 120 h immersion in 3.5 wt.% NaCl solution, respectively. rGO particles provide barrier properties to the rGO@APTES. A number of studies showed that rGO can improve the anticorrosion properties of the sol–gel film by the modulation of diffusion path length for corrosives and by the sealing the micro-defects.[157,173]

Nowadays, magnesium alloys attract more and more interest as green-engineering materials.[174–176] However, magnesium can be easily corroded. Thus, there are a lot of attempts to improve the anticorrosion properties of magnesium alloys using different corrosion inhibitors.[177–180] Among them 2D materials show a high potential for future applications.[181] For instance, Fan *et al.*[182] reported an anticorrosion coating consisted of a GO layer, a cerium-based conversion layer, and a branched poly (ethylene imine) (PEI)/poly (acrylic acid) (PAA) multilayer on AZ31. PEI/PAA multilayers provided the rapid self-healing ability, while GO increased polarization resistance by two orders as compared with the uncoated AZ31 alloy.[183,184] Yan *et al.*[185] utilized GO flakes with oxide/zinc–aluminum layered double hydroxide (LDH) to produce a rGO/zinc–aluminum LDH (rGO/Zn–Al LDH) coating on a magnesium alloy. LDH nanosheets grown on both the GO surface and the magnesium substrate effectively prevented the agglomeration of GO. The corrosion current density of the bare substrate and Zn–Al LDH film was 33.2 and 4.33 $\mu A/cm^2$, respectively. The addition of GO to the LDH coating resulted in more improved corrosion resistance with a very low corrosion current density of 0.546 $\mu A/cm^2$. The organic combination of the Zn–Al LDH coating with GO dramatically increased the penetration resistance to the corrosive media.

Jiang *et al.*[186] fabricated a composite coating composed of N-doped graphene quantum dots (N-GQDs) and polymethyltrimethoxysilane (PMTMS) and coated it on the surface of AZ91D via electrodeposition with further silane treatment. After coating the N-GQDs/PMTMS composite, the R_s was increased by six times in comparison with the bare alloy. The N-GQDs/PMTMS composite film exhibited considerable improvement in anticorrosion properties thanks to the interface-free intermediate layer formed by the firm chemical bonding of N-GQDs with both PMTMS and magnesium substrate besides the physical barrier provided by PMTMS coating.

To date, 2D materials have depicted distinctly promising anticorrosion properties for protecting the materials used in automotive and aerospace industries such as steel, aluminum, and magnesium alloys. Thus, currently, there is a great interest in using them in anticorrosion coatings on industrial scale and trying them in novel combinations and systems for better performance.

Metals are widely used as biomedical materials such as stents,[187,188] bone and joint constructs,[189,190] dental implants,[191] and pacemaker generators.[192,193] One of the main problems is uncontrollable corrosion of metal implants in the harsh body environment, leaching of ions, and also chemical reactions with some acids and enzymes. Complex *in vivo* degradation of implants is not well understood yet and cannot be controlled. It has been discovered that soluble metal ions induce a natural monocytes/macrophage response, which stimulates immune responses, leading to toxic, allergic, inflammatory, allergic, or mutagenic reactions in patients.[194] Metallic debris, created by corrosion, can accumulate in periprosthetic soft tissues and cause metallosis.[195,196] For instance, Mg is used for the engineering of biodegradable implants. However, it corrodes *in vivo* with a production of hydrogen gas.[197–199] A composite of poly (3,4-ethylenedioxythiophene) and GO (PEDOT/GO) on magnesium ribbon was proposed for biomedical applications.[200] The Tafel scans displayed that the E_{corr} of the magnesium ribbon coated with PEDOT/GO underwent a positive shift and reached $-1.55\,\mathrm{V}$ higher than that of the bare metal ($-1.705\,\mathrm{V}$). Furthermore, the anticorrosion PEDOT/GO coating reduced the corrosion current by 75% down to $12.3\,\mu\mathrm{A}$ from $49.6\,\mu\mathrm{A}$ of the uncoated samples. Mg ions and OH^- release to the corrosive solution declined by 41.1% and 43% on an average daily basis after being coated with PEDOT/GO. Despite the bare alloy, the PEDOT/GO coatings depicted a larger radius in the Nyquist arc over the 22 test days. When the samples were coated with PEDOT/GO and exposed to 100% corrosion media, they exhibited a 40% decrease in toxicity. Large GO sheets subsist trapped in the polymer film after reduction of the polymer backbone, causing a net negative charge in the coating,[201] which can ameliorate resistance to the diffusion of the corrosion boosting Cl^-.[202] In addition, the GO interfaces provide good implant–tissue interaction.[203]

Zhu *et al.*[204] coated carbon steel with a modified GO composited with waterborne acrylic modified alkyd resin (AMAR) and evaluated its corrosion in simulated body fluid (SBF). The coating in the optimum conditions resisted against corrosion in SBF by increasing E_{corr} to $-0.1807\,\mathrm{mV}$ versus SCE (in comparison with $-0.541\,\mathrm{mV}$ of the uncoated steel) and decreasing the high electrical conductivity of the neat steel and its i_{corr} to the negligible numbers of 1.709×10^{-11} S cm^{-1} and 2.8511×10^{-8} A cm^{-2}, respectively.

In general, 2D materials are superior to bulk materials in anticorrosion applications due to their electrical and mechanical properties, surface area,[205] impermeability,[206] thermal stability,[207] bandgap tunability,[208] not affecting the physical properties of the substrates,[209–212] and transparency.[213–215] The most promising superiority of some 2D materials for anticorrosion applications is their barrier properties for small molecules and ions.

9. Conclusion and Perspectives

The robust graphene-based body structure and the tunable abundant functional groups such as hydroxyl, epoxy, carboxyl on the surface make GO and GO membranes great candidates for multiple applications on ion/molecular separation, new generation sensors, biomedical devices and applications, energy related applications, and anticorrosion. In these applications, GO membranes play a number of different roles, such as a *d*-spacing controllable block (in ion/water separation), mechanical enforcement additive (in biomedical test), separator (in secondary battery), solid-state electrolyte (in FC), and coating (in anticorrosive protection). In addition, GO membranes can form composites with other materials to fulfill their applications via synergetic effects.

However, there are still some issues to be addressed. First, even though some basic principles for ion and water transport in GO have been revealed, their better understanding and control of the performance of corresponding applications is very important. Second, despite the good performance, the toxicology and environment impact of GO membranes need to be further researched especially when applied in biomedical fields. Third, majority of existing applications related to GO membranes is limited to laboratory, and thus more works need to be done to pave the way for the laboratory-to-industry transfer. Last but not least, we need to find an efficient way of precise control of the functionalization level of GO to be able to gain both from the high conductivity of graphene and the chemical versatility of the functional groups of GO. A relevant question is: the need of standardization for GO and rGO materials.

With potentially good biocompatibility and flexible nature, fascinating and novel applications of GO membranes can be expected such as *in vivo* sensing and intelligent foldable electronic devices. We can expect with further studying, GO membranes could exhibit more attractive applications which might contribute to solving the up-to-date issues.

Acknowledgment

We thank NRF (Singapore) for financial support through the project Medium-sized centre programme R-723-000-001-281. KSN also acknowledges support from EU Flagship Programs (Graphene CNECTICT-604391 and 2D-SIPC Quantum Technology), European Research Council Synergy Grant Hetero2D, the Royal Society, EPSRC grants EP/N010345/1, EP/P026850/1, EP/S030719/1. E. E.V was supported by the State Task No. 075-00355-21-00. This work was supported by the Ministry of Education (Singapore) through the Research Centre of Excellence program (grant EDUN C-33-18-279-V12, I-FIM).

References

1. D. R. Dreyer, S. Park, C. W. Bielawski and R. S. Ruoff, *Chem. Soc. Rev.* **39** (2010) 228.
2. J. W. Suk, R. D. Piner, J. An and R. S. Ruoff, *ACS Nano* **4** (2010) 6557.
3. D. A. Dikin, S. Stankovich, E. J. Zimney, R. D. Piner, G. H. B. Dommett, G. Evmenenko, S. T. Nguyen and R. S. Ruoff, *Nature* **448** (2007) 457.
4. T. Tsuchiya, K. Terabe and M. Aono, *Adv. Mater.* **26** (2014) 1087.
5. M. Hadadian, E. K. Goharshadi and A. Youssefi, *J. Nanopart. Res.* **16** (2014) 2788.
6. J. Kim, L. J. Cote, F. Kim, W. Yuan, K. R. Shull and J. Huang, *J. Am. Chem. Soc.* **132** (2010) 8180.
7. X. Cai, Y. Luo, B. Liu and H. M. Cheng, *Chem. Soc. Rev.* **47** (2018) 6224.
8. W. S. Hummers and R. E. Offeman, *J. Am. Chem. Soc.* **80** (1958) 1339.
9. L. J. Cote, R. Cruz-Silva and J. Huang, *J. Am. Chem. Soc.* **131** (2009) 11027.
10. T. Zhang, J. Liu, C. Wang, X. Leng, Y. Xiao and L. Fu, *Biosens. Bioelectron.* **89** (2017) 28.

11. C. Chen, Q. H. Yang, Y. Yang, W. Lv, Y. Wen, P. X. Hou, M. Wang and H. M. Cheng, *Adv. Mater.* **21** (2009) 3007.
12. K. Huang, G. Liu, Y. Lou, Z. Dong, J. Shen and W. Jin, *Angew. Chem. Int. Ed.* **53** (2014) 6929.
13. X. Hu, Y. Yu, W. Hou, J. Zhou and L. Song, *Appl. Surf. Sci.* **273** (2013) 118.
14. X. Wang, H. Bai and G. Shi, *J. Am. Chem. Soc.* **133** (2011) 6338.
15. C.-J. Shih, S. Lin, R. Sharma, M. S. Strano and D. Blankschtein, *Langmuir* **28** (2012) 235.
16. X. Yang, Y. Wang, X. Huang, Y. Ma, Y. Huang, R. Yang, H. Duan and Y. Chen, *J. Mater. Chem.* **21** (2011) 3448.
17. Y. Oh, D. L. Armstrong, C. Finnerty, S. Zheng, M. Hu, A. Torrents and B. Mi, *J. Membr. Sci.* **541** (2017) 235.
18. D. V. Andreeva, M. Trushin, A. Nikitina, M. C. F. Costa, P. V. Cherepanov, M. Holwill, S. Chen, K. Yang, S. W. Chee, U. Mirsaidov, A. H. Castro Neto and K. S. Novoselov, *Nat. Nanotechnol.* (2021), https://doi.org/10.1038/s41565-020-00795-y.
19. J. Abraham, K. S. Vasu, C. D. Williams, K. Gopinadhan, Y. Su, C. T. Cherian, J. Dix, E. Prestat, S. J. Haigh, I. V. Grigorieva, P. Carbone, A. K. Geim and R. R. Nair, *Nat. Nanotechnol.* **12** (2017) 546.
20. R. K. Joshi, P. Carbone, F. C. Wang, V. G. Kravets, Y. Su, I. V. Grigorieva, H. A. Wu, A. K. Geim and R. R. Nair, *Science* **343** (2014) 752.
21. K. G. Zhou, K. S. Vasu, C. T. Cherian, M. Neek-Amal, J. C. Zhang, H. Ghorbanfekr-Kalashami, K. Huang, O. P. Marshall, V. G. Kravets, J. Abraham, Y. Su, A. N. Grigorenko, A. Pratt, A. K. Geim, F. M. Peeters, K. S. Novoselov and R. R. Nair, *Nature* **559** (2018) 236.
22. Q. Yang, Y. Su, C. Chi, C. T. Cherian, K. Huang, V. G. Kravets, F. C. Wang, J. C. Zhang, A. Pratt, A. N. Grigorenko, F. Guinea, A. K. Geim and R. R. Nair, *Nat. Mater.* **16** (2017) 1198.
23. Y. Su, V. G. Kravets, S. L. Wong, J. Waters, A. K. Geim and R. R. Nair, *Nat. Commun.* **5** (2014) 4843.
24. H. M. Hegab and L. Zou, *J. Membr. Sci.* **484** (2015) 95.
25. Y. You, V. Sahajwalla, M. Yoshimura and R. K. Joshi, *Nanoscale* **8** (2016) 117.
26. G. Zhao, J. Li, X. Ren, C. Chen and X. Wang, *Environ. Sci. Technol.* **45** (2011) 10454.
27. Q. Xu, H. Xu, J. Chen, Y. Lv, C. Dong and T. S. Sreeprasad, *Inorg. Chem. Front.* **2** (2015) 417.
28. H. W. Kim, H. W. Yoon, S. M. Yoon, B. M. Yoo, B. K. Ahn, Y. H. Cho, H. J. Shin, H. Yang, U. Paik and S. Kwon, *Science* **342** (2013) 91.
29. Y. P. Tang, D. R. Paul and T. S. Chung, *J. Membr. Sci.* **458** (2014) 199.
30. E. Halakoo and X. Feng, *Chem. Eng. Sci.* **216** (2020).
31. G. Liu, J. Shen, Q. Liu, G. Liu, J. Xiong, J. Yang and W. Jin, *J. Membr. Sci.* **548** (2018) 548.
32. K. Toda, R. Furue and S. Hayami, *Anal. Chim. Acta* **878** (2015) 43.
33. S. Goenka, V. Sant and S. Sant, *J. Control Release* **173** (2014) 75.
34. F. Ordikhani, M. R. Farani, M. Dehghani, E. Tamjid and A. Simchi, *Carbon* **84** (2015) 91.

35. B. Zhang, Y. Wang and G. Zhai, *Mater. Sci. Eng. C* **61** (2016) 953.
36. S. Chen, K. Yang, X. Leng, M. Chen, K. S. Novoselov and D. V. Andreeva, *Polym. Int.* **69** (2020) 1173.
37. M. Ioniţă, L. E. Crică, S. I. Voicu, S. Dinescu, F. Miculescu, M. Costache and H. Iovu, *Carbohydr. Polym.* **183** (2018) 50.
38. M. Majumder and A. K. Thakur, in *Surface Engineering of Graphene* (Springer, Cham, 2019), pp. 113–138.
39. W. K. Chee, H. N. Lim, Z. Zainal, N. M. Huang, I. Harrison and Y. Andou, *J. Phys. Chem. C* **120** (2016) 4153.
40. R. R. Nair, H. A. Wu, P. N. Jayaram, I. V. Grigorieva and A. K. Geim, *Science* **335** (2012) 442.
41. S. Hong, C. Constans, M. V. Surmani Martins, Y. C. Seow, J. A. Guevara Carrio and S. Garaj, *Nano Lett.* **17** (2017) 728.
42. D. W. Boukhvalov, M. I. Katsnelson and Y. W. Son, *Nano Lett.* **13** (2013) 3930.
43. J. Dix, Understanding the mechanism of permeation through graphene-based membranes using molecular dynamics simulations, Ph. D. thesis, University of Manchester, United Kingdom (2017).
44. L. Chen, G. Shi, J. Shen, B. Peng, B. Zhang, Y. Wang, F. Bian, J. Wang, D. Li, Z. Qian, G. Xu, G. Liu, J. Zeng, L. Zhang, Y. Yang, G. Zhou, M. Wu, W. Jin, J. Li and H. Fang, *Nature* **550** (2017) 380.
45. J. Abraham, K. S. Vasu, C. D. Williams, K. Gopinadhan, Y. Su, C. T. Cherian, J. Dix, E. Prestat, S. J. Haigh, I. V. Grigorieva, P. Carbone, A. K. Geim and R. R. Nair, *Nat. Nanotechnol.* **12** (2017) 546.
46. W. S. Hung, T. J. Lin, Y. H. Chiao, A. Sengupta, Y. C. Hsiao, S. R. Wickramasinghe, C. C. Hu, K. R. Lee and J. Y. Lai, *J. Mater. Chem. A* **6** (2018) 19445.
47. A. V. Talyzin, V. L. Solozhenko, O. O. Kurakevych, T. Szabo, I. Dekany, A. Kurnosov and V. Dmitriev, *Angew Chem. Int. Ed. Engl.* **47** (2008) 8268.
48. H. Huang, Y. Mao, Y. Ying, Y. Liu, L. Sun and X. Peng, *Chem. Commun.* **49** (2013) 5963.
49. J. Yang, X. Hu, X. Kong, P. Jia, D. Ji, D. Quan, L. Wang, Q. Wen, D. Lu, J. Wu, L. Jiang and W. Guo, *Nat. Commun.* **10** (2019) 1171.
50. Y. Feng, H. Dai, J. Chen, X. Kong, J. Yang and L. Jiang, *J. Mater. Chem. A* **7** (2019) 20182.
51. S. Zheng, Q. Tu, M. Wang, J. J. Urban and B. Mi, *ACS Nano* **14** (2020) 6013.
52. Z. Zhang, S. Li, B. Mi, J. Wang and J. Ding, *Sci. Adv.* **6** (2020) eaba9471.
53. S. Homaeigohar and M. Elbahri, *NPG Asia Mater.* **9** (2017) e427.
54. M. Hu and B. Mi, *Environ. Sci. Technol.* **47** (2013) 3715.
55. A. Boretti, S. Al-Zubaidy, M. Vaclavikova, M. Al-Abri, S. Castelletto and S. Mikhalovsky, *NPJ Clean Water* **1** (2018) 5.
56. L. Jin, Z. Wang, S. Zheng and B. Mi, *J. Membr. Sci.* **545** (2018) 11.
57. H. R. Chae, J. Lee, C. H. Lee, I. C. Kim and P. K. Park, *J. Membr. Sci.* **483** (2015) 128.
58. Q. Nan, P. Li and B. Cao, *Appl. Surf. Sci.* **387** (2016) 521.

59. J. Ran, L. Wu, Y. He, Z. Yang, Y. Wang, C. Jiang, L. Ge, E. Bakangura and T. Xu, *J. Membr. Sci.* **522** (2017) 267.
60. S. M. Hosseini, E. Jashni, M. Habibi, M. Nemati and B. Van der Bruggen, *J. Membr. Sci.* **541** (2017) 641.
61. S. Gahlot, P. P. Sharma, H. Gupta, V. Kulshrestha and P. K. Jha, *RSC Adv.* **4** (2014) 24662.
62. S. M. Hosseini, E. Jashni, S. Amani and B. Van der Bruggen, *J. Colloid Interface Sci.* **505** (2017) 763.
63. R. Hu, G. Zhao, Y. He and H. Zhu, *Desalination* **477** (2020) 114271.
64. L. Cui, Y. Wang, L. Gao, L. Hu, L. Yan, Q. Wei and B. Du, *Chem. Eng. J.* **281** (2015) 1.
65. G. Z. Kyzas, E. A. Deliyanni and K. A. Matis, *J. Chem. Technol. Biotechnol.* **89** (2014) 196.
66. W. Peng, H. Li, Y. Liu and S. Song, *J. Mol. Liq.* **230** (2017) 496.
67. C. J. Madadrang, H. Y. Kim, G. Gao, N. Wang, J. Zhu, H. Feng, M. Gorring, M. L. Kasner and S. Hou, *ACS Appl. Mater. Interfaces* **4** (2012) 1186.
68. Y. Qi, M. Yang, W. Xu, S. He and Y. Men, *J. Colloid Interface Sci.* **486** (2017) 84.
69. G. Zhao and H. Zhu, *Adv. Mater.* **32** (2020) 1905756.
70. C. Xu, A. Cui, Y. Xu and X. Fu, *Carbon* **62** (2013) 465.
71. H. Li, Z. Song, X. Zhang, Y. Huang, S. Li, Y. Mao, H. J. Ploehn, Y. Bao and M. Yu, *Science* **342** (2013) 95.
72. K. Zahri, K. C. Wong, P. S. Goh and A. F. Ismail, *RSC Adv.* **6** (2016) 89130.
73. L. Cheng, K. Guan, G. Liu and W. Jin, *J. Membr. Sci.* **595** (2020).
74. X. Jin, T. Foller, X. Wen, M. B. Ghasemian, F. Wang, M. Zhang, H. Bustamante, V. Sahajwalla, P. Kumar, H. Kim, G.-H. Lee, K. Kalantar-Zadeh and R. Joshi, *Adv. Mater.* **32** (2020) 1907580.
75. F. Malekian, H. Ghafourian, K. Zare, A. A. Sharif and Y. Zamani, *Eur. Phys. J. Plus* **134** (2019).
76. C. Sun, B. Wen and B. Bai, *Sci. Bull.* **60** (2015) 1807.
77. D.-E. Jiang, V. R. Cooper and S. Dai, *Nano Lett.* **9** (2009) 4019.
78. S. M. Fatemi, M. Arabieh and H. Sepehrian, *Carbon Lett.* **16** (2015) 183.
79. Q. Liu, K. M. Gupta, Q. Xu, G. Liu and W. Jin, *Sep. Purif. Technol.* **209** (2019) 419.
80. R. R. Nair, H. A. Wu, P. N. Jayaram, I. V. Grigorieva and A. K. Geim, *Science* **335** (2012) 442.
81. G. Li, L. Shi, G. Zeng, Y. Zhang and Y. Sun, *RSC Adv.* **4** (2014) 52012.
82. S. Gupta, G. B. Thorat and Z. V. P. Murthy, *Macromol. Res.* **28** (2020) 587.
83. Y. Mao, M. Zhang, L. Cheng, J. Yuan, G. Liu, L. Huang, M. Barboiu and W. Jin, *J. Membr. Sci.* **595** (2020) 117545.
84. C. T. Lee and Y. S. Wang, *J. Alloys Compd.* **789** (2019) 693.
85. J. Yan, T. Wei, B. Shao, Z. Fan, W. Qian, M. Zhang and F. Wei, *Carbon* **48** (2010) 487.
86. J. Peng, J. Liao, X. Yang and W. Feng, *Chin. Chem. Lett.* **31** (2019) 2145.
87. A. Miyamoto, Y. Kuwaki, T. Sano, K. Hatakeyama, A. Quitain, M. Sasaki and T. Kida, *ACS Omega* **2** (2017) 2994.

88. X. Huang, T. Leng, T. Georgiou, J. Abraham, R. Raveendran Nair, K. S. Novoselov and Z. Hu, *Sci. Rep.* **8** (2018) 43.
89. M. R. R. Khan and S. W. Kang, *IEEE Sens. J.* **18** (2018) 2667.
90. B. Liu, X. D. Chen, N. Li, X. Y. Li and Y. Yao, *J. Electron. Lett.* **50** (2014) 1447.
91. L. Jia, Y. Zhou, Y. Jiang, A. Zhang, X. Li and C. Wang, *J. Alloys Compd.* **685** (2016) 167.
92. J. H. Jeon, R. K. Cheedarala, C. D. Kee and I. K. Oh, *Adv. Funct. Mater.* **23** (2013) 6007.
93. N. Mahmoudi and A. Simchi, *Mater. Sci. Eng. C* **70** (2017) 121.
94. L. Chen, H. Jiang, Y. Li, B. L. Zimba, X. Yu, C. Chen, G. Xiong and Q. Wu, *Mater. Res. Express* **6** (2019) 075404.
95. I. S. Yahia, M. Shkir and S. M. A. S. Keshk, *Results Phys.* **16** (2020) 102990.
96. L. Jin, Z. Zeng, S. Kuddannaya, D. Yue, J. Bao, Z. Wang and Y. Zhang, *J. Mater. Chem. B* **3** (2015) 4338.
97. C. Yang, X. Huang, Y. Huang, Y. Chen, L. Wang, X. Zheng, H. Wen, N. Dan and W. Dan, *Int. J. Polym. Anal. Charact.* **24** (2019) 741.
98. G. Zou, J. Shen, P. Duan, X. Xia, R. Chen and B. Jin, *Int. J. Polym. Sci.* **2018** (2018) 7906747.
99. Z. Yang, C. Sun, L. Wang, H. Chen, J. He and Y. Chen, *J. Colloid Interface Sci.* **507** (2017) 344.
100. Z. Mao, J. Li, W. Huang, H. Jiang, B. L. Zimba, L. Chen, J. Wan and Q. Wu, *RSC Adv.* **8** (2018) 16619.
101. S. Thampi, V. Muthuvijayan and R. Parameswaran, *J. Appl. Polym. Sci.* **132** (2015) 41809.
102. C. He, Z. Q. Shi, C. Cheng, H. Q. Lu, M. Zhou, S. D. Sun and C. S. Zhao, *Biomater. Sci.* **4** (2016) 1431.
103. A. Eftekhari, Y. M. Shulga, S. A. Baskakov and G. L. Gutsev, *Int. J. Hydrog. Energy* **43** (2018) 2307.
104. I. Nicotera, C. Simari, M. Agostini, A. Enotiadis and S. Brutti, *J. Phys. Chem. C* **123** (2019) 27406.
105. F. Zhang, F. Shen, Z. Y. Fan, X. Ji, B. Zhao, Z. T. Sun, Y. Y. Xuan and X. G. Han, *Rare Metals* **37** (2018) 510.
106. J. Y. Kim, D. O. Shin, K. M. Kim, J. Oh, J. Kim, S. H. Kang, M. J. Lee and Y. G. Lee, *Sci. Rep.* **9** (2019) 2464.
107. L. Wang, F. Liu, W. Shao, S. Cui, Y. Zhao, Y. Zhou and J. He, *Compos. Commun.* **16** (2019) 150.
108. S. H. Park, T. H. Lee, Y. J. Lee, H. B. Park and Y. J. Lee, *Small* **14** (2018) e1801456.
109. Z. Guo, J. Li, Y. Xia, C. Chen, F. Wang, A. G. Tamirat, Y. Wang, Y. Xia, L. Wang and S. Feng, *J. Mater. Chem. A* **6** (2018) 6022.
110. H. Song, C. Zuo, X. Xu, Y. Wan, L. Wang, D. Zhou and Z. Chen, *RSC Adv.* **8** (2018) 429.
111. D. K. Lee, S. J. Kim, Y. J. Kim, H. Choi, D. W. Kim, H. J. Jeon, C. W. Ahn, J. W. Lee and H. T. Jung, *Adv. Mater. Interfaces* **6** (2019) 1801992.
112. M. A. Aziz and S. Shanmugam, *J. Mater. Chem. A* **6** (2018) 17740.

113. X. Lou, D. Yuan, Y. Yu, Y. Lei, M. Ding, Q. Sun and C. Jia, *Chem. Asian J.* **15** (2020) 1.
114. S. C. Park, T. H. Lee, G. H. Moon, B. S. Kim, J. M. Roh, Y. H. Cho, H. W. Kim, J. Jang, H. B. Park and Y. S. Kang, *ACS Appl. Energy Mater.* **2** (2019) 4590.
115. W. Du, J. Xiao, H. Geng, Y. Yang, Y. Zhang, E. H. Ang, M. Ye and C. C. Li, *J. Power Sources* **450** (2020) 227716.
116. Z. Xie, B. Liu, C. Xie, B. Yang, Y. Jiao, D. Cai, L. Yang, Q. Shu and A. Shi, *Mater. Chem. Phys.* **220** (2018) 208.
117. W. Ren, W. Ma, S. Zhang and B. Tang, *Energy Storage Mater.* **23** (2019) 707.
118. T. Bayer, S. R. Bishop, M. Nishihara, K. Sasaki and S. M. Lyth, *J. Power Sources* **272** (2014) 239.
119. H. Tateishi, K. Hatakeyama, C. Ogata, K. Gezuhara, J. Kuroda, A. Funatsu, M. Koinuma, T. Taniguchi, S. Hayami and Y. Matsumoto, *J. Electrochem. Soc.* **160** (2013) F1175.
120. D. Ege, A. R. Kamali and A. R. Boccaccini, *Adv. Eng. Mater.* **19** (2017) 1700627.
121. U. R. Farooqui, A. L. Ahmad and N. A. Hamid, *Renew. Sustain. Energy Rev.* **82** (2018) 714.
122. R. P. Pandey, G. Shukla, M. Manohar and V. K. Shahi, *Adv. Colloid Interface Sci.* **240** (2017) 15.
123. R. Yadav, A. Subhash, N. Chemmenchery and B. Kandasubramanian, *Ind. Eng. Chem. Res.* **57** (2018) 9333.
124. M. Vinothkannan, A. R. Kim, G. Gnana kumar and D. J. Yoo, *RSC Adv.* **8** (2018) 7494.
125. E. Abouzari-Lotf, M. Zakeri, M. M. Nasef, M. Miyake, P. Mozarmnia, N. A. Bazilah, N. F. Emelin and A. Ahmad, *J. Power Sources* **412** (2019) 238.
126. K. Ben Liew, J. X. Leong, W. R. Wan Daud, A. Ahmad, J. J. Hwang and W. Wu, *J. Power Sources* **449** (2020) 227490.
127. H. Lee, J. Han, K. Kim, J. Kim, E. Kim, H. Shin and J. C. Lee, *J. Ind. Eng. Chem.* **74** (2019) 223.
128. W. Rehman, K. Liaqat, S. Fazil, S. Saeed, M. Waseem, M. Shakeel, S. Mir, I. Bibi and C. Y. Guo, *J. Polym. Res.* **26** (2019).
129. L. Liu and G. Sun, *J. Electrochem. Soc.* **12** (2017) 9262.
130. M. H. Gouda, W. Gouveia, M. L. Afonso, B. Šljukić, N. A. El Essawy, A. A. A. Nassr and D. M. F. Santos, *J. Power Sources* **432** (2019) 92.
131. X. Peng, L. Peng, C. Wu and Y. Xie, *Chem. Soc. Rev.* **43** (2014) 3303.
132. L. Feng, K. Wang, X. Zhang, X. Sun, C. Li, X. Ge and Y. Ma, *Adv. Funct. Mater.* **28** (2018) 1704463.
133. X. Yang, F. Zhang, L. Zhang, T. Zhang, Y. Huang and Y. Chen, *Adv. Funct. Mater.* **23** (2013) 3353.
134. H. J. Kim, S. Y. Lee, L. H. Sinh, C. S. Yeo, Y. R. Son, K. R. Cho, Y. Song, S. Ju, M. K. Shin, S. J. Park and S. Park, *J. Power Sources* **346** (2017) 113.
135. C. Ogata, R. Kurogi, K. Awaya, K. Hatakeyama, T. Taniguchi, M. Koinuma and Y. Matsumoto, *ACS Appl. Mater. Interfaces* **9** (2017) 26151.

136. C. Ogata, R. Kurogi, K. Hatakeyama, T. Taniguchi, M. Koinuma and Y. Matsumoto, *Chem. Commun.* **52** (2016) 3919.
137. Z.-S. Wu, Y. Zheng, S. Zheng, S. Wang, C. Sun, K. Parvez, T. Ikeda, X. Bao, K. Müllen and X. Feng, *Adv. Mater.* **29** (2017) 1602960.
138. J. D. Roy-Mayhew and I. A. Aksay, *Chem. Rev.* **114** (2014) 6323.
139. T. Mahmoudi, Y. Wang and Y. B. Hahn, *Nano Energy* **47** (2018) 51.
140. K. Patil, S. Rashidi, H. Wang and W. Wei, *Int. J. Photoenergy* **2019** (2019) 1812879.
141. M. Acik and S. B. Darling, *J. Mater. Chem. A* **4** (2016) 6185.
142. E. L. Lim, C. C. Yap, M. H. H. Jumali, M. A. M. Teridi and C. H. Teh, *Nanomicro Lett.* **10** (2018) 27.
143. A. P. Litvin, X. Zhang, K. Berwick, A. V. Fedorov, W. Zheng and A. V. Baranov, *Renew. Sustain. Energy Rev.* **124** (2020) 109774.
144. T. A. Amollo, G. T. Mola and V. O. Nyamori, *Crit. Rev. Solid State Mater. Sci.* (2019) 1.
145. A. Iwan, F. Caballero-Briones, K. A. Bogdanowicz, J. D. O. Barceinas-Sánchez, W. Przybyl, A. Januszko, J. A. Baron-Miranda, A. P. Espinosa-Ramirez and J. Guerrero-Contreras, *Adv. Mater. Lett.* **9** (2018) 58.
146. K. Wee Shing, G. Choon How, P. Wee Kee, Y. A. Akimov and B. Ping, *IEEE J. Sel. Top. Quantum Electron.* **20** (2014) 36.
147. S. J. Sung, J. Park, Y. S. Cho, S. H. Gihm, S. J. Yang and C. R. Park, *Carbon* **150** (2019) 275.
148. L. Phan, W. G. T. Walkup, D. D. Ordinario, E. Karshalev, J. M. Jocson, A. M. Burke and A. A. Gorodetsky, *Adv. Mater.* **25** (2013) 5621.
149. Z. Ahmad, in *Principles of Corrosion Engineering and Corrosion Control*, ed. Z. Ahmad (Butterworth-Heinemann, Oxford, 2006), pp. 1–8.
150. S. Böhm, *Nat. Nanotechnol.* **9** (2014) 741.
151. X. Ye, Z. Lin, H. Zhang, H. Zhu, Z. Liu and M. Zhong, *Carbon* **94** (2015) 326.
152. X. Ye, J. Long, Z. Lin, H. Zhang, H. Zhu and M. Zhong, *Carbon* **68** (2014) 784.
153. C. Cui, A. T. O. Lim and J. Huang, *Nat. Nanotechnol.* **12** (2017) 834.
154. Y. P. Hsieh, M. Hofmann, K. W. Chang, J. G. Jhu, Y. Y. Li, K. Y. Chen, C. C. Yang, W. S. Chang and L. C. Chen, *ACS Nano* **8** (2014) 443.
155. J. Hu, Y. Ji, Y. Shi, H. Duan and M. Lanza, *Ann. J. Mater. Sci. Eng.* **3** (2014) 1.
156. B. Ramezanzadeh, M. H. M. Moghadam, N. Shohani and M. Mahdavian, *Chem. Eng. J.* **320** (2017) 363.
157. B. Ramezanzadeh, A. Ahmadi and M. Mahdavian, *Corros. Sci.* **109** (2016) 182.
158. Y. Zhan, J. Zhang, X. Wan, Z. Long, S. He and Y. He, *Appl. Surf. Sci.* **436** (2018) 756.
159. N. Parhizkar, B. Ramezanzadeh and T. Shahrabi, *Appl. Surf. Sci.* **439** (2018) 45.
160. M. Mo, W. Zhao, Z. Chen, Q. Yu, Z. Zeng, X. Wu and Q. Xue, *RSC Adv.* **5** (2015) 56486.
161. N. S. Arul, J. I. Han and P. C. Chen, *ChemElectroChem* **5** (2018) 2747.
162. E. György, A. Pérez del Pino, C. Logofatu, C. Cazan and A. Duta, *J. Am. Ceram. Soc.* **97** (2014) 2718.

163. S. Chhetri, S. Ghosh, P. Samanta, N. C. Murmu and T. Kuila, *ChemistrySelect* **4** (2019) 13446.
164. B. Song, Y. Shi and Q. Liu, *Polym. Adv. Technol.* **31** (2020) 309.
165. V. Srimaneepong, D. Rokaya, P. Thunyakitpisal, J. Qin and K. Saengkiettiyut, *Sci. Rep.* **10** (2020) 3247.
166. B. Ramezanzadeh, G. Bahlakeh and M. Ramezanzadeh, *Corros. Sci.* **137** (2018) 111.
167. S. Zhou, Y. Wu, W. Zhao, J. Yu, F. Jiang, Y. Wu and L. Ma, *Mater. Des.* **169** (2019) 107694.
168. R. R. Nair, H. A. Wu, P. N. Jayaram, I. V. Grigorieva and A. K. Geim, *Science* **335** (2012) 442.
169. Q. Chen, Z. Jiang, S. Tang, W. Dong, Q. Tong and W. Li, *Appl. Surf. Sci.* **423** (2017) 939.
170. G. S. Hikku, K. Jeyasubramanian, A. Venugopal and R. Ghosh, *J. Alloys Compd.* **716** (2017) 259.
171. B. Xue, M. Yu, J. Liu, S. Li, L. Xiong and X. Kong, *J. Electrochem. Soc.* **163** (2016) C798.
172. S. Nezamdoust and D. Seifzadeh, *Prog. Organ. Coat.* **109** (2017) 97.
173. J. Li, J. Cui, J. Yang, Y. Ma, H. Qiu and J. Yang, *Prog. Organ. Coat.* **99** (2016) 443.
174. B. L. Mordike and T. Ebert, *Mater. Sci. Eng. A* **302** (2001) 37.
175. X. Guo, K. Chang, L. Q. Chen and M. Zhou, *Eng. Fract. Mech.* **96** (2012) 401.
176. G. Chen, J. Gao, Y. Cui, H. Gao, X. Guo and S. Wu, *J. Alloys Compd.* **735** (2018) 536.
177. R. C. Zeng, Z. G. Liu, F. Zhang, S. Q. Li, H. Z. Cui and E. H. Han, *J. Mater. Chem. A* **2** (2014) 13049.
178. G. L. Song and Z. Shi, *Corros. Sci.* **85** (2014) 126.
179. Q. Liu, D. Chen and Z. Kang, *ACS Appl. Mater. Interfaces* **7** (2015) 1859.
180. R. Xu, X. Yang, P. Li, K. W. Suen, G. Wu, and P. K. Chu, *Corros. Sci.* **82** (2014) 173.
181. J. H. Chu, L. B. Tong, M. Wen, Z. H. Jiang, K. S. Wang and H. J. Zhang, *Carbon* **145** (2019) 389.
182. F. Fan, C. Zhou, X. Wang and J. Szpunar, *ACS Appl. Mater. Interfaces* **7** (2015) 27271.
183. Y. Li, Z. Yang, H. Qiu, Y. Dai, Q. Zheng, J. Li and J. Yang, *J. Mater. Chem. A* **2** (2014) 14139.
184. W. Sun, L. Wang, T. Wu, Y. Pan and G. Liu, *J. Mater. Chem. A* **3** (2015) 16843.
185. L. Yan, M. Zhou, X. Pang and K. Gao, *Langmuir* **35** (2019) 6312.
186. B. K. Jiang, A. Y. Chen, J. F. Gu, J. T. Fan, Y. Liu, P. Wang, H. J. Li, H. Sun, J. H. Yang and X. Y. Wang, *Carbon* **157** (2020) 537.
187. T. H. Baron, *N. Engl. J. Med.* **344** (2001) 1681.
188. M. Kaw, S. Singh, H. Gagneja and P. Azad, *Surg. Endosc.* **17** (2003) 646.
189. P. Korovessis, G. Petsinis, M. Repanti and T. Repantis, *JBJS* **88** (2006) 1183.
190. F. S. Haddad, R. R. Thakrar, A. J. Hart, J. A. Skinner, A. V. F. Nargol, J. F. Nolan, H. S. Gill, D. W. Murray, A. W. Blom and C. P. Case, *J. Bone Joint. Surg. Br.* **93-B** (2011) 572.

191. M. C. Cortizo, M. F. L. de Mele and A. M. Cortizo, *Biol. Trace Elem. Res.* **100** (2004) 151.
192. K. Ishii, E. Kodani, S. Miyamoto, T. Otsuka, M. Hosone, K. I. Ogata, W. Sato, S. Matsumoto, T. Tadera, C. Ibuki, Y. Kusama and H. Atarashi, *Pacing Clin. Electrophysiol.* **29** (2006) 1299.
193. F. Tondato, D. W. Ng, K. Srivathsan, G. T. Altemose, M. Y. Halyard and L. R. Scott, *Expert Rev. Med. Devices* **6** (2009) 243.
194. M. S. Caicedo, P. H. Pennekamp, K. McAllister, J. J. Jacobs and N. J. Hallab, *J. Biomed. Mater. Res. A* **93A** (2010) 1312.
195. J. W. Romesburg, P. L. Wasserman and C. H. Schoppe, *J. Radiol. Case Rep.* **9** (2010) 7.
196. J. J. Jacobs and N. J. Hallab, *JBJS* **88** (2006) 1171.
197. J. Kuhlmann, I. Bartsch, E. Willbold, S. Schuchardt, O. Holz, N. Hort, D. Höche, W. R. Heineman and F. Witte, *Acta Biomater.* **9** (2013) 8714.
198. F. Witte, *Acta Biomater.* **6** (2010) 1680.
199. F. Witte, N. Hort, C. Vogt, S. Cohen, K. U. Kainer, R. Willumeit and F. Feyerabend, *Curr. Opin. Solid State Mater. Sci.* **12** (2008) 63.
200. K. Catt, H. Li and X. T. Cui, *Acta Biomater.* **48** (2017) 530.
201. C. L. Weaver, J. M. LaRosa, X. Luo and X. T. Cui, *ACS Nano* **8** (2014) 1834.
202. G. Song, A. Atrens, D. S. John, X. Wu and J. Nairn, *Corros. Sci.* **39** (1997) 1981.
203. X. Luo, C. L. Weaver, S. Tan and X. T. Cui, *J. Mater. Chem. B* **1** (2013) 1340.
204. K. Zhu, X. li, H. Wang, J. Li and G. Fei, *J. Appl. Polym. Sci.* **134** (2017).
205. H. K. Chae, D. Y. Siberio-Pérez, J. Kim, Y. Go, M. Eddaoudi, A. J. Matzger, M. O'Keeffe, O. M. Yaghi, D. Materials and G. Discovery, *Nature* **427** (2004) 523.
206. J. S. Bunch, S. S. Verbridge, J. S. Alden, A. M. van der Zande, J. M. Parpia, H. G. Craighead and P. L. McEuen, *Nano Lett.* **8** (2008) 2458.
207. H. Y. Nan, Z. H. Ni, J. Wang, Z. Zafar, Z. X. Shi and Y. Y. Wang, *J. Raman Spectrosc.* **44** (2013) 1018.
208. T. Chu, H. Ilatikhameneh, G. Klimeck, R. Rahman and Z. Chen, *Nano Lett.* **15** (2015) 8000.
209. B. Ramezanzadeh, S. Niroumandrad, A. Ahmadi, M. Mahdavian and M. H. M. Moghadam, *Corros. Sci.* **103** (2016) 283.
210. D. Prasai, J. C. Tuberquia, R. R. Harl, G. K. Jennings and K. I. Bolotin, *ACS Nano* **6** (2012) 1102.
211. B. Jin, D. B. Xiong, Z. Tan, G. Fan, Q. Guo, Y. Su, Z. Li and D. Zhang, *Carbon* **142** (2019) 482.
212. M. Cui, S. Ren, S. Qin, Q. Xue, H. Zhao and L. Wang, *Corros. Sci.* **131** (2018) 187.
213. A. Ashraf, Y. Wu, M. C. Wang, K. Yong, T. Sun, Y. Jing, R. T. Haasch, N. R. Aluru and S. Nam, *Nano Lett.* **16** (2016) 4708.
214. J. Rafiee, X. Mi, H. Gullapalli, A. V. Thomas, F. Yavari, Y. Shi, P. M. Ajayan and N. A. Koratkar, *Nat. Mater.* **11** (2012) 217.
215. C. J. Shih, M. S. Strano and D. Blankschtein, *Nat. Mater.* **12** (2013) 866.

https://doi.org/10.1142/9789811247859_0012

Chapter 12

Enhancement of Reduced Graphene Oxide Bolometric Photoresponse via Addition of Graphene Quantum Dots

S. Lai[*,||], M. Chen[*,||], Yu. N. Khanin[†], K. S. Novoselov[*,‡,§,¶,||,**] and D. V. Andreeva[*,‡,||,††]

[*]*Department of Materials Science and Engineering, National University of Singapore, Singapore 117575, Singapore*

[†]*Institute for Microelectronics Technology, Russian Academy of Sciences, Chernogolovka 142432, Russia*

[‡]*Centre for Advanced 2D Materials, National University of Singapore, Singapore 117546, Singapore*

[§]*National Graphene Institute, University of Manchester, Oxford Road, Manchester M13 9PL, UK*

[¶]*Chongqing 2D Materials Institute, Liangjiang New Area, Chongqing 400714, China*

[||]*Institute for Functional Intelligent Materials, National University of Singapore, 117544, Singapore*

[**]*kostya@nus.edu.sg*

[††]*daria@nus.edu.sg*

Reduced graphene oxide (rGO) has attracted interest in its potential application in large area photodetectors owing to its ease of manufacture and wideband optical absorbance. Here, we report that thin rGO films produced via vacuum filtration of GO followed by reduction by immersion in L-ascorbic acid are capable of sensing light through a bolometric mechanism. The photoresponse of these rGO thin films can be further enhanced by dropcasting graphene quantum dots (GQDs) on the rGO

[*]Corresponding author.

surface. These GQDs were observed to increase the opacity of the rGO film and hence its absorptivity of light, thereby enabling a significant increase in the photoresponse of the device.

Keywords: Graphene; graphene oxide; quantum dots; photodetector.

1. Introduction

Owing to its superlative electronic and physical properties,[1–4] graphene has attracted considerable interest for a wide range of applications. For example, the high charge carrier mobilities featured by graphene, which exceed 200,000 $cm^2/V \cdot s$[5] have been exploited by Xia *et al.* in constructing extremely fast photodetectors capable of detecting light modulated to a frequency of 40 GHz without degradation in photoresponse.[6] Graphene and graphene-based materials may therefore show great promise in photonic and optoelectronic applications[7] such as light-sensing and photodetection.[8]

In order to achieve photodetectors that are open to more cost-effective production methods, several researchers have also studied graphene oxide (GO) as an alternative to graphene in photodetector applications.[9–11] Unlike graphene, the oxygen-containing functional groups in GO enable their dispersion in water and hence allow for easier in-solution processing.[12,13] However, since GO features much poorer carrier mobilities than graphene, an additional reduction step to produce reduced graphene oxide (rGO) is necessary to improve its conductivity before it can be integrated into an electronic device.[14,15]

Several studies have thus far explored the use of rGO as a potential photodetector, with several researchers suggesting that the photoresponse of rGO can be attributed to a variety of different mechanisms depending on the design and active material of the device.[9,11,16,17] For example, when reduced using a thermal annealing process in which heat is used to remove oxygen-containing groups from GO, Abid *et al.*[16] observed that the photoresponse of the resulting rGO photodetector could be best explained by a photoconductive effect. In another study, Tian *et al.*[11] utilized a chemical reduction pathway using hydrazine to synthesize rGO and proposed that the resulting photoresponse was instead dominated by a photothermal effect due to its much slower exponential photoresponse behavior. By removing the supporting substrate, Tian *et al.*,[11] removed a cooling pathway for the rGO film, thus allowing for the more effective accumulation of thermal energy by the film when irradiated by a light source resulting in a more pronounced photoresponse.

One advantage of rGO photodetectors over monolayer graphene is that they usually comprise multiple stacked layers of RGO flakes instead of a single monolayer sheet resulting in increased light absorptivity.[11,17] This is especially advantageous considering the weak light absorption of monolayer graphene which only absorbs about 2.3% of incident light[18,19] and has been correlated to the poor responsivity of pure graphene devices.[20] However, increased thicknesses may have a detrimental effect on the material cost of the photodetector, as well as decreased sensitivity for rGO photodetectors that utilize photothermal mechanisms due to the increased heat capacity of the thicker film.

One strategy that has been utilized to improve the photodetection characteristics of graphene is to combine it with a quantum dot layer to produce a hybrid photodetector with improved responsivity.[20–24] For example, Konstantatos *et al.*[20] constructed a hybrid quantum dot/graphene photodetector which utilized a spin-coated top layer of Lead Sulfide (PbS) quantum dots which demonstrated an electron–photon gain mechanism where charge carriers were transferred from the PbS quantum dots to the underlying graphene sheet, which enabled a photogating effect that allowed for high responsivities of around 10^7 A W^{-1} to be achieved.

In recent years, the viability of integrating PbS quantum dots with rGO in producing large-area photodetectors has also been explored. Litvin *et al.*[25] adapted the MPTS-functionalized rGO developed by Martin-Garcia *et al.*[26] in linking PbS quantum dots to rGO sheets. The Langmuir–Blodgett (LB) technique was then used to fabricate thin films composed of these PbS-MPTS-rGO hybrid materials which could be used to detect light. In a similar vein, Lara-Canche *et al.*[27] found that a dip coating technique could be used to coat a substrate with alternating layers of rGO and PbS quantum dots that could then be used as a photodetector.

To our knowledge, most studies incorporating quantum dots with rGO or graphene utilize PbS quantum dots[25–28] which are generally sensitive in the infrared (IR) region of light. There may therefore exist a research gap in integrating rGO with other types of quantum dots which may have different spectral sensitivities and characteristics. For example, graphene quantum dots (GQDs) can be visualized as nanoflakes of graphene typically less than 10 nm in lateral dimension.[29] In recent years, GQDs have attracted research interest due to their low toxicity compared to inorganic quantum dots, coupled with their unique optical and physical characteristics.[29,30] Several researchers have also noted that GQDs are capable of efficient light-to-heat conversion.[31] These properties may make them excellent candidates for enhancing the photoresponse of rGO photodetectors which utilize a photothermal effect.

In this work, we demonstrate the fabrication of a simple rGO bolometric photodetector via a simple vacuum filtration process on an aqueous graphene oxide dispersion to form a thin film of GO followed by reduction through immersion in saturated L-ascorbic acid solution. We further report that the addition of small amounts of GQDs via dropcasting on the rGO film surface was capable of enhancing the photoresponse observed from the rGO thin film photodetectors by enhancing the amount of light absorbed by the rGO film.

2. Materials and Methods

2.1. *Preparation of rGO films and fabrication of rGO-based photodetector device*

Standard methodology[32–34] was adapted to prepare thin GO films. Briefly, a 2 mg/ml GO dispersion was first prepared by diluting a GO stock solution purchased from Graphenea. Scanning electron microscopy (SEM) analysis of a diluted dropcasted sample of this dispersion before sonication showed that it contained GO flakes with lateral dimensions of less than 10 μm, while atomic force microscopy (AFM) analysis of our sample showed that these flakes generally had a thickness of around 1.2 nm, which is in agreement with literature on the thickness of monolayer graphene oxide flakes.[35] This 2 mg/ml GO dispersion was sonicated for 1 h to maximize the exfoliation of GO into monolayer sheets. 0.2 ml of the sonicated dispersion, which was 0.4 mg of GO, was then vacuum filtered through a 0.03 μm alumina filter membrane to give a thin paper-like GO film.

To reduce the GO film to rGO, the films were immersed in a saturated L-ascorbic acid solution (30 mg/ml), a mild reducing agent shown to be a safe and environmentally-friendly method to chemically reduce GO,[14,35,36] for 1 week. We note here that the mode of reduction may be an important factor in determining the photoresponse characteristics of the rGO device[11,16] as it may affect the optical and electrical properties of the rGO film. Based on surface profilometer measurements, these rGO films had a thickness of around 177 nm compared to a thickness of around 254 nm in the unreduced GO sheet. The successful reduction of the GO was confirmed via a change in colour of the film from a largely transparent brown (Fig. 1(a)) to a semi-translucent dark brown (Fig. 1(b)) as well as an increase in the conductivity of the film. We noted that before reduction, the as-filtered GO films exhibit insulating behavior, whereas a 177 nm thick film of the rGO film exhibits a resistance of 1.9 k$\Omega/\square$. From this rGO film, a small segment, approximately 1 cm^2 in area, was cut out and placed on a glass slide. Silver glue was then used to form contacts on the rGO film which were then connected to copper cables to form a simple device.

To evaluate the effect of GQDs on this device, 10 μl of a 1 mg/ml aqueous dispersion of GQDs purchased from Sigma Aldrich (S/N: 900712) was dropcasted on the surface of the rGO film and left to dry in a desiccator for 1 h. The GQDs used in this process had lateral dimensions of less than 5 nm, with an excitation maxima at 485 nm, an emission maxima at 530 nm $\pm$ 10 nm, and a quantum yield $\geq$17%. We note here that while the pure rGO was significantly darker and less transparent compared to the GO before reduction, the addition of GQDs on the surface of the rGO resulted in the further increase in opacity of the rGO film at the spot where GQDs were dropcasted as seen in Fig. 1. This suggests an increase in the light absorptivity of the device. The photoresponse of the device before and after the GQD was added was measured to evaluate the effect that the GQD had on the response.

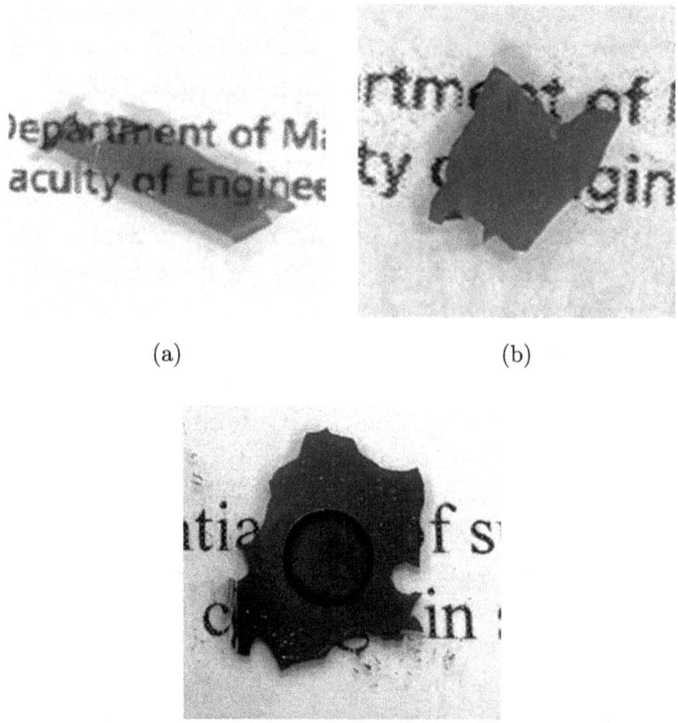

(a) (b)

(c)

Fig. 1. (Color online) Appearance of rGO film before reduction (a); after reduction (b); and after the addition of GQDs (c).

2.2. *Photoresponse characterization of device*

To study the photoresponse of the photodetector, the device was connected to a Keithley 6430 sourcemeter and illuminated using different light cubes of a fluorescence microscope. The fluorescence microscope and the light cubes used in this experiment provided a convenient platform to adjust the wavelength of the incident light, its intensity, and the spot size incident on the device. The sourcemeter was then used to detect changes in potential difference across the device over time as the sample was illuminated while a source current was maintained across the device. The photoresponse of each device was characterized by noting the change in its resistance over time when illuminated.

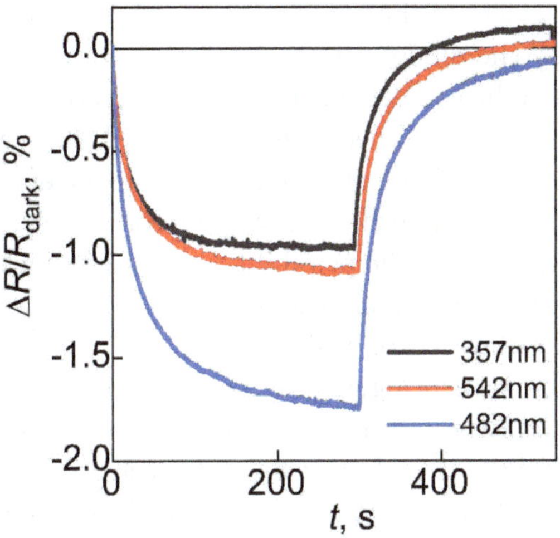

Fig. 2. (Color online) Normalized photoresponse of rGO device in response to different wavelengths of light.

3. Results and Discussion

3.1. *Photoresponse of pure rGO photodetector*

The photoresponse of a roughly 1 cm^2 pure rGO device is depicted in Fig. 2, which shows the change in the normalized resistance ($\Delta R/R$) of the device over time when illuminated with 357, 482, and 542 nm light sources where a 100 μA source current was applied. This pure rGO device was fabricated via the process described in Sec. 2.1 in which vacuum filtration is used to produce graphene oxide films which were then reduced by immersion in L-ascorbic acid.

As noted from Fig. 2, the resistance across the device was observed to drop upon illumination. This is consistent with the results from pure rGO devices studied by other sources.[11,16,17] Furthermore, we measured a photoresponse across different wavelengths of light ranging from 357 nm to 542 nm suggesting that the photodetector is capable of broad spectrum photodetection. Unlike the device studied by Abid *et al.*,[16] the potential difference across the pure rGO photodetector featured a slow exponential decay when the light was turned on and off, similar to the behavior of pure rGO devices studied by Tian *et al.*[11] Differences in the photoresponse behavior of the different rGO devices in these studies, and our pure-rGO devices may be accounted by the different reduction processes used. This suggests that a bolometric heating effect may be responsible for the observed change in resistance. Furthermore, we noticed that when the bias voltage was set to 0 V, no short-circuit photocurrent was observed under illumination, this precludes any photovoltaic effect in this device.

To further investigate the possibility of a bolometric heating effect resulting in the observed change in resistance, we evaluated the temperature dependence of resistance of the rGO film. This was accomplished by measuring the resistance of the rGO film with a sourcemeter at different temperatures which we controlled using a Linkam LTS420 hot stage. Figure 3 depicts the normalized resistance ($\Delta R/R$) of the rGO film from 30°C to 150°C. As noted from Fig. 3, the resistance of the device saw a drop in resistance with increased temperature, similar to what was observed when it was illuminated. Here, we note that similar temperature sensitivity in the resistance of rGO has been noted by other studies[37] as well. This, together with the slow exponential photoresponse observed in Fig. 2, strongly suggests that a bolometric heating effect was responsible for the observed photoresponse.

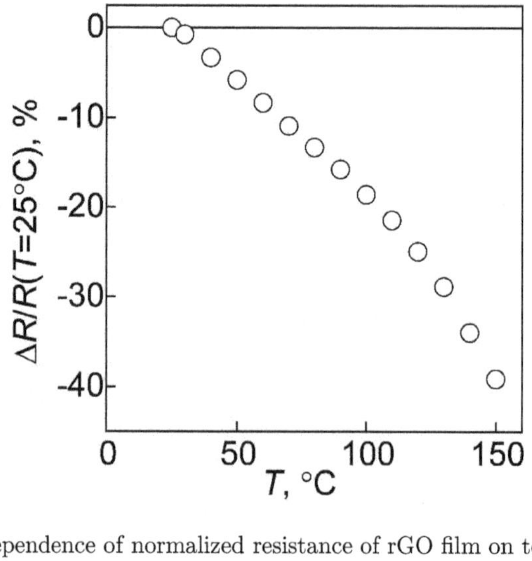

Fig. 3. Dependence of normalized resistance of rGO film on temperature.

3.2. *Photoresponse of rGO photodetector with dropcasted GQDs*

As mentioned in Sec. 2.1, dropcasting GQDs provided a relatively simple means of producing hybrid GQD/rGO photodetectors. Essentially, a pure-rGO device, similar to that described in the previous section, is first fabricated before 10 μl of GQDs are dropcasted on top of the rGO thin film and left to dry in a desiccator. The hybrid device displayed strong fluorescence that was detectable by the fluorescence microscope corresponding to the emission wavelengths of the GQDs. We also noted that this hybrid GQD/rGO device was noticeably darker and more opaque than the pure rGO device, suggesting increased light absorption.

The photoresponse of the device before and after GQDs was added, as shown in Fig. 4 which depicts the normalized resistance ($\Delta R/R_{\mathrm{dark}}$) of the device under a 482 nm light source while a 100 μA source current is applied (here R_{dark} is the sample resistance without illumination). As evident from Fig. 4, the addition of GQDs resulted in a significant increase in the change of voltage of the device upon

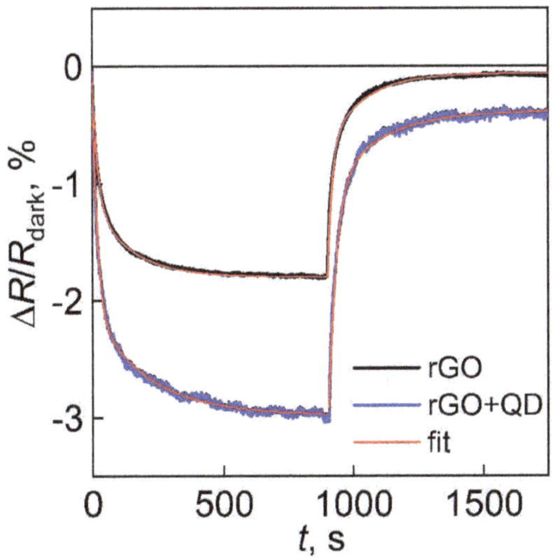

Fig. 4. (Color online) Normalized photoresponse of rGO device before (black) and after (blue) GQDs are added, with double exponential fit applied (red).

illumination. A similar increase in photoresponse was recorded for measurements taken under wavelengths of light at 357 nm and 542 nm.

Similar to work by Chitara *et al.*[17] and Khan *et al.*[10] on pure rGO-photodetectors, we found that the normalized photoresponse curves could be well described ($r^2 > 0.99$) using the double exponential function of the following form:

$$R_{\text{norm}} = R_0 + A\exp(-t/\tau_s) + B\exp(-t/\tau_f), \tag{1}$$

$$R_{\text{norm}} = R_0 - \left[A\exp\left(-\frac{t}{\tau_s}\right) + B\exp(-t/\tau_f)\right], \tag{2}$$

where R_0, A, B, τ_S and τ_f are fitted constants, and t is the time after illumination is turned on or off. Here, Eqs. (1) and (2) describes the normalized resistance behavior when the light is turned on and off respectively as depicted in Fig. 4 which shows the fitting function against the raw results from the device before and after GQDs are added. The applicability of this equation to the photoresponse of our devices suggests that the photoresponse can be expressed as the sum of a faster component associated with a characteristic time τ_f and pre-exponential B, and a slower component associated with τ_s and A.

Table 1 lists the different fitted parameters associated with each of the curves. The characteristic times here are relatively long ($\tau_S \sim 200$ s and $\tau_f \sim 20$ s) but are consistent with other studies on rGO photodetectors.[9,17] We note here that the addition of GQDs was associated with the increase in the pre-exponential B (corresponding

Table 1. Fitted parameters to double exponential function.

	A (with τ_s)	B (with τ_f)	A/B	τ_s	τ_f
Light On — ($R_0 + A\exp(-t/\tau_s) + B\exp(-t/\tau_f)$)					
GQD	0.0089	0.0203	0.397	206.6	17.6
No GQD	0.0064	0.0107	0.598	141.8	19.5
Light Off — ($R_0 - A\exp(-t/\tau_s) + B\exp(-t/\tau_f)$)					
GQD	0.0067	0.0215	0.311	196.3	30.1
No GQD	0.0062	0.0113	0.549	137.9	17.7

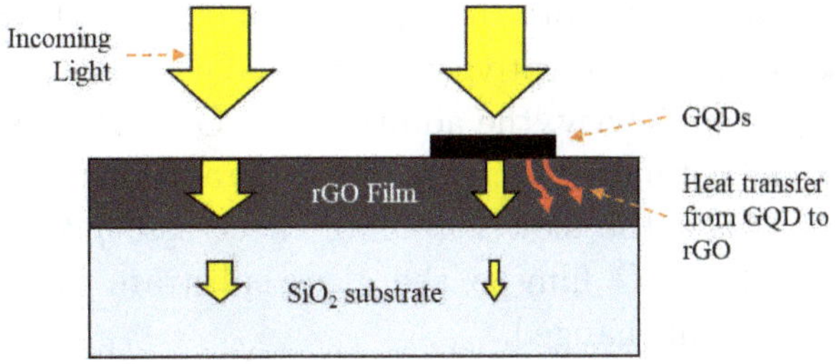

Fig. 5. (Color online) Proposed mechanism for GQD enhancement of photoresponse; the GQDs increase the light absorption of the device, transferring the additional absorbed light in the form of heat to the underlying rGO film.

to the fast component of the decay), as well as an increase in the slow characteristic time τ_S for both the 'light on' and 'light off'. The 10–90% rise time for the device with GQDs was 198 s which was a slight increase over the 184 s for the device before GQDs were added.

As mentioned prior, the slow exponential behavior of the photoresponse suggests that a heating effect may be responsible for the observed change in resistance of the pure-rGO photodetector. Here light incident on the rGO film is first converted into heat which then results in a change in the resistance of the rGO film which can be detected and correlated to the presence of light. The relatively long period of time needed for the rGO to achieve a steady-state temperature when exposed to light additionally accounts for the extended time needed for photoresponse saturation of the device.

We propose that the increase in the photoresponse of the device by the addition of GQDs can be attributed to an enhanced bolometric effect. We note here that the efficient conversion of light into heat by GQDs has already been documented by other studies.[31] Hence, we propose that the GQDs act to increase the amount of light absorbed by the device, while efficiently converting this incident radiation into heat. The heat produced by the GQDs is then transferred via thermal contact to the rGO film, resulting in an increase in the recorded photoresponse. This process is described schematically in Fig. 5.

This explanation also accounts for why the addition of GQDs enhances the pre-exponential B associated with the fast component of photoresponse decay for both the 'light-on' and 'light-off' cases. Since the increase in light absorptivity by the GQDs also results in an increase in light emissivity, the addition of GQDs also results in the rGO film losing heat more effectively through radiation when the light is turned off, while the slower cooling effect associated with heat transfer from the rGO film to the glass substrate via conduction remains relatively unchanged.

While we note that light absorption may also be improved by increasing the thickness of the rGO film, there may be several advantages to using GQD via dropcasting to improve the light absorption of the rGO layer. First, since relatively small amounts of GQDs are necessary, the usage of GQD over increased amounts of rGO may contribute to lower material costs. Second, as the mass of dropcasted GQDs is small compared to that of the rGO, the increase in the heat capacity from the dropcasted GQDs is likely to be relatively low which would enable a more pronounced change in the normalized photoresponse since this correlates with the temperature of the rGO film.

4. Conclusion

A simple method involving the vacuum filtration of graphene oxide to form a GO film, followed by subsequent reduction of the film via immersion in saturated L-ascorbic acid solution was used to fabricate photosensitive rGO films. Under illumination, the resistance of these rGO films was observed to decrease with slow exponential behavior suggesting that a bolometric effect is responsible for the observed photoresponse. Our study also examined the viability of integrating GQDs into this rGO-based photodetector via a simple dropcasting method. After the addition of GQDs, an appreciable increase in photoresponse over a pure-rGO photodetector control experiment was observed. We attributed this increase in photoresponse to an enhanced bolometric effect that arises from the ability of the dropcasted GQDs to efficiently absorb light and convert it into heat which is then transferred to the underlying rGO film.

Acknowledgments

We thank NRF (Singapore) for financial support through the project Medium-Sized Centre programme R-723-000-001-281. KSN acknowledges support from EU Flagship Programs (Graphene CNECTICT-604391 and 2D-SIPC Quantum Technology), European Research Council Synergy Grant Hetero2D, the Royal Society, EPSRC grants EP/N010345/1, EP/P026850/1, EP/S030719/1. Yu.N.Kh. was supported by the RFBR (No. 20-02-00601). This work was supported by the Ministry of Education (Singapore) through the Research Centre of Excellence program (grant EDUN C-33-18-279-V12, I-FIM).

References

1. K. S. Novoselov *et al.*, *Science* **306** (2004) 666, doi: 10.1126/science.1102896.
2. A. K. Geim and K. S. Novoselov, *Nat. Mater.* **6** (2007) 183.
3. A. K. Geim and I. V. Grigorieva, *Nature* **499** (2013) 419, doi: 10.1038/nature12385.
4. K. S. Novoselov, A. Mishchenko, A. Carvalho and A. H. Castro Neto, *Science* **353** (2016) aac9439, doi: 10.1126/science.aac9439.
5. K. S. Novoselov *et al.*, *Nature* **490** (2012) 192, doi: 10.1038/nature11458.
6. F. Xia, T. Mueller, Y.-M. Lin, A. Valdes-Garcia and P. Avouris, *Nat. Nanotechnol.* **4** (2009) 839, doi: 10.1038/nnano.2009.292.
7. F. Bonaccorso, Z. Sun, T. Hasan and A. C. Ferrari, *Nat. Photon* **4** (2010) 611, doi: 10.1038/nphoton.2010.186.
8. A. De Sanctis, J. D. Mehew, M. F. Craciun and S. Russo, *Materials* **11** (2018) 1762, doi: 10.3390/ma11091762.
9. H. Chang *et al.*, *ACS Nano* **7** (2013) 6310, doi: 10.1021/nn4023679.
10. M. A. Khan, K. K. Nanda and S. B. Krupanidhi, *Mater. Res. Exp.* **4** (2017) 085603, doi: 10.1088/2053-1591/aa8042.
11. H. Tian, Y. Cao, J. Sun and J. He, *RSC Adv.* **7** (2017) 46536, doi: 10.1039/C7RA09826J.
12. Y. Zhu *et al.*, *Adv. Mater.* **22** (2010) 3906, doi: 10.1002/adma.201001068.
13. D. Konios, M. M. Stylianakis, E. Stratakis and E. Kymakis, *J. Colloid Interf. Sci.* **430** (2014) 108, doi: 10.1016/j.jcis.2014.05.033.
14. J. Gao *et al.*, *Chem. Mater.* **22** (2010) 2213, doi: 10.1021/cm902635j.
15. A. T. Smith, A. M. LaChance, S. Zeng, B. Liu and L. Sun, *Nano Mater. Sci.* **1** (2019) 31, doi: 10.1016/j.nanoms.2019.02.004.
16. Abid, P. Sehrawat, S. S. Islam, P. Mishra and S. Ahmad, *Sci. Rep.* **8** (2018) 3537, doi: 10.1038/s41598-018-21686-2.
17. B. Chitara, S. B. Krupanidhi and C. N. R. Rao, *Appl. Phys. Lett.* **99** (2011) 113114, doi: 10.1063/1.3640222.

18. R. R. Nair *et al.*, *Science* **320** (2008) 1308, doi: 10.1126/science.1156965.
19. S.-E. Zhu, S. Yuan and G. C. A. M. Janssen, *Europhys. Lett.* **108** (2014) 17007, doi: 10.1209/0295-5075/108/17007.
20. G. Konstantatos *et al.*, *Nat. Nanotechnol.* **7** (2012) 363, doi: 10.1038/nnano.2012.60.
21. Q. Nian *et al.*, *ACS Appl. Mater. Interfaces* **9** (2017) 44715, doi: 10.1021/acsami.7b14468.
22. D. De Fazio *et al.*, *ACS Nano* **14** (2020) 11897, doi: 10.1021/acsnano.0c04848.
23. A. J. K. Al-Alwani *et al.*, *Appl. Surf. Sci.* **424** (2017) 222, doi: 10.1016/j.apsusc.2017.03.235.
24. A. Black *et al.*, *ACS Omega* **4** (2019) 15824, doi: 10.1021/acsomega.9b01449.
25. A. P. Litvin *et al.*, *Nanomaterials* **10** (2020) 723, doi: 10.3390/nano10040723.
26. B. Martín-García, A. Polovitsyn, M. Prato and I. Moreels, *J. Mater. Chem. C* **3** (2015) 7088, doi: 10.1039/C5TC01137J.
27. A. R. Lara-Canche *et al.*, *Nanotechnology* **32** (2020) 055604, doi: 10.1088/1361-6528/abc209.
28. A. Mnoyan, K. Kim, J. Y. Kim and D. Y. Jeon, *Solar Energy Mater. Solar Cells* **144** (2016) 181, doi: 10.1016/j.solmat.2015.09.001.
29. Y. Li *et al.*, *Adv. Mater.* **23** (2011) 776, doi: 10.1002/adma.201003819.
30. Y. Chong *et al.*, *Biomaterials* **35** (2014) 5041, doi: 10.1016/j.biomaterials.2014.03.021.
31. X. Bao *et al.*, *Light. Sci. Appl.* **7** (2018) 91, doi: 10.1038/s41377-018-0090-1.
32. D. A. Dikin *et al.*, *Nature* **448** (2007) 457, doi: 10.1038/nature06016.
33. D. V. Andreeva *et al.*, *Nat. Nanotechnol.* **16** (2021) 174, doi: 10.1038/s41565-020-00795-y.
34. S. Y. Chen *et al.*, *Polym. Int.* **69** (2020) 1173, doi: 10.1002/pi.6080.
35. J. Zhang *et al.*, *Chem. Commun.* **46** (2010) 1112, doi: 10.1039/B917705A.
36. M. J. Fernández-Merino *et al.*, *J. Phys. Chem. C* **114** (2010) 6426, doi: 10.1021/jp100603h.
37. S. Afroj *et al.*, *ACS Nano* **13** (2019) 3847, doi: 10.1021/acsnano.9b00319.

https://doi.org/10.1142/9789811247859_index

Index

CPSIA information can be obtained
at www.ICGtesting.com
Printed in the USA
JSHW041202181021
19643JS00001B/3

9 789811 247842